METHODS IN CELL BIOLOGY

VOLUME 33

Flow Cytometry

Series Editor

LESLIE WILSON

Department of Biological Sciences
University of California
Santa Barbara, California

METHODS IN CELL BIOLOGY

BIOLOGY

Prepared under the Auspices of the American Society for Cell Biology

VOLUME 33
Flow Cytometry

Edited by

ZBIGNIEW DARZYNKIEWICZ

CANCER RESEARCH INSTITUTE OF THE
NEW YORK MEDICAL COLLEGE AT VALHALLA
VALHALLA, NEW YORK 10595

HARRY A. CRISSMAN

BIOMEDICAL SCIENCES DIVISION
LOS ALAMOS NATIONAL LABORATORY
LOS ALAMOS, NEW MEXICO 87545

ACADEMIC PRESS, INC.
Harcourt Brace Jovanovich, Publishers

San Diego   New York   Boston
London   Sydney   Tokyo   Toronto

Academic Press, Inc.
San Diego, California 92101

United Kingdom Edition published by
Academic Press Limited
24-28 Oval Road, London NW1 7DX

Library of Congress Catalog Card Number: 64-14220

ISBN   0-12-564133-8   (alk. paper)
ISBN   0-12-203050-8   (pbk. : alk. paper)

Printed in the United States of America
90 91 92 93   9 8 7 6 5 4 3 2 1

CONTENTS

CONTRIBUTORS

Numbers in parentheses indicate the pages on which the authors' contributions begin.

HEINZ BAISCH, Institut für Biophysik und Strahlenbiologie, Universität Hamburg, D-2000 Hamburg 20, Federal Republic of Germany (217)

MARTY F. BARTHOLDI, Life Sciences Division, Los Alamos National Laboratory, University of California, Los Alamos, New Mexico 87545 (369)

KENNETH D. BAUER, Department of Pathology, Northwestern University School of Medicine, Chicago, Illinois 60611 (235)

WOLFGANG BEISKER, Biomedical Sciences Division, Lawrence Livermore National Laboratory, Livermore, California 94550 (207)

CATHERINE BERGOUNIOUX, Laboratoire de Physiologie Végétale Moléculaire, Faculté des Sciences, Université d'Orsay, 91405 Orsay, France (563)

RALPH M. BÖHMER, Ludwig Institute for Cancer Research, Melbourne Tumour Biology Branch, Melbourne, Victoria 3050, Australia (173)

ERIK BOYE, Department of Biophysics, Institute for Cancer Research, Montebello, 0310 Oslo, Norway (519)

SPENCER C. BROWN, Cytométrie, Institut des Sciences Végétales, Centre National de la Recherche Scientifique, 91198 Gif-sur-Yvette, France (563)

MARY CAMPBELL, Life Sciences Division, Los Alamos National Laboratory, University of California, Los Alamos, New Mexico 87545 (377)

ROBERT CERRA, Department of Surgery, Henry Ford Hospital, Detroit, Michigan 48202 (1)

DAVID J. CHAPLIN, Medical Biophysics Unit, B. C. Cancer Research Centre, Vancouver, British Columbia, V5Z 1L3, Canada (509)

YUHCHYAU CHEN, Department of Pathology, School of Medicine, University of Washington, Seattle, Washington 98195 (185)

IB JARLE CHRISTENSEN, The Finsen Laboratory, Rigshospitalet, University Hospital, DK-2100 Copenhagen, Denmark (127)

GAETANO CIANCIO, Department of Surgery, Jackson Memorial Hospital, Miami, Florida 33101 (19)

L. SCOTT CRAM, Life Sciences Division, Los Alamos National Laboratory, University of California, Los Alamos, New Mexico 87545 (369, 377)

HARRY A. CRISSMAN, Cell Biology Group, Los Alamos National Laboratory, Los Alamos, New Mexico 87545 (89, 97, 105, 199, 305)

JOHN D. CRISSMAN, Department of Pathology, Wayne State University, Detroit, Michigan 48202 (1)

ZBIGNIEW DARZYNKIEWICZ, Cancer Research Institute of the New York Medical College at Valhalla, Valhalla, New York 10595 (285, 305, 337, 655)

LARRY L. DEAVEN, Life Sciences Division, Los Alamos National Laboratory, University of California, Los Alamos, New Mexico 87545 (377)

FRANK DOLBEARE, Biomedical Sciences Division, Lawrence Livermore National Laboratory, Livermore, California 94550 (81, 207)

LYNN G. DRESSLER, Division of Molecular and Cellular Diagnostics, University of New Mexico Cancer Center, University of New Mexico School of Medicine, Albuquerque, New Mexico 87131 (157)

RALPH E. DURAND, Medical Biophysics Unit, B. C. Cancer Research Centre, Vancouver, British Columbia V5Z 1L3, Canada (509, 647)

DONALD P. EVENSON, Olson Biochemistry Laboratories, Department of Chemistry, South Dakota State University, Brookings, South Dakota 57007 (401)

JOHN J. FAWCETT, Life Sciences Division, Los Alamos National Laboratory, University of California, Los Alamos, New Mexico 87545 (377)

OSKAR S. FRANKFURT,[1] Grace Cancer Drug Center, Roswell Park Memorial Institute, Buffalo, New York 14263 (13, 299)

MACK J. FULWYLER, Laboratory for Cell Analysis, Department of Laboratory Medicine, University of California, San Francisco, California 94143 (613)

DAVID W. GALBRAITH, Department of Plant Sciences, University of Arizona, Tucson, Arizona 85721 (527, 549)

ELIO GEIDO, IST, Istituto Nazionale per la Ricerca sul Cancro, Laboratorio di Biofisica, 16132 Genova, Italy (149)

JOHANNES GERDES, Division of Molecular Immunology, Forschungsinstitut Borstel, D-2061 Borstel, Federal Republic of Germany (217)

WALTER GIARETTI, IST, Istituto Nazionale per la Ricerca sul Cancro, Laboratorio di Biofisica, 16132 Genova, Italy (149)

JOE W. GRAY, Biomedical Sciences Division, Lawrence Livermore National Laboratory, Livermore, California 94550 (207)

ANGELIKA GROSSMANN, Department of Comparative Medicine, School of Medicine, University of Washington, Seattle, Washington 98195 (185)

DAVID W. HEDLEY, Departments of Medicine and Pathology, Princess Margaret Hospital, Toronto, Ontario M4X 1K9, Canada (59, 139)

HANS HERWEIJER, TNO Institute of Applied Radiobiology and Immunology, 2280 HV Rijswijk, and Daniel den Hoed Cancer Center, Rotterdam, The Netherlands (631)

RYUJI HIGASHIKUBO, Section of Cancer Biology, Radiation Oncology Center, Washington University School of Medicine, St. Louis, Missouri 63108 (325, 353)

HOLGER HOEHN, Department of Human Genetics, University of Würzburg Biocenter, Am Hubland, 87 Würzburg, Federal Republic of Germany (185)

MARIANNE H. HOFLAND, Cell Biology Group, Los Alamos National Laboratory, Los Alamos, New Mexico 87545 (89)

PAUL KARL HORAN, Zynaxis Cell Science, Inc., Malvern, Pennsylvania 19355 (469)

BRUCE D. JENSEN, Zynaxis Cell Science, Inc., Malvern, Pennsylvania 19355 (469)

RICHARD JONKER, TNO Institute of Applied Radiobiology and Immunology, 2280 HV Rijswijk, The Netherlands (631)

CARL H. JUNE, Immune Cell Biology Program, Naval Medical Research Institute, Bethesda, Maryland 20814 (37)

JAN KAPUSCINSKI, Cancer Research Institute of the New York Medical College at Valhalla, Valhalla, New York 10595 (655)

MAREK KIMMEL,[2] Investigative Cytology Laboratory, Memorial Sloan-Kettering Cancer Center, New York, New York 10021 (249)

[1]*Present address:* Oncology Laboratory, Cedars Medical Center, Miami, Florida 33136.
[2]*Present address:* Department of Statistics, Rice University, Houston, Texas 77251.

AWTAR KRISHAN, Department of Oncology, University of Miami School of Medicine, Miami, Florida 33101 (121, 491)

MANFRED KUBBIES, Boehringer Mannheim Research Center, 8122 Penzberg, Federal Republic of Germany (185)

WEN-LIN KUO, Biomedical Sciences Division, Lawrence Livermore National Laboratory, Livermore, California 94550 (207)

JUDITH LAFFIN, Department of Microbiology and Immunology, Albany Medical College, Albany, New York 12208 (271)

JØRGEN K. LARSEN, The Finsen Laboratory, Rigshospitalet, University Hospital, DK-2100 Copenhagen, Denmark (227)

JOHN M. LEHMAN, Department of Microbiology and Immunology, Albany Medical College, Albany, New York 12208 (271)

THOMAS M. McHUGH, Department of Laboratory Medicine, University of California, San Francisco, California 94143 (613)

MERYLE J. MELNICOFF, Zynaxis Cell Science, Inc., Malvern, Pennsylvania 19355 (469)

A. DUSTY MILLER, Fred Hutchinson Cancer Research Center, Seattle, Washington 98104 (71)

ELIZABETH A. MUSGROVE, Garvan Institute of Medical Research, St. Vincent's Hospital, Darlinghurst, Sydney, N.S.W. 2010, Australia (59)

KEES NOOTER, TNO Institute of Applied Radiobiology and Immunology, 2280 HV Rijswijk, The Netherlands (631)

MICHAEL NÜSSE, GSF, Gesellschaft für Strahlen- und Umweltforschung, Institut für Biophysikalische Strahlenforschung, D-6000 Frankfurt-am-Main, Federal Republic of Germany (149)

MARY J. O'CONNELL, Analytical Cytology Unit, Department of Pathology, University of Rochester Medical Center, Rochester, New York 14642 (501)

PEGGY L. OLIVE, Medical Biophysics Unit, B. C. Cancer Research Centre, Vancouver, British Columbia V5Z 1L3, Canada (509)

FRIEDRICH OTTO, Fachklinik Hornheide, D-4400 Münster, Federal Republic of Germany (105)

DAN PINKEL, Biomedical Sciences Division, Lawrence Livermore National Laboratory, Livermore, California 94550 (383)

ALAN POLLACK, Department of Radiation Therapy, U.T./M.D. Anderson Cancer Center, Houston, Texas 77030 (19, 315)

MARTIN POOT, Department of Human Genetics, University of Würzburg Biocenter, Am Hubland, 87 Würzburg, Federal Republic of Germany (185)

PETER S. RABINOVITCH, Department of Pathology, School of Medicine, University of Washington, Seattle, Washington 98195 (37, 185)

FRANK A. RAY, Life Sciences Division, Los Alamos National Laboratory, University of California, Los Alamos, New Mexico 87545 (369)

JAY E. REEDER, Analytical Cytology Unit, Department of Pathology, University of Rochester Medical Center, Rochester, New York 14642 (501)

JOSEPH L. ROTI ROTI, Section of Cancer Biology, Radiation Oncology Center, Washington University School of Medicine, St. Louis, Missouri 63108 (325, 353)

HOWARD M. SHAPIRO, Howard M. Shapiro, M.D., P.C., West Newton, Massachusetts 02165 (25)

KIRSTEN SKARSTAD, Department of Biophysics, Institute for Cancer Research, Montebello, 0310 Oslo, Norway (519)

SUE E. SLEZAK, Zynaxis Cell Science, Inc., Malvern, Pennsylvania 19355 (469)

HARALD B. STEEN, Department of Biophysics, Institute for Cancer Research, Montebello, 0310 Oslo, Norway (519)

JOHN A. STEINKAMP, Cell Biology Group, Los Alamos National Laboratory, Los Alamos, New Mexico 87545 (199, 305)

ANITA P. STEVENSON, Cell Biology Group, Los Alamos National Laboratory, Los Alamos, New Mexico 87545 (89)

CARLETON C. STEWART, Laboratory of Flow Cytometry, Roswell Park Memorial Institute, Buffalo, New York 14263 (411, 427)

RICHARD A. THOMAS, RATCOM, Inc., Miami, Florida 33193 (111)

JERRY T. THORNTHWAITE, Immuno-Oncology Laboratories, Baptist Hospital of Miami, Miami, Florida 33176 (111)

ROBERT A. TOBEY, Biochemistry and Biophysics Group, Los Alamos National Laboratory, Los Alamos, New Mexico 87545 (89, 97, 305)

FRANK TRAGANOS, Cancer Research Institute of the New York Medical College at Valhalla, Valhalla, New York 10595 (249)

BARBARA TRASK, Biomedical Sciences Division, Lawrence Livermore National Laboratory, Livermore, California 94550 (363, 383)

GER VAN DEN ENGH, Biomedical Sciences Division, Lawrence Livermore National Laboratory, Livermore, California 94550 (363, 631)

MARTIN VANDERLAAN, Biomedical Sciences Division, Lawrence Livermore National Laboratory, Livermore, California 94550 (207)

LARS VINDELØV, Departments of Hematology L and Oncology ONK, Rigshospitalet, University Hospital, DK-2100 Copenhagen, Denmark (127)

JAN W. M. VISSER, Radiobiological Institute TNO, Rijswijk, The Netherlands (451)

PETER DE VRIES, Radiobiological Institute, TNO, Rijswijk, The Netherlands (451)

LEON L. WHEELESS, JR., Analytical Cytology Unit, Department of Pathology, University of Rochester Medical Center, Rochester, New York 14642 (501)

MARK E. WILDER, Cell Biology Group, Los Alamos National Laboratory, Los Alamos, New Mexico 87545 (89)

WILLIAM D. WRIGHT, Section of Cancer Biology, Radiation Oncology Center, Washington University School of Medicine, St. Louis, Missouri 63108 (325, 353)

CLARICE M. YENTSCH, J. J. MacIsaac Flow Cytometry/Sorting Facility, Bigelow Laboratory for Ocean Sciences, West Boothbay Harbor, Maine 04575 (613)

RICHARD J. ZARBO, Department of Pathology, Henry Ford Hospital, Detroit, Michigan 48202 (1)

PREFACE

Progress in cell biology has been closely associated with the development of quantitative analytical methods applicable to individual cells or cell organelles. Three distinctive phases characterize this development. The first started with the introduction of microspectrophotometry, microfluorometry, and microinterferometry. These methods provided a means to quantitate various cell constituents such as DNA, RNA, or protein. Their application initiated the modern era in cell biology, based on quantitative—rather than qualitative, visual—cell analysis. The second phase began with the birth of autoradiography. Applications of autoradiography were widespread and this technology greatly contributed to better understanding of many functions of the cell. Especially rewarding were studies on cell reproduction; data obtained with the use of autoradiography were essential in establishing the concept of the cell cycle and generated a plethora of information about the proliferation of both normal and tumor cells.

The introduction of flow cytometry initiated the third phase of progress in methods development. The history of flow cytometry is short, with most advances occurring over the past 15 years. Flow cytometry (and, associated with it electronic cell sorting) offers several advantages over the two earlier methodologies. The first is the rapidity of the measurements. Several hundred, or even thousands of cells can be measured per second, with high accuracy and reproducibility. Thus, large numbers of cells from a given population can be analyzed and rare cells or subpopulations detected. A multitude of probes have been developed that make it possible to measure a variety of cell constituents. Because different constituents can be measured simultaneously and the data are recorded by the computer in list mode fashion, subsequent bi- or multivariate analysis can provide information about quantitative relationships among constituents either in particular cells or between cell subpopulations. Still another advantage of flow cytometry stems from the capability for selective physical sorting of individual cells, cell nuclei, or chromosomes, based on differences in the variables measured. Because some of the staining methods preserve cell viability and/or cell membrane integrity, the reproductive and immunogenic capacity of the sorted cells can be investigated. Sorting of individual chromosomes has already provided the basis for development of chromosomal DNA libraries, which are now indispensable in molecular biology and cytogenetics.

Flow cytometry is a new methodology and is still under intense development, improvement, and continuing change. Most flow cytometers are quite complex and not yet user friendly. Some instruments fit particular applications better than others, and many proposed analytical applications have not been extensively tested on different cell types. Several methods are not yet routine and a certain degree of artistry and creativity is often required in adapting them to new biological material, to new applications, or even to different instrument designs. The methods published earlier often undergo modifications or improvements. New probes are frequently introduced.

This volume represents the first attempt to compile and present selected flow cytometric methods in the form of a manual designed to be of help to anyone interested in their practical applications. Methods having a wide immediate or potential application were selected, and the chapters are written by the authors who pioneered their development, or who modified earlier techniques and have extensive experience in their application. This ensures that the essential details are included and that readers may easily master these techniques in their laboratories by following the described procedures.

The selection of chapters also reflects the peculiarity of the early phase of method development referred to previously. The most popular applications of flow cytometry are in the fields of immunology and DNA content–cell cycle analysis. While the immunological applications are now quite routine, many laboratories still face problems with the DNA measurements, as is evident from the poor quality of the raw data (DNA frequency histograms) presented in many publications. We hope that the descriptions of several DNA methods in this volume, some of them individually tailored to specific dyes, flow cytometers, and material (e.g., fixed or unfixed cells or isolated cell nuclei from solid tumors), may help readers to select those methods that would be optimal for their laboratory setting and material. Of great importance is the standardization of the data, which is stressed in all chapters and is a subject of a separate chapter.

Some applications of flow cytometry included in this volume are not yet widely recognized but are of potential importance and are expected to become widespread in the near future. Among these are methods that deal with fluorescent labeling of plasma membrane for cell tracking, flow microsphere immunoassay, the cell cycle of bacteria, the analysis and sorting of plant cells, and flow cytometric exploration of organisms living in oceans, rivers, and lakes.

Individual chapters are designed to provide the maximum practical information needed to reproduce the methods described. The theoretical bases of the methods are briefly presented in the introduction of most chapters. A separate section of each chapter is devoted to applicability of the described method to different biological systems, and when possible, references are provided to articles that review the applications. Also discussed under separate subheads are the critical points of procedure, including the experience of the authors with different instruments, and the appropriate controls and standards. Typical results, often illustrating different cell types, are presented and discussed in the "Results" section. The "Materials and Methods" section of each chapter is the most extensive, giving a detailed description of the method in a cookbook format.

Flow cytometry and electronic sorting have already made a significant impact on research in various fields of cell and molecular biology and medicine. We hope that this volume will be of help to the many researchers who need flow cytometry in their studies, stimulate applications of this methodology to new areas, and promote progress in many disciplines of science.

Zbigniew Darzynkiewicz
Harry A. Crissman

Chapter 1

Dissociation of Cells from Solid Tumors

ROBERT CERRA

Department of Surgery
Henry Ford Hospital
Detroit, Michigan 48202

RICHARD J. ZARBO

Department of Pathology
Henry Ford Hospital
Detroit, Michigan 48202

JOHN D. CRISSMAN

Department of Pathology
Wayne State University
Detroit, Michigan 48202

I. Introduction

The ability or, more correctly, the inability to easily dissociate solid tumors into single-cell or nuclear suspensions has remained one of the major obstacles for analysis by flow cytometry (FCM). Lymphomas, which lack intercellular attachments, were easily adapted to FCM analysis as a result of the ease with which hematological neoplasms can be dissociated into single-cell suspensions. The ability to identify phenotypically diverse subsets of lymphocyte populations was the second and equally important stimulus for application of FCM technology to lymphomas. In contrast, solid tumors have varying numbers of intercellular desmosome attachments and adhesive contacts with the basal lamina and host stromal components. Squamous cell carcinomas contain numerous well-developed desmosomes and are extremely difficult to dissociate. Most adenocarcinomas contain both

1

desmosomes and tight circumferential junctional complexes, as well as attachment to basal lamina and host stromal components. In general, the ability to dissociate solid tumors into single-cell suspensions is dependent on the type of neoplasm and the degree of differentiation. Well-differentiated squamous carcinomas will have numerous well-developed desmosomes and be extremely difficult to dissociate. In contrast, poorly differentiated adenocarcinomas will have fewer poorly formed desmosomes and patchy irregular fragments of basal lamina, and predictably are much easier to dissociate.

The major application of FCM technology in solid tumors is quantitation of cellular DNA content. At this time there are few cytoplasmic or membrane antigens in solid tumors that are important for subclassification. The difficulty in producing intact cells and the paucity of interest in cytoplasmic markers has led to the development of enucleation techniques. Nuclear suspensions for DNA measurements are reported to be easier to prepare than intact cell suspensions, although our laboratory is not convinced. The application of nuclear extraction protocols is relatively straightforward and requires cytoplasmic disruption and stripping, leaving a suspension of "bare" nuclei. However, this approach precludes conjoint cytoplasmic or membrane antigen indentification and restricts FCM analysis to nuclear parameters.

II.   Solid-Tumor Dissociation: Theoretical Considerations

Tumor dissociation techniques are broadly grouped into two major categories, mechanical and enzymatic, each with numerous different methodological approaches. Dissociation protocols may apply combinations of mechanical and enzymatic techniques, resulting in an infinite number of possible conditions. Therefore, an investigator must clearly establish what parameters are to be measured by FCM. The dissociation protocol must be optimized to yield a population of cells (or nuclei) that is representative of the solid tumor and in sufficient quantity for analysis. If cell surface markers are to be examined, a dissociation protocol must be designed that results in viable single cells that retain cell surface antigens or other markers. This generally requires careful and gentle approaches, because the disruption of desmosomes often causes cell membrane injury that must be repaired by the cell to preserve viability and restore surface markers. In contrast, enucleation procedures can mechanically tear or enzymatically dissolve the desmosomes from the cytoplasm, which is subsequently stripped from the nuclei by enzymes.

The major goal of any dissociation procedure is to obtain a suspension of whole cells or nuclei that adequately represent the solid-tumor cell population. Therefore, an assessment of cell representation is required to document the cell suspension content. The simplest approach of analyzing for cell population is to maximize cell viability and the total cell yield. Methods based on trypan blue dye exclusion or fluorescein acetate enzymatic hydrolysis are standard techniques for assessing cell viability. In addition, functional studies such as uptake of tritiated thymidine are even better indicators of cell viability and more important, since preservation of cell function is measured by continued DNA synthesis (McDivitt *et al.*, 1984). The theoretical cell yield can be calculated by the quantitation of DNA in a known weight of tumor tissue and in a documented number of dispersed cells, providing a denominator for determining efficiency of dissociation. Suspensions with over 90% viable cells and cell yields approaching the theoretical maximum are indicative of representative cell population in any dissociation protocol.

A second approach in evaluation of cell representation is to determine the percentage of the various cell populations (e.g., inflammatory, stromal, tumor). Cytospin preparations of the single-cell suspension can be compared to histological sections of the original tumor. This is a relatively crude approach but may alert the investigator if tumor cells are preferentially being lost in the dissociation protocol. This is especially pertinent in that selected tumor cell populations are commonly lost in numerous dissociation procedures. We have noted in two separate studies that aneuploid colon tumor cells are sensitive to enzymatic cell dissociation, being preferentially lost during the dissociation procedure (Ensley *et al.*, 1987a; Crissman *et al.*, 1988).

Ideally, protocol development should include repetitive dissociation experiments to appraise cell yields, efficiency of dispersion, and representation of all cell populations. These experiments often require extensive amounts of comparable tumor tissue. For this reason we have used murine tumor models to compare dissociation protocols (Ensley *et al.*, 1987b; Crissman *et al.*, 1988). This approach allows adequate comparison of the numerous experimental variables, but also allows documentation of preservation of the characteristic cell or parameter to be measured (e.g., preservation of aneuploid tumor cells). Repeated use of animal tumor models has confirmed that for DNA quantitation, mechanical dissociation methods are optimum for adenocarcinomas, sarcomas, lymphomas, and transitional carcinomas. Enzymatic or enucleation procedures are optimum for producing suspensions of intact squamous carcinoma cells. Obviously, enucleation procedures work equally well on all forms of solid tumors.

III.   Solid-Tumor Dissociation: Practical Considerations

Fresh solid tumors may be dissociated by a variety of mechanical, enzymatic, or combination of mechanical and enzymatic methods to yield a viable single-cell suspension for cell surface, cytoplasmic, and nuclear DNA content measurement. Mechanical methods usually give adequate cell yields with fair viability, especially in the more easily dissociated solid tumors. Mechanical dissociation has the advantage of preserving cell surface antigens. However, mechanical dissociation often causes irreversible damage to the cell membrane that is presumably due to tearing out of desmosomal attachments. We have found that initial cell viability is greater than viability measured after 24-hour incubation at 37°C. This would support the hypothesis of irreversible latent damage occurring during the dissociation procedure. However, short-term and especially long-term viability is important only in the study of functional activities or in attempts to culture dissociated cells. Enzymatic dissociation procedures usually give a better cell yield with increased viability, although alteration of cell surface markers may occur as a result of enzyme digestion. In some instances, enzymatic dissociation has been shown to result in a nonrepresentative population (Engelholm *et al.*, 1985; Ensley *et al.*, 1987b). In general, the highest yields and most representative populations require testing both mechanical and enzymatic methods to determine empirically the optimum method of dissociation-specific types of solid tumors.

IV.   Mechanical Cell Dispersion

Mechanical dispersion is most effective in tumors without well-developed desmosomes, such as colon, breast, bladder, and sarcoma. This method results in mechanical release of cells from neoplastic tissue into media; the cells are then harvested and analyzed.

Isolation Procedure

1. Fresh tumor (or normal tissue) is placed in standard tissue culture medium, such as RPMI–1640 (GIBCO Laboratories) supplemented with 5% fetal bovine serum (FBS) and 1% penicillin–streptomycin, and processed immediately.

Storage of specimen for as long as 72 hours at 4° C in media is possible and does not appear to affect DNA quantitation procedures.

2. Trim tissue of fatty, necrotic, and other extraneous tissue.
3. Place trimmed tissue in Petri dish and mince using two scalpel blades in a criss cross motion. Scissors should be avoided because they tend to "mash" the tissue rather than slicing it. Soft or medullary tumors will literally fall apart and the medium will become cloudy.
4. Sieve solution through 100-mesh cell sieve and rinse.
5. Count cells and determine viability; vital stain tests such as 0.5% trypan blue dye exclusion are relatively easy to perform.
6. Centrifuge for 10 minutes at 1000 rpm and resuspend cell pellet in 1 ml media.
7. Process as desired for FCM studies.

For scirrhous tumors (e.g., breast adenocarcinomas), disaggregation can be accomplished (substituting in step 3 above) by scraping the tissue surface (instruments such as a glass slide or scalpel blade work well) and rinsing the scraper in culture medium. Scirrhous tumors consist primarily of fibrous tissue (desmoplastic tumor response) with a minority of the tumor composed of infiltrating malignant cells. Enzymatic dissociation techniques are extremely inefficient, because the extensive collagen develops into an amorphous aggregate restricting release of the malignant cells. We have found that scraping is effective in some scirrhous neoplasms, while mechanical mincing provides better DNA histograms in other scirrhous breast carcinomas. These two methods of mechanical cell release are influenced by unknown parameters, as one method will often appear to work better in a specific tumor. However, neither technique has resulted in many viable cells, although mechanical mincing dissociation appears more favorable. (Table I). Mucosal tissue also may be easily prepared by opening a section of intestine longitudinally and scraping the mucosal surface

TABLE I

COMPARISON OF THE SCRAPING VERSUS MECHANICAL MINCING PROCEDURES FOR
SCIRRHOUS BREAST ADENOCARCINOMAS

	Scraped			Mechanically minced		
Case	Cell yield (10^6)	Viability (%)	Quality of DNA histogram	Cell yield (10^6)	Viability (%)	Quality of DNA histogram
1	10	2	Unacceptable	30	85	Acceptable
2	3	0	Good	16	3	Good
3	0.04	0	Good	0.44	0	Good
4	2.2	0	Good	8.8	10	Unacceptable
5	0.1	0	Unacceptable	.21	0	Good

with the edge of a glass slide; the slide is then rinsed in media. Eade *et al.* (1981) have immobilized a section of intestine on a wine cork attached to a mechanical engraver and used the vibration to produce single cells as an alternative to scraping.

Another acceptable technique for harvesting cells for single-cell suspensions is fine-needle aspiration samples. Small needles (22–24 gauge) can be inserted into the tumor (*in vivo* or *in vitro*); after producing negative pressure, the needle is moved through the tumor. The needle (and syringe) is rinsed with medium and the suspension strained to disaggregate any small cell clumps. This sampling technique appears to increase the yield of nondiploid cells, although the total cell yields are low, especially in scirrhous neoplasms (Greenebaum *et al.*, 1984; Remvikos *et al.*, 1988).

V.   Enzymatic Cell Dispersion

Enzymatic dissociation procedures are very diverse, since the possible choices of enzyme(s), concentrations, digestion time, and temperature are infinite and must be optimized for each procedure. In general, enzyme choices may be classified into two categories: proteases and collagenases. The former is primarily used to disrupt desmosome structure, while the latter serves to break down extracellular matrix material. Use of a protease in an enzyme disaggregation procedure risks damage to cell surface components and should be avoided if any cell surface markers are to be measured. If this enzyme is necessary, the isolated cells should be washed free of enzyme and placed in tissue culture media at 37°C for 1–2 hours or sufficient time to permit regeneration of the cell surface constituent to be measured. In addition, DNase is routinely used in all enzymatic procedures because it prevents reaggregation of the suspended cells by digesting the extremely sticky free DNA strands released by damaged cells.

Trypsin, a serine protease with specificity for lysine and arginine residues, is the most commonly used protease and has been shown to produce good cell yields in solid-tumor dissociation protocols. However, the enzyme is harsh and may reduce cell viability. Ensley *et al.* (1987b) have reported, for head and neck squamous carcinomas, maximum viable cell yields are obtained with 0.25% trypsin concentrations with incubations of 2 hours. After 2 hours they reported $>90\%$ cell viability (by trypan blue exclusion) with cell yields of 4.5×10^7 cells/g tissue. Bijman *et al.* (1985) also reported similar cell yields for head and neck tumors using overnight incubations at 4°C. However, their viability dropped to about 60–70%. Low viability (55–72%) was also reported by Sitar *et al.* (1987) with tryspin treatment alone.

Neutral protease from *Bacillus polymyxa* has also been used for tumor dispersion in place of tryspin (Allalunis-Turner and Siemann, 1986; Ensley *et al.*, 1987b; Keng *et al.*, 1984). These studies indicated that a 4-fold increase in cell yield is obtained with this protease compared to trypsin. Pronase, a proprietary trade name for a protease from *Streptomyces griseus*, is a broad-spectrum protease that will digest proteins to amino acid residues. This enzyme has also been used in place of trypsin and does appear to give better yields of cells with good preservation of tumor cell representation (Allalunis-Turner, and Siemann, 1986; Ensley *et al.*, 1987b; Keng *et al.*, 1984), but cell surface changes are likely.

The collagenase family of enzymes are widely used to dissociate tissue and are primarily used in tumors containing large amounts of extracellular matrix material. One set of collagenase enzymes selectively cleaves one or all three chains of the collagen triple helix (reviewed in Liotta *et al.*, 1982) of collagens I, II, and III (stromal components). A second family of collagenases that degrade type IV or type V collagen (basement membrane components) has also been reported (reviewed in Liotta *et al.*, 1982). McDivitt *et al.* (1984) have found that the dissociation of breast cancer by type IV collagenase results in good correlations between S-phase cells determined by FCM and [^{3}H]thymidine labeling.

Solid-tumor dissociation may be enhanced by using an enzyme "cocktail" composed of several enzymes. The most commonly used combinations to produce single-cell suspensions are trypsin, neutral protease, or pronase, to disrupt intercellular junctions, along with collagenase, to degrade basement membrane components, and DNase to prevent cell aggregation by DNA fragments. Ensley *et al.* (1987b) has reported an extensive comparison between several individual enzymes and cocktails using a murine tumor model of squamous carcinoma. They found that a combination of trypsin and collagenase over a short period of time produces the best yield with highest viability. Allalunis-Turner and Siemann (1986) also report the optimum enzymatic dissociation methods using both collagenase and pronase.

Most commercially available enzymes contain contaminating proteases, which may or may not adversely affect tumor dissociation. For example, collagenase type I, that which cleaves type I, II, or III collagen, contains clostripain, a protease with trypsinlike activities that has been reported to damage plasma membranes (Hedley *et al.*, 1983). Only highly purified enzymes should be used for a dissociation protocol.

Isolation Procedure

The following method of enzymatic disaggregation has been used successfully for squamous cell carcinoma.

1. Prepare enzyme cocktail consisting of 0.25% trypsin, 0.5 mg/ml collagenase (Worthington, type II, Freehold, NJ) and 0.002% DNase. All enzymes must be of high quality and are purchased from Worthington Biochemicals.
2. Perform mechanical disaggregation procedure (given earlier) and rinse leftover tumor pieces with medium without any additives to remove serum.
3. Place tissue pieces in fresh tube containing 10 ml/1 g tissue of enzyme cocktail warmed to 37°C.
4. Incubate on rocker at 37°C for 1 hour.
5. Sieve solution, add 1–2 ml FBS to released cells to stop enzyme reaction and wash cells.

Disaggregation procedure may be repeated on tumor pieces retained on sieve.

VI. Enucleation Procedure

Analysis of cellular DNA content does not require viable or even intact single-cell populations. Tissue, either fresh, fixed, or frozen, may be used to extract intact nuclei for flow cytometric DNA analysis. Several methods of "one-step" staining and isolation of nuclei have been reported (Thornthwaite *et al.*, 1980; Vindeløv *et al.*, 1983) and are useful for fresh tissue, such as fine-needle aspirates. For archival samples, various proteases have been used (pepsin, collagenase, pronase, trypsin) to extract nuclear suspensions from formalin-fixed, paraffin-embedded tissue blocks (Oud *et al.*, 1986; Schutte *et al.*, 1985; Stephenson *et al.*, 1986; Takamatsu *et al.*, 1981). At present, the most widely accepted is the Hedley enucleation technique (Hedley *et al.*, 1983) utilizing pepsin digestion. Numerous variables, such as enzyme purity and activity units, digestion time, enzyme concentration, and other modifications to the Hedley method, have been attempted to optimize nuclear yields and aneuploid nuclear recovery (Crissman *et al.*, 1988). When compared to dissociated fresh tissues, the nuclear suspensions derived from thick paraffin sections yield inferior histograms or poor resolution (wide CV) due to increased amounts of background debris (Crissman *et al.*, 1988). This remains a serious problem with enucleation procedures and results in loss of sensitivity in identifying near-diploid cell populations and inability to calculate SPF in some tumors.

VII. Paraffin-Embedded Specimens[1]

Nuclei Isolation

1. Trim paraffin block of excess necrotic and stromal tissue.
2. Cut 5–10 sections measuring 50-μm and place into nylon mesh bags prepared by (a) folding a strip of 50-μm nylon mesh (14 × 6 cm) in half, (b) folding tumor long edges over on themselves (0.7 cm overlap), and (c) sewing pleats with cotton thread.
3. Deparaffinize as follows:
   3 × Xylene washes, 10 minutes each
   2 × 100% Ethanol washes, 5–10 minutes each
   2 × 95% Ethanol washes, 5–10 minute each
   2 × 70% Ethanol washes, 5–10 minutes each
   2 × 50% Ethanol washes, 5–10 minutes each
   3 × Distilled-water washes, 5 minutes each
4. Scrape out hydrated tissue with Pasteur pipet into a 5-ml polypropylene tube and gently mash tissue to increase surface area. To prevent drying, 100–200 μl Hanks' balanced salt solution (HBSS) is added.
5. Add 1 ml of warm (36°C) 0.5% pepsin in 0.9% saline, pH 1.5, to tube and vortex. Incubate for 45–60 minutes at 36° C and vortex at 10-minute intervals.
6. Remove residual tissue with forceps or glass rod and save residual tissue. Add equal volume of HBSS to remaining suspension and filter through a 5-cm^2 50-μm nylon mesh using a plastic funnel. Any fragments retained on the nylon are combined with residual tissue.
7. Repeat step 6 two more times.
8. Collect all filtered nuclei and wash three times with HBSS at 1000 rpm for 5 minutes.
9. Adjust nuclei concentration to 1–2 × 10^6 per milliliter.

Staining Procedure

1. Centrifuge 1 ml nuclei (0.5–2 × 10^6 nuclei/ml) at 750 g for 5 minutes.
2. Remove all but 100 μl HBSS and resuspend nuclei by vortexing.

---

[1] A more detailed description is given by Hedley (Chapter 15, this volume).

3. Add 1 ml propidium iodide (Calbiochem–Behring Corp., La Jolla, CA) (0.05 ml/1 ml in 0.9% saline). Stock solution may be stored at 4°C for 3 months in the dark.
4. Add 100 μl, 1 mg/ml 0.9% saline, RNase A and incubate for 30 minutes in the dark at room temperature (RT).
5. Filter samples through 53-μm nylon mesh.

VIII.   Optimizing Procedures

Reduction of background debris, generated by the cell dissociation technique and consisting of broken cells and cell constituents, is of prime importance for flow analysis. Density gradient centrifugation with Ficoll-Paque gradient medium is widely used to collect a live-cell population because it is the simplest and most economical method. Approximately 5 ml of a cell suspension in complete medium is layered onto the top of 5 ml Ficoll-Paque (Pharmacia, Piscataway, NJ) and centrifuged at 500 g for 15 minutes. Viable cells that band at the interface are removed with a pipet and washed several times in complete medium.

More complex techniques for debris removal have also been employed, including cell elutriation (Keng *et al.*, 1984; Moore and Mortari, 1983; Onoda *et al.*, 1988) and immunological methods (Hunt *et al.*, 1985). Most of these techniques are time-consuming and technically difficult to perform.

A second approach to optimize flow studies is to use tissue-specific markers to gate the flow cytometer selectively to measure only epithelial or other cells. Crissman *et al.* (1989) and Feitz *et al.* (1985) have taken advantage of this two-color multiparametric technique to eliminate background and increase sensitivity. Tumor cell suspensions are stained for cytokeratin and analysis gates are created to generate DNA histograms only from those cells that are stained positive for cytokeratin and DNA, eliminating normal and inflammatory cells from analysis.

A two-color multiparameter analysis of carcinomas using cytokeratin–DNA staining provides the following advantages in the analysis of FCM DNA histograms.

1. Limitation of DNA analysis to urothelial, colonic, and breast epithelial cells by generation of cytokeratin-gated histograms.
2. Separation of near-diploid epithelial G_0/G_1 peaks from the intrinsic diploid standard, thereby increasing sensitivity of identifying near-diploid stem lines.

3. Clarification of tetraploid tumor populations that overlap G_2/M regions by excluding dilutional nonepithelial peaks and enhancing identification of corresponding octoploid peaks.
4. Elimination of overlapping histograms of FCM aneuploid tumors and cell cycle contribution of diploid nonepithelial cells for more accurate cell cycle calculations.

In addition, use of a two-color staining method with FITC-labeled antibody to the leukocyte common antigen (LCA) enables correct identification of a patient-specific intrinsic diploid (LCA^+) reference standard obtained from leukocytes within each individual solid tumor (Zarbo and Crissman, 1989).

REFERENCES

Allalunis-Turner, M. J., and Siemann, D. W. (1986). *Br. J. Cancer* **54**, 615–622.

Bijman, J. T., Wagener, D. J. T., van Rennes, H., Wessels, J. M. C., and van den Broek, P. (1985). *Cytometry* **6**, 334–341.

Crissman, J. D., Zarbo, R. J., Niebylski, C. D., Corbett, T., and Weaver, D. D. (1988). *Mod. Pathol.* **1**, 198–204.

Crissman, J. D., Zarbo, R. J., MA, C. K., and Visscher, D. W., (1989). *Pathol. Annu.* **24**, in press.

Eade, O. E., Andre-Ukena, S. S., and Beeken, W. L. (1981). *Digestion* **21**, 25–32.

Engelholm, S. A., Spang-Thomsen, M., Brunner, N., Nohr, I., and Vindelov, L. L. (1985). *Br. J. Cancer* **51**, 93–98.

Ensley, J., Maciorowski, Z., Pietraszkiewicz, H., Hassan, M., Crissman, J., and Valdivieso, M. (1987a). *Am. Assoc. Cancer Res.* **28**, 33 (Abstr. 130).

Ensley, J. F., Maciorowski, Z., Pietraszkiewicz, H., Klemic, G., Kukuruga, M., Sapareto, S., Corbett, T., and Crissman, J. (1987b). *Cytometry* **8**, 479–487.

Feitz, W. F., Beck, H. L. M., Smeets, A. W. G. B., Debruyne, F. M. J., Vooiis, G. P., Herman, C. J., and Ramaekers, F. C. S. (1985). *Int. J. Cancer* **36**, 349–356.

Greenebaum, E., Koss, L. G., Sherman, A. B., and Elequin, F. (1984). *Am. J. Clin. Pathol.* **82**, 559–564.

Hedley, D. W., Friedlander, M. L., Taylor, I. W., Rugg, C. A., and Musgrove, E. A. (1983). *J. Histochem. Cytochem.* **31**, 1333–1335.

Hefley, T. J., Stern, P. H., and Brand, J. S., (1983). *Exp. Cell Res.* **149**, 227–236.

Hunt, J. M., Buckley, M. T., Laishes, B. A., and Dunsford, H. A. (1985). *Cancer Res.* **45**, 2226–2233.

Keng, P. C., Rubin, P., Constine, L. S., Frantz, C., Narissa, N., and Gregory, P. (1984). *Int. J. Radiat. Oncol. Biol. Phys.* **10**, 1913–1922.

Liotta, L. A., Thorgeirsson, U. P., and Garbisa, S. (1982). *Cancer Metastasis. Rev.* **1**, 277–288.

McDivitt, R. W., Stone, K. R., and Meyer, J. S. (1984). *Cancer Res.* **44**, 2628–2633.

Moore, K., and Mortari, F. (1983). *Br. J. Exp. Pathol.* **64**, 354–360.

Onoda, J. M., Nelson, K. K., Grossi, I. M., Umberger, I. A., Taylor, J. D., and Honn, K. V. (1988). *Proc. Soc. Exp. Biol. Med.* **187**, 250–255.

Oud, P. S., Hanselaar, T. G. J. M., Reubsaet-Veldhuizen, J. A. M., *et al.* (1986). *Cytometry* **7**, 595–600.

Remvikos, Y., Magdelenat, H., and Zajdela, A. (1988). *Cancer* **61**, 1629–1634.
Schutte, B., Reynders, M. M. J., Bosman, F. T., and Blijham, G. H. (1985). *Cytometry* **6**, 26.
Sitar, G., Brusamolino, E., Scivetti, P., and Borroni, R. (1987). *Haematologica* **72**, 23–28.
Stephenson, R. A., Gay, H., Fair, W. R., and Melamed, M. R. (1986). *Cytometry* **7**, 41–44.
Takamatsu, T., Nakanishi, K., Fukada, M., and Fujita, S., (1981). *Histochemistry* **71**, 161–170.
Thornthwaite, J. T., Sugarbaker, E. V., and Temple, W. I. (1980). *Cytometry* **1**, 229–237.
Vindelov, L. L., Christensen, I. J., and Nissen, N. I. (1983). *Cytometry* **3**, 323–327.
Zarbo, R. J., and Crissman, J. D. (1989). *Lab. Invest.* **60**, 109A.

Chapter 2

Flow Cytometric Measurement of Cell Viability Using DNase Exclusion

OSKAR S. FRANKFURT[1]

Grace Cancer Drug Center
Roswell Park Memorial Institute
Buffalo, New York 14263

I. Introduction

Cells that have lost plasma membrane integrity and become permeable to external compounds, including dyes and enzymes, are considered to be nonviable and do not function metabolically. Measurements of cell viability may be of importance for characterization of cell populations exhibiting spontaneous cell death such as solid tumors or cells affected by cytotoxic agents. Nonviable cells may bind antibodies or other cellular markers nonspecifically, and therefore should be excluded from flow cytometric (FCM) measurements.

The method described here is a procedure to obtain selective DNA distributions for viable cells only and to determine viability for different cell subsets, including cell cycle phases. The method is based on the treatment of nonfixed cells with DNase to eliminate DNA from nonviable cells, removal of DNase, and staining of cells with DNA fluorochrome in the presence of a detergent (Frankfurt, 1982, 1983).

---

[1] Present address: Oncology Laboratory, Cedars Medical Center, Miami, Florida 33136.

13

II.   Materials and Methods

A.   Reagents

Solution SMT (sucrose–magnesium–Tris)

1. Dissolve 2.43 g Tris (hydroxymethyl) aminomethane in 800 ml distilled water (dH$_2$O).
2. Adjust to pH 6.5 with concentrated HCl.
3. Add 85.6 g sucrose and 1.016 g MgCl$_2$.
4. Add dH$_2$O to 1 l.

DNase Solution

Dissolve 100 mg DNase (Deoxyribonuclease I, Worthington) in 200 ml SMT solution. Store DNase solution frozen at −20°C in small portions.

Staining Solution

Prepare RPMI-1640 medium containing 10% calf serum, 0.2% Triton X-100, 1 μg/ml 4′,6-diamidino-2-phenylindole (DAPI), and 2 × 10^5 cells/ml chicken red blood cells (CRBC).

B.   Procedure

DNase Treatment

1. Distribute cell suspension into two tubes, each containing 6 × 10^6 cells.
2. Centrifuge cells at 1000 rpm for 5 minutes.
3. Resuspend cells in one tube in 1 ml SMT solution (tube S) and resuspend cells in the other tube in 1 ml DNase solution (tube D).
4. Incubate both tubes at 37°C for 15 minutes.

Staining

1. Place tubes S and D on ice for 5–10 minutes.
2. Rinse cells with RPMI-1640 medium containing 10% fetal or newborn bovine serum.
3. Resuspend cells in 2.5 ml staining solution.

Flow Cytometry

1. Measure DNA distributions from samples S and D until 10^4 CRBC are counted. This is accomplished by integrating CRBC peak during measurement.

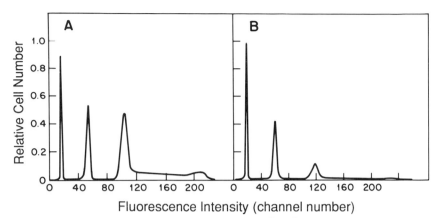

Fig. 1. DNA histograms of Lewis lung carcinoma (LLC) cells mixed with chicken red blood cells (CRBC). Suspensions of cells were obtained by mechanical disaggregation and stained with DAPI without (A) or with (B) DNase treatment. Left-hand peak represents CRBC. G_1 peak of diploid host cells appears in channel 55 and G_1 peak of tetraploid tumor cells in channel 110.

2. Integrate the number of cells other than CRBC in the DNA histograms from samples S and D. These data show a number of nonviable plus viable cells per 10^4 CRBC in sample S and a number of viable cells per 10^4 CRBC in sample D.

Examples of Calculations

1. In the DNA histogram of Lewis lung carcinoma (LLC: Fig. 1), the number of diploid cells in sample S (A) was 2×10^4; in sample D (B) it was 1.5×10^4. Viability of these cells is $(1.5 \times 10^4/2 \times 10^4) \times 100 = 75\%$. For aneuploid cells viability was 40% (2×10^4 in sample D versus 5×10^4 in sample S).
2. The proportion of cells in S phase among aneuploid tumor cells was 35% in sample S and 37% in sample D. Viability of S-phase cells was determined by multiplying the viability of all aneuploid cells (40%) by the ratio of the S-phase index in sample D to the S-phase index in sample S: $40\% \times 37/35 = 42.3\%$.

III. Results and Discussion

A. DNase Treatment

Two cell types were seen in DNA histograms of mechanically disaggregated solid LLC (Fig. 1A). Diploid cells were represented by cells in G_1

phase and presumed to contain only nontumor host cells. Tumor cell population consisted of actively proliferating tetraploid cells. After treatment with DNase the population of diploid cells was reduced only slightly, whereas most of the tetraploid cells became nonstainable (Fig. 1B). We assume that all nonviable cells that lost membrane integrity are permeable to DNase and define the proportion of viable cells (i.e., cells excluding DNase) as a percentage of cells that retained DNA after DNase treatment, stainable by DNA fluorochrome and registered by FCM. DNA content was similar in viable cells and in the cells from which DNA was eliminated by DNase, as relative position of the peaks was the same for both populations (Fig. 1).

To estimate optimal conditions of enzyme treatment in the DNase exclusion test, suspensions of LLC cells were treated with 0.5 mg/ml DNase for 5–40 minutes and cell viability was estimated by FCM analysis. Viability of tetraploid LLC cells was 15.9% after 5 minutes of DNase treatment. Treatment with DNase for an additional 35 minutes did not change significantly the percentage of viable cells. Increase of DNase concentration to 1 mg/ml did not affect the results. Treatment with 0.1–0.2 mg/ml DNase was not effective, as DNA-containing cell debris obscured peaks in the DNA histograms.

B.   Spontaneous Cell Death

It is known that spontaneous cell death (i.e., cell death without effects of extracellular physical and chemical factors) occurs in normal and tumor tissues and in aging tissue cultures. We characterized spontaneous cell death in LLC tumor by the DNase permeability test with regard to cell type and cell cycle position. DNase treatment selectively eliminated tetraploid tumor cells in comparison with diploid host cells (Fig. 1). Viability of diploid host cells was in the range of 80–100% and that of tetraploid cells 10–50% in 1- to 2-week-old tumors. Cell cycle distribution was similar for total and viable populations of tetraploid cells, with S-phase index $35 \pm 6\%$ and $32 \pm 5\%$, respectively. These data show that spontaneous cell death was more common in tetraploid tumor cells compared to normal cells in the same tumor and that cell death was not selective with regard to cell cycle.

C.   Selective Death of Cells Arrested in S/G_2 Phases

DNA distributions of tetraploid LLC cells 24 hours after injection of the alkylating agent cyclophosphamide showed that the proportion of cells

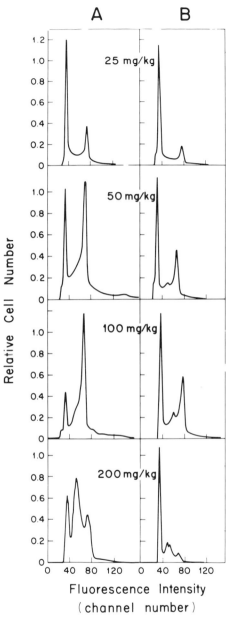

FIG. 2. DNA histograms of tetraploid LLC cells obtained 24 hours after i.p. injection of 25–200 mg/kg cyclophosphamide in mice 7 days after i.m. transplantation of 5×10^5 tumor cells. Cells were stained with DAPI without (A) or with (B) DNase treatment. Threshold was adjusted to eliminate diploid cells during FCM measurement.

blocked in S/G_2 phases was significantly higher in total (viability 3–10%) than in viable cell populations (Fig. 2). Thus, selective death of cells arrested in S/G_2 phases was apparent.

IV. Discussion

The DNase exclusion test permits the characterization of cell death and its relation to cell type and cell cycle phase (Frankfurt, 1982; 1983). This test may also be used in double-parameter FCM assays for selective measurements of viable cells, which retain DNA after DNase treatment. The test may also have broader applications, for example, in biochemical and autoradiographic analysis of radiolabeled thymidine incorporation after selective elimination of DNA from nonviable cells.

ACKNOWLEDGMENTS

This work was supported by NIH grants CA21071, CA13038, and CA24538.

REFERENCES

Frankfurt, O. S. (1982). *Proc. Am. Assoc. Cancer Res.* **23,** 208.
Frankfurt, O. S. (1983). *Exp. Cell Res.* **144,** 478–483.

Chapter 3

Cell Cycle Phase-Specific Analysis of Cell Viability Using Hoechst 33342 and Propidium Iodide after Ethanol Preservation

ALAN POLLACK

Department of Radiation Therapy
U.T./M.D. Anderson Cancer Center
Houston, Texas 77030

GAETANO CIANCIO

Department of Surgery
Jackson Memorial Hospital
Miami, Florida 33101

I. Introduction

Several techniques have been developed to quantitate cell viability utilizing the differential staining of fluorescent dyes. Propidium iodide (PI), ethidium bromide (EB), fluorescein diacetate (FDA), acridine orange (AO), erythrosin B, and diamidinophenylindole (DAPI) have been used to quantitate cell viability, and some of these have been used in flow cytometric (FCM) assays. Horan and Kappler in 1977 described an FCM assay based on differential PI staining. The permeability of dying cells to PI or EB is analogous to the staining by trypan blue (Edidin, 1970); PI is excluded from viable cells.

Fluorescein diacetate has also been applied to FCM analysis (Hamori *et al.*, 1980). A nonpolar nonfluorescent fatty acid ester, FDA accumulates intracellularly in viable cells after it is hydrolyzed by esterases, yielding free fluorescein (Rotman and Papermaster, 1966). The free fluorescein

19

is polar and is trapped intracellularly. Dead and dying cells remain non-fluorescent or weakly fluorescent. Hamori *et al.* (1980) used FDA in combination with Hoechst 33342 to quantitate DNA content and viability. Another method described by Jones and Senft (1985) combines FDA staining of viable cells with PI staining of dead cells.

A dual-staining technique described by Böhmer (1984, 1985) uses the vital staining of AO in combination with the dead-cell staining of EB. The EB suppresses the DNA-specific green AO fluorescence and permits earlier identification of damaged cells. Using this technique, two steps were observed in the kinetics of cell death following treatment with various chemotherapeutic agents and hyperthermia.

In 1980, Stohr and Vogt-Schaden described an FCM method for the determination of cell cycle phase-specific cell death using Hoechst 33662 and PI. Wallen *et al.* (1983) modified the technique slightly using Hoechst 33342 (HO342) and PI. Hoechst stains all of the cells (live and dead), while PI stains only the dead cells. Hoechst fluorescence is quenched in the dead cells by PI, resulting in an increase in red fluorescence from PI. The resulting bivariate histograms show the equivalent of two single-parameter DNA histograms; one with high blue–low red fluorescence (live) and one with high red–low blue fluorescence (dead). We have subsequently modified the method such that the cells can be preserved for as long as 3 days using 25% ethanol (Ciancio *et al.*, 1988). The method for HO342–PI staining in 25% ethanol will be described.

II.   Materials and Methods

A.   Reagents and Materials for HO342–PI Method

1. HO342 stock solution: 1 mM Hoechst 33342 in distilled water.
2. HO342 working solution: Dilute the Hoechst stock solution 1 : 4 in Dulbecco's phosphate-buffered saline (PBS, calcium and magnesium free).
3. PI working solution: 20 μg/ml PI (Calbiochem–Behring Corp., La Jolla, CA) in PBS.
4. 25% ethanol: Dilute 100% or 95% ethanol to 25% in PBS.

B.   Staining Procedure for HO342–PI Method

During the entire procedure the cells are kept at 4°C. The cells are washed in PBS.

1. Pellet 3×10^5–7.5×10^5 cells in a tube, decant medium, and vortex-mix the pellet.
2. Add 0.1 ml PI working solution, vortex-mix, and keep on ice for 30 minutes.
3. Add 1.9 ml 25% ethanol and vortex-mix.
4. Add 50 μl of HO342 working solution, vortex-mix, and incubate on ice for at least 30 minutes. Samples are stable for 72 hours.
5. Examine under phase-contrast fluorescence microscopy for clumping.

C. Instrumentation for HO342/PI Procedure

A Coulter Electronics (Hialeah, FL) EPICS V flow cytometer with a 5-W argon ion laser at 351 nm (50 mW) was used. The fluorescence emission was first filtered through a 488-nm dichroic filter, with the <488 nm fluorescence filtered through a 450-nm long-pass filter [blue photomultiplier tube (PMT) for HO342 fluorescence]. The >488-nm fluorescence is filtered through a 515-nm long-pass filter, and a 560-nm dichroic filter with the >560-nm fluorescence filtered through a 630-nm long-pass filter (red PMT for PI fluorescence). Samples were kept at 4°C using the Coulter Viable-Cell-Handling-System.

III. Results

Figure 1 shows HEP-2 cells stained with HO342–PI after fixation in 25% ethanol and incubation on ice for 24 hours. Shown are untreated control and freeze-thawed cell samples containing 45% dead cells. The bivariate histogram shows the equivalent of two single-parameter histograms, one for the live and one for the dead cells. The terms "live" and "dead" cells used herein refer to the status of the cells prior to staining and assumes that cells staining with PI or trypan blue are dead or dying cells. Quantitation of the percentage of cells in G_1, S, and G_2/M in the HO342–PI histogram correlates with estimates using PI–FITC to quantitate DNA and nuclear protein (Pollack and Ciancio, 1989). The proportion of dead cells in the HO342–PI histograms correlated closely with proportion of cells staining with trypan blue.

During the first 12 hours after staining there is some skewing of the histograms, particularly the live G_1 peak (Ciancio *et al.*, 1988). After 12 hours there is little skewing. Also, the relationship of the live-cell histogram to the dead-cell histogram remained stable from 12 to 72 hours after staining.

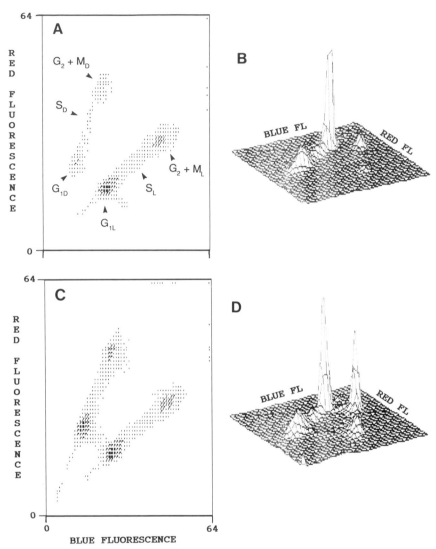

FIG. 1. Flow cytometric analysis of cell viability in Hoechst 33342- and propidium iodide-stained cells preserved with 25% ethanol for 24 hours. (A) and (B) Control HEP-2 cells in exponential growth *in vitro*. The live (G_{1L}, S_L, $G_2 + M_L$) and dead (G_{1D}, S_D, $G_2 + M_D$) cells are labeled in panel (A). (C) and (D) HEP-2 cells after freeze-thawing twice (45% dead by both trypan blue and HO342–PI).

The addition of 25% ethanol after PI staining serves three purposes. First, the cells are permeabilized, facilitating HO342 staining without incubation at 37°C. Second, fixation in ethanol results in prolonged stability of the staining over 72 hours. Third, the PI concentration in the dead cells remains at significantly higher concentrations than the live cells despite permeabilization. Cells treated with 25% ethanol show 100% trypan blue uptake, yet the differential PI staining between live and dead cells remains. In contrast, cells permeabilized with Triton X-100 after PI staining show no differential staining of the live and dead cells. These data suggest that the treatment with ethanol results in the decreased intercalation of PI into those cells that are living at the time of staining as compared to the cells that are dead at the time of staining (Ciancio et al., 1988).

IV.   Standards

The results of the FCM assay should be compared to another method of determining cell viability, such as trypan blue. The comparison of the results of dye exclusion assays with colony-forming assays will result in significant differences, primarily because of the intrinsic differences in what is being measured. Dye exclusion assays quantitate immediate cell death in a defined window of time, while colony-forming assays quantitate the cumulative effect and are concerned only with reproductive ability of the cell (Ciancio et al., 1988; Weisenthal et al., 1983). Thus, to assess cell loss accurately over time, both cell number and cell viability must be determined at multiple time points.

The major live-cell G_1 peak was used to align the histograms. There is some shifting in the relationship of the live and dead cell populations to one another and there is some skewing of the histograms (mostly in G_1), particularly during the first 12 hours after staining. Therefore, an unperturbed control sample should be run with each group of samples when the analysis is at different times after staining (e.g., 12 hours vs. 24 hours).

V.   Summary of Critical Aspects of the Procedure

Several different concentrations of HO342, PI and ethanol were tested and the concentrations used herein were derived empirically. The final optimal PI concentration, after dilution with ethanol, was found to be

1 μg/ml. This concentration appears to be close to saturation of DNA-binding sites (Darzynkiewicz et al., 1984), although this may vary with the cell type used. Some adjustment in the concentrations of these reagents may be required.

The initial procedures described by Stohr and Vogt-Schaden (1980) and by Wallen et al. (1983) for cell cycle phase-specific viability measurements using Hoechst and PI involved incubation of the cells at 37°C in the presence of Hoechst. In the procedure described herein, all staining is performed at 4°C and Hoechst is added after PI staining. The principal difference is the use of ethanol as a preservative. The cells are initially stained with PI and then permeabilized with 25% ethanol. The cells that were viable prior to permeabilization, and hence excluded PI, continue to have less PI staining than the cells that were dead prior to permeabilization. Hoechst 33342 staining after ethanol permeabilization is rapid, even at 4°C.

References

Böhmer, R. M. (1984). *Cell Tissue Kinet.* **17,** 593–600.

Böhmer, R. M. (1985). *Cytometry* **6,** 215–218.

Ciancio, G., Pollack, A., Taupier, M. A., Block, N. L., and Irvin, G. L., III (1988). *J. Histochem. Cytochem.* **36,** 1147–1152.

Darzynkiewicz, Z., Traganos, F., Kapuscinski, J., Staiano-Coico, L., and Melamed, M. R. (1984). *Cytometry* **5,** 355–363.

Edidin, M. (1970). *J. Immunol.* **104,** 1303–1306.

Hamori, E., Arndt-Jovin, D. J., Grimwade, B. G., and Jovin, T. M. (1980). *Cytometry* **1,** 132–135.

Horan, P. K., and Kappler, J. W. (1977). *J. Immunol. Methods* **18,** 309–316.

Jones, K. H., and Senft, J. A. (1985). *J. Histochem. Cytochem.* **33,** 77–79.

Pollack, A., and Ciancio, G., (1989). *In* "Flow Cytometry II: Advanced Research and Clinical Applications" (A. Yen, ed.), pp 29–48. CRC Press, Boca Raton, FL.

Rotman, B., and Papermaster, B. W. (1966). *Proc. Natl. Acad. Sci. U.S.A.* **55,** 134–141.

Stohr, M., and Vogt-Schaden, M. (1980). *In* "Flow Cytometry IV" (O. D. Laerum, T. Lindmo, and E. Thorud, eds.), pp. 96–99. Bergen, Universitetsforlaget.

Wallen, C. A., Higashikubo, R., and Roti Roti, J. L. (1983). *Cell Tissue Kinet.* **16,** 357–365.

Weisenthal, L. M., Dill, P. L., Kurnick, N. B., and Lippman, M. E. (1983). *Cancer Res.* **43,** 258–264.

Chapter 4

Cell Membrane Potential Analysis

HOWARD M. SHAPIRO

Howard M. Shapiro, M.D., P.C.
West Newton, Massachusetts 02165

I. Introduction

Cytoplasmic and mitochondrial membrane potential (MP) changes may occur during the early stages of surface receptor-mediated activation processes related to the development, differentiated function, and pathology of a large number of cell types, and are thought to play a role in signal transduction between the cell surface and the interior. Investigations in this area have been facilitated by the development of methods for flow cytometric (FCM) estimation of MP in single cells (Shapiro *et al.*, 1979; Shapiro, 1981, 1982, 1988). This chapter discusses the principles, practical aspects, and limitations of such methods.

A. The Basis of Membrane Potentials

Electrical potential differences across eukaryotic cell membranes are due in part to concentration gradients of Na^+, K^+, and Cl^- ions across the cell membrane, and in part to the operation of electrogenic pumps. The potential differences across the cytoplasmic membranes of resting mammalian cells range in magnitude from ~ 10 to 90 mV, with the cell interior negative with respect to the exterior. There is also a potential difference of ≥ 100 mV across the membranes of energized mitochondria, with the mitochondrial interior negative with respect to the cytosol; this potential is dependent on energy metabolism. In prokaryotes, the enzymes responsible for energy metabolism are located on the inner surface of the cytoplasmic membrane, and most of the potential difference across this membrane, which is typically 100–200 mV, is generated by energy metabolism.

25

B.   Potential Measurement by Indicator Distribution

Membrane potential can be estimated from the distribution of lipophilic cationic indicators or dyes between cells and the suspending medium. Lipophilicity, or hydrophobicity—that is, a high lipid/water partition coefficient—enables dye molecules to pass freely through the lipid portion of the membrane; the concentration gradient of a cationic dye, $[C^+]$, across the membrane is determined by the transmembrane potential difference according to the Nernst equation:

$$[C^+]_i/[C^+]_o = e^{-F\Delta\Psi/RT}$$

where $[C^+]_i$ is the concentration of C^+ ions inside the cell, $[C^+]_o$ is the concentration of C^+ ions outside the cell, $\Delta\Psi$ is the MP, R is the gas constant, T is the temperature in degrees Kelvin, and F is the Faraday. Indicators or dyes used in this fashion are referred to as distributional probes of MP.

Once cells have been equilibrated with a cationic dye, electrical depolarization of the cells (i.e., a decrease in MP) will cause release of dye from cells into the medium, while hyperpolarization (i.e., an increase in MP) will make cells take up additional dye from the medium. The dye distribution will not adequately represent the new value of potential until equilibrium has again been reached; this process requires periods ranging from a few seconds to several minutes. Thus, while distributional probes may be suitable for detection of slow potential changes, they cannot be used to monitor the rapidly changing action potentials in excitable tissues such as nerve and muscle.

The use of cyanine dyes as MP probes began with the study of Hoffman and Laris (1974), who used 3,3'-dihexyloxacarbocyanine [DiOC$_6$(3) in the common notation introduced by Sims et al. (1974)] to estimate MP in red blood cell (RBC) suspensions based on the partitioning of the dye into the cells. They noted that addition of RBC to a micromolar dye solution in a spectrofluorometer cuvet produced a suspension with lower fluorescence than that of the original solution, indicating that—at these concentrations—the fluorescence of cyanine dyes taken into cells is quenched. When extracellular ion concentrations were manipulated so as to hyperpolarize the cells, increasing cellular uptake of dye, the fluorescence of the suspension decreased further; when the cells were depolarized, releasing dye into the medium, the fluorescence of the suspension increased. Membrane potential measurements of giant *Amphiuma* RBC using cyanine dyes compared favorably with results obtained from microelectrode measurements.

Under normal circumstances, the intracellular concentration of K^+ ions is considerably higher than the extracellular concentration, while the intracellular concentration of Na^+ ions is considerably lower than the extracellular concentration. Valinomycin (VMC), a lipophilic potassium-selective ionophore, forms complexes with K^+ ions and thus readily transports them across cell membranes, effectively increasing the potassium permeability of cells to the point at which MP is determined almost entirely by the transmembrane K^+ gradient. Addition of VMC hyperpolarizes cells if the external K^+ concentration is lower than the internal concentration and depolarizes them if it is higher. The ionophore gramicidin A, which forms channels passing Na^+, K^+, and other ions, depolarizes cells in solutions with approximately physiologic ionic concentrations.

C.   Single-Cell Measurements with Distributional Probes: Principles and Problems

Potential estimation by fluorometry of cell suspensions in cuvets requires dye concentrations sufficiently high that the fluorescence of intracellular dye is largely quenched, so that most of the measured fluorescence comes from free dye in solution. When cells are hyperpolarized, they take up more dye from the solution, decreasing total fluorescence. When cells are depolarized, dye molecules that were quenched when inside the cell are released into solution, increasing total fluorescence.

In cytometric estimation of MP, it is the fluorescence of intracellular dye that is measured. The dye concentrations used are lower than those used for bulk measurements, to minimize quenching of intracellular dye. In principle, MP could be calculated accurately from measurements of intracellular and extracellular dye concentrations using the Nernst equation (Ehrenberg *et al.*, 1988). However, the accuracy of the FCM procedure is limited, for several reasons.

The flow cytometer measures the amount, not the concentration, of dye in each cell. To find the intracellular concentration, it would be necessary to divide the fluorescence value for each cell by the cell's volume, obtained by an electronic (Coulter) volume sensor or estimated from forward-scatter or extinction measurements. Measuring the extracellular concentration of dye is more problematic; this would require signal-processing electronics of a type not normally used in flow cytometers. However, the large variances of fluorescence distributions obtained from conventional FCM measurements of cyanine dye fluorescence in cells at a known MP suggest that accuracy would not be significantly increased by the instrumental refinements just discussed.

One possible source of fluorescence variance is mitochondrial uptake of dye. Since the mitochondrial MP is typically $\geqslant 100$ mV negative with respect to the cytosol, dye should be present in energized mitochondria at almost 100 times the concentration found in the cytoplasm. The mitochondrial MP, in fact, provides the basis for the accumulation in mitochondria of cationic dyes such as rhodamine 123 (Johnson *et al.*, 1980, 1981; Darzynkiewicz *et al.*, 1981), pinacyanol, and Janus green. However, the high concentration of dye in mitochondria should result in substantial quenching. Indeed, in at least some cell types, treatment with metabolic inhibitors and uncouplers that abolish the mitochondrial MP gradient, which should eliminate concentration of dye in mitochondria, reduces neither the intensity nor the variance of cyanine dye fluorescence.

In cells depolarized with gramicidin A, the MP should be zero; the intracellular and extracellular dye concentrations predicted by the Nernst equation should therefore be equal, and the very fact that the flow cytometer can detect the cells' fluorescence above background requires some explanation. One reason for the cells' increased fluorescence is that the fluorescence of cyanine dyes is enhanced [~6-fold in the case of $DiOC_6(3)$, less for other dyes] when the dye is in a hydrophobic environment (Sims *et al.*, 1974) such as the membranous structures in which it can be observed in cells.

Perhaps even more important, the lipophilic, hydrophobic character of cyanine dyes causes them to be concentrated in cells even in the absence of a potential gradient. Similar problems with such "non-Nernstian" probe binding are also encountered when membrane potential is estimated with radiolabeled lipophilic cations such as [^{3}H]triphenylmethylphosphonium ($TPMP^+$) or with less hydrophobic cationic dyes (Ehrenberg *et al.*, 1988). In order to get accurate values of cytoplasmic MP using these indicators, it is necessary to inhibit the mitochondria and to correct for probe uptake by cells in the absence of a potential gradient. The affinity of hydrophobic cyanine dyes for cells is sufficiently high that cells may take up most of the dye molecules even in a suspension in which the cells occupy only a fraction of a percentage of the total volume. This makes it necessary to keep cell concentrations relatively constant from sample to sample in order to obtain reproducible results.

Flow cytometry of cells exposed to increasing concentrations of cyanine dyes shows that saturation occurs; that is, eventually, further increasing the dye concentration does not increase fluorescence in the cells. For $DiOC_6(3)$, this happens when cells at a concentration of 10^6/ml are incubated with 2 μM dye. The variance of the fluorescence distribution remains large. It is likely that most of the fluorescence measured in cells comes from dye in hydrophobic regions, which represent both the highest affinity binding sites and the sites in which dye fluorescence yield would be highest; the variance of fluorescence would then be explained by cell-to-cell variations in the number of binding sites. It has been observed (Shapiro, 1981) that, when dye-binding sites are saturated, cellular fluorescence does not change when cells are depolarized or hyperpolarized by ionophores; however, such potential changes would be detectable by bulk fluorometry in a cell suspension exposed to the 2 μM saturating concentration of $DiOC_6(3)$. Thus, at this concentration, the cells contain dye that is essentially nonfluorescent as a result of quenching, as well as dye bound to the hydrophobic sites in which fluorescence is enhanced.

The dye concentration at which saturation of binding sites occurs is determined primarily by the hydrophobicity of the dye; the fluorescence of cells equilibrated (at 10^6/ml) with 2 μM diethyloxacarbocyanine [$DiOC_2(3)$], which is less hydrophobic than $DiOC_6(3)$, is less than the fluorescence of cells equilibrated with the C_6 dye, and does change when MP is changed by ionophore addition or by manipulation of ion concentrations in the medium, indicating that the saturating concentration for the C_2 dye is higher than for the C_6 dye. When cells in 2 μM $DiOC_2(3)$ in NaCl (normal MP, higher fluorescence) are mixed with an equal volume of cells in 2 μM $DiOC_2(3)$ in KCl (depolarized, lower fluorescence), the cells and dye reequilibrate within a few minutes, yielding a fluorescence distribution that reflects the intermediate value of MP resulting from the ionic composition of the mixed suspending medium. Thus, the fluorescence of cell-associated cyanine dye can provide a reasonably rapid indication of substantial changes in MP, even if fluorescence variance limits overall accuracy and precision.

Fluorescence variance also limits the capability of FCM measurements to detect heterogeneous responses within cell populations; two subpopulations of equal size with a mean difference of 50% in fluorescence intensity will be clearly resolved, while a 5% subpopulation with a mean 10% above the population mean might be undetectable.

II. Materials and Methods

A. Dyes and Reagents

1. CYANINE DYES

The choice of dye is determined primarily by the excitation wavelength(s) available and the emission wavelengths desired. Dihexyl- or dipentyloxacarbocyanine [$DiOC_6(3)$ or $DiOC_5(3)$] or hexamethylindocarbocyanine [$DiIC_1(3)$] are both suitable for blue-green (488 nm) excitation. $DiIC_1(3)$ can also be used with green (515–546 nm) excitation. $DiOC_6(3)$

and $DiOC_5(3)$ are green fluorescent and can be used with the same detector–filter combination used for fluorescein; $DiIC_1(3)$ is orange fluorescent and works well with filters designed for phycoerythrin. Hexamethylindodicarbocyanine [$DiIC_1(5)$; sometimes also known as HIDC] can be excited with a red (633 nm) He–Ne laser and measured through 665-nm glass filters using an R928 or other red-sensitive photomultiplier tubes. The dyes are available from Molecular Probes (Eugene, OR) and other sources.

Stock solutions with a 1 mM concentration of any of the dyes mentioned in dimethyl sulfoxide (DMSO) may be kept for several weeks in the dark at room temperature (RT). Working solutions are made by diluting the stock solution with absolute ethanol or DMSO to allow the desired final dye concentration to be obtained by adding 5 μl of working solution to each 1 ml of cell suspension. For $DiOC_5(3)$ and $DiOC_6(3)$, a 10 μM working solution and a 50 nM final concentration are appropriate; for $DiIC_1(3)$ and $DiIC_1(5)$, a 20 μM working solution yields a 100 nM final concentration.

2.   IONOPHORES

Valinomycin is used to hyperpolarize cells in high-Na^+, low-K^+ media and to depolarize cells in high-K^+ media. A 1 mM stock solution in DMSO is stable for several weeks at RT; adding 5 μl of stock solution to a 1-ml cell sample produces a 5 μM concentration of VMC.

Gramicidin D is used to depolarize cells; this material is a mixture containing a variable percentage of gramicidin A, which is the active ingredient. A stock solution of 1 mg/ml gramicidin D in DMSO is stable for several weeks at RT; addition of 5 μl stock solution to 1 ml of cell suspension yields a final concentration of $\sim$ 20 μM gramicidin A.

The proton ionophore carbonyl cyanide chlorophenylhydrazone (CCCP), an uncoupler and mitochondrial inhibitor, is an effective depolarizing agent for both aerobic and anaerobic bacteria. A final concentration of 5 μM is obtained by adding 5 μl of a 1 mM stock solution in DMSO to 1 ml of dilute bacterial suspension.

B.   Staining Procedures

Cyanine dye equilibration with cells in protein-free media is usually complete after 15 minutes at RT. For cells in media containing protein, 30 minutes' incubation at 37°C generally suffices. The cell concentration should be kept relatively constant from sample to sample, since the high affinity of the dye for cell constituents otherwise produces variations in fluorescence intensity per cell with cell concentration. Concentrations in the range of 10^6 cells/ml give satisfactory results. Cells are not washed prior to analysis.

C. Flow Cytometry and Data Analysis

The incubation temperature and the interval between dye addition and introduction of samples into the flow cytometer should be kept as nearly constant as possible; it is also advisable to run all samples in an experiment at the same flow rate to avoid artifacts due to dye diffusion.

While cyanine dye fluorescence signals are generally strong enough to be used as gating signals, it is preferable to use a forward-scatter signal to avoid missing depolarized cells. Since MP is meaningful only in viable cells, although cyanine dyes will stain dead cells and debris, using the scatter signal for gating allows extraneous signals to be excluded during data collection. Pronounced changes in scatter signals following exposure of cells to some stimulus to be tested suggest that an apparent MP effect may be secondary to a more obvious change such as membrane lysis.

When using $DiOC_5(3)$ or $DiOC_6(3)$ as an MP probe, it is convenient to add propidium iodide (5–20 μg/ml final concentration) to the cell suspension, allowing dead (i.e., membrane-damaged) nucleated cells to be gated out of analyses on the basis of their strong red fluorescence.

The effects of many agents on MP of homogeneous cell populations may be appreciated by simple comparison of single-parameter fluorescence distributions such as those shown in Fig. 1. In other situations such as observation of stimulated lymphocytes, multiparameter analysis that, for

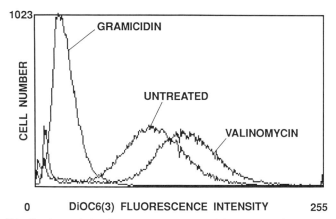

FIG. 1.   Distributions of the fluorescence of $DiOC_6(3)$ in CCRF-CEM lymphoblasts in phosphate-buffered saline (pH 7.4), initially incubated with 50 nM dye for 15 minutes. Fluorescence histograms are shown for untreated cells, for cells hyperpolarized after an additional 10 minutes' incubation with 5 μM valinomycin (VMC), and for cells depolarized after a further 10 minutes' exposure to 50 μg/ml gramicidin following the VMC treatment. Each histogram represents 25,000 cells; fluorescence measurements were made in a "Cytomutt" flow cytometer (Shapiro, 1988) with 21-mW excitation at 488 nm, using forward-scatter signals for gating.

example, relates MP to nuclear DNA content in cells vitally stained with Hoechst 33342 (Shapiro, 1988), may be preferable. Rapid changes in MP are best appreciated when time is used as a measurement parameter.

D. Controls

When an experiment extends over a period of a few hours, control samples should be run during the course of the experiment as well as at the beginning and end. Controls should include an untreated cell sample, a sample of cells hyperpolarized by addition of 5 μM VMC, and another sample depolarized by addition of 50 $\mu g/ml$ gramicidin.

A more dynamic control procedure, which verifies that cells are capable of response to hyperpolarizing and depolarizing stimuli, can be implemented by examining one aliquot of a sample of cells after they have equilibrated with dye, then adding VMC and analyzing a second aliquot after 5–10 minutes, and finally adding gramicidin and analyzing a third aliquot after an additional 5–10 minutes. The data shown in Fig. 1 are typical of the results of such a procedure. Valinomycin should increase the fluorescence of the cells; gramicidin should decrease the fluorescence well below the initial value (and, incidentally, render cells unresponsive to subsequent treatment with VMC or other hyperpolarizing agents).

While the large variance (CV values typically $\sim 30\%$) of cyanine dye fluorescence distributions speaks against attempting to calibrate flow MP distributions to absolute voltage, the nature and magnitude of changes observed by different investigators in many cell systems have been fairly consistent. On several occasions, cyanine dye-stained lymphocytes from different donors, run on the same flow cytometer on successive days, have yielded distributions with peaks within 5% of one another. This suggests that both MP and the amount of dye-binding sites per cell are relatively well regulated.

The control procedures just described make it possible to establish that the cells under study have a detectable MP and that the dye in use will respond to potential changes in either direction from the control value. All of this is prerequisite to any investigation of the effects of biologic, chemical, and/or physical agents on cell MP.

E. Pitfalls and Cautions

Addition of any substantial amount of protein to a cell suspension that has been equilibrated with dye in a protein-free medium will decrease the dye concentration in cells, because the dye will bind to the protein in solution. This artifactual apparent depolarization can be avoided by working in a medium with added protein (e.g., 1–10% albumin) when trying to determine effects of adding specific proteins to cells. Similar cautions apply to studies of the effects of solvents such as DMSO, which alter the partitioning of dye between cells and the medium. The total amount of added solvent should be maintained relatively constant from sample to sample and should not exceed 2% of sample volume.

Other problems with cyanine dyes have been discussed at length elsewhere (Shapiro, 1988). At micromolar concentrations, cyanines have been observed to be toxic to bacteria and mammalian cells. The dyes themselves may perturb MP directly by altering membrane conductivity; their inhibition of energy metabolism might also result in potential changes. When used to monitor neutrophil responses to chemotactic peptides, $DiOC_6(3)$ and $TPMP^+$ were reported to give contradictory results, while the thiocyanine dye $DiSC_3(5)$ was found to be destroyed by oxidation following neutrophil activation.

Toxicity is a liability shared by the cyanines with other families of cationic dyes such as acridines, safranins, oxazines, pyronins, rhodamines, and triarylmethanes, and with lipophilic cations such as $TPMP^+$. When the dyes are used in flow cytometry, at concentrations of 5–100 nM, toxicity is less than when radiolabeled cations or dyes are used at micromolar concentrations for bulk measurements. Different cell types appear to have different degrees of susceptibility to cyanine dye toxicity. Crissman, *et al.*, (1988) found that simultaneous staining with $DiOC_5(3)$ improved Hoechst 33342 staining of live CHO cells, presumably by interfering sufficiently with energy metabolism to block efflux of the dye from cells via a pump mechanism. However, the cyanine dye did not affect cell viability following sorting, which remained ~90%. In this instance, at least, cyanine dye toxicity is evidently entirely reversible.

Oxonols, which are negatively charged lipophilic dyes, bind to the cytoplasmic membrane but do not accumulate in intact cells; this probably explains why they are much less toxic than cyanines and other cationic dyes. Because oxonols do not enter cells, oxonol fluorescence is largely unaffected by mitochondrial potentials. These desirable characteristics of oxonols are offset somewhat by their much weaker fluorescence, as compared to cyanines; the variance of oxonol fluorescence distributions is no better than that of cyanine fluorescence distributions, and oxonol and cyanine dyes produce comparable results in most cases.

F. Mitochondrial Staining with Rhodamine 123

The lipophilic cationic dye rhodamine 123 has been used for investigations of mitochondrial structure and function (Johnson *et al.*, 1980, 1981; Darzynkiewicz *et al.*, 1981); it accumulates in energized mitochondria as a result of their MP. Cells are equilibrated for 30 minutes with 10 μg/ml

($\sim 25~\mu M$) dye, then washed and examined; most of the retained dye is found in mitochondria. Safranin and the less hydrophobic cyanine dyes show similar potential-dependent mitochondrial staining when cells are prepared as they are for the rhodamine 123 procedure.

III. Improving Cytometry of Membrane Potentials

The limits of accuracy and precision with which cell membrane potentials and changes can be measured by FCM are due largely to the population variance of fluorescence intensity, presumably reflecting cell-to-cell differences in numbers of dye-binding sites. This problem is largely eliminated when repeated measurements of the fluorescence of a cyanine dye or other distributional probe of MP are made of the same cell at intervals using a static low-resolution or image cytometer.

Even under the best of circumstances, however, distributional probes could only be used to monitor relatively slow MP changes, occurring over periods of seconds to minutes. Better results would be expected using faster responding dyes, which sense MP by different mechanisms. A number of dyes developed for use by neurophysiologists respond to MP changes by changing their position and/or orientation in the membrane; most do not penetrate to the cell interior. Since all of the transmembrane potential difference is developed across the thickness of the membrane, the electric field strength in the membrane itself can be quite high. The fluorescence or absorption changes seen with fast-response dyes are typically small, on the order of a few percentage points for a 60-mV change in potential; yet such changes, which are difficult to detect reliably in a flow cytometer, are readily measured in a static apparatus.

At present, interest among investigators studying transmembrane signaling has shifted from cytometry of MP to cytometry of intracellular pH and calcium. This is due in part to the relative inconstancy and variable magnitude of MP changes associated with surface ligand–receptor interactions, as compared to changes in calcium concentration and distribution and pH, and in part to better probe technology. Both pH and calcium probes suitable for ratiometric measurements of fluorescence using two excitation or emission wavelengths are now available; a ratiometric procedure eliminates the contribution of cell-to-cell differences in dye binding to measurement variance, yielding narrow distributions and making it easier to identify heterogeneity in populations.

Bacterial membrane potentials are primarily dependent on metabolic activity, of which they are a sensitive indicator. Membrane potential changes substantially and rapidly in response to the availability or lack thereof of suitable energy sources, and is rapidly dissipated when the organism is killed by drugs or other agents. It is thus possible to exploit potential-sensitive dyes in rapid FCM procedures for bacterial detection, indentification, and antibiotic susceptibility testing (Shapiro, 1988). This application, however, could also be facilitated by the availability of ratiometric potential probes.

Loew and co-workers have developed electrochromic probes of MP, which undergo spectral changes in responses to changes in the electric field in the membrane. One such dye, di-4-ANEPPS, has been used for dual-excitation beam ratiometric MP measurements in an image cytometer; it yields a 10% change in fluorescence ratio for a 90-mV potential change (Montana *et al.*, 1989).

Even if probes usable for ratiometric measurements are available, it is obviously more appropriate to assess effects of stimuli on cell MP by measuring the same cell at different times, as is done in a static system, than by measuring different cells at different times and assuming their behavior to be similar, a compromise forced on the experimenter by the nature of FCM. It thus seems likely that further progress in understanding transmembrane signaling will come from the refinement of static cytometric techniques for measurement of MP and of other functional parameters such as intracellular pH, calcium flux, and redox state.

REFERENCES

Crissman, H. A., Hofland, M. H., Stevenson, A. P., Wilder, M. E., and Tobey, R. A. (1988). *Exp. Cell Res.* **174**, 388–396.

Darzynkiewicz, Z., Staiano-Coico, L., and Melamed, M. R. (1981). *Proc. Natl. Acad. Sci. U.S.A.* **78**, 6696–6698.

Ehrenberg, B., Montana, V., Wei, M.-D., Wuskell, J. P., and Loew, L. M. (1988). *Biophys. J.* **53**, 785–794.

Hoffman, J. F., and Laris, P. C. (1974). *J. Physiol. (London)* **239**, 519–552.

Johnson, L. V., Walsh, M. L., and Chen, L. B. (1980). *Proc. Natl. Acad. Sci. U.S.A.* **77**, 990–994.

Johnson, L. V., Walsh, M. L., Bockus, B. J., and Chen, L. B. (1981). *J. Cell Biol.* **88**, 526–535.

Montana, V., Farkas, D. O., and Loew, L. M. (1989). *Biochemistry* **28**, 4536–4539.

Shapiro, H. M. (1981). *Cytometry* **1**, 301–312.

Shapiro, H. M. (1982). U.S. Patent No. 4, 343, 782.

Shapiro, H. M. (1988). "Practical Flow Cytometry" (2nd Ed.). Liss/Wiley, New York.

Shapiro, H. M., Natale, P. J., and Kamentsky, L. A. (1979). *Proc. Natl. Acad. Sci. U.S.A.* **76**, 5728–5730.

Sims, P. J., Waggoner, A. S., Wang, C.-H., and Hoffman, J. F. (1974). *Biochemistry* **13**, 3315–3330.

Chapter 5

Flow Cytometric Measurement of Intracellular Ionized Calcium in Single Cells with Indo-1 and Fluo-3

CARL H. JUNE

Immune Cell Biology Program
Naval Medical Research Institute
Bethesda, Maryland 20814

PETER S. RABINOVITCH

Department of Pathology
School of Medicine
University of Washington
Seattle, Washington 98195

I. Introduction

Measurement of intracellular ionized calcium concentration ($[Ca^{2+}]_i$) in living cells is of considerable interest to investigators over a broad range of cell biology. Calcium has an important role in a number of cellular functions and, perhaps most interestingly, can function to transmit information from the cell membrane to regulate diverse cellular functions. An optimal indicator of $[Ca^{2+}]_i$ should span the range of calcium concentrations found in quiescent cells (~100 nM) to levels measured in stimulated cells (micromolar free Ca^{2+}), with greatest sensitivity to small changes at the lower end of that range. The indicator should freely diffuse throughout the cytoplasm and be easily loaded into small cells. The response of the indicator to transient changes in $[Ca^{2+}]_i$ should be rapid. Finally, the indicator itself should have little or no effect on $[Ca^{2+}]_i$ itself or on other cellular functions.

37

Until 1982 it was not possible to measure $[Ca^{2+}]_i$ in small intact cells, and attempts to measure cytosolic free calcium were restricted mostly to large invertebrate cells where the use of microelectrodes was possible. Bioluminescent indicators such as aequorin, a calcium-sensitive photoprotein, are well suited for certain applications (Blinks *et al.*, 1982). Their greatest limitation is the necessity for loading into cells by microinjection or other forms of plasma membrane disruption. $[Ca^{2+}]_i$ was first measured in diverse populations of cells with the development of quin2 (Tsien *et al.*, 1982). The indicator was easily loaded into small intact cells using a chemical technique developed by Tsien (1981). Cells are incubated in the presence of the acetoxymethyl ester of quin2. This uncharged form is cell permeant and diffuses freely into the cytoplasm, where it serves as a substrate for esterases. Hydrolysis releases the tetraanionic form of the dye, which is trapped inside the cell. Unfortunately, quin2 has several disadvantages that limit its application to flow cytometry (FCM) (Ransom *et al.*, 1986). A relatively low extinction coefficient and quantum yield have made detection of the dye at low concentrations difficult; at higher concentrations, quin2 itself buffers the $[Ca^{2+}]_i$. Grynkiewicz *et al.* (1985) have described a new family of highly fluorescent calcium chelators that overcome most of the aforementioned limitations. One of these dyes, indo-1, has spectral properties that make it especially useful for analysis with FCM. In particular, indo-1 exhibits large changes in fluorescence emission wavelength upon calcium binding (Fig. 1). As described later, use of the

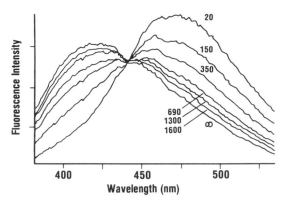

FIG. 1. Emission spectra for indo-1 as a function of ionized calcium. Indo-1 (3 μM, Molecular Probes) was added to a buffer consisting of 21.64 mM K_2H_2–EGTA, 100 mM KCl, 20 mM HEPES, pH 7.20. Small aliquots of a buffer that contained equimolar calcium and EGTA as K_2Ca–EGTA and was otherwise identical to the first buffer were added. Fluorescence excited at 356 nm was measured in a spectrofluorimeter (Tracor Northern Fluoroplex III).

ratio of intensities of fluorescence at two wavelengths allows calculation of $[Ca^{2+}]_i$ independent of variability in cellular size or intracellular dye concentration. For the first time, single-cell measurements of $[Ca^{2+}]_i$ are possible in large numbers of cells. The use of a flow cytometer allows the correlation of $[Ca^{2+}]_i$ with other cell parameters such as surface antigen expression, and furthermore, allows one to sort cells electronically based on $[Ca^{2+}]_i$.

II. Flow Cytometric Assay with Indo-1

A. Loading of Cells with Indo-1

Uptake and retention of indo-1 [(1-(2-amino-5-(6-carboxylindol-2-yl)-phenoxy)-2-(2′-amino-5′-methylphenoxy) ethane $N, N, N′N′$-tetraacetic acid)] is facilitated by the use of the pentaacetoxymethyl ester of indo-1, using the scheme already described. Approximately 20% of the total dye is trapped in this manner during typical loadings. After loading, the extracellular indo-1 should be diluted 10- to 100-fold before FCM analysis (Rabinovitch et al., 1986). One incidental benefit of this loading strategy is that this procedure, like the more familiar use of fluorescein diacetate or carboxyfluorescein diacetate, allows one to distinguish between live and dead cells. The latter will not retain the hydrophilic impermeant dye and can be excluded during subsequent analysis.

The lower limit of useful intracellular loading concentrations of indo-1 is determined by the sensitivity of fluorescence detection of the flow cytometer, and the upper limit is determined by avoidance of buffering of $[Ca^{2+}]_i$ by the presence of the calcium-chelating dye itself. In practice, one should use the least amount of indo-1 that is necessary to quantitate the fluorescence signal reliably. Fortunately, indo-1 has excellent fluorescence characteristics [30-fold greater quantum yield than quin2 (Grynkiewicz et al., 1985)], and useful ranges of indo-1 loading are much lower than the millimolar amounts required with quin2. For human peripheral blood T cells, we have found adequate detection at $\geqslant 1$ μM indo-1 ester, under conditions that achieve ~ 5 μM intracellular indo-1. Buffering of $[Ca^{2+}]_i$ in human T cells was observed as a slight delay in the rise in $[Ca^{2+}]_i$, and a retarded rate of return of $[Ca^{2+}]_i$ to baseline values when loading concentrations >3 μM (22 μM intracellular concentration) were used. A reduction in peak $[Ca^{2+}]_i$ occurred at even higher indo-1 concentrations (Rabinovitch et al., 1986). Chused et al. (1987) have observed a much greater sensitivity of murine B cells to indo-1 buffering, recommending a

TABLE I

Proliferation of Peripheral Blood Lymphocytes Loaded with Indo-1[a]

Day of culture	Indo-1/AM (μM)	Percentage of cells			
		Noncycling	Cycle 1	Cycle 2	Cycle 3
2	0	69.8 ± 1.3	28.4 ± 0.8	1.9 ± 0.7	—
	3	70.7 ± 1.8	27.5 ± 1.2	1.5 ± 0.8	—
3	0	58.1 ± 2.5	14.3 ± 0.5	26.5 ± 2.7	1.2 ± 0.4
	3	59.5 ± 2.6	14.9 ± 1.1	24.7 ± 3.4	0.9 ± 0.3
4	0	43.8 ± 0.7	8.4 ± 0.5	29.5 ± 1.0	18.4 ± 1.7
	3	44.6 ± 0.8	9.5 ± 0.9	28.6 ± 1.4	17.3 ± 3.0

[a] Peripheral blood lymphocytes were either loaded or not loaded with indo-1 ester and cultured with phytohemagglutinin (10 μg/ml) and bromodeoxyuridine (1×10^{-4} M). The cells were harvested on the indicated days, stained with Hoechst 33258, and analyzed by flow cytometry. The percentage of cells (mean $\pm$ SEM, $n = 4$) in each cell cycle was quantitated as described (Kubbies and Rabinovitch, 1985).

loading concentration of no greater than 1 μM. In side-by-side comparisons, we have found that calcium transients in B cells are much more sensitive to the effects of buffering by indo-1 than T cells. For human platelets, a 2 μM loading concentration has been reported (Davies et al., 1988). Rates of loading of the indo-1 ester can be expected to vary between cell types, perhaps as a consequence of variations in intracellular esterase activity. In peripheral human blood, more rapid rates of loading are seen in platelets and monocytes than in lymphocytes. Even within one cell type, donor- or treatment-specific factors may affect loading; for example, lower rates of indo-1 loading are seen in splenocytes from aged than from young mice (Miller et al., 1987).

Indo-1 has been found to be remarkably nontoxic to cells subsequent to loading. Analysis of the proliferative capacity of human T lymphocytes loaded with indo-1 (Table I) has shown no adverse effects on the ability of cells to enter and complete three rounds of the cell cycle. Similar results have been obtained with murine B lymphocytes (Chused et al., 1987). This is especially pertinent to the sorting of indo-1-loaded cells based on $[Ca^{2+}]_i$, as described subsequently.

B. Instrumental Technique.

The absorption maximum of indo-1 is between 330 and 350 nm, depending on the presence of calcium (Grynkiewicz et al., 1985); this is well suited to excitation at either 351–356 nm from an argon ion laser, or to 337–356 nm excitation from a krypton ion laser. Laser power requirements depend on the choice of emission filters and optical efficiency of the instrument; however, it is our experience that although 100 mW is often routinely employed, virtually identical results can be obtained with 10 mW. This should allow, in principle, the use of helium–cadmium lasers. Stability of the intensity of the excitation source is less important in this application than many others, because of the use of the ratio of fluorescence emissions. Analysis with indo-1 has also been performed using excitation by a mercury arc lamp (FACS Analyzer, Becton Dickinson). For details on the practice of instrument alignment, see Ransom *et al.* (1987).

An increase in $[Ca^{2+}]_i$ is detected with indo-1 as an increase in the ratio of fluorescence intensity at a lower to a higher emission wavelength. The optimal strategy is to select bandpass filters so that one minimizes the collection of light near the isosbestic point and maximizes collection of fluorescence that exhibits the largest variation in calcium-sensitive emission. The choice of filters used to select these wavelengths is dictated by the spectral characteristics of the shift in indo-1 emission upon binding to calcium (Fig. 1). The original spectral curves published for indo-1 (Grynkiewicz *et al.*, 1985) did not depict the large amounts of indo-1 emission in the blue-green and green wavelengths; in practice, on flow cytometers, we find that there is more light available in the blue region than in the violet region, although the dynamic range of the calcium-sensitive changes in the violet region exceeds that of the blue region (Fig. 1). Commercially available bandpass filters for analysis with indo-1 are often centered on the "violet" emission of the calcium-bound indo-1 dye (405 nm) and calcium-free indo-1 dye "blue" emission (485 nm). However, we have found these wavelengths to be suboptimal, and that a larger dynamic range in the ratio of wavelengths is obtained if "blue" emission <485 nm is not collected and the center of the "blue"-emission bandpass filter is moved upward. Similarly, the "violet" bandpass filter should be chosen to minimize the collection of wavelengths >405 nm. Thus, in order to optimize the calcium signal, it is important to exclude light from analysis that is near or at the isosbestic point. This effect is summarized in Table II by values of R_{max}/R, which indicates the range of change in indo-1 ratio observed from resting intracellular calcium to saturated calcium.

C. Calibration of Ratio to $[Ca^{2+}]_i$

Prior to the development of indo-1, $[Ca^{2+}]_i$ determination with quin2 fluorescence was sensitive to cell size, and intracellular dye concentration

TABLE II

EFFECT OF WAVELENGTH CHOICE ON CALCIUM-SENSITIVE INDO-1 RATIO SHIFTS[a]

Wavelength pair (nm)	R_{max}	R_{min}	R	R_{max}/R_{min}	R_{max}/R
475/395	2.33	0.040	0.352	58.2	6.62
475/405	2.38	0.100	0.410	23.8	5.80
495/395	3.51	0.048	0.429	73.1	8.18
495/405	3.59	0.119	0.501	30.2	7.17
515/395	5.75	0.070	0.644	82.1	8.93
515/405	5.88	0.176	0.752	33.4	7.82
530/395	9.68	0.117	1.073	82.7	9.02
530/405	9.89	0.292	1.252	33.9	7.90

[a] By spectrofluorimetry, slit width 2 nm; uncorrected fluorescence.

as well as $[Ca^{2+}]_i$, This made necessary calibration at the end of each individual assay by determination of the fluorescence intensity of the dye at zero and saturating Ca^{2+}. In contrast, with indo-1 use of the Ca^{2+}-dependent shift in dye emission wavelength allows the ratio of fluorescence intensities of the dye at the two wavelengths to be used to calculate $[Ca^{2+}]_i$

$$[CA^{2+}]_i = K_d \cdot \frac{(R - R_{min})S_{f2}}{(R_{max} - R)S_{b2}} \tag{1}$$

where K_d is the effective dissociation constant (250 nM), R, R_{min}, and R_{max} are the fluorescence intensity ratios of violet/blue fluorescence at resting, zero, and saturating $[Ca^{2+}]_i$, respectively; and S_{f2}/S_{b2} is the ratio of the blue florescence intensity of the calcium-free and bound dye, respectively (Grynkiewicz et al., 1985). Because this calibration is independent of cell size and total intracellular dye concentration as well as instrumental variation in efficiency of excitation or emission detection, it is not necessary to measure the fluorescence of the dye in the calcium-free and saturated states for each individual assay. In principle, it is sufficient to calibrate the instrument once after setup and tuning by measurement of the constants R_{max}, R_{min}, S_{f2}, and S_{b2}, after which only R is measured for each subsequent analysis on that occasion. It is important to note that the apparent K_d of indo-1 was measured at 37°C at an ionic strength of 0.1 at pH 7.08; this value will change significantly at different temperatures, pH, and ionic strength. For example, the K_d of fura-2 changes from 225 nM to 760 nM when ionic strength is changed from 0.1 to 0.225. Similar values for indo-1 have not been published.

One strategy to obtain the R_{max} and R_{min} values for indo-1 is to lyse cells in order to release the dye to determine fluorescence at zero and saturating

$[Ca^{2+}]_i$, as is carried out in fluorimeter-based assays with quin2 and fura-2. However, this is not possible with flow cytometry because of the loss of cellular fluorescence. Another strategy is the use of an ionophore to saturate or deplete $[Ca^{2+}]_i$, in order to allow approximation of the true endpoints without cell lysis. For this approach the ionophore ionomycin is best suited because of its specificity for calcium and low fluorescence. When FCM quantitation of fluorescence from intact cells treated with ionomycin or ionomycin plus EGTA was compared with spectrofluorimetric analysis of lysed cells in medium with or without EGTA, the indo-1 ratio of unstimu-lated cells (R) and the ratio at saturating amounts of Ca^{2+} (R_{max}) were similar by both techniques (Rabinovitch et al., 1986). The latter indicates that ionomycin-treated cells reach near-saturating levels of $[Ca^{2+}]_i$. The value of R_{min} that is obtained by treatment of intact cells with ionomycin in the presence of EGTA, however, is substantially higher than either that predicted from the spectral emission curves (Fig. 1) or that obtained by cell lysis and spectrofluorimetry. Spectrofluorimetric quantitation with either quin2 or indo-1 indicates that $[Ca^{2+}]_i$ remains at ~ 50 nM in intact cells treated with ionomycin and EGTA. Thus, because of the inability to obtain a valid FCM determination with calcium-free dye, we have used for calibration the values of R_{min} and S_{f2} or S_{b2} derived from spectro-fluorimetry, either of the indo-1 pentapotassium salt or after lysis cells loaded with indo-1 acetoxymethyl ester in the presence of EGTA. It is essential that the same optical filters be used for FCM and spectro-fluorimetry, since the standardization is very sensitive to the wavelengths chosen. Typical values of R_{max}, R_{min}, R, and S_{f2}/S_{b2} are shown for different emission wavelength combinations in Table II. Even if careful calibration of the fluorescence ratio to $[Ca^{2+}]_i$ is not being performed for a particular experiment, ordinary quality control should include a daily determination of the value of R_{max}/R. A limited range of R_{max}/R values should be obtained, since unperturbed cells are routinely found to have a reproducible value of $[Ca^{2+}]_i$, and day-to-day optical variations in the flow cytometer are usually minimal (with the same filter set). The R_{max}/R values obtained on the Ortho Cytofluorograph are typically in the range of 6 to 8; for unexplained reasons, several groups have reported that the value obtained on Becton Dickinson instruments is only 4.

As an alternative to the combined use of a flow cytometer with a spectrofluorimeter, a procedure that may be cumbersome or for which the instrument may be unavailable, Parks et al. (1987) have proposed that with minor modification, the flow cytometer may be used as a spectrofluori-meter. In essence, the fluorescence of a steady stream of dye is measured by the photomultiplier and the photomultiplier voltage is analyzed as in a standard spectrofluorimeter, or converted to a pulse for processing by the flow cytometer. Preliminary experience with this technique suggests that it should be possible to determine all constants for indo-1 necessary for calibration directly on the flow cytometer.

Chused *et al.* (1987) have suggested that metabolic inactivation of cells by the use of a cocktail of nigericin, high concentrations of potassium, 2-deoxyglucose, azide, and carbonyl cyanide *m*-chlorophenylhydrazone and the calcium ionophore ionomycin can collapse the calcium gradient to zero, and therefore, that $[Ca^{2+}]_i = [Ca^{2+}]_o$. To avoid the apparent impossibility of assessing R_{min} in intact cells, the calibration is based on a regression formula that relates R to ionomycin-treated cells suspended in a series of precisely prepared calcium buffers. Thus, this technique allows one to estimate $[Ca^{2+}]_i$ without the need to determine R_{min}, S_{f2}, or S_{b2}, although it is subject to the inability to quantitate calcium concentrations lower than those found in resting cells as well as the precision with which one can prepare a series of calcium buffers that yield known and reproducible free-calcium concentrations.

Accuracy of prediction of ionized Ca^{2+} concentration in buffer solutions is dependent on a variety of interacting factors, so that care must be exercised in formulating Ca^{2+} standards. The ionized calcium concentration in an EGTA buffer system is dependent on the magnesium concentration; other metals such as aluminum, iron, and lanthanum also bind avidly to EGTA (Martell and Smith, 1968). In addition, the dissociation constant of Ca^{2+}–EGTA is a function of pH, temperature, and ionic strength (Harafuji and Ogawa, 1980; reviewed in Blinks *et al.*, 1982). For example, in an EGTA buffer (total EGTA 2 mM, total Ca^{2+} 1 mM, ionic strength 0.1 at 37°C), changing the pH from 7.4 to 7.0 can result in the ionized calcium increasing by ∼ 200 nM, a change that is approximately twice the magnitude of that found in resting cells, and is easily measured on a flow cytometer. Finally, it is important to prepare the buffers using the "pH metric technique" (Moisescu and Pusch, 1975), in part because of the varying purity of commercially available EGTA (Miller and Smith, 1984).

D. Display of Results

It is possible to display the data as a bivariate plot of the inversely correlated "violet" versus "blue" signals derived from each cell. Thus, the increase in ratio seen with increased $[Ca^{2+}]_i$ will be observed as a rotation about the axis through the origin. This method of ratio analysis is cumbersome, and fortunately, commercial flow cytometers all have some provision for a direct calculation of the value of the fluorescence ratio itself, either by analog circuitry or by digital computation.

Plotted as a histogram of the ratio values, quiescent cell populations show narrow distributions of ratio, even when cellular loading with indo-1 is very heterogeneous, and *CV* values < 10% are not uncommon (Fig. 2). The effects of perturbation of $[Ca^{2+}]_i$ by agonists can be noted by changes in the ratio histogram profiles by storing histograms sequentially with subsequent analysis of data. This approach utilizes a linear display of indo-1 blue and violet fluorescence. If cellular indo-1 loading is extremely heterogeneous, it may be desirable to work with a logarithmic conversion of "violet" and "blue" emission intensities in order to observe a broader range of cellular fluorescence. In this case, the hardware must permit the logarithm of the ratio to be calculated by *subtraction* of the log "blue" from the log "violet" signals (Rabinovitch *et al.*, 1986). Use of pluronic F-127 (see later) can often result in more homogeneous loading with indo-1.

A more informative and elegant display is obtained by a bivariate plot of ratio versus time. The bivariate histogram can be stored and the data displayed as "dot plots" on which the indo-1 ratio of each cell (proportional to $[Ca^{2+}]_i$) is plotted on the *y* axis versus time on the *x* axis (Fig. 3A). Alternatively, the data can be subjected to further analysis for presentation as "isometric plots" in which the *x* axis represents time, the *y* axis $[Ca^{2+}]_i$, and the *z* axis, number of cells (Fig. 3B). In these bivariate plots, kinetic changes in $[Ca^{2+}]_i$ are seen with much greater resolution, limited only by the number of channels on the time axis, the interval of time between each channel and the rate of cell analysis. Changes in the fraction of cells responding, in the mean magnitude of the response, and in the heterogeneity of the responding population are best observed with these displays. For example, it can be seen in Fig. 3 that not all murine thymocytes respond to stimulation by anti-CD3 and of those that do, the values of $[Ca^{2+}]_i$ are quite heterogeneous.

Calculation of the mean *y*-axis value for each *x*-axis time interval allows presentation of the data as mean ratio versus time (Rabinovitch *et al.*, 1986). Calibration of the ratio to $[Ca^{2+}]_i$ (see earlier) allows data presentation in the same manner as traditionally displayed by spectrofluorimetric analysis (i.e., mean $[Ca^{2+}]_i$ versus time). While this presentation yields much of the information of interest in an easily displayed format, data relating to heterogeneity of the $[Ca^{2+}]_i$ response is lost. Some of this information can be displayed by a calculation of the "proportion of responding cells." If a threshold value of the resting ratio distribution is chosen (e.g., one at which only 5% of control cells are above), the proportion of cells responding by ratio elevations above this threshold versus time yields a presentation informative of the heterogeneity of the response (Fig. 3C and 3D). In some instances, however, plots of mean response and percentage responding cells versus time do not provide an

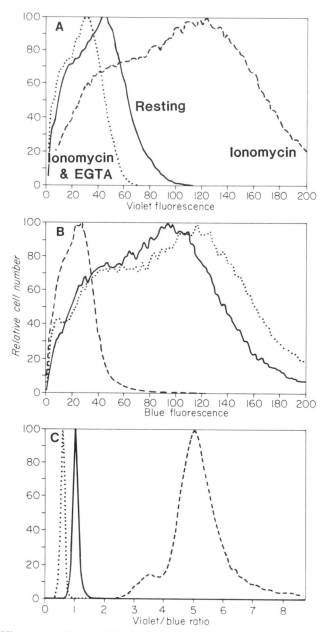

FIG. 2.   Histograms of violet (405 nm) fluorescence (A), blue (485 nm) fluorescence (B), and the ratio of violet/blue fluorescence (C) of peripheral blood lymphocytes loaded with indo-1. Cells were analzyed in a basal state (————), in the presence of 3 μM ionomycin (----), or in the presence of ionomycin and EGTA, used to reduce free calcium in the medium to ~20 nM (....). Data are plotted on linear scales; the violet/blue ratio was normalized to unity for resting cells. Reproduced from Rabinovitch *et al.*, the *Journal of Immunology*, 1986, Vol. 137, p. 952, by copyright permission of the American Association of Immunologists.

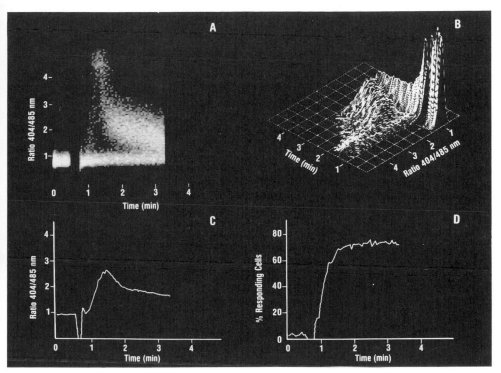

Fig. 3. Methods to express calcium signaling measured in single cells as a function of time. Thymocytes from a C57BL/6 mouse were loaded with indo-1 and stimulated with anti-CD3 antibody 145-2C11. The cells were analyzed at 300 cells/second. In (A) the results are displayed on a "dot plot" in which $[Ca^{2+}]_i$ is plotted for each cell analyzed on a 100×100 pixel grid. The number of cells per pixel is displayed by intensity that ranges over 16 levels. In (B) isometric plots of the same experiment in (A) are shown. Sequential histograms are plotted in which the x axis represents time, the y axis $[Ca^{2+}]_i$, and the z axis number of cells. In (C) and (D) the mean calcium versus time and percentage cells responding to two standard deviations above the mean of the cells before antibody stimulation are plotted for the same experiment depicted in (A) and (B).

adequate representation of the complexity of the data. For example, in Fig. 3, it can be seen from the "dot plot" and the "isometric" plot that the early response consists of small population (~10%) of cells with an extremely high-magnitude response of brief duration, while the later response is composed of ~50% of cells that have a low-magnitude response. Thus, in this example, the time of the occurrence of the *peak* response is dissociated from the time of the maximal *mean* response (Fig. 3A and 3B vs Fig. 3C). Note that this heterogeneity of the pattern of response is not apparent in the displays of mean $[Ca^{2+}]_i$ versus time and the percentage responding cells versus time.

E.   Simultaneous Analysis of $[Ca^{2+}]_i$ and Other Fluorescence Parameters

Although the broad spectrum of indo-1 fluorescence emission will likely preclude the simultaneous use of a second UV-excited dye, the use of two or more excitation sources allows additional information to be derived from visible-light-excited dyes. Perhaps the most usual application will be determination of cellular immunophenotype simultaneously with the indo-1 assay, allowing alterations in $[Ca^{2+}]_i$ to be examined in, and correlated with, specific cell subsets.

Combining the use of FITC and PE-conjugated antibodies with indo-1 analysis allows determination of $[Ca^{2+}]_i$ in complex immunophenotypic subsets. On instruments without provision for analysis of four separate fluorescence wavelengths, detection of temporally delayed signals from both FITC and the higher indo-1 wavelength with the same filter element may allow successful implementation of these experiments. Gating the analysis of indo-1 fluorescence upon windows of FITC versus PE fluorescence allows information relating to each identifiable cellular subset to be derived from a single sample.

Using other probes excited by visible light, it may be possible to analyze additional physiologic responses in cells simultaneously with $[Ca^{2+}]_i$. The simultaneous analysis of membrane potential and $[Ca^{2+}]_i$ has been accomplished by several groups (Lazzari et al., 1986; Ishida and Chused, 1988). Similarly, the simultaneous analysis of intracellular pH and $[Ca^{2+}]_i$ has been made possible with the use of SNARF-1 and indo-1.

F.   Sorting on the Basis of $[Ca^{2+}]_i$ Responses

The ability of the FCM analysis with indo-1 to observe small proportions of cells with different $[Ca^{2+}]_i$ responses than the majority of cells suggests that the flow cytometer may be useful to identify and sort variants in the population for their subsequent biochemical analysis or growth. Results of artificial mixing experiments with Jurkat (T cell) and K562 (myeloid cell) leukemia lines indicated that subpopulations of cells with variant $[Ca^{2+}]_i$ comprising <1% of total cells could be accurately identified (Rabinovitch et al., 1986). Goldsmith and Weiss (1987, 1988) have reported the use of sorting on the basis of indo-1 fluorescence to identify mutant Jurkat cells that fail to mobilize $[Ca^{2+}]_i$ in response to CD3 stimulus, in spite of the expression of structurally normal CD3/Ti complexes. These experiments suggest that sorting on the basis of indo-1 fluorescence can be an important tool for the selection and identification of genetic variants in the biochemical pathways leading to Ca^{2+} mobilization and cell growth and differentiation.

III. Use of Flow Cytometry and Fluo-3 to Measure $[Ca^{2+}]_i$

Fluo-3 is a fluorescein-based, calcium-sensitive probe developed by Minta and co-workers (Minta *et al.*, 1989). The principal advantage of this compound is that it is the first calcium indicator that does not require UV excitation, and therefore it can be used on all flow cytometers that have the capability to measure fluorescence emission from fluorescein. This compound is less sensitive than quin2 and indo-1 at detecting small changes in $[Ca^{2+}]_i$, in part because the K_d is higher (400 nM); this may be an advantage or disadvantage, depending on the experimental situation. An additional advantage with fluo-3 is that it can be used with other probes such as caged calcium chelators that may themselves require UV excitation (Tsien, 1989). Furthermore, the use of fluo-3 permits, for the first time, the simultaneous determination of cell cycle or $[Na^+]_i$ and correlation with calcium responses.

Practical experience with fluo-3 remains limited. The primary disadvantage of this probe is that it does not have fluorescence properties that allow ratiometric determinations. Thus, calibration on a flow cytometer is particularly difficult because the signal is proportional to cell size and dye concentration as well as $[Ca^{2+}]_i$. This is not a particular problem with fluorimeter-based assays but represents a substantial limitation for FCM because destructive calibration techniques are not possible or are cumbersome. Furthermore, the ability to measure responses in subsets of cells is limited because the discrimination of fluorescence changes is usually less than with indo-1. This is primarily attributable to the broad distribution of fluorescence intensities of unstimulated cells, which often results in an overlapping distribution of the values from stimulated and unstimulated cells. This problem can be minimized by loading cells in the presence of pluronic F-127 (available from Molecular Probes, Eugene OR), which minimizes the cell-to-cell variation in loading with fluo-3 (Vandenberghe and Ceuppens, 1990). However, the ability of fluo-3 to distinguish responding cells from nonresponding cells will never be as satisfactory as with indo-1. This is due to the lack of ratiometric properties that, in the case of indo-1, permits one to distinguish fluorescence signals attributable to cell size and dye content from those due to $[Ca^{2+}]_i$ (Fig. 3).

IV. Pitfalls and Critical Aspects

A. Instrumental Calibration and Display of Data

In some instruments there may be difficulty in displaying the ratio of indo-1 fluorescence so that increases in Ca^{2+} are depicted as increased ratio values (Breitmeyer et al., 1987). In particular, the analog ratio circuits of some Coulter instruments are limited in their range of acceptable inputs, for example, that the "violet" signal never be greater than the "blue," yielding a ratio of >1. Rather than reversing the ratio (blue/violet) so that a rise in calcium results in a counterintuitive decline in ratio, the violet/blue ratio can be used as long as the signal gains are initially set such that subsequent rises in the ratio will not exceed the permitted value.

Some instruments may have nonlinearity in signal amplification or introduce errors into the calculation of the indo-1 ratio.By either analog or digital calculation, it is important that no artifactual offset be introduced in the ratio; this can be quickly tested by altering the excitation power over a broad range of values in an analysis of a nonperturbed indo-1-loaded cell population: a correctly calculated ratio will not show any dependence on excitation intensity. It can similarly be shown that loading of cells with a broad range of indo-1 concentrations results in a constant value of the violet/blue ratio (Rabinovitch et al., 1986).

B. Difficulties with Loading Indo-1

Several problems may be encountered in cells loaded with indo-1. These include compartmentalization, leakage, secretion, quenching by heavy metals, and incomplete deesterification of indo-1 ester. The analysis of $[Ca^{2+}]_i$ using indo-1 is predicted on achieving uniform distribution of the dye within the cytoplasm. In several cell types, the related dye fura-2 has been reported to be compartmentalized within organelles (Di Virgilio et al., 1988; Malgawli et al., 1987). In bovine aortic endothelial cells, fura-2 has been reported to be localized to mitochondria; however, under those conditions, indo-1 remained diffusely cytoplasmic (Steinberg et al., 1987). We have observed that indo-1 may become compartmentalized in cells. Some cell types such as neutrophils and monocytes, and some cell lines rather than primary cells are more susceptible to compartmentalization. In addition, compartmentalization is enhanced by prolonged incubation of cells at 37°C. Thus, it is possible that there will be fewer problems with compartmentalization of indo-1 than with fura-2; however, it is advisable to examine the cellular distribution of indo-1 microscopically, and in each new application to confirm the expected behavior of the dye. This is done by determining the ratio of R_{max} to R as a control for each experiment as described later. In addition, one should store indo-1-loaded cells at room temperature (RT) after loading and use the cells promptly after loading. In prolonged experiments, it is preferable to discard cells after several hours and to reload fresh aliquots of cells. In experiments in which it is necessary to incubate loaded cells at 37°C with various treatments, it is necessary to include control cells incubated under identical conditions. The use of the membrane-permeant heavy-metal chelator diethylenetri-aminepentaacetic acid (TPEN) for cell lines that contain increased amounts of heavy metals has been described (Arslan et al., 1985); these cells appear to have artefactually low concentrations of $[Ca^{+2}]_i$ due to the interaction of the heavy metal with the calcium indicator. It is possible that probenecid, a blocker of organic anion transport, may be useful in cells that actively secrete fluo-3 or indo-1 (Di Virgilio et al., 1989).

If, for a particular cell type loaded with indo-1, the magnitude of change between R and R_{max} is in good agreement with the values predicted from spectral curves of indo-1 in a cell-free buffer, then it would be unlikely that the dye is in a compartment inaccessible to cytoplasmic Ca^{2+}, in a form unresponsive to $[Ca^{2+}]_i$ (e.g., still esterified), or in a cytoplasmic environment in which the spectral properties of the dye were altered. With regard to the second condition, it has been proposed that since indo-1 fluorescence, but not that of the indo-1 ester, is quenched in the presence of millimolar concentrations of Mn^{2+}, that Mn^{2+} in the presence of ionomycin can be used as a further test of complete hydrolysis of the indo-1 ester within cells (Luckhoff, 1986).

It has been suggested that both fura-2 and indo-1 may be incompletely deesterified within some cell types (Luckhoff, 1986; Scanlon et al., 1987; Owen, 1988). Since the fluorescence of the ester has little spectral dependence on changes in Ca^{2+}, the presence of this dye form could lead to false estimates of $[Ca^{2+}]_i$. Again, results of calibration experiments are helpful in excluding this possibility. In some circumstances, the use of Pluronic F-127, a nonionic, high molecular weight surfactant, may aid in the loading of fura-2, fluo-3, and indo-1 into cells that are otherwise dificult to load (Cohen et al., 1974; Poenie et al., 1986; Vandenberghe et al., 1990). Pluronic F-127 may be obtained from Molecular Probes.

C.   Unstable Baseline

Under typical conditions, the baseline indo-1 ratio should show little ($<3\%$) variation from sample to sample. Some cell lines may have altered mean values of "resting" $[Ca^{2+}]_i$, which can often be ascribed to a subpopulation of cells with elevated $[Ca^{2+}]_i$. This may result from impaired viability of some cells or, presumably, may be due to cells traversing certain phases of the cell cycle. In circumstances where the baseline is not stable from sample to sample, the following situations should be considered. The cells should be equilibrated to 37°C for at least 5 minutes before analysis. Regulation of the temperature of the cell sample is essential, as transmembrane signaling and calcium mobilization are temperature-dependent and active processes. Most applications will require analyses at 37°C. If cells are allowed to cool before they flow past the laser beam, calcium signals will often become impaired, so that either the sample input tubing should be warmed, or narrow-gauge tubing and high flow rates (e.g., ≥ 50 μl sample/minute, in the case of the Ortho Cytofluorograph) should be used to keep transit times from warmed sample to flow cell minimized. As noted before, it is necessary to maintain the working stock of indo-loaded cells at RT and to warm the cells just prior to the assay. It is not clear to what mechanism the variation in basal $[Ca^{2+}]_i$ with temperature variation may be ascribed. It is possible that the changes reported by indo-1 are real and reflect strict temperature requirements of the cell for the maintenance of calcium homeostasis. Alternatively, the changes of calcium reported by indo-1 may in part be due to temperature-dependent changes in the effective dissociation constant of indo-1 for calcium. Occasionally the baseline will start at a normal level and then rise with time. This may be due to the failure to remove completely an agonist from the sample lines from a previous experiment. The most common problem has been residual calcium ionophore; this can be efficiently removed by first washing the sample lines with dimethyl sulfoxide and then scavenging residual ionophore by washing with a buffer containing 1% bovine serum albumin or fetal calf serum.

D.   Sample Buffer

The choice of medium in which the cell sample is suspended for analysis can be dictated primarily by the metabolic requirements of the cells, subject only to the presence of millimolar concentrations of calcium (to enable calcium agonist-stimulated calcium influx) and reasonable pH buffering. The use of phenol red as a pH indicator does not impair the FCM detection of indo-1 fluorescence signals. Although the new generation of

Ca^{2+} indicator dyes are not highly sensitive to small fluctuations of pH over the physiologic range (Grynkiewicz et al., 1985), unbuffered or bicarbonate-buffered (in low-pCO_2 environments) solutions can impart large and uncontrolled pH shifts. Finally, if analysis of release of Ca^{2+} from intracellular stores is desired, independent of extracellular Ca^{2+} influx, addition of 5 mM EGTA to the cell suspension (final concentration) will reduce Ca^{2+} from several millimolar to <20 nM, thus abolishing the usual extracellular to intracellular gradient.

E.   Poor Cellular Response

When one encounters cells that are poorly responsive to various treatments, it is necessary first to determine if there is difficulty with the cells or with the instrument. The cells should be stimulated with the calcium ionophore ionomycin and the magnitude of R_{max} to R determined. If the ratio increases by ~6-fold, then the instrument is functioning properly. If the increase is less than expected, then one should obtain an independent preparation of cells, such as murine thymocytes or human peripheral blood lymphocytes (PBL). Aliquots of cryopreserved cells are convenient for this purpose. If these cells load properly and also respond poorly, then the instrument alignment should be checked. Not uncommonly, the violet or blue signals may not be properly focused, or perhaps there is interference from a second laser. This problem can be pinpointed by analyzing separately the blue and violet signals after ionophore treatment; the violet signal should increase ~3-fold and the blue signal should decrease ~2-fold (Figs. 1 and 2).

If the instrument is functioning properly, then the problem may be in the cells. The cells must be loaded with sufficient indo-1 to be easily detected. This should be checked independently with fluorescence microscopy. If the cells are too dim or excessively bright, or if the indo-1 is compartmentalized, the ability to detect calcium signals will be impaired. For unknown reasons, the calcium signaling of B cells and not T cells is particularly sensitive to overloading with indo-1 (Rabinovitch et al., 1986; Chused et al., 1987). The cells must be suspended in media that contains calcium; occasionally, responses will appear blunted because of the inadvertent resuspension of cells in medium than contains no added calcium. In the simultaneous analysis of $[Ca^{2+}]_i$ and immunofluorescence, consider that the use of the antibody probe can itself alter the cellular $[Ca^{2+}]_i$. It is becoming increasingly clear that binding of monoclonal antibodies (mAb) to cell surface proteins can alter $[Ca^{2+}]_i$, even when these proteins are not previously recognized as part of a signal-transducing pathway (Anasetti et al., 1987; Geppert et al., 1988; June et al., 1987; Pezutto et al., 1987,

1988; Rabinovitch *et al.*, 1986; Wilson *et al.*, 1987b). For example, antibody binding to CD4 will reduce CD3-mediated $[Ca^{2+}]_i$ signals; if the anti-CD4 mAb is crosslinked to the CD3 complex, as with a goat antimouse mAb, the CD3 signals are augmented (Ledbetter *et al.*, 1987, 1988a).

As a consequence of these concerns, a reciprocal staining strategy should be used whenever possible, so that the celluar subpopulation of interest is unlabeled while undesired cell subsets are identified by mAb staining. The CD4$^+$ subset in PBL may be identified, for example, by staining with a combination of CD8, CD20, and CD11 mAb (Rabinovitch *et al.*, 1986), and the CD5$^+$ subset can be identified by staining with CD16, CD20, and HLA-DR mAb (June *et al.*, 1987). Finally, it is important when staining cells with mAb for functional studies that the antibodies be azide-free, in order that metabolic processes be uninhibited. Commercial antibody preparations may thus require dialysis before use.

V.   Limitations

There are several limitations to the FCM assay of cellular calcium concentration. First, certain problems attributable to the use of fluorescent indicators have been mentioned. In some cells indo-1 will not load uniformly into cells, or may not be uniformly hydrolyzed to the calcium-sensitive moiety. Quin2 at intracellular concentrations often used may block plasma membrane sodium–calcium transport (Allen and Baker, 1985), and quin2 may be quenched by heavy metals that are found in the cytoplasm of some cell lines (Arslan *et al.*, 1985). Similarly, quin2 has been reported to be mitogenic in certain cells and to alter cellular functions (Lanza *et al.*, 1987; Hesketh *et al.*, 1983), although this has not as yet been observed in indo-1-loaded cells (Chused *et al.*, 1987; Rabinovitch *et al.*, 1986). It is possible that indo-1 may have similar limitations, although this would be expected to be less of a problem as a result of the much lower concentrations of indo-1 that are required to attain a satisfactory fluorescent signal from cells. Second, there are limitations imposed by the nature of the FCM assay system. Flow cytometry is unable to detect heterogeneity of cellular calcium concentrations within a single cell, and there are reports from assays using digital videomicroscopy that in some situations, calcium transients may be compartmentalized (Poenie *et al.*, 1987; Williams *et al.*, 1987). The use of the photoprotein aequorin may in some circumstances detect changes in cytosolic calcium not reported by indo-1 (Ware *et al.*, 1987), although use of the two indicators is complementary because aequorin cannot measure Ca^{2+} in single cells (Cobbald and Rink, 1987). In addition, there is evidence that calcium elevations occurring after cellular stimulation may be oscillatory rather than sustained (Ambler et al., 1988; Wilson et al., 1988; Wilson et al., 1987a), thus raising the possibility that some cellular processes controlled by calcium may be frequency modulated as well as amplitude modulated. Since FCM cannot measure the calcium concentration inside a single cell as a function of time, it is not possible to distinguish between a subpopulation of cells that is responding with a sustained response or alternatively, whether there are two populations of cells, one that has elevated calcium concentration and one that has basal levels.

In spite of these limitations, determination of $[Ca^{2+}]_i$ in large numbers of single cells using FCM with indo-1 offers great practical advantages and allows measurements of a kind not possible by alternative techniques that are currently available.

VI.   Results

The FCM assay of cellular calcium concentration has already been applied to a wide variety of cells, providing interesting and sometimes unexpected results. Examples of the initial applications of the technique are presented in several reviews (June and Rabinovitch, 1988; Rabinovitch and June, 1990). A small CV (often <10%) was found in lymphocytes for the distribution of the basal calcium level; this value appears to reflect physiologic variation because the instrumental variation was only 4.5% (Fig. 2C). Surprisingly, there is heterogeneity in the response of lymphocytes to doses of calcium ionophores that are capable of stimulating physiologic responses in cells. Ischida and Chused (1988) found that the response of murine splenic B cells was lower than that of splenic or lymph node T cells. After CD2 (T11) stimulation of PBL, it was found that both $CD3^+$ T cells and $CD16^+CD3^-$ large granular lymphocytes (LGL) mobilized calcium. Following CD3 (antigen-specific) stimulation, >90% of T cells responded, and as was expected, LGL did not respond (June et al., 1986). The pattern of the calcium signal after CD2 stimulation differed in that calcium mobilization in LGL was early in onset and low in magnitude while in T cells, the calcium signal was delayed and high in magnitude (Fig. 4). Thus, mechanisms of calcium homeostasis in T lymphocytes appear to differ among T-cell subsets and in several respects from B cells. Recently, infection with the HIV-1 retrovirus was found to impair signal transduction in CD4 cells, the major target of HIV-1 (Linette et al., 1988). There are many potentially exciting clinical applications of the FCM assay

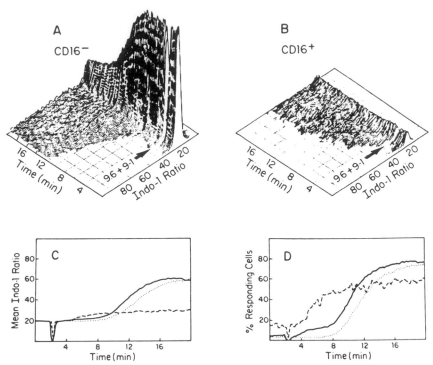

FIG. 4.   Effects of CD2 stimulation on $[Ca^{2+}]_i$ of CD16$^-$ or CD16$^+$ lymphocytes. Indo-1-loaded peripheral blood T cells were stained with phycoerythrin—conjugated anti-CD16 antibody, which identifies a population of large granular lymphocytes that do not express the CD3 antigen, and stimulated with CD2 antibodies 9.6 plus 9-1. The cells were analyzed at 500 cells/second and electronic "gating" used to display the phycoerythrin-negative and positive cells as iometric plots in (A) and (B), respectively. The time course of the mean indo-1 ratio (C) and the percentage responding cells (D) are displayed (———, total cell population analyzed; ---, CD16$^+$ cells; ..., CD16$^-$ cells). Separate analysis showed that these cells were 99% CD2$^+$, 95% CD3$^+$, and 2% CD16$^+$. Reproduced from June *et al.*, the *Journal of Clinical Investigation*, 1986, vol. 77, p. 1224, by copyright permission of the American Society for Clinical Investigation.

of cellular calcium concentration (June and Rabinovitch, 1988). Demonstration of such heterogeneity in $[Ca^{2+}]_i$ signals would have been impossible to discern in conventional assays carried out in a fluorimeter where only the mean calcium response is recorded.

Acknowledgments

We thank J. A. Ledbetter for many valuable comments. This work was supported in part by the Naval Medical Research and Development Command, and by National Institutes of Health Grant AG01751.

REFERENCES

Allen, T. J., and Baker, P. F. (1985). *Nature (London)* **315,** 755.

Ambler, S. K., Poenie, M., Tsien, R. Y., and Taylor, P. (1988). *J. Biol. Chem.* **263,** 1952.

Anasetti, C., Martin, P .J., June, C. H., Hellstrom, K. E., Ledbetter, J. A., Rabinovitch, P. S. Morishita, Y., Hellstrom, I., and Hansen, J. A. (1987). *J. Immunol.* **139,** 1772.

Arslan, P., Di Virgilio, F., Beltrame, M., Tsien, R. Y., and Pozzan, T. (1985). *J. Biol. Chem.* **260,** 2719.

Blinks, J. R., Wier, W. G., Hess, P., and Prendergast, F. G. (1982). *Prog. Biophys. Mol. Biol.* **40,** 1.

Breitmeyer, J., Daley, J. F., Levine, H. B., and Schlossman, S. F. (1987). *J. Immunol.* **139,** 2899.

Chused, T. M., Wilson, H. A., Greenblatt, D., Ishida, Y., Edison, L. J., Tsien, R. Y., and Finkelman, F. D. (1987). *Cytometry* **8,** 396.

Cobbold, P. H., and Rink, T. J. (1987). *Biochem. J.* **248,** 313.

Cohen, L. B., Salzberg, B. M., Davila, H. V., Ross, W. N., Landowne, D., Waggoner, A. S., and Wang, C. H. (1974). *J. Membr. Biol.* **19,** 1.

Davies, T. A., Drotts, D., Weil, G. J., and Simons, E. R. (1988). *Cytometry* **9,** 138.

Di Virgilio, F., Steinberg, T. H., and Silverstein, S. C. (1989). *Methods Cell Biol.* **31,** 453.

Geppert, T. D., Wacholtz, M. C., Davis, L. S., and Lipsky, P. E. (1988). *J. Immunol.* **140,** 2155.

Goldsmith, M. A., and Weiss, A. (1987). *Proc. Natl. Acad. Sci. U.S.A.* **84,** 6879.

Goldsmith, M. A., and Weiss, A. (1988). *Science* **240,** 1029.

Grynkiewicz, G., Poenie, M., and Tsien, R. Y. (1985). *J. Biol. Chem.* **260,** 3440.

Harafuji, H., and Ogawa, Y. (1980). *J. Biochem.* **87,** 1305.

Hesketh, T. R., Smith, G. A., Moore, J. P., Taylor, M. V., and Metcalfe, J. C. (1983). *J. Biol. Chem.* **258,** 4876.

Ishida, Y., and Chused, T. M. (1988). *J. Exp. Med.* **168,** 839.

June, C. H., and Rabinovitch, P. S. (1988). *Pathol. Immunopathol. Res.,* **7,** 409.

June, C. H., Ledbetter, J. A., Rabinovitch, P. S., Martin, P. J., Beatty, P. G., and Hansen, J. A. (1986). *J. Clin. Invest.* **77,** 1224.

June, C. H., Rabinovitch, P. S., and Ledbetter, J. A. (1987). *J. Immunol.* **138,** 2782.

Kubbies, M., and Rabinovitch, P. S. (1985). *Cell Tissue Kinet.* **18,** 551.

Lanza, F., Beretz, A., Kubina, M., and Cazenave, J. P. (1987). *Thromb. Haemostasis* **58,** 737.

Lazzari, K. G., Proto, P. J., and Simons, E. R. (1986). *J. Biol. Chem.* **261,** 9710.

Ledbetter, J. A., June, C. H., Grosmaire, L. S., and Rabinovitch, P. S. (1987). *Proc. Natl. Acad. Sci. U.S.A.* **84,** 1384.

Ledbetter, J. A., June, C. H., Rabinovitch, P. S., Grossmann, A., Tsu, T. T., and Imboden, J.B. (1988a). *Eur. J. Immunol.* **18,** 525.

Ledbetter, J. A., Rabinovitch, P. S., Hellstrom, I., Hellstrom, K. E., Grosmaire, L. S., and June, C. H. (1988b). *Eur. J. Immunol.,* **18,** 1601.

Linette, G. P., Hartzman, R. J., Ledbetter, J. A., and June, C. H. (1988). *Science* **241,** 573.

Luckhoff, A. (1986). *Cell Calcium* **7,** 233.

Malgawli, A., Milani, D., Meldolesi, J., and Pozzan, T. (1987). *J. Cell Biol.* **105,** 2145.

Martell, A. E., and Smith, R. M. (1968). "Critical Stability Constants. Vol. 1: Amino Acids, pp. 269–272." Plenum, New York.

Miller, R. A., Jacobson, B., Weil, G., and Simons, E. R. (1987). *J. Cell. Physiol.* **132,** 337.

Minta, A., Kao, J. P.Y., and Tsien, R. Y. (1989). *J. Biol. Chem.* **264,** 8171.

Moisescu, D. G., and Pusch, H. (1975). *Pflugers Arch.* **355,** R122.

Owen, C. S. (1988). *Cell Calcium* **9,** 141.

Parks, D. R., Nozaki, T., Dunne, J. F., and Peterson, L. L. (1987). *Cytometry Suppl.* **1**, 104.

Pezzutto, A., Dorken, B., Rabinovitch, P. S., Ledbetter, J. A., Moldenhauer, G., and Clark, E. A. (1987). *J. Immunol.* **138**, 2793.

Pezzutto, A., Rabinovitch, P. S., Dorken, B., Moldenhauer, G., and Clark, E. A. (1988). *J. Immunol.* **140**, 1791.

Poenie, M., Alderton, J., Steinhardt, R., and Tsien, R. (1986). *Science* **233**, 886.

Poenie, M., Tsien, R. Y., and Schmitt-Verhulst, A. M. (1987). *EMBO J.* **6**, 2223.

Rabinovitch, P. S., and June, C. H., (1990). *In* "Flow Cytometry and Cell Sorting" (M. R. Melamed, T. Lindmo, and M. L. Mendelsohn, eds.) Wiley-Liss, New York.

Rabinovitch, P. S., June, C. H. Grossmann, A., and Ledbetter, J. A. (1986). *J. Immunol.* **137**, 952.

Ransom, J. T., DiGusto, D. L., and Cambier, J. C. (1986). *J. Immunol.* **136**, 54.

Ransom, J. T., DiGiusto, D. L., and Cambier, J. (1987). *Methods Enzymol.* **141**, 53.

Scanlon, M., Williams, D. A., and Fay, F. S. (1987). *J. Biol. Chem.* **262**, 6308.

Steinberg, S. F., Bilezikian, J. P., and Al-Awqati, Q. (1987). *Am. J. Physiol.* **253** (Pt. 1), C744.

Tsien, R. Y. (1981). *Nature (London)* **290**, 527.

Tsien, R. Y., Pozzan, T., and Rink, T. J., (1982). *J. Cell Biol.* **94**, 325.

Tsien, R. Y. (1989). *Methods Cell Biol.* **30**, 127.

Vandenberghe, P. A., and Ceuppens, J. L. (1990). *J. Immunol. Methods* **127**, 197.

Ware, J. A., Smith, M., and Salzman, E. W. (1987). *J. Clin. Invest.* **80**, 267.

Williams, D. A., Becker, P. L., and Fay, F. S. (1987). *Science* **235**, 1644.

Wilson, H. A., Greenblatt, D., Poenie, M., Finkelman, F. D., and Tsien, R. Y. (1987a). *J. Exp. Med.* **66**, 601.

Wilson, H. A., Greenblatt, D., Taylor, C. W., Putney, J. W., Tsien, R. Y., Finkelman, F. D., and Chused, T. M. (1987b). *J. Immunol.* **138**, 1712.

Chapter 6

Measurement of Intracellular pH

ELIZABETH A. MUSGROVE

Garvan Institute of Medical Research
St. Vincent's Hospital
Darlinghurst, Sydney, N. S. W. 2010
Australia

DAVID W. HEDLEY

Departments of Medicine and Pathology
Princess Margaret Hospital
Toronto, Ontario M4X 1K9
Canada

I. Introduction

A. Intracellular pH Regulation

Normal functioning of cell metabolism occurs within a restricted pH range. Regulation of intracellular pH (pH_i) is accomplished via active extrusion of H^+ by a Na^+/H^+ antiport localized in the plasma membrane and driven by the energy of the inward Na^+ electrochemical gradient. These mechanisms are reviewed by Seifter and Aronson (1986) and Grinstein and Rothstein (1986). The Na^+/H^+ antiport is present in a diversity of cell types, but in some (e.g., nerve and muscle cells) there is evidence for an additional pH-regulatory mechanism using HCO_3^-/Cl^- exchange. The electroneutral exchange of Na^+ and H^+ by the antiport occurs on a $1:1$ basis, at a rate which depends on the pH_i. The "set point" is typically $pH_i \sim 7.2$ for resting cells, but stimuli, including several hormones and growth factors, as well as tumor-promoting phorbol esters, apparently shift the set point to a pH_i of 7.4–7.5. These observations suggest a role for pH_i regulation in the control of cell processes.

59

Intracellular pH can be experimentally manipulated, by preloading with NH_4^+ and other weak acids, or by using ionophores. A method for preloading with NH_4^+ is described here, and sources for other methods can be found in the review by Grinstein and Rothstein (1986). The activity of the antiport can be modulated by alterations in the ionic composition of the extracellular medium. For example, since the antiport is driven by the free-energy gradients for Na^+ and H^+, under some conditions (e.g., extracellular Na^+ concentrations <12 mM) the exchange may reverse direction. In addition, the antiport can be reversibly inhibited by the diuretic amiloride, which binds to the external Na^+ site with an inhibition constant of 7–30 mM (Seifter and Aronson, 1986, and references therein).

B. Measurement of pH with Trapped Indicators

The current fluorimetric methods for measuring pH_i rely on the use of membrane-impermeant dyes. These dyes are cleaved by intracellular esterases, from membrane-permeant, nonfluorescent precursors. The charged nature of the cleaved products then inhibits their diffusion from the cell. The wavelength and/or intensity of emission and/or excitation of the dye is pH sensitive. These spectral changes are monitored by ratiometric techniques, typically for flow cytometry (FCM) by measuring the ratio of emission intensities at two wavelengths. Ratio techniques have the significant advantage of making the pH measurement independent of cell-to-cell variables, including the intracellular dye concentration.

1.   Fluorescein Derivatives: BCECF

The first fluorochromes utilized for pH_i measurements in FCM were fluorescein derivatives. These early probes suffered from limitations that included poor intracellular localization, and excessive dye leakage. Currently the most appropriate such compound is BCECF (2′,7′-bis(2-carboxyethyl)-5,6-carboxyfluorescein), which is confined relatively specifically to the cytoplasm (Paradisio et al., 1984), and well retained within the cell, so that stable ratio measurements can be made over at least 2 hours (Musgrove et al., 1986). The pK of BCECF is 7.0, which is in the range of expected physiological pH values (Rink et al., 1982). The absorption maximum is 500 nm, and the dye can be readily used in flow cytometers equipped either with argon ion lasers or with mercury arc lamps. The emission intensity, maximum at 520 nm, increases with increasing pH.

Related probes that may offer further improvements are still being developed: recently Molecular Probes has released details of the pH indicators SNAFL (seminaphthofluorescein) and SNARF (seminaphtho-rhodafluor). These dyes have yet to be fully evaluated but have the advantage of two separable, pH-dependent emissions.

2. ULTRAVIOLET-EXCITED PROBES: DCH

In contrast to the fluorescein derivatives, where the intensity of emission is pH sensitive, the pH indicator DCH (2,3-dicyano-hydroquinone) has a shift in emission wavelength and decrease in emission intensity with increasing pH (Valet *et al.*, 1981). It has a pK of 8.0 and an absorbance maximum in the UV range, requiring either high-power argon ion laser or mercury arc lamp-based flow cytometers. Although up to 10% of the dye is not confined to the cytoplasm (Valet *et al.*, 1981), and dye efflux from the cell leads to a marked reduction in fluorescence within minutes (Musgrove *et al.*, 1986), DCH offers resolution superior to BCECF (see Section I, B, 4). The choice of fluorochrome will depend on the instrumentation available and the particular experimental requirements. The dyes in common use were compared by Musgrove *et al.* (1986). Since that publication, Cook and Fox (1988b) have examined in some detail the requirements for good staining with DCH and their procedure offers advantages over the original method of Valet *et al.* (1981). In addition, Alabaster and colleagues have published modifications of the original protocol (Alabaster *et al.*, 1984).

3. CALIBRATION OF pH

A calibration curve is constructed by using cells loaded with the dye in parallel with the experimental sample(s), then resuspended in high $[K^+]$ buffers containing the proton ionophore nigericin. If extracellular and intracellular K^+ concentrations ($[K^+]_e$ and $[K^+]_i$, respectively) are approximately equal, intracellular pH follows extracellular pH, since in the presence of nigericin,

$$\frac{[K^+]_i}{[K^+]_e} = \frac{[H^+]_i}{[H^+]_e}$$

(Thomas *et al.*, 1979). The ratio measurements of cells in nigericin-containing buffers of a range of pH_e can be used to relate histogram channel numbers to pH_i. Note that since $[K^+]_i$ and $[K^+]_e$ are not usually determined experimentally, the calibration is more accurate for relative pH values than absolute pH measurement. However, values of pH_i obtained are consistently within the range of those obtained with independent methods.

4. Measurement Accuracy

There are two components to the accuracy of pH_i measurements. The first relates to the ability to detect differences in the mean of a population normally distributed in pH_i—that is, the sensitivity of the measurement. Our calibration curves with BCECF have slopes of ~25 channels per pH unit, whereas Cook and Fox, using DCH, obtain slopes of −50 channels per pH unit. This suggests a maximum sensitivity of 0.02–0.04 units. In practice this sensitivity is only useful if the measurement is reproducible within similar limits. For both dyes reproducibility is excellent. In repeated, independent estimates of pH_i in PMC-22 amelanotic melanoma cells we obtained a value of 7.47 ± 0.03 (mean $\pm$ SEM, $n = 8$), while Cook and Fox (1988b) similarly claim reproducibility within <0.05 pH units on the basis of replicate measurements. Measurements of changes of ~0.05 pH unit should therefore be possible.

The second component of measurement accuracy is the capacity to

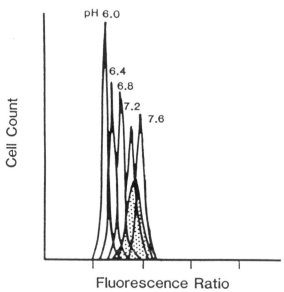

Fluorescence Ratio

Fig. 1. Representative fluorescence ratio histograms. Histograms of 515–525 nm fluorescence/615–625 nm fluorescence ratio, using CCRF-CEM cultured leukemic cells loaded with BCECF and measured in high-$[K^+]$ buffers with nigericin, at the indicated pH. Stippled histogram is that of a parallel sample in pH 7.2 Hanks' balanced salt solution without nigericin. The ratio increases linearly with channel number over the 256-channel histogram, with channel 100 representing a ratio of 1. Reproduced from Musgrove et al. (1986) with permission of the publisher.

resolve subpopulations of different pH_i within the measured sample—that is, the resolution of the technique. This can be estimated by examination of the overlap between the ratio distributions of separate samples differing in pH_i. Clearly the resolution will depend on the coefficient of variation (CV) of the ratio histogram, which in turn depends on the inherent cell-to-cell variation in pH_i as well as instrument and dye-related measurement uncertainties. These components can be separated by comparison of histograms run in the presence and absence of nigericin, since in the presence of nigericin cell-to-cell variation in pH_i should be negligible. The resolution of DCH is superior to that of BCECF: compare Fig. 4A of Cook and Fox (1988b) with Fig. 1 here.

In summary, measurements with both dyes are highly reproducible but DCH is more sensitive and has a higher resolution.

II.   Materials

A.   Stock Solutions

BCECF

The acetoxymethyl ester of BCECF (BCECF–AM) is available from Molecular Probes (Eugene, OR). Stock solutions, at a concentration of 1 mM in dimethyl sulfoxide, are stored at $-20°C$. The same source can provide BCECF.

DCH

Both DCH and its ester, 1,4-diacetoxy-2,3-dicyanobenzol (ADB) are also available from Molecular Probes. Stock solutions of ADB are stored at 4°C, dissolved in dimethyl formamide at a concentration of 2 mg/ml. In aqueous solution ADB undergoes hydrolysis (Cook and Fox, 1988b). Therefore, stock solutions should be stored dessicated or renewed frequently.

Other Chemicals

Sigma Chemical Co., (St. Louis, MO) (in common with a number of other chemical supply companies) supply both nigericin and amiloride. Nigericin stock solutions, of 1 mM in absolute ethanol, are stored at 4°C and amiloride, 18.8 mM (i.e., 5 mg/ml) at 4°C in distilled water. Amiloride is not freely soluble in aqueous solution and may need warming to dissolve, both initially and after prolonged storage at 4°C.

B. Buffers

Earl's Balanced Salt Solution (EBSS)

 140 mM NaCl
 5.4 mM KCl
 1.8 mM CaCl$_2$
 0.8 mM MgSO$_4$
  5 mM Glucose
 25 mM HEPES

Adjust pH to 7.2–7.4 with 1 M NaOH or 1 M Tris.

High-[K$^+$] Buffer

Mix appropriate proportions of 135 mM KH$_2$PO$_4$, 20 mM NaCl, and 110 mM K$_2$HPO$_4$, 20 mM NaCl to give buffers with a range of pH between 6.0 and 8.0.

Alternate High-[K$^+$] Buffer (Method of Cook and Fox)

  30 mM NaCl
 120 mM KCl
   1 mM CaCl$_2$
 0.5 mM MgSO$_4$
   1 mM NaHPO$_4$
   5 mM Glucose
  10 mM HEPES
  10 mM PIPES

Adjust pH with 1 M NaOH.

III. Methods

A. Instrumentation

Either hardware or software capable of generating a signal proportional to the ratio of two fluorescences on a cell-by-cell basis is required for pH$_i$ measurement. Such hardware is standard on some flow cytometers and on others, compensation circuitry can be readily adapted (since logFL1 − logFL2 = FL1/FL2). Alternately, some software packages offer the ability to calculate a ratio from list mode fluorescence intensity data, or the more useful option of real-time ratio calculation.

1. BCECF

The 488-nm line of an argon ion laser, commonly at a power of 400 mW, or mercury arc lamp emission filtered to give blue light of similar wavelength, is used to excite the dye. The resulting fluorescence is first filtered through a 520-nm long-pass filter, then separated into high and low wavelengths by a 560-nm dichroic filter. These are further selected by a 520-nm bandpass filter (10-nm bandwidth) and a 580-nm long-pass filter followed by a 620-nm bandpass filter (10-nm bandwidth). The ratio of 515–525 nm fluorescence/615–625 nm fluorescence is measured and increases with increasing pH. Figure 1 shows examples of the histograms obtained. The expected narrowing of the distribution in the presence of nigericin is apparent.

2. DCH

Either a suitably filtered mercury arc lamp emission or the UV doublet lines of an argon ion laser, at up to 200 mW, can be used. The resulting fluorescence is then filtered through a 418-nm long-pass filter and 50% beam splitter before detection of the two fluorescences (see Cook and Fox, 1988b, for details of a suitable filter set). In this case, the ratio 418–440 nm fluorescence/469–484 fluorescence is measured, and decreases with increasing pH.

For both fluorochromes, some flexibility in both the choice of filters and precise emission bands is possible. Variation in the wavelengths measured, in filter sets used, and even in individual filters of the same specifications, though still yielding useful data, may affect either the sensitivity or the resolution of the pH_i estimation. Additional filters may be necessary to block leakage of high-intensity light at the emission maximum into the selected wavelength band.

B. Dye Loading

1. BCECF

Harvest cells and wash once in EBSS, then resuspend at a concentration of 10^7 cells/ml in the same buffer. Add BCECF–AM to a final concentration of 10 μM (10 μl stock solution/ml cells + buffer). Hanks' balanced salt solution (HBSS) or serum-free medium can be used instead of EBSS. Medium containing serum is unsuitable as a buffer for loading esterified probes, since serum contains high levels of nonspecific esterase activity.

Incubate cells for at least 5 minutes at 37°C to allow cleavage of the esterified dye. The cell fluorescence will increase with longer incubation times, but the ratio measurement is essentially independent of intracellular dye concentration. An incubation of 30 minutes is recommended.

Remove aliquots of ~10^6 cells, comprising the experimental sample(s) and those for the calibration curve, and pellet by centrifugation. Aspirate the supernatant and hold the pellets on ice (where fluorescence will be maintained for 1–2 hours). Immediately before running on the flow cytometer, resuspend the cell pellet in 1–2 ml EBSS (experimental sample) or high-[K^+] buffer with 10 μM nigericin (calibration samples).

2.   DCH

This follows the protocol of Cook and Fox (1988b).

Harvest cells and wash with saline buffer (the published method uses saline buffer of similar composition to their high-[K^+] buffer, but with 5 mM KCl and 145 mM NaCl replacing 120 mM KCl and 30 mM NaCl, respectively) or with alternate high-[K^+] buffer (calibration samples), then resuspend at a concentration of 4–6 × 10^6 cells/ml in the same buffer.

Add a 5-μl aliquot of the stock solution of ADB to each milliliter of cells in buffer, to give a final concentration of 10 μg/ml (41 μM). Incubate at room temperature for 20 minutes, then run on flow cytometer immediately.

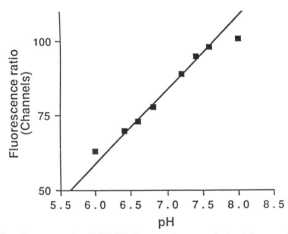

Fig. 2.   Calibration curve for BCECF. Data points are derived from calculations of the mean fluorescence ratio of the histograms presented in Fig. 1. The line is fitted, by least-squares linear regression, to data points from pH 6.4 to 7.6 inclusive. Correlation coefficient, $r = .997$. Data redrawn from Musgrove et al. (1986) with permission of the publisher.

C.   Calibration

A range of high-[K$^+$] buffers, with pH in 0.2-unit increments from 6.0 to 8.0, is required. Stained cells resuspended in these buffers are equilibrated to the extracellular pH by the addition of nigericin, 10 μM final concentration, 2–5 minutes before running on the flow cytometer. Cook and Fox dye-load cells for the calibration curve in the presence of nigericin in high-[K$^+$] buffer.

The fluorescence ratio histograms are normally distributed, typically with CV values of ≤5%. A calibration curve is obtained by plotting the calculated mean ratio of samples run in high-[K$^+$] buffers with nigericin, as a function of external pH. Over a wide range of pH, this plot will yield a sigmoid curve, which can be fitted by some computer algorithms. However, the curve is linear near the pK and for that range can be fitted by least-squares linear-regression analysis. An example is shown in Fig. 2.

D.   Manipulation of pH$_i$

1.   ACID LOADING

A fall in pH$_i$ of up to 1 pH unit can be generated by exposing dye-loaded cells to 10 mM NH$_4$Cl for 30 minutes at 37°C (Moolenaar *et al.*, 1984), then resuspending the cells in ammonium-free buffer. During the initial incubation, the ammonium ion enters the cells and equilibrates with NH$_3$ and H$^+$. A rapid efflux of NH$_3$ on changing to ammonium-free conditions then leaves an excess of H$^+$ within the cells. Recovery of pH$_i$ is temperature-dependent, and occurs within minutes at 37°C (see Hedley and Jorgensen, 1989, for examples of recovery at different temperatures).

2.   pH$_i$ CLAMPING

The protocols just described assume that the pH$_i$ is at equilibrium and will be held constant by the cell throughout the dye-loading and running procedure. In some situations (e.g. for tumor cells held below equilibrium pH$_i$ by an acidic microenvironment), it is necessary to measure pH$_i$ under nonequilibrium conditions. Means for achieving this have been examined (Hedley and Jorgensen, 1989); the definitive protocol from that study is reproduced here.

Collect cells on ice in EBSS containing 0.5 mM amiloride (an inhibitor of the antiport). Resuspend at 10^7 cells/ml in choline-based EBSS containing 10 μM BCECF–AM and dye-load by incubating at 37°C for 5 minutes only. Choline-based EBSS is prepared by substitution of 140 mM choline chloride for the NaCl in the EBSS formulation given earlier.

At the end of the incubation, pellet the cells and resuspend in either ice-cold Na-based EBSS containing 0.5 mM amiloride, pH 7.2 (experimental sample), or in high-[K$^+$] buffers with nigericin (calibration samples). Less than 30 minutes should elapse between collection of the sample and the experimental ratio measurement.

IV.   Applications and Future Directions

Publications demonstrating application of FCM pH$_i$ measurement in several areas of cell biology have appeared. These include studies linking changes in pH$_i$ with transition to quiescence (Musgrove *et al.*, 1987) and others investigating the cytotoxicity of hyperthermia in combination with low pH (Cook and Fox, 1988a).

It should be possible to correlate pH$_i$ measurements using BCECF with other parameters for which UV-excited, membrane-permeable probes are available, such as intracellular [Ca^{2+}] (measured using indo-1) or DNA content (measured using Hoechst 33342), using dual-laser instruments. Indeed, Shapiro (1983) has shown that it is possible to measure independently Hoechst 33342 fluorescence and pH-dependent, fluorescein derivative fluorescence. There are few, if any, published studies using this approach despite the potential applications, for example in work on cell activation.

Acknowledgments

We would like to thank Helle Jorgensen for her useful comments on the manuscript. Elizabeth Musgrove is supported by a scholarship from the Government Employees Assistance to Medical Research Fund.

References

Alabaster, O., Clagget Carr, K., and Leonaridis, L. (1984). *Methods Achiev. Exp. Pathol.* **11,** 96–110.
Cook, J. A., and Fox, M. H. (1988a). *Cancer Res.* **48,** 496–502.
Cook, J. A., and Fox, M. H. (1988b). *Cytometry* **9,** 441–447.
Grinstein, S., and Rothstein, A. (1986). *J. Membr. Biol.* **90,** 1–12.
Hedley, D. W., and Jorgensen, H. B. (1989). *Exp. Cell Res.* **180,** 106–116.
Moolenaar, W. H., Tertoolen, G. J., and de Laat, S. W. (1984). *J. Biol. Chem.* **259,** 7563–7569.

Musgrove, E. A., Rugg, C. A., and Hedley, D. W. (1986). *Cytometry* **7,** 347–355.

Musgrove, E. A., Seaman, M., and Hedley, D. W. (1987). *Exp. Cell Res.* **172,** 65–75.

Paradisio, A. M., Tsien, R. Y., and Machen, T. E. (1984). *Proc. Natl. Acad. Sci. U. S. A.* **81,** 7436–7440.

Rink, T. J., Tsien, R. Y., and Pozzan, T. (1982). *J. Cell Biol.* **95,** 189–196.

Seifter, J. L., and Aronson, P. S. (1986). *J. Clin. Invest.* **78,** 859–864.

Shapiro, H. M. (1983). *Cytometry* **3,** 227–243.

Thomas, J. A., Buschbaum, R. N., Zimniak, A., and Racker, E. (1979). *Biochemistry* **18,** 2210–2218.

Valet, G., Raffael, A., Moroder, L., Wunsch, E., and Ruhenstroth-Bauer, G. (1981). *Naturwissenschaften* **68,** 265–266.

Chapter 7

Flow Cytometric Techniques for Measurement of Cytochrome P-450 Activity in Viable Cells

A. DUSTY MILLER

Fred Hutchinson Cancer Research Center
Seattle, Washington 98104

I. Introduction

Cytochrome P-450-containing monooxygenases initiate the metabolism of a wide variety of lipophilic substances. Exogenous substrates include drugs, carcinogens, and pesticides, and endogenous substrates include steroids, fatty acids, and prostaglandins. Metabolism of these compounds is mediated by a family of P-450 enzymes having different substrate specificities. P-450 activity is often regulated in response to substrate levels, and this response is mediated by specific protein receptors.

One of the best-studied members of the P-450 family is cytochrome P_1-450, which metabolizes polycyclic aromatic hydrocarbons, including environmental pollutants and carcinogens. For example, cytochrome P_1-450 can initiate metabolism of benzo[a]pyrene (BP, Fig. 1), a widespread carcinogen produced during combustion. Other enzymes continue the conversion of BP to water-soluble forms that can be excreted. However, the initial oxidized products of BP can bind to cellular macromolecules and are mutagenic and carcinogenic. Thus metabolism of BP involves an initial step that converts BP to a more toxic form followed by detoxification and elimination steps. P_1-450 is induced by exposure to BP and other related compounds, which results in more rapid metabolism and elimination following exposure of cells to increased concentrations of these compounds.

Flow cytometric (FCM) techniques have been developed for measurement of cytochrome P-450 activity in intact viable cells, which allows quantitation of enzyme activity in individual cells in a population and isolation

71

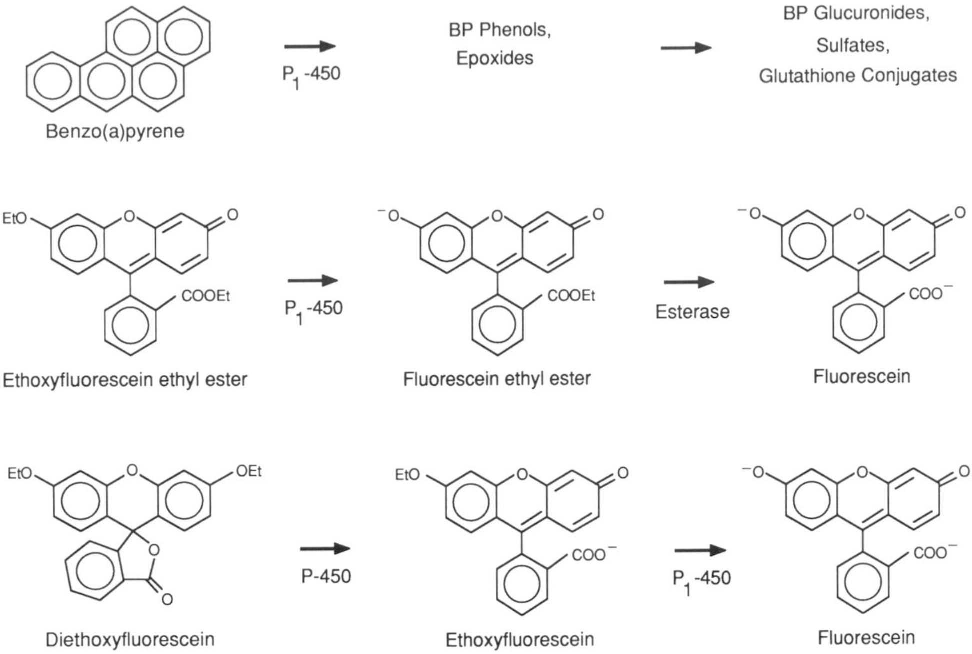

FIG. 1.   *P*-450 substrates and their metabolism.

of subpopulations with defined activity. Three analytical techniques are presented here. The first is based on oxidation of the fluorescent substrate BP (Fig. 1) by cytochrome P_1-450 to nonfluorescent derivatives (Miller and Whitlock, 1981). The other two techniques measure *O*-deethylation of relatively nonfluorescent derivatives of fluorescein to highly fluorescent fluorescein (Miller, 1983; White *et al.*, 1987). Two derivatives of fluorescein have been used (Fig. 1), ethoxyfluorescein ethyl ester (EFEE) and diethoxyfluorescein (DEF). Metabolism of EFEE is mediated by P_1-450 (Miller, 1983), while metabolism of DEF is mediated by an uncharacterized member of the *P*-450 family (White *et al.*, 1987).

II.   Applications

The techniques described here allow quantitation of *P*-450 activity in individual cells in a population. This allows study of heterogeneity of enzyme activity in cells, including cells obtained directly from animals.

P_1-450 is induced in response to enzyme substrates; thus the response of cells to enzyme inducers can also be examined. Because the cells remain viable during the procedure, somatic cell genetic approaches to the study of this enzyme system are also possible. Hepatoma cell line variants having abnormally high or low P_1-450 enzyme activity have been isolated (Miller and Whitlock, 1981), and these were found to have defects in different components of this inducible enzyme system (Miller *et al.*, 1983; Jones *et al.*, 1984). Rapid complementation analysis can be performed on mutant cells without the need to isolate, propagate, and analyze hybrid cell clones by analysis of P-450 activity in heterokaryons formed by cell fusion (Miller *et al.*, 1983; Jones *et al.*, 1984). This capability reduces the potential for chromosome loss during propagation and subsequent artifactual effects on complementation analysis.

In principle, these techniques can be applied to any cell type. Most studies have analyzed P-450 activity associated with hepatic cells, either primary hepatocytes or hepatoma cell lines. The liver contains the highest levels of P-450 enzymes, and is the primary organ for metabolism of many exogenous P-450 substrates. The techniques described here readily measure P-450 activity in hepatic cells, but may not be sensitive enough to detect low levels of activity present in other cell types.

III. Measurement of Benzo[a]pyrene Metabolism

Benzo[a]pyrene (BP) is extremely hydrophobic, and when added to cells in suspension, rapidly moves from the suspension medium into the cells (Miller and Whitlock, 1982b). It is also highly fluorescent and can be excited using the ultraviolet (UV) spectral band of an argon laser. The only way for BP to leave the cell is by metabolism, the first step of which is catalyzed by cytochrome P_1-450. Thus, one can measure P-450 activity by FCM measurement of BP disappearance from single cells. The cells remain viable during the procedure, and cells with various activities can be isolated from a population with heterogeneous activity (Miller and Whitlock, 1981).

A. Materials

Materials can be obtained as follows: BP from Aldrich Chemical Co., Dulbecco's phosphate-buffered saline (PBS) without calcium and magnesium from Grand Island Biological Co. (GIBCO), 2,3,7,8-tetrachlorodibenzo-p-dioxin (TCDD) from the Chemical Carcinogen Repository of the

National Cancer Institute (NCI, Bethesda, MD). The emission filter for BP fluorescence measurement is a 10-nm bandpass filter with center wavelength of 405 nm (Ditric Optics, Hudson, MA).

B. Procedures

1. CELL PREPARATION

Cytochrome P_1-450 can be induced by using a variety of compounds, including BP itself. In general, residual amounts of inducers remain after induction. This can interfere with the assay in two ways: by inhibiting enzyme activity (most of the inducers are also metabolized by the induced P-450) and by contributing to the background fluorescence of the cells (Miller and Whitlock, 1982b). An ideal inducer for this assay is TCDD, because it induces the enzyme to very high levels, does not interfere with P-450 enzyme activity, and is nonfluorescent (Miller and Whitlock, 1982b). Exposure of cells to 1 nM TCDD (achieved by using a 1 μM stock solution in dimethyl sulfoxide, DMSO) for 18 hours provides maximum induction.

Following growth in the presence or absence of inducers, cells that were grown in suspension are centrifuged at 500 g for 4 minutes and resuspended in PBS at 5×10^5 cells/ml. Cells grown in monolayer culture are treated with trypsin (0.5%)–EDTA (0.02%), diluted 5-fold with PBS, centrifuged at 500 g for 4 minutes, and resuspended in fresh PBS at 5×10^5 cells/ml. Cells are kept on ice until assay. Even though P-450 enzyme activity is dependent on NADPH, the cells do not require an exogenous energy source for this assay. Thus, the cells need not be suspended in cell culture medium for assay. This is an advantage for reduction of background fluorescence.

Cells isolated from animal tissue, particularly liver, can also be examined. Perfusion of organs with collagenase and other proteases is useful to help disaggregate the cells since a single-cell suspension is required for analysis. Animals can be pretreated with P-450 inducers prior to cell harvest to study effects of these inducers on enzyme activity.

2. CELL STAINING

Cell suspensions can be kept on ice for at least 1 hour prior to assay. For assay, the cells are warmed to 37°C. At the same time, a 50 nM solution of BP in warm PBS is made using a 10 μM stock solution of BP in acetone. A Hamilton syringe is useful for measurement of the BP stock solution. It is important to make this solution shortly before addition of cells because BP is hydrophobic and may tend to crystallize or adhere to the walls of the tube upon standing. To start the reaction, equal volumes of warm 50 nM BP in PBS and cells suspended at 5×10^5 per milliliter are mixed and incubated at 37°C. Fluorescence of BP is then measured as a function of time. Glass or plastic tubes convenient for attachment to the flow cytometer may be used. While attached to the flow cytometer for analysis, the cells should be kept at 37°C to avoid temporary reduction of enzyme activity. For situations where large numbers of cells must be analyzed, such as in screening for mutants, the cell density can be increased. However, since most of the added BP partitions into the cells (Miller and Whitlock, 1982b), the BP concentration must also be increased proportionally to provide the same starting BP concentration per cell.

3. INSTRUMENT SETUP

For use with an argon laser, tune the laser to the band providing 350/360 nm UV. The fluorescence emission peak of BP in cells is 406 nm, so a 405-nm emission filter with a 10-nm bandpass works well as an emission filter. Cellular fluorescence is best displayed on a logarithmic scale. Cells take up BP in direct proportion to cell size, thus fluorescence/cell size should ideally be displayed. An alternative for cell sorters without this feature is to analyze only those cells within a narrow size window, so as to minimize the effect of varying cell size on fluorescence (Miller and Whitlock, 1982a).

C. Interpretation of Data

Useful qualitative information can easily be obtained using this assay. Cells with the highest enzyme activity lose fluorescence most rapidly as a function of time, and cells with low activity remain highly fluorescent. A population of cells with similar activity tend to migrate as a single peak in fluorescence histograms taken as a function of time, and cells with widely varying activity show wide fluorescence histograms. Its hydrophobicity prevents BP from migrating between cells, so low-activity cells do not lose fluorescence even in the presence of high-activity cells (Miller and Whitlock, 1981). In addition, quantitative analysis of enzyme activity distributions within cell populations can be obtained from fluorescence histograms taken at various times after addition of BP; for a complete description, see Miller and Whitlock (1982a).

D. Other Considerations

Benzo[*a*]pyrene is carcinogenic and TCDD is toxic, so stock solutions should be handled with care. Rubber gloves should be used while handling the stock solutions and treated cells. However, BP is also an ubiquitous pollutant and possible experimental exposure levels are far below those normally encountered in the environment. Benz[a]anthracene can be substituted for TCDD as an inducer (Miller and Whitlock, 1981, 1982b), although it is not ideal and care must be taken that is completely metabolized before P-450 measurement. At high concentrations BP is toxic to cells (Hankinson, 1979), but it is nontoxic at concentrations employed here. Toxicity should be tested for cell lines not previously analyzed.

IV. Measurement of Ethoxyfluorescein Ethyl Ester Metabolism

The fluorogenic compound ethoxyfluorescein ethyl ester (EFEE) is metabolized by the cytochrome P-450 induced by polycyclic aromatic hydrocarbons. The product of metabolism in intact cells is fluorescein, which can be measured using standard FCM detection systems (Miller, 1983; White et al., 1987). The utility of this method is primarily for measurement of P-450 activity in cells with relatively high activity.

A. Materials

One can obtain EFEE from Molecular Probes (Eugene, OR) TCDD from the Chemical Carcinogen Repository of the NCI (Bethesda, MD) and Dulbecco's PBS without calcium and magnesium from GIBCO.

B. Procedures

1. CELL PREPARATION

Procedures are described in Section III. Fluorescence of compounds used to induce P-450 does not interfere appreciably with measurement of fluorescein fluorescence, in contrast to their ability to interfere with BP fluorescence measurement.

2. CELL STAINING

Cells at a density of 5×10^5 per milliliter in PBS are incubated at 37°C with 50 nM EFEE and fluorescence measured after 5 or 10 minutes. The EFEE can be added directly to the cell suspension from an aqueous stock solution. The concentration of EFEE and time of incubation are for cultured hepatoma (Hepa-lclc7) cells, and may need modification for other cell types.

3.   INSTRUMENT SETUP

Standard excitation and emission wavelengths for detection of fluorescein are used. For use with an argon laser, the excitation wavelength is 488 nm and the emission filter is a long-pass filter with half-maximal transmission at 515 nm. Cellular fluorescence is best displayed on a logarithmic scale.

C.   Interpretation of Data

This assay relies on detection of fluorescein accumulation in cells following metabolism of EFEE. However, fluorescein can leak from the cells, thus the amount present in a cell is not a direct function of metabolism. Also, EFEE contributes some background fluorescence at wavelengths used to measure fluorescein, so that the concentration of EFEE employed is a compromise between acceptable background fluorescence and that for maximal enzyme activity. Thus, results should be interpreted in a qualitative sense. Fluorescence should be measured at a time when cellular fluorescence has stabilized, representing an equilibrium between fluorescein production and leakage out of the cell. The additional complication of possible fluorescein leakage into cells with low enzyme activity does not occur (White *et al.*, 1987).

D.   Other Considerations

This assay complements the BP disappearance assay for measurement of P_1-450 activity. The BP disappearance assay is relatively sensitive, but very high-activity cells metabolize BP too rapidly for convenient analysis. The EFEE assay allows measurement and discrimination between cells with very high activity.

V.   Measurement of Diethoxyfluorescein Metabolism

Diethoxyfluorescein (DEF) is nonfluorescent. O-Deethylation of DEF results in the production of highly fluorescent ethoxyfluorescein and fluorescein, a proportion of which remains in cells and can be measured by standard methods for detection of fluorescein fluorescence. The particular P-450 involved in DEF metabolism is not known, although it is not induced in rats pretreated with methylcholanthrene, which induces cytochrome P_1-450, or with phenobarbitone, another standard P-450 enzyme inducer (White *et al.*, 1987).

A. Materials

One can obtain DEF from Molecular Probes (Eugene, OR), TCDD from the Chemical Carcinogen Repository of the NCI (Bethesda, MD), and Dulbecco's PBS without calcium and magnesium from GIBCO.

B. Procedures

1. CELL PREPARATION

Procedures are described in Section III.

2. CELL STAINING

Cells at a density of 5–10×10^5 per milliliter in PBS are incubated at 37°C with 50 μM DEF and fluorescence measured after 4–10 minutes. The DEF can be added directly to the cell suspension from a 10 mM stock in DMSO. The concentration of DEF and time of incubation are for rat hepatocytes, and may need modification for other cell types.

3. INSTRUMENT SETUP

Setup is described in Section IV.

C. Interpretation of Data

This assay relies on detection of fluorescein accumulation in cells following metabolism of DEF. However, fluorescein can leak from the cells, thus the amount present in a cell is not a direct function of metabolism. Thus, results should be interpreted in a qualitative sense. Fluorescence should be measured at a time when cellular fluorescence has stabilized, representing an equilibrium between fluorescein production and leakage out of the cell. An advantage of this assay in comparison with that using EFEE as a substrate, is that while EFEE is slightly fluorescent, DEF is not, and thus concentrations of DEF can be used that saturate the enzyme responsible for its metabolism and result in maximal rates of fluorescein production.

D. Other Considerations

It is not known what member of the *P*-450 family of enzymes is responsible for DEF metabolism, nor are possible inducers of the enzyme known. Induction of P_1-450, which results in increased metabolism of BP and EFEE, does not result in a similar increase in DEF metabolism (White *et al.*, 1987).

REFERENCES

Hankinson, O. (1979). *Proc. Natl. Acad. Sci. U.S.A.* **76**, 373–376.
Jones, P. B. C., Miller, A. G., Israel, D., Galeazzi, D. R., and Whitlock, J. P. (1984). *J. Biol. Chem.* **259**, 12357–12363.
Miller, A. G. (1983). *Anal. Biochem.* **133**, 46–57.
Miller, A. G., and Whitlock, J. P. (1981). *J. Biol. Chem.* **256**, 2433–2437.
Miller, A. G., and Whitlock, J. P. (1982a). *Mol. Cell. Biol.* **2**, 625–632.
Miller, A. G., and Whitlock, J. P. (1982b). *Cancer Res.* **42**, 4473–4478.
Miller, A. G., Israel, D., and Whitlock, J. P. (1983). *J. Biol. Chem.* **258**, 3523–3527.
White, I. N. H., Green, M. L., and Legg, R. F. (1987). *Biochem. J.* **247**, 23–28.

Chapter 8

Fluorescent Staining of Enzymes for Flow Cytometry

FRANK DOLBEARE

Biomedical Sciences Division
Lawrence Livermore National Laboratory
Livermore, California 94550

I. Introduction

Fluorogenic substrates for quantitative evaluation of enzymes in single cells by flow cytometry (FCM) range from natural substrates for the dehydrogenases (e.g., NAD and NADP) to exotic naphthol and coumarin derivatives (Dolbeare and Smith, 1979). Many of these substrates can be used in direct analysis of specific activities but have to be used with limited incubation periods followed by immediate flow analysis because of diffusion artifacts (Dolbeare and Phares, 1979). Diffusion artifacts can mostly be overcome by using simultaneous coupling techniques. Uncoupled reactions with substrates like fluorescein diacetate (Hulett *et al.*, 1969), fluorescein diphosphate (Watson *et al.*, 1977), and naphthol esters (Dolbeare and Phares, 1979) have been quantified on unfixed cells by FCM. Fluorescent measurements were done in the first 5–10 minutes of reaction to avoid diffusion losses of the fluorescent products.

The following protocols will describe methods for coupled histochemical reactions for assaying proteases such as cathepsin B (Dolbeare and Smith, 1977), and esterases such as alkaline phosphatase (Dolbeare *et al.*, 1980). Chemically blocked amino acid and peptide derivatives of 4-methoxy-2-naphthylamine (MNA) or 2-naphthylamine (NA) have been used for a

METHODS IN CELL BIOLOGY, VOL. 33

number of years as chromogenic and fluorogenic substrates for a class of proteases called arylamidases (Smith and Van Frank, 1975). The protease splits the arylamide bond, releasing the peptide and naphthylamine. Diazonium salts react with the protease-liberated naphthylamine, forming an insoluble colored product in the cell. Many of the diazonium complexes with naphthylamine are deeply colored and appear not to be fluorescent, at least by visible excitation. Others form colored complexes that fluoresce red.

The liberated naphthylamine can also react with 5-nitrosalicylaldehyde (2-hydroxy-5-nitrobenzaldehyde) (5 NSA) to form an insoluble fluorescent Schiff base that fluoresces orange (Dolbeare and Smith, 1977).

Naphthol esters are used as synthetic substrates for a number of esterases (e.g, acid and alkaline phosphatases, β-galactosidase, β-glucuronidase, and arylsulfatase). The naphthol liberated during esterase hydrolysis is complexed with a diazonium salt to form an insoluble colored and sometimes fluorescent product. The fluorescent product does not diffuse from the cell. This permits the investigator to stain the cells for a specific enzyme, wash the cells, and analyze the fluorescence at a later time.

II. Applications

These methods can be applied to viable and fixed cells where the fixative does not inactivate the enzyme to be assayed. Flow cytometric enzyme assays have been used to: (1) discriminate viable from dead or damaged cells, (2) distinguish transformed from nontransformed cells, (3) discriminate and quantify differential leukocyte populations, (4) follow the progression of cellular differentiation, (5) quantify subcellular organelles (e.g., lysosomes and nuclei), and (6) measure enzyme kinetics on heterogeneous cell populations. Enzyme staining may be done in combination with a second enzyme stain, or may be used in combination with a DNA stain, RNA stain, or with fluorescent antibodies to specific cellular antigens.

Other enzymes that may be stained by the alkaline phosphatase protocol by varying the phenolic ester group of the naphthol AS-MX include β-galactosidase, β-glucuronidase, and aryl sulfatase. Enzymes that may be stained by the cathepsin B protocol include all acid proteases with arylamide activity. Alkaline proteases (e.g., dipeptidyl aminopeptidase IV and γ-glutamyl transpeptidase) may be stained by using diazonium salts, which couple arylamines at alkaline pH.

III.   Materials

A.   Alkaline Phosphatase

1. Tris buffer, pH 8.1, containing 5 mM MgCl$_2$. This buffer may be made as a stock solution and stored in the refrigerator.
2. Naphthol AS-MX phosphate sodium salt (Sigma Chemical Co., St. Louis, MO) should be a white powder and when dissolved in the Tris buffer should not fluoresce or produce a red color when mixed with the diazonium salt. This reagent should be prepared freshly before the experiment as a 10 mg/ml stock solution in distilled water (dH$_2$O).
3. Fast red TR salt (Sigma) should also be a white crystalline material and should not be used if colored. This reagent should be stored at $-10°$ to $-20°C$, desiccated. A stock solution containing 10 mg/ml in H$_2$O should be prepared fresh before each experiment.
4. Reaction solution contains Tris buffer, MgCl$_2$ plus 1 mg of Naphthol AS-MX phosphate, and 0.5 mg fast red TR. Any red coloration is an indication of decomposition of the reaction solution.

B.   Cathepsin B

1. Morpholineethanesulfonic acid (MES), pH 6.1, containing 5 mM dithiothreitol (DTT). This buffer may stored frozen or at 4°C without the DTT.
2. CBZ-Ala-Arg-Arg-methoxynaphthylamine (Enzyme Systems Products, Dublin, CA) should be stored below freezing with desiccation and should not show any discoloration as a powder or give any blue fluorescence when dissolved in the reaction buffer. A stock solution of 10 mg/ml in dimethyl formamide or dimethyl sulfoxide may be stored frozen. β-Naphthylamine may be substituted for the MNA in the protease substrates.
3. 5-Nitrosalicylaldehyde (2-hydroxy-5-nitrobenzaldehyde) (5NSA) may be obtained from Aldrich Chemical Co. (Milwaukee, WI). A stock solution of 20 mg/ml in 95% ethanol may be prepared and stored refrigcrated for use over a period of several weeks.
4. Reaction solution contains 0.1 M MES buffer, 5 mM DTT, 0.5 mg/ml CBZ-Ala-Arg-Arg-MNA, and 0.5 mg 5NSA.

IV. Staining Procedure

A. Alkaline Phosphatase

1. Cells should be fixed in monodisperse suspension with 1% paraformaldehyde in PBS for 15 minutes at room temperature (RT).
2. Cells are then washed with 5 ml of PBS.
3. 500,000 cells are then incubated in 1 ml of staining solution for periods up to 30 minutes at RT. Overincubation will lead to crystalline product formation.
4. Cells should be washed with H_2O for 5 minutes prior to FCM analysis.

B. Cathepsin B

1. Cells may be stained unfixed or fixed in 0.1% paraformaldehyde in PBS for 30 minutes at RT. Wash cells with MES buffer prior to staining. If unfixed cells are used, incorporate 10% ethanol in the reaction solution to increase permeability of the cells for the CBZ-Ala-Arg-Arg-MNA.
2. Incubate cells in the substrate reagent at 37°C for 5–30 minutes depending on the level of cathepsin B activity in the cells.
3. Centrifuge at 500 g for 1 minute. Pour off supernatant. Vortex cell pellet. Resuspend cells in 2 ml MES buffer pH 6.1 and centrifuge again.
4. Remove wash buffer, and resuspend the pellet in 2 ml of 1% paraformaldehyde in PBS for 15 minutes to stabilize the fluorescence product.
5. Centrifuge to remove fixative. Resuspend the pellet in 1.5 ml of MES buffer, pH 6.1.
6. Analyze by FCM. Product is excited at either 457 nm or 488 nm, and fluoresces at 580 nm.

V. Critical Aspects of These Procedures

A. Alkaline Phosphatase

Diazonium salts are reactive and generally unstable. Yellow discoloration of the fast red salts indicates decomposition and may contribute to increased nonspecific fluorescence. Diazonium compounds tend to inhibit esterase and protease activity. Avoid high concentrations in the staining mixture. For this reason many histochemical methods involving diazo salts call for post coupling; that is, the substrate incubation with the tissue is followed by incubation with a higher concentration of diazo salt to trap the reaction product.

B.   Cathepsin B

Since cathepsin B is a thiol protease and is inhibited by aldehydes, minimize paraformaldehyde fixation. Do not use glutaraldehyde as a fixative, since high levels of yellow autofluorescence are produced. Avoid any aldehyde fixation with hepatocytes, since the levels of autofluorescence may exceed the fluorescence of the reaction product.

Schiff base formation from the 5NSA reaction with aromatic amines is faster at lower pH values, so that acid proteases (pH optima between 5 and 6.5) can be measured. At alkaline pH, the reaction is too slow with an unfavorable equilibrium, so that most of the reaction product will diffuse from the cell before coupling. For alkaline proteases like dipeptidyl aminopeptidase IV and γ-glutamyl transpeptidase, use fast red violet LB salt as a substitute for 5NSA as the coupling agent.

Crystal formation is undesirable in any histochemical reaction, since product is lost from the cell and poor localization results. If cathepsin B levels are high, decrease the incubation time or do the reaction at RT. Smaller clusters of reaction product are formed during shorter incubation periods.

VI.   Controls and Standards

1. Controls without substrate are used to compensate for autofluorescence or fluorescence induced by the coupling reagents. Omit the substrate for the alkaline phosphatase (naphthol AS-MX phosphate) and for cathepsin B (CBZ-Ala-Arg-Arg-MNA). Include the coupling reagent indicated in the individual protocols.
2. To test specificity of the staining reaction, use cells without enzyme activity. This may be accomplished by inactivating the enzyme by heat or fixatives. An alternate is to use specific inhibitors for the enzyme activity to determine if the specific enzyme is being stained. Alkaline phosphatase inhibitors include 20 mM L-homoarginine or L-phenylalanine or 1 mM levamisole depending on the source of the

enzyme. Cathepsin B1 is inhibited completely by 1 μM leupeptin. Specificity of the proteases can be demonstrated by using substrates containing specific peptide sequences.

VII.   Instruments

Flow cytometers equipped with an argon ion laser can be used to measure the fluorescence of the histochemical products for both the alkaline phosphatase and cathepsin B1 reactions. Both reaction products can be excited at 488 nm.

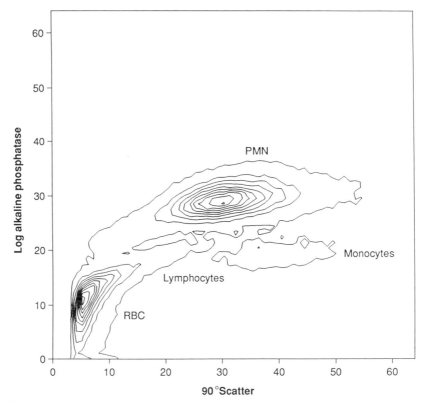

FIG. 1.   Alkaline phosphatase activity in rat peritoneal leukocytes after 3 hours *in vivo* stimulation with 2% BSA in PBS. The leukocyte population consists of neutrophils (90%), monocytes (5%), and lymphocytes (5%). *x* Axis, right-angle light scatter; *y* axis, log red fluorescence (alkaline phosphatase).

VIII. Results

A. Alkaline Phosphatase

Figure 1 is a bivariate plot of alkaline phosphatase/light-scatter distribution of rat peritoneal leukocytes harvested 3 hours after an i.p. injection of 2% bovine serum albumin (BSA) in isotonic saline. The leukocyte suspension contains red blood cells (RBC) and 5% lymphocytes, 5% monocytes, and 90% polymorphonuclear cells (PMN).

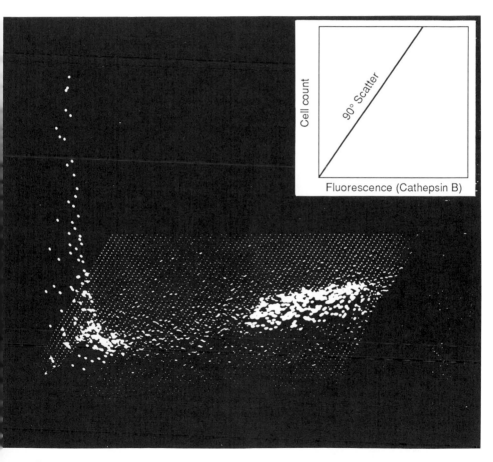

FIG. 2. Cathepsin B activity in human lung diploid fibroblasts (WI-38) and in SV-40-transformed diploid fibroblasts (VA-13). x Axis, cathepsin B activity; y axis, right-angle light scatter.

B.   Cathepsin B

Figure 2 is an isometric plot of the cathepsin B and 90° light scatter for a mixture of WI-38 human lung diploid fibroblasts and VA-13 (a SV-40 transformation of the WI-38). The transformed cell line has very low cathepsin B1 activity compared to the WI-38 cell line.

REFERENCES

Dolbeare, F. A., and Phares, W. (1979). *J. Histochem. Cytochem.* **27**, 120–124.
Dolbeare, F. A., and Smith, R. E. (1977). *Clin. Chem.* **23**, 1485–1491.
Dolbeare, F. A., and Smith, R. E. (1979). *In* "Flow Cytometry and Sorting" (M. R. Melamed, F. F. Mullaney, and M. L. Mendelsohn, eds.), pp. 317–333. Wiley, New York.
Dolbeare, F., Vanderlaan, M., and Phares W. (1980). *J. Histochem. Cytochem.* **28**, 419–426.
Hulett, H. R. Bonner, W. A., Barret, J., and Herzenberg, L. A. (1969). *Science* **166**, 747–749.
Smith, R. E., and Van Frank, R. (1975). *In* "Lysosomes in Biology and Pathology" (J. T. Dingle and R. T. Dean, eds.). North Holland Publ., Amsterdam.
Watson, J. V., Chambers, S. H., Workman, P., and Horsnell, T. S. (1977). *FEBS Lett.* **81**, 179–182.

Chapter 9

Supravital Cell Staining with Hoechst 33342 and DiOC₅(3)

HARRY A. CRISSMAN, MARIANNE H. HOFLAND,
ANITA P. STEVENSON, AND MARK E. WILDER

Cell Biology Group
Los Alamos National Laboratory
Los Alamos, New Mexico 87545

ROBERT A. TOBEY

Biochemistry and Biophysics Group
Los Alamos National Laboratory
Los Alamos, New Mexico 87545

I. Introduction

The Hoechst (HO342) dyes are benzimidazole derivatives that have a high specificity for double-helical DNA and bind preferentially to A-T base regions, but do not intercalate. They emit blue fluorescence when excited by ultraviolet (UV) light at ~350 nm. Hilwig and Gropp (1973) used Hoechst 33258 in mouse chromosome-banding studies. Latt (1973) showed that this dye was quenched when bound to bromodeoxyuridine (BrdUrd)-substituted DNA and developed a method for detecting regions of sister chromatid exchange in metaphase chromosomes labeled with BrdUrd. Arndt-Jovin and Jovin (1977) using flow cytometry (FCM), first demonstrated the use of HO258 and preferably Hoechst 33342 (HO342) for quantitative DNA staining and sorting of viable cells.

89

Although this technique has been useful for a number of cell types (Arndt-Jovin and Jovin, 1977), unfortunately HO342 can be limited in usefulness in many cell systems by (a) poor cell cycle resolution due to limited uptake of the dye through the viable plasma membrane, (b) dye toxicity, and (c) loss of long-term viability following UV excitation and cell sorting.

Dye uptake and/or retention is cell type-dependent and many cell types including line CHO cells are particularly refractory in this regard. The quality of the DNA histogram for CHO cells stained with HO342 alone is extremely poor (coefficient of variation, CV = 10%). Such results are generally not acceptable when attempting to obtain a high degree of accuracy in sorting cells from specific phases of the cell cycle. However, the cell type variability in dye uptake has been exploited in some studies. Loken (1980) demonstrated that B cells had a greater capacity for dye uptake than T cells. Visser (1979) showed that large myeloid progenitor cells were deficient in HO342 uptake compared to the various other cell types in mouse bone marrow samples. Lalande *et al.* (1981) also demonstrated a correlated variability in HO342 uptake within colcemid-resistant clones of CHO cells, in which HO342 stainability decreased with decreasing colcemid uptake into cells. Unfixed cells permeabilized with nonionic detergents yielded excellent DNA histograms, indicating that the cell type variability in dye content was not due to cell type differences in DNA–dye binding.

Effects of HO342 staining on cell viability were studied by Fried *et al.* (1982) and others (Durand and Olive, 1982). Results from these studies indicated that the degree of toxicity was more dependent on the type of cell than on the amount of dye taken up by the cells. Pallavicini *et al.* (1979) also observed a synergistic killing effect of X-ray and HO342 staining.

In this chapter, we provide a technique for combining the membrane potential (MP)-modifying fluorochrome $DiOC_5(3)$ (Waggoner, 1979) with HO342 to yield a superior CV of 3.0% for CHO cells under conditions where dye toxicity is eliminated (Crissman *et al.*, 1988). DNA content resolution, compared well to ethanol-fixed and mithramycin-stained CHO populations (Crissman and Tobey, 1974), and there is no evidence of cytotoxicity in cells sorted following laser excitation settings ranging from 25 to 500 mW. The technique was also shown to yield viable, sorted cells from stained cultures of normal human fibroblast cells and CHO-K1 cells. Furthermore, since excellent DNA resolution and viable cell sorting was achieved with a laser excitation power as low as 25 mW, this new staining procedure is suitable for analysis with relatively inexpensive, low-power laser FCM systems. In some cell types, such as human skin fibroblasts, where HO342 alone provides good DNA stainability, the addition of

$DiOC_5(3)$ (DiO) does not improve the resolution in FCM DNA content analysis. However, DiO does appear to reduce dye toxicity in these cell types (Crissman *et al.*, 1988).

II. Applications

Except for the Hoechst dyes, and preferably HO342, most DNA-specific fluorochromes are not easily transported into and/or retained in viable cells, at least in concentrations satisfactory for stoichiometric DNA content analysis. If cell membranes are perforated with nonionic detergents (e.g., 1% Triton X-100 or Nonidet P-40, NP-40) or ethanol, HO342 staining in most cell types is rapid (i.e., 5–10 minutes) at dye concentrations as low as 0.5 μg/ml. Also, viable cells with damaged membranes stain rapidly. These results indicate that the cell membrane of different cell types, and not DNA binding per se, is responsible for observed variations in the quality of DNA staining. Our study using DiO and that of Krishan (1987) with calcium channel-blocking agents show that treatment of viable cells with these membrane-interacting compounds improve HO342 stainability and possibly cell viability. Our FCM study showed that DiO, a MP mitochondrial stain, increased uptake of HO342 in CHO cells 2-fold and provided CV values of 3.0% compared to values of 8.3% for cells treated with HO342 alone as judged by analysis using a laser power of 500 mW. Cell viability was not impaired as based on cell sorting and plating efficiency studies. Even at powers of 25 mW, CV values of 4.0% were obtained. By comparison, the more commonly used mitochondrial stain, rhodamine 123 (R123, Eastman Kodak, Rochester, NY) (Johnson *et al.*, 1981), did not increase HO342 uptake by cells. Since DiO is not excited in the UV range, energy transfer from DiO to HO342 cannot account for the improvement in the HO–DNA content resolution. Also cells permeabilized by detergents and treated with HO342 with or without DiO provided the same HO342 intensity, indicating that the DiO did not increase HO342 binding to chromatin.

III. Materials

Stock solutions of HO342 (Calbiochem, San Diego, CA) are prepared in distilled water (dH_2O) at 1.0 mg/ml. The solution is stable for at least 1 month when stored at refrigerated temperatures in tubes wrapped with aluminum foil. HO342 tends to precipitate in phosphate-buffered saline (PBS) at concentrations $> 50 \mu g/ml$.

DiOC$_5$(3) (3, 3-dipentyloxacarbocyanine, Molecular Probes, Eugene, OR) is initially dissolved in dimethyl sulfoxide (DMSO) at a concentration of 1.0 mg/ml. This solution is stable for at least 1 year when stored at refrigerated temperatures. At this temperature DMSO will freeze but the solution can be thawed rapidly at 37°C prior to dilution in dH$_2$O (100 $\mu g/ml$) as a stock solution for cell staining.

The aqueous stain solutions just described can be sterilized by filtration (0.22-μm filter) when being used in cell sorting and long-term growth studies.

IV.   Staining Procedure

For viable-cell staining, HO342 and DiO are added directly to the cells actively growing in the culture medium (37°C) at concentrations ranging from 1.0 to 5.0 $\mu g/ml$ and 0.1 to 0.3 $\mu g/ml$, respectively. Cells should remain under normal incubating conditions for a staining period of 30–60 minutes. The dye concentrations and the length of the staining period, which vary according to the cell type, can best be judged from results obtained by FCM analysis, and when appropriate, by dye cytotoxicity studies if long-term cell culturing is involved. Cells from suspension cultures may be removed and analyzed directly in the culture medium containing the dyes. Cells on monolayers should be treated with trypsin containing the same dye concentrations as used for staining, and the trypsin-neutralization culture medium should contain these same dye concentrations also. This ensures that equilibrium staining is continuously maintained, since in the absence of dye, in some cell types, including line CHO, the HO342 tends to diffuse out rapidly. Cells may remain in this condition for ~ 2 hours at room temperature during extensive cell sorting experiments. After that time period a freshly stained cell sample should be substituted. This may help to ensure better cell viability.

V.   Critical Aspects

Hoechst 33342 remains the most valuable DNA-binding fluorochrome for viable DNA cell staining. Optimal conditions for viable cell staining with HO342 and DiO vary according to the cell type, and even when these conditions are obtained, cell viability may need to be assessed. Cells cultured for long periods of time in the presence of HO342 will accumulate in G_2 phase and lose both division capacity and long-term viability. However, with the design of the appropriate control experiments, one may arrive at conditions that satisfy some aspects of a study that would not be available except through the use of a viable DNA stain. We have obtained excellent staining and sorting results with CHO and CHO-K1 cells as well as with human skin fibroblasts (HSF). Good HO342–DNA stainability has also been obtained with L1210 cells, but viability was not examined in that study. The addition of DiO did not improve HO342 stainability of HSF; however, there was some indication of DiO cytoprotection against HO342 toxicity (Crissman et al., 1988). The use of DiO in conjunction with HO342 needs to be evaluated in studies with other cell types.

The DNA content histogram obtained by viable-cell staining should compare well to that obtained from a duplicate sample fixed in 70% ethanol and stained with 0.5–1.0 μg/ml HO342 in PBS. The CV values will be superior for the fixed sample. Krishan (1987) has demonstrated that there are molecular mechanisms in the membranes of some viable-cell populations that actively pump HO342 out of the cell. This results in a decrease in fluorescence intensity and an increase in the CV values observed during FCM analysis.

VI. Instrumentation

Stained cells are excited in the UV range (i.e., 350 nm) and analyzed in a flow system equipped with either a laser or a mercury arc lamp. A barrier filter (GG 400, Schott Glass Technologies Inc., Duryea, PA) is used and light from about 400 to 500 nm is collected. The use of excessive laser power for excitation of HO342-stained cells may damage chromatin; therefore, the power of the excitation source should be maintained as low as possible when cell viability is of consideration. Cell sorting rates should be adjusted to ensure good purity of cell subpopulations sorted from different phases of the cell cycle.

VII. Results

The resolution of DNA content in HO342–DiO, viably stained CHO cells, gauged by CV of G_1 subpopulations, is far superior to the resolution of DNA content obtained with HO342-stained cells (Fig. 1). Computer-fit

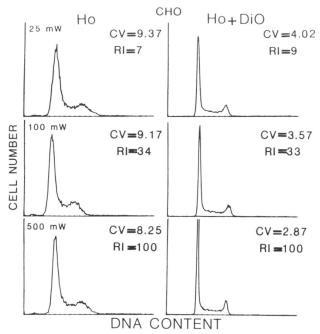

FIG. 1. DNA content frequency histograms obtained for viable CHO cells stained with Hoechst 33342 (HO342) either with or without DiOC$_5$(3) (DiO) and excited with a UV laser adjusted to 25, 100, or 500 mW power. RI, Relative intensity of G$_1$ cells based on a value of 100 for cells excited at 500 mW. The absolute intensity of cells stained with both HO342 and DiO was twice that obtained for cells stained only with HO342. CV, Coefficient of variation of the G$_1$ subpopulations.

analysis of the fraction of cells in G$_1$, S, and G$_2$ + M in HO342–DiO stained CHO populations was also in excellent agreement with results obtained in mithramycin-stained cells (Crissman *et al.*, 1988). Figure 1 shows DNA histograms for viable cells analyzed at 25, 100, and 500 mW laser power following staining with HO342 or with HO342–DiO. The quality of the histograms of HO342–DiO-stained populations is extremely good even at 25 mW laser power (CV = 4.02) compared to those of the HO342-stained cells (CV = 9.37). The survival (plating efficiency) values obtained for HO342–DiO-stained CHO cells excited with 500 mW laser power and sorted over the entire DNA content range was 100% compared to 90% for HO342-stained cells stained and sorted over the same range (Crissman *et al.*, 1988). These data indicated that DiO provides a significant degree of cytoprotection in HO342 staining and analysis of CHO cells. This is particularly interesting in view of the fact that the HO342-

DiO-stained cells have twice the fluorescence intensity of HO342-stained cells indicating a 2-fold cellular HO342 dye concentration in the HO342-DiO-stained cells compared to the HO342-stained cells. Cells stained with DiO only then excited in the UV range and analyzed over the 400–500 nm range yielded no detectable fluorescence. These data show that DiO fluorescence cannot be contributing to the HO342 fluorescence detected in that range. Aso cells with permeabilized membranes stained equally as well with HO342 as with the HO342–DiO combination, demonstrating that DiO does not increase HO342–DNA binding. Comparisons of the effects of the two treatments on cell proliferation and cell survival can be found elsewhere (Crissman et al., 1988).

ACKNOWLEDGMENTS

This work was supported by the Los Alamos National Flow Cytometry and Sorting Research Resource funded by the Division of Research Resources of NIII (grant P41-RR01315) and the Department of Energy.

REFERENCES

Arndt-Jovin, D. J., and Jovin, T. M. (1977). J. Histochem. Cytochem. 25, 585–589.
Crissman, H. A., and Tobey, R. A. (1974). Science 184, 1297–1298.
Crissman, H. A., Hofland, M. H., Stevenson, A. P., Wilder, M. E., and Tobey, R. A. (1988). Exp. Cell. Res. 174, 388–396.
Durand, R., and Olive, J. (1982). J. Histochem. Cytochem. 30, 111–122.
Fried, J., Doblin, J., Takamoto, S., Perez, A., Hansen, H., and Clarkson, B. (1982). Cytometry 3, 42–47.
Hilwig, I., and Gropp, A. (1973). Exp. Cell. Res. 81, 474–477.
Johnson, L. V., Walsh, M. L., Bockus, B. J., and Chen, L. B. (1981). J. Cell Biol. 88, 526–535.
Krishan, A. (1987). Cytometry 8, 642–645.
Lalande, M. E., Ling, V., and Miller, R. G. (1981). Proc. Natl. Acad. Sci. U.S.A. 78, 363–367.
Latt, S. A. (1973). Proc. Natl. Acad. Sci. U.S.A. 70, 3395–3399.
Loken, M. R. (1980). J. Histochem. Cytochem. 28, 36–39.
Pallavicini, M. G., Lalande, M. E., Miller, R. G., and Hill, R. P. (1979). Cancer Res. 39, 1891–1897.
Visser, J. W. M. (1979). In "Flow Cytometry IV" (O. D., Laerum, T., Lindmo, and E., Thorud, eds.), pp. 86–90. Norway Universitetsforlaget, Bergen.
Waggoner, A. S. (1979). In "Methods in Enzymology" (S. Fleischer and L., Packer, eds.), Vol. 55, p. 689–695. Academic Press, New York.

Chapter 10

Specific Staining of DNA with the Fluorescent Antibiotics, Mithramycin, Chromomycin, and Olivomycin

HARRY A. CRISSMAN

Cell Biology Group
Los Alamos National Laboratory
Los Alamos, New Mexico 87545

ROBERT A. TOBEY

Biochemistry and Biophysics Group
Los Alamos National Laboratory
Los Alamos, New Mexico 87545

I. Introduction

Mithramycin (MI), chromomycin A3 (CH), and olivomycin (OL) are closely related, fluorescent antibiotics that show a high specificity for double-helical DNA but not RNA (Ward *et al.*, 1965). When complexed with magnesium these compounds have an affinity for the 2-amino group of guanine in DNA (Kersten *et al.*, 1966). Binding of the metal–antibiotic complex to DNA occurs by nonintercalating mechanisms and is noncovalent in nature.

Spectrofluorometric analysis of MI– and CH–Mg complexes bound to calf thymus DNA in phosphate-buffered saline (PBS) shows two excitation peaks: a minor peak at 320 nm and a major peak at ~445 nm (Crissman *et al.*, 1979). A broad emission peak is observed at ~575 nm, the green–yellow region of the spectrum. Olivomycin has a slightly different chromophore than MI or CH and by comparison has excitation and emission peaks ~10 nm lower than those mentioned for these compounds.

97

The ease of application of these antibiotics for staining cells in suspension coupled with the high specificity for DNA made these compounds especially attractive for use in flow cytometry (FCM; Crissman and Tobey, 1974). Also the 457.9-nm line for fluorescence excitation was available in most argon ion lasers in FCM. Together these features provided considerable advantages over the more cumbersome and lengthy Feulgen procedure that had been previously employed in FCM.

II. Applications

In addition to usage in FCM for cell cycle analysis, MI has been used to stain and analyze DNA content of cells on microscope slides (Johannisson and Thorell, 1977) and to quantitate DNA in sonicated cell preparations (Hill and Whatley, 1975). Swarzendruber (1977a,b) showed that bromodeoxyuridine (BrdUrd) incorporation into cellular DNA significantly increased the availability of MI-binding sites. Cell populations cultured in BrdUrd for the duration of one cell cycle had MI fluorescence values elevated 25% above untreated populations. Larsen et al. (1986), using FCM, showed that mitotic apparatuses could be distinguished from G_2 nuclei in preparations of nuclear suspensions fixed in formalin and stained with MI. In another FCM study, van Kroonenburgh et al. (1985) distinguished eight different cell populations in MI-stained rat testes cell suspensions. Mithramycin has also been used in combination with Hoechst 33342 (HO342) and propidium iodide (PI) for analysis and correlation of the availability of binding sites of the different DNA bases in cells by FCM (Crissman and Steinkamp, 1986). Zante et al. (1976) and Barlogie et al. (1976) have used MI in combination with ethidium bromide (EB) to improve cellular DNA content resolution by taking advantage of energy transfer from MI to EB in flow systems employing mercury arc excitation. Hedley et al. (1985) also used the combination MI–EB to stain and analyze nuclei prepared from paraffin-embedded tissue. Chromomycin A3 (CH) was used by Jensen (1977) in FCM cell cycle analysis studies, and it is used most frequently in combination with Hoechst 33258 (HO258) to stain chromosomes for FCM analysis (Gray et al., 1975). In some of the earliest FCM studies on plant material we used CH as a vital DNA stain in algal cultures of Chlamydomonas for simultaneous analysis of cell cycle and chlorophyll content (Crissman et al., 1975; Price et al., 1976). Olivomycin (OL), though not so commonly used in FCM, was shown to produce reverse-banding patterns (R-banding) in untreated chromosomes (Van de Sande et al., 1977).

III.   Materials

Vials of MI obtained from Pfizer Company (Groton, CN) are soluble in aqueous solutions up to at least 1.0 mg/ml. Solutions of MI in Tris-HCl buffer containing $MgCl_2$ show no deterioration with regard to DNA stainability when stored for at least 1 month at refrigerated temperatures in glass or plastic containers, wrapped in foil. The same solubility conditions apply for CH and OL unless the antibiotics have not been prepared in aqueous soluble salt form, in which case a few drops of dimethyl sulfoxide (DMSO) can be used as a solvent followed by dilution with the appropriate aqueous solution.

These DNA-binding antibiotics do not sufficiently permeate viable cells for DNA-staining purposes; therefore, a fixative, such as ethanol, that perforates cell membranes, is routinely employed prior to staining. For ethanol fixation, cells are centrifuged and the tissue dispersal solution and/or culture medium removed by aspiration. The cell pellet is thoroughly resuspended by pipetting in one part cold saline GM (in grams per liter: glucose 1.1, NaCl 8.0, KCl 0.4, $Na_2HPO_4 \cdot 12 H_2O$ 0.39, KH_2PO_4 0.15 containing 0.5 mM EDTA). Three parts of cold 95% ethanol are added with pipetting. Volumes of saline GM and ethanol are adjusted so that the final cell density is $\sim 10^6$ cells/ml of 70% ethanol fixative. Cell staining for DNA content may be performed after as short a time as 1 hour fixation, and cells can remain in fixative for at least 1 year at refrigerated temperatures prior to staining.

Alternatively, glutaraldehyde (1%) or paraformaldehyde (1%) can be used for fixation of some cell types; however, the coefficient of variation (CV) is somewhat larger for the G_1 peak of the DNA histograms, and the fluorescence intensity of stained cells is only about one-half that of ethanol-fixed cells (Crissman et al., 1979). However, these fixatives preserve cell morphology better than ethanol.

IV.   Staining Procedure

We have previously studied the effects of pH, ionic strength, magnesium concentration, and dye concentration on MI staining of ethanol-fixed cells (Crissman et al., 1979). In summary, cell staining is optimal if performed over a pH range of 5.5–8.5, at ionic strength 0.15–0.40 M NaCl, with MI concentrations of 5.0–100.0 μg/ml and at $MgCl_2$ concentration of 15.0–30.0 mM.

For routine DNA staining, one vial containing 2.5 mg of MI is dissolved in 50 ml of Tris-HCl buffer (pH 7.2) containing 20 mM MgCl$_2$ and 0.15 M NaCl. Final MI concentration is 50 μg/ml. The shelf life of this solution is at least 1 month at refrigerated temperatures.

Cells are removed from ethanol fixative by centrifugation and aspiration, and the cell pellet is resuspended to a final concentration of ~7.5×10^5 cells/ml in the dye solution. Cells can remain in stain at room temperature for 30 minutes or even overnight prior to analysis in the dye solution (i.e., equilibrium staining). Stained CHO cells repeatedly produce good DNA histograms over at least a 1-week period if stored in a refrigerator, and such samples are often retained as standards for instrument alignment.

In our original procedure, viable cells were stained with MI in 0.15 M saline containing 20% ethanol and 15 mM MgCl$_2$ for 20 minutes. Although results obtained were adequate for cell cycle analysis, better CV values and increased cellular fluorescence have been obtained using the procedure described earlier. It appears that ethanol in the stain solution interfered with the stability of the sample stream in systems employing saline sheath fluid.

V. Critical Aspects

Formalin fixative should be avoided for cell fixation. Ethanol fixatives appear best, but a brief (1–2 hours) fixation period in glutaraldehyde or paraformaldehyde followed by resuspension and storage in PBS at refrigerated temperatures may be desirable if cell morphology is to be preserved.

Cells should be analyzed in the dye solution, since fluorescence is decreased ~90% when cells are removed from the stain and analyzed in 0.15 M NaCl. Cellular fluorescence is only decreased 60% when cells are analyzed in 0.15 M NaCl containing 15 mM MgCl$_2$ (Crissman et al., 1979).

Stoichiometric DNA staining can be somewhat affected in cells that have been treated with chemotherapeutic agents. Alabaster et al. (1978) noted differences in MI–DNA stainability of L1210 cells treated with a combination of cytosine arabinoside and adriamycin compared to stainability of untreated control cells.

Compounds such as the DNA-specific, fluorescent antibiotics, like other DNA-binding reagents, should be handled with some precautions. Although these antibiotics were originally designed for administration to patients as chemotherapeutic agents, contact with these compounds should be avoided. Dye solutions should be immediately washed from the skin if accidental contact is encountered.

VI.   Controls and Standards

In testing for DNA specificity of staining, ethanol-fixed cells are treated with DNase (Worthington) in PBS containing 3 mM MgCl$_2$ for 1 hour at 37°C. Staining with the antibiotics should be negative as determined by FCM and/or microscopic analysis. Also computer-fit analysis (Dean and Jett, 1974) of the DNA content histogram by FCM should yield results similar to those obtained by analysis of PI- or HO342-stained cells.

VII.   Instrumentation

Analysis of cells in the stain solution is performed in an FCM equipped with an argon ion laser tuned to 457.9 nm and using a 495-nm long-pass barrier filter that will remove laser light scattered by cells at that wavelength. The sample should be allowed to run for about 1–2 minutes until the dye solution and the sheath fluid are equilibrated. The position of the G$_1$ peak will stabilize and the CV will generally decrease over this initial time period prior to analysis and collection of data. The time required for stabilization will vary from instrument to instrument, depending on the hydrodynamics of the flow system.

VIII.   Results

The DNA content frequency histograms shown in Fig. 1 demonstrate the similarity in DNA specificity of the three antibiotics. Likewise, analysis of the same cell populations stained by each of the three dyes, yielded percentages of cells in the different phases of the cell cycle that are quite comparable. CHO cells fixed in ethanol and stained with the DNA dyes PI or HO342 have provided DNA content histograms similar to those shown in Fig. 1. These data verify the accuracy of DNA staining of the three antibiotics.

The staining technique using any one of the three antibiotics is simple and highly reproducible. It has been applied successfully to a wide variety of cell types in cell cycle analysis studies. Also, DNA stainability of these dyes is unaffected when used in combination with other DNA fluorochromes (Crissman and Steinkamp, 1986).

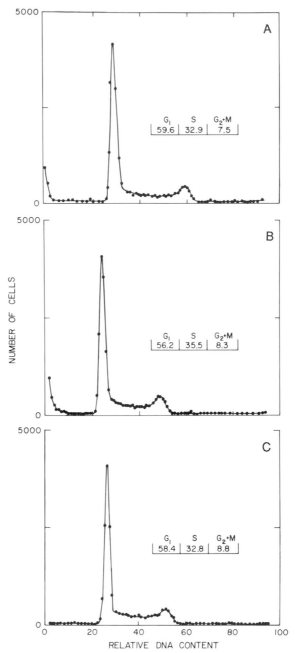

Fig. 1 DNA distribution patterns obtained for Chinese hamster cells (line CHO) stained with (A) olivomycin, (B) chromomycin A3, or (C) mithramycin (100 μg/ml in 0.15 M NaCl containing 15 mM MgCl$_2$). The CV values of the G$_1$ peaks were 4.3%, 4.7% and 4.5%, respectively, for OL-stained, CH-stained, and MI-stained cell populations. Computer-fit analysis (Dean and Jett, 1974) of the DNA profiles was used to determine the percentages of cells in G$_1$, S, and G$_2$ + M.

ACKNOWLEDGMENTS

This work was supported by the Los Alamos National Flow Cytometry and Sorting Research Resource funded by the Division of Research Resources of NIH (Grant P41-RR01315) and the Department of Energy.

REFERENCES

Alabaster, O., Tannenbaum, E., Habbersett, M. C., Magrath, I., and Herman, C. (1978). *Cancer Res.* **38**, 1031–1035.

Barlogie, B., Spitzer, G., Hart, J. S., Johnston, D., Buchner, T., and Schumann, J. (1976). *Blood* **48**, 245–258.

Crissman, H. A., and Steinkamp, J. A. (1986). *In* "Techniques in Cell Cycle Analysis" (J. E. Gray and Z. Darzynkiewicz, eds.). pp. 163–206. Humana Press, Clifton, New Jersey.

Crissman, H. A., and Tobey, R. A. (1974). *Science* **184**, 1297–1298.

Crissman, H. A., Kollman, V. H., and Martin, J. C. (1975). *J. Cell Biol.* **67**, 81a (Abstr. 161).

Crissman, H. A., Stevenson, A. P., Kissane, R. J., and Tobey, R. A. (1979). *In* "Flow Cytometry and Sorting" (M. R. Melamed, P. F. Mullaney, and M. L. Mendelsohn, eds.), Chap. 14, pp. 243–261. Wiley, New York.

Dean, P. N., and Jett, J. H. (1974). *J. Cell Biol.* **60**, 523–527.

Gray, J. W., Carrano, A. V., Steinmetz, L. L. Van Dilla, M. A., Moore, D. H., Mayall, B. H., and Mendelsohn, M. L. (1975). *Proc. Nat. Acad. Sci. U. S. A.* **72**, 1231–1234.

Hedley, D. W., Friedlander, M. L., and Taylor, I. W. (1985). *Cytometry* **6**, 327–333.

Hill, B. T., and Whatley, S. (1975). *FEBS Lett.* **56**, 20–23.

Jensen, R. H. (1977). *J. Histochem. Cytochem.* **25**, 573–579.

Johannisson, E., and Thorell, B. (1977). *J. Histochem. Cytochem.* **25**, 122–128.

Kersten, W., Kersten, H., and Szybalski, W. (1966). *Biochemistry* **5**, 236–244.

Larsen, J. K., Munch-Petersen, B., Christiansen, J., and Jorgensen, K. (1986). *Cytometry* **7**, 54–63.

Price, R. L., Crissman, H. A., Martin, J. C., and Kollman, V. H. (1976). *In* "Proceedings of the Second International Conference on Stable Isotopes" (E. R. Klein and P. F. Klein, eds.), pp. 71–85. U. S. Energy Research and Development Administration. CONF-751027, Oak Ridge, Tennessee.

Swartzendruber, D. E. (1977a). *Exp. Cell Res.* **109**, 439–443.

Swartzendruber, D. E. (1977b). *J. Cell. Physiol.* **90**, 445–454.

van Kroonenburgh, M. J., Beck, J. L., Scholtz, J. W., Hacker-Klom, V., and Herman, C. J. (1985). *Cytometry* **6**, 321–326.

Van de Sande, J. H., Lin, C. C., and Jorgenson, K. F. (1977). *Science* **195**, 400–402.

Ward, D. C., Reich, E., and Goldberg, I. H. (1965). *Science* **149**, 1259–1263.

Zante, J., Schumann, J., Barlogie, B., Gohde, W., and Buchner, Th. (1976). *In* "Pulse Cytophotometry" (W. Gohde, J. Schumann, and Th. Buchner, eds.). pp. 97–106. European Press, Ghent, Belgium.

Chapter 11

DAPI Staining of Fixed Cells for High-Resolution Flow Cytometry of Nuclear DNA

FRIEDRICH OTTO

Fachklinik Hornheide
D-4400 Münster
Federal Republic of Germany

I. Introduction

The use of DAPI (4′, 6-diamidino-2-phenylindole) for flow cytometry (FCM) was proposed by Göhde *et al.* (1978). In the following years, DAPI has been used by many authors. It proved to be a specific, highly fluorescent stain, very well suited for FCM of DNA in whole cells, in nuclei, and in chromosomes.

The staining protocol presented here was developed for use in cell suspensions after fixation with 70% ethanol. It is applicable in specimens that have to be stored before measurement.

The procedure consists of a pretreatment of cells with citric acid plus Tween 20 in order to produce a suspension of isolated nuclei, followed by addition of a solution of disodium phosphate or trisodium citrate with DAPI, which raises the pH and stains the DNA.

This protocol is recommended for accurate and reproducible measurement of the nuclear DNA, especially for the assessment of DNA content variations in cell populations affected by chemical or physical mutagens and for the discrimination of spermatogenic cells.

II. Application

This method can be applied to a large variety of cell types. Flow cytometric histograms with excellent resolution were obtained from fibroblasts,

105

leukocytes, bone marrow cells, thymocytes, lung, liver, and testicular cells of humans and experimental animals, from chicken and trout erythrocytes, and from various cultured cell lines.

In mammalian spermatozoa the method does not yield quantitative DNA staining because of the highly condensed chromatin. These cells need a pretreatment with strong decondensing agents such as dithioery-thritol and papain (Otto *et al.*, 1979).

Using bone marrow of experimental animals it is possible to establish an *in vivo* mutagenicity test, taking the increased dispersion of the nuclear DNA content as a measure of the mutagenic effect (Otto *et al.*, 1981, 1984; Otto and Oldiges, 1983). An example is shown in Fig. 1.

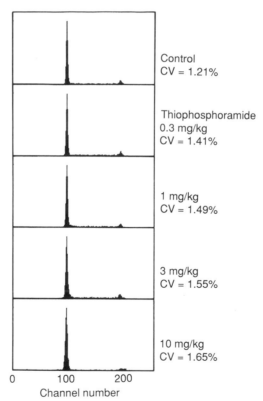

FIG. 1.   Flow cytometric histograms of the DNA content distribution of mouse bone marrow cells treated *in vivo* with increasing doses of the chemical mutagen thiophosphor-amide. The coefficient of variation (CV) of the G_1 peak is calculated to quantify the mutagenic effect.

In testicular cells the method can be used for the assessment of sper-matogenic and spermiogenic cell kinetics and for the detection of so-called diploid spermatozoa, which may be used as an indicator of genotoxic effects (Otto and Oldiges, 1986; Otto, 1987).

It must be kept in mind that DAPI preferentially stains AT-rich DNA. Thus, the procedure described here is not suited to yield absolute DNA values or to compare cells with differing proportions of AT. Despite this reservation, the addition of chicken or trout erythrocytes as internal stan-dard to one and the same cell type permits the assessment of peak posi-tion and coefficient of variation (CV) of the mean cell fluorescence of that cell type.

III.   Materials

The following solutions are used.

Pretreatment Solutions

100 ml Distilled water (dH$_2$O)
2.1 g Citric acid · H$_2$O (0.1 M)
0.5 ml Tween 20 (SERVA 37470)

100 ml dH$_2$O
4.2 g Citric acid · H$_2$O (0.2 M)
0.5 ml Tween 20

Staining Solution (Phosphate)

100 ml dH$_2$O
7.1 g Na$_2$HPO$_4$ · 2 H$_2$O
0.2 mg DAPI (4′, 6-diamidino-2-phenylindole · 2 HCl (SERVA 18860)

Staining Solution (Citrate)

100 ml dH$_2$O
11.8 g Citric acid trisodium salt · 2 H$_2$O
0.2 mg DAPI

The solutions are stable at room temperature (RT) for 1–2 weeks. The DAPI staining solutions should be stored in the dark.

IV.   Cell Separation Methods

Cells from suspension cultures and other suspensions like chicken and trout erythrocytes and human leukocytes are harvested without dispersing pretreatment and fixed with 70% ethanol.

Cells from bone marrow and other suitable tissues are isolated by flushing with EDTA containing Ca- and Mg-free PBS. The cell suspension is then fixed with 70% ethanol.

Testicular cells are obtained by mincing the tubuli contorti of testes and incubating in 0.1 M citric acid pretreatment solution for 10–20 minutes. The resulting single-cell suspension is then fixed with 70% ethanol. Cells from other solid tissues are tested in the same way.

All fixed cells can be stored at 4°C for as long as several months.

V.   Staining Procedure

The fixed cells are centrifuged at 200 g for 10 minutes and the fixative is removed completely.

Cells from suspension cultures, other suspensions, bone marrow, and similar tissues are resuspended in one volume of the pretreatment solution consisting of 0.2 M citric acid and 0.5% Tween 20 and incubated at RT for 20 minutes with gentle shaking. Subsequently, nine volumes of the staining solution containing 0.4 M sodium hydrogen phosphate and 5 μM DAPI are added to raise the pH to ~7.0 and to stain the DNA.

Testicular cells and cells from other solid tissues are resuspended in one volume of the 0.1 M pretreatment solution and incubated for 10 minutes. Then six volumes of the staining solution containing 0.4 M trisodium citrate and 5 μM DAPI are added.

The stained cells are stable for 24–48 hours.

VI.   Instrument Setup

The excitation maximum of DAPI is in the near-ultraviolet (UV) range at ~360 nm and the emission maximum in the blue at 460 nm.

Using the PAS II flow cytometer (Partec AG, Switzerland) the following filters are recommended: KG 1, BG 38, and UG 1 for excitation; TK 420 as dichroic mirror; and GG 435 as barrier filter.

In order to guarantee a smooth and reproducible sample flow, it is recommended to use a syringe pump for the injection of the cell suspension.

VII.   Results

The treatment of ethanol-fixed cells with citric acid plus Tween 20 produces isolated nuclei free of cytoplasm. The addition of phosphate or citrate solution raises the pH and results in a buffer system, in which DAPI is working as an effective and specific DNA stain.

The procedure yields FCM histograms of the nuclear DNA content with excellent resolution. Coefficients of variation in the range of 1% can be obtained in most specimens routinely. This allows the detection of small increases of DNA content variability induced by chemical or physical mutagens. The increased variability can be quantified by calculating the CV of the G_1 peak (Fig. 1). The protocol also allows the separation of cell

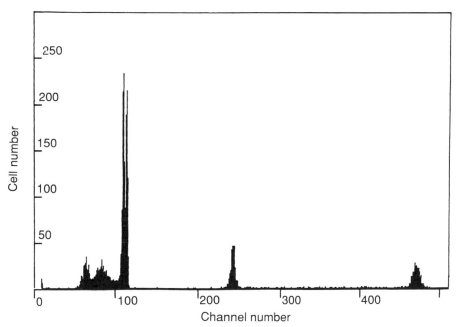

FIG. 2.   Flow cytometric histograms of mouse testicular cells showing separation of the X- and Y-chromosome-bearing round spermatids (at channel 110 and 114) and representing the elongated spermatids and spermatozoa as two distinct peaks to the left of the round spermatids (below channel 100).

populations with small differences in DNA content such as the X- and Y-chromosome-bearing round spermatids in testicular tissue (Fig. 2).

Using this protocol, double staining with DAPI and sulforhodamine 101 is possible to yield two-parameter histograms of DNA and protein. In this case it must be taken into account that only nuclear proteins that are resistant to the acid–detergent treatment are measured.

REFERENCES

Göhde, W., Schumann, J., and Zante, J. (1978). *In* "Pulse Cytophotometry" (D. Lutz, ed.), European Press, Ghent, pp. 229–232.

Otto, F. J. (1987). *In* "Clinical Cytometry and Histometry" (G. Burger, J. S. Ploem, and K. Goerttler, eds.), Academic Press, London, pp. 297–299.

Otto, F. J., and Oldiges, H. (1983). *Wiss. Umwelt* **83,** 109–121.

Otto, F. J., and Oldiges, H. (1986). *Wiss. Umwelt* **86,** 15–30.

Otto, F. J., Hacker, U., Zante, J., Schumann, J., Göhde, W., and Meistrich, M.L. (1979). *Histochemistry* **61,** 249–254.

Otto, F. J., Oldiges, H., Göhde, W., and Jain, V. K. (1981). *Cytometry* **2,** 189–191.

Otto, F. J., Oldiges, H., and Jain, V. K. (1984). *In* "Biological Dosimetry" (W. Eisert and M. L. Mendelsohn, eds.), Springer-Verlag, Berlin, pp. 37–49.

Chapter 12

High-Resolution DNA Measurements Using the Nuclear Isolation Medium, DAPI, with the RATCOM Flow Cytometer

JERRY T. THORNTHWAITE

Immuno-Oncology Laboratories
Baptist Hospital of Miami
Miami, Florida 33176

RICHARD A. THOMAS

RATCOM, Inc.
Miami, Florida 33193

I. Introduction

Fluorescent DNA-specific dyes have been used for rapid determination of DNA content. These include the guanine–cytosine (GC) specific binding fluorochrome, mithramycin (Williams, *et al.*, 1980), and the adenine–thymine (AT) specific dyes, Hoechst 33258 (Cesarone *et al.*, 1979; Labarca and Paigen, 1980) and 4′, 6-diamidino-2-phenylindole (DAPI) (Brunk *et al.*, 1979; Dann *et al.*, 1971; Lin *et al.*, 1977; Kapuscinski and Skoczylas, 1977). Unlike intercalating fluorochromes, however, these compounds can be used directly on crude tissue homogenates, without interference from RNA (Coleman *et al.*, 1981; Taylor and Milthorpe, 1980).

A very suitable fluorochrome for measuring DNA in cells by flow cytometry (FCM) is DAPI. This dye has a very high quantum efficiency and is stable in ultraviolet (UV) light (Coleman *et al.*, 1981). It passes through the nuclear envelope to bind stoichiometrically with the AT-rich regions of the intact DNA molecule (Dann *et al.*, 1971). The DNA–DAPI

111

complex is maximally excited at 365 nm. This complex fluoresces at 465 nm with about a 20-fold increase in fluorescence as compared to DAPI alone (Lin *et al.*, 1977; Kapuscinski and Skoczylas, 1977). RNase treatment has no effect on DAPI–DNA fluorescence (Coleman *et al.*, 1981; Taylor and Milthorpe, 1980).

The use of a one-step nuclear isolation medium (NIM) combined with a DNA fluorochrome keeps tissue preparation to a minimum while avoiding losses of DNA due to extraction steps (Thornthwaite *et al.*, 1980). Using this method, Thornthwaite *et al.* (1980) have obtained coefficients of variation (CV) $\sim 1\%$ for the mean DNA content of the G_0/G_1 populations of cells from a variety of normal and neoplastic tissues. We present data here showing that FCM of cells prepared with a NIM–DAPI solution and combined with trout red blood cells (TRBC) as the biological standard is a rapid, precise method for quantitating DNA in nuclei.

In this report, comparisons will be made between DAPI- and propidium iodide (PI)-stained human tissue samples such as breast, prostate, and bladder washings. The methodologies that work best in our laboratory will be presented with recommendations for obtaining high-resolution DNA information.

II.   Materials and Methods

A.   Medium and Sample Preparation

1.   NIM–DAPI SOLUTION

The *NIM–DAPI solution* contains a new nuclear isolation medium (NIM-II) with 10 μg/ml of DAPI (Accurate Scientific and Chemical Co., Hicksville, NY). The solution can be kept at 4°C for as long as a year with no appreciable change in staining efficiency. One may use NIM-II with any nucleic acid stain. For example, PI (Calbiochem) may be substituted for DAPI except at 50 μg/ml in the protocol already given. Staining kits using the improved NIM-II formulation with a variety of fluorochromes are now commercially available (RATCOM, Inc.).

2.   SAMPLE PREPARATION

To prepare the sample, 2.5 ml NIM–DAPI are added to ~ 0.1 g of mammalian tissue, either fresh or frozen (-90°C). The tissue is minced for 30 seconds with a scalpel and filtered through a disposable 40-μm filter assembly (RATCOM, Inc.). Tissue culture cells and leukocytes were prepared by resuspending a centrifuged pellet of 10^6 viable cells in 1 ml of NIM–DAPI.

3. Pepsin HCL–NIM–RNase Technique

In order to extract histones and study their contribution to DAPI staining, the following procedure (modified from Drs. K. Bauer and V. Shankey, personal communication) is used.

1. Prostate tissue (0.1–0.2 g) is minced in 2 ml of 0.5% crystallized pepsin (w/v) (Boehringer-Mannheim) in 0.14 M NaCl, 3% PEG 8000 (polyethylene glycol, Sigma), adjusted to pH 1.5 with HCl.
2. The tissue pieces are incubated at 37°C for 30 minutes, after which 0.2 ml of pepstatin A (0.5 mg/ml, Boehringer-Mannheim) is added to stop the pepsin activity.
3. After filtration (40-μm mesh), the nuclei are centrifuged in 15 ml of Hanks' balanced salt solution with calcium and magnesium (HBSS) at 200 g, 10 minutes at 4°C.
4. The pellet is resuspended in 2 ml of 0.1% Triton X-100 with 10 mM $MgCl_2$ and 180 Kunz units of RNase (Worthington: Chromatographically pure).
5. After incubation at 37°C for 30 minutes, the nuclei are centrifuged as before, resuspended in 1 ml NIM–DAPI, and analyzed after 10 minutes in an ice bath.

B.   Instrumentation

The RATCOM Flow Cytometer is ideally suited for the DAPI-stained nuclear measurements. A unique, triangular orifice allows a very stable flow. A 100-W mercury arc lamp with special power supply circuitry for long, uniform lamp life (>300 hours) is used with a UG1 2-mm filter to isolate the 365-nm emission line. A 410-nm dichroic mirror allows the 365-nm light to be reflected on the DAPI-stained nuclei, while the resulting fluorescence 450 nm light passes through the dichroic mirror and a long-pass 410-nm filter to a photodetector. The nuclei are analyzed at rates of 150–300 per second. The stabilized TRBC standard routinely obtains CV in the 1.2–1.6 range using DAPI–NIM. The FACScan Flow Cytometer (Becton Dickinson) was used to analyze NIM–PI prepared samples. Excitation was at 488 nm from a 15-mW argon laser with resulting fluorescence measured above 580 nm. The CV range of the stabilized TRBC standard stained with NIM–PI was 1.4–1.8 at flow rates of 50–100 nuclei/second.

III.   Results

Using the NIM in combination with a stabilized TRBC suspension (RATCOM, INC.), high-precision measurements yielding CV values of DNA content of G_0/G_1 cells in the 0.9–1.8% range, of the mean are routinely obtained for either DAPI or PI. Figure 1a shows a typical standard TRBC DNA histogram using NIM–DAPI with the RATCOM Flow Cytometer, while Fig. 1b reveals with almost equal precision a high-resolution TRBC DNA histogram using NIM–PI with the FACSCAN Flow Cytometer (shown in black in the figures). The stabilized TRBC form less than 1% doublets (unless stained while frozen) and may be stored refrigerated.

However, when larger nuclei are analyzed, as in the case of cancer, the nonspecific component of the PI nuclear stainability and varying sensitivity of laser-based flow cytometers with respect to nuclear size variation prevent accurate determination of aneuploidy in near-diploid tumors. Examples of this observation are shown in Figs. 1–3. Figure 1c reveals discrimination of near-diploid lymphomas with NIM–DAPI (DNA index, DI = 1.06), but not with NIM–PI (Fig. 1d). Other examples of solid tissue-prepared NIM–DAPI nuclei are displayed in Fig. 1e,f. The breast tumor tissue (Fig. 1e) reveals a narrow distribution (CV = 1.42), with a G_0/G_1 peak channel at 79, small percentage of S-phase nuclei (3.2%), and a G_2 + M peak at channel 158. When the TRBC peak distribution is placed in channel 50, the diploid human G_1/G_0 peak is in channel 79. Other examples of peak channel 79, diploid G_0/G_1 human populations are shown in near-diploid breast cancer (Fig. 1f), prostate tissues with (Fig. 2d,f) and without (Fig. 2a,c,e) prior treatment with pepsin–RNase, and positive bladder cancer washings (Fig. 3a,c,e).

Figure 2 is a composite of DNA ploidy histograms of nuclei from prostate autopsy tissues from individuals without a history of cancer. The NIM–DAPI-prepared nuclei reveal aneuploid DNA Histograms (Fig. 2a,c–f) obtained with (Fig. 2d,f) or without (Fig. 2a,c,e) prior treatment with pH 1.5 pepsin–RNase before NIM–DAPI. The prostate nuclei from a 40-year-old clearly showed aneuploidy (A = 72.0%), while the NIM–PI sample with (not shown) or without (Fig. 2b) RNase treatment could not clearly separate the two DNA populations, even though a broad CV was suggestive of two populations being present. In order to see if the aneuploidy was due to the DAPI staining, a study was conducted to see if the removal of histones (pH 1.5-pepsin) and even nuclear RNA (subsequent RNase treatment) could eliminate the aneuploidy and reveal a single DNA G_0/G_1 population. In a study of six aneuploid prostates,

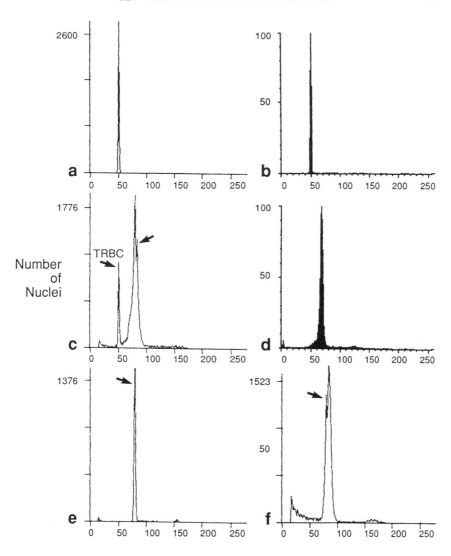

DNA Ploidy Content

FIG. 1. DNA ploidy histograms prepared with NIM–DAPI or NIM–PI. (a) and (b) Trout red blood cells (TRBC) preserved in a special refrigerated storage medium for 3 months and prepared with (a) NIM–DAPI (CV = 1.23) or (b) NIM–PI (CV = 1.47). (c) Lymphoma cells stained with NIM–DAPI (DI = 1.06) or (d) NIM–PI. (e) Diploid, benign (CV = 1.42) and (f) near-diploid, malignant breast masses prepared with NIM–DAPI (DI = 1.05). Arrows denote human diploid G_0/G_1 nuclei (channel 79) versus (TRBC) G_0/G_1 nuclei (channel 50).

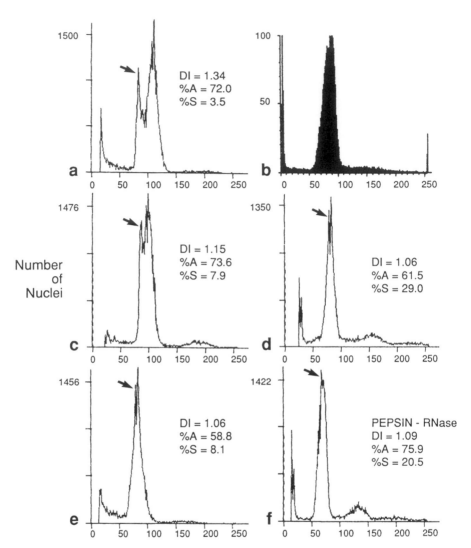

DNA Ploidy Content

FIG. 2.   DNA ploidy histograms of frozen control human prostates prepared using various methodologies. Prostate from a 40-year-old patient prepared using (a) NIM–DAPI or (b) NIM–PI. Prostate from a 16-year-old prepared using (c) NIM–DAPI or (d) pepsin (pH 1.5), followed by RNase, and then NIM–DAPI. Prostate from an 11-month-old prepared using (e) NIM–DAPI or (d) pepsin (pH 1.5)–RNase/NIM–DAPI. The arrows denote the diploid G_0/G_1 population (peak channel 79) versus TRBC G_0/G_1 nuclei (channel 50).

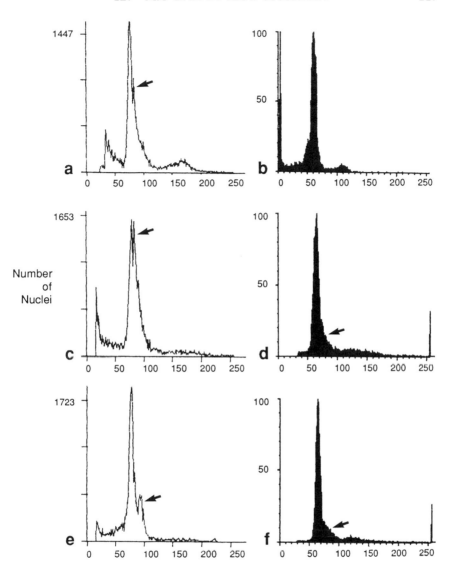

DNA Ploidy Content

FIG. 3.   DNA ploidy histograms of human bladder washings that were positive for tumor cells by cytology. Cells from patient 1 (a, b), patient 2 (c, d), and patient 3 (e, f) were prepared by staining 10^6 cells/ml with NIM–DAPI (a, c, e) or NIM–PI (b, d, f). Arrows denote aneuploid populations.

with and without pH 2-pepsin–RNase treatment, all showed aneuploidy after histone removal. However, the DI decreased in four of the cases as illustrated in Fig. 2c,d (DI = 1.15 − 1.06, respectively), and in all cases there was an apparent increase in the S-phase. Examples of this increase are shown as 3.5% (Fig. 2a) to 25.0% (not shown), 7.9% (Fig. 2c) to 29.0% (Fig. 2d), and 8.1% (Fig. 2e) to 20.5% (Fig. 2f). The increase in both S and $G_2 + M$ nuclei is probably artifactual, since very few mitotic figures are present in normal prostate tissues.

Finally, Fig. 3 illustrates the importance of high-resolution DNA measurements in bladder washings. Figure 3a,c,e reveal aneuploidy present (55.5%, 63.3%, 31.2%, respectively) in cytologically positive samples in a double blind study. However, the NIM–PI/FACSCAN preparations could not detect aneuploidy in these samples (Fig. 3b,d,f) except for suspicious shoulders in Fig. 3d,f.

IV. Discussion

This chapter has shown that both sample preparation and instrumentation are important in obtaining high-resolution DNA FCM histograms. To accomplish these goals in the clinical environment, the following recommendations may be helpful:

1. Use a DNA standard that is stable and will allow the lowest CV for your instrument. Human normal tissues of the same histological type (preferably from the same patient), are ideal but are not easily obtained and do not establish a universal standard for all samples. Stabilized TRBC allow the maximum resolution for a DNA standard and have been shown in this study and our other publications (Lee *et al.*, 1984; Thornthwaite *et al.*, 1985) to be a reliable DNA standard.

2. DAPI is to be preferred as the DNA stain over PI (and possibly all other stains) because of its high DNA specificity and relatively low sensitivity to chromatin changes.

3. The NIM—especially the newer formulation, NIM II—is a stable, easy-to-use sample preparation medium in which nuclear isolation and DAPI staining occurs in the same medium. Since no centrifugation steps are necessary, sample preparation time is < 5 minutes. The pepsin–RNase procedure appears not to be superior over using NIM–DAPI with fresh or frozen tissues or cells. It is the method of choice for paraffin-embedded tissues or, as shown in this study, investigating the effects of histone binding on DAPI staining.

4. The use of DAPI necessitates using a UV source. While UV light lasers can excite DAPI, their cost and short life due to running in the UV light mode do not recommend them for this purpose. The mercury arc lamp is ideally suited for DAPI and high-resolution (CV = 1–2%) histograms can be obtained as shown in Figs. 1–3.

ACKNOWLEDGMENTS

We thank Drs. Riemer and Moskowitz for the pathology evaluations and lucid suggestions in this manuscript. We thank Diane Ream, director of the hospital library, Jo Baxter and Joan Gisselberg in the marketing department, and Diane Pruitt in the preparation of this chapter. Finally, we thank the Baptist Hospital Auxilary for their commitment to the Immuno-Oncology Laboratories in conducting cancer research.

REFERENCES

Brunk, C. F., Jones, K. C., and James, T. W. (1979). *Anal. Biochem.* **92,** 497–500.
Cesarone, C. F., Bolognesi, C., and Santi, L. (1979). *Anal. Biochem.* **100,** 188–197.
Coleman, A. W., Maguire, M. J., and Coleman, J. R. (1981). *J. Histochem. Cytochem.* **29,** 959–968.
Dann, O., Bergen, G., Demant, E., and Voltz, G. (1971). *Justus Liebigs Ann. Chem.* **748,** 68–79.
Kapuscinski, J., and Skoczylas, B. (1977). *Anal. Biochem.* **83,** 252–257.
Labarca, C., and Paigen, K. (1980). *Anal. Biochem.* **102,** 344–352.
Lee, G. M., Thornthwaite, J. T., and Rasch, E. M. (1984). *Anal. Biochem.* **137,** 221–226.
Lin, M. S., Comings, D. E., and Alfi, O. S. (1977). *Chromosoma* **60,** 15–25.
Taylor, I. W., and Milthorpe, B. K. (1980). *J. Histochem. Cytochem.* **28;** 1224–1232.
Thornthwaite, J. T., Sugarbaker, E. V., and Temple, W. J. (1980). *Cytometry* **1,** 229–237.
Thornthwaite, J. T., Thomas, R. A., Russo, J., Ownby, H. *et al.* (1985). *In* "Immunochemistry in Tumor Diagnosis" (J., Russo, ed.), pp. 380–398. Martinus Nijhoff, Boston.
Williams, S. K., Sasaki, A. W., Matthews, M. A., and Wagner, R. C. (1980). *Anal. Biochem.* **107,** 17–20.

Chapter 13

Rapid DNA Content Analysis by the Propidium Iodide–Hypotonic Citrate Method

AWTAR KRISHAN

Department of Oncology
University of Miami School of Medicine
Miami, Florida 33101

I. Introduction

Following publication of the Crissman and Steinkamp method (1973) for rapid quantitation of DNA and protein content of fixed cells by flow cytometry (FCM), we experimented with direct staining of isolated nuclei (without prior fixation and enzyme digestion) with various DNA-binding fluorochromes. The propidium iodide (PI)–citrate method (Krishan, 1975) was particularly developed for rapid determination of cell cycle distribution in leukemic peripheral blood and bone marrow samples of patients on high-dose methotrexate rescue protocols (Krishan *et al.*, 1976). Fried *et al.* (1976) further employed this method and found it suitable for rapid cell cycle analysis. Subsequently, it was used by several workers to study cell cycle phase distribution of mammalian tissue culture cells, and a variety of tumor cells. Modifications have involved inclusion of RNase and a detergent in the staining solution (Vindeløv, 1977; see Chapter 14, this volume). A brief description of this simple staining method follows.

II. Application

This rapid method is especially suitable for small samples such as a drop of blood from a fingerprick, tissue culture cells from multiwell plates, and

121

small bone marrow aspirates. For solid-tumor material, mincing of tissue with scissors or crossed scalpels and filtering through nylon cloth are essential.

III. Materials

1. Cell pellets from tissue culture, leukemic blood, or a drop of blood from a fingerprick, bone marrow aspirates or biopsy samples, minced tissue, or solid tumors
2. PI (Calbiochem catalog no. 537059)
3. Sodium citrate
4. Nonidet P-40 (NP-40), Shell trademark (obtainable from Bethesda Research Labs, Rockville, MD)
5. Pasteur pipets
6. Nylon cloth, 40-μm mesh
7. Preparation of staining solution:

> PI, 5 mg
> Sodium citrate, 100 mg
> Glass-distilled water, 100 ml

Addition of 30 μl of NP-40 per 100 ml of final staining solution may help in isolation of nuclei.

Prepare the staining solution and store in an amber-colored bottle. We prefer a 1-liter Repipet bottle, which allows for rapid transfer of premeasured volumes of staining solution directly into the specimen vial. The staining solution can be stored indefinitely in a cold room or at room temperature without any measurable loss of activity.

IV. Staining Procedure

1. *For normal or leukemic cells from the peripheral blood:* Add a drop of the specimen directly to 1–2 ml of stain. Mix vigorously with a syringe/pipet and run on a flow analyzer.
2. *For monolayer cultures:* Decant growth medium and wash cell monolayer with normal saline or balanced salt solution. Add staining solution ($\sim 10^6$ cells/ml), scrape with a rubber "policeman," and shake vigorously. Transfer to a test tube and pipet to break open cells

and dislodge any cytoplasmic fragments sticking to isolated nuclei. Filter through nylon cloth and run on a flow analyzer.

3. *For solid-tumor surgical biopsy specimens:* Remove excess fat and necrotic tissue, and wash with saline to remove blood. Cover tissue in a Petri dish with staining solution. Dice or mince with crossed scalpels or scissors, or grate on a metal sieve. Remove large pieces and fat, and collect cell suspension in a test tube. Use a narrow-gauge needle to syringe the specimen vigorously and isolate single nuclei. Filter through a nylon cloth and run on an analyzer.

4. *For needle biopsy specimens:* The syringe and needle can be directly flushed into a 15-ml centrifuge tube containing 1–2 ml of staining solution. Syringe vigorously to isolate nuclei.

V. Caution

It is important that the specimen be vigorously pipetted to dislodge cytoplasmic fragments from nuclei.

Do not use trypsin for removal of cell monolayers or heparin as an anticoagulant for blood or bone marrow aspirates. Both of these agents may interfere with binding of the fluorochrome and may not yield good DNA distribution histograms (Krishan *et al.*, 1978).

It is important to remember that certain DNA-intercalating agents may interfere with binding of PI to DNA. This is especially true of clinical specimens from patients on chemotherapy. Readers are referred to Krishan *et al.* (1978) for further consideration of this artifact.

Stained specimen can be run within 5 minutes of staining and often can be stored as long as 24 hours in an ice bucket or refrigerator at 4°C. Longer storage leads to swelling of nuclei.

Propidium iodide may be a potential health hazard and care should be taken in handling, especially avoiding spills on hands. Proper disposal of hazardous waste is also recommended.

VI. Instrument Setup

Although peak excitation (540 nm) of PI with the 488- or 514-nm argon ion laser line is ideal, the PI–citrate method can also be used in analyzers equipped with a mercury HBO lamp light source (e.g., Ortho 1CP/22 system).

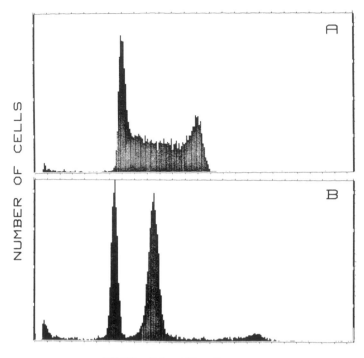

DNA CONTENT

Fɪɢ. 1.   (A) Histogram of murine leukemic cells (P388) growing in a log-phase suspension culture. (B) Histogram of a human solid-tumor biopsy specimen showing a prominent peak with 2N and a near 3N peak. Note the presence of a small G_2/M (6N) peak of the tumor cells. The 2N peak may consist of normal cells infilterating the tumor.

In the Coulter EPICS 753 unit (argon ion 488-nm laser line), we use peak emission fluorescence signal (>590 nm) and following approximate instrument settings to get low coefficients of variation (CV values): laser power 500 MW, HV 700, amplifier gain 2×.

In the Ortho 1CP/22 system with a mercury HBO source, we use an HV setting of 2.2–2.5, red PMT, and area (under the curve) for data analysis.

Sample histograms of mouse leukemic cells and a human solid tumor stained with the PI–hypotonic citrate and analyzed on an EPICS 753 (Coulter Electronics) are illustrated in Fig. 1.

VII.   Comments

Several modifications of the original PI–citrate method have been subsequently published. In cells with low nucleocytoplasmic ratio (e.g., lymphoid cells), the original PI–citrate method gives excellent results. In cells with large cytoplasmic mass or RNA content, use of detergents and/or RNase may be indicated.

VIII. Results

Histograms in Fig. 1 are typical of cells stained by the PI–hypotonic citrate method.

REFERENCES

Crissman, H. A., and Steinkamp, J. A. (1973). *J. Cell Biol.* **59,** 766–771.
Fried, J., Perez, A. G., and Clarkson, B. D. (1976). *J. Cell Biol.* **71,** 172–181.
Krishan, A. (1975). *J. Cell Biol.* **66,** 188–193.
Krishan, A., Pitman, S. W., Tattersall, M. H. N., Paika, K. D., Smith, D. C., and Frei, III, E. (1976). *Cancer Res.* **36,** 3813–3820.
Krishan, A., Ganapathi, R. N., and Israel, M. (1978). *Cancer Res.* **38,** 3656–3662.
Vindeløv, L. L. (1977) *Virchows Arch. Abt. B Cell Pathol.* **24,** 227–242.

Chapter 14

An Integrated Set of Methods for Routine Flow Cytometric DNA Analysis

LARS VINDELØV

Departments of Hematology L and Oncology ONK
Rigshospitalet, University Hospital
DK-2100 Copenhagen, Denmark

IB JARLE CHRISTENSEN

The Finsen Laboratory
Rigshospitalet, University Hospital
DK-2100 Copenhagen, Denmark

I. Introduction

The amount of information that can be extracted from a DNA distribution depends heavily on the technical quality of the analysis and the methods used for standardization and statistical analysis of the histogram. The aim of the methods presented in this chapter has been to secure a constant output of reproducible and reliable endpoint results, based on good-quality DNA histograms. The latter are characterized by a minimal amount of debris and symmetrical G_1 peaks with low coefficients of variation (CV). The data reduction obtained by statistical analysis yields the desired endpoint results that will adequately and exhaustively describe the DNA distribution for most purposes. These are (1) the number of subpopulations with different DNA content present in the sample, and (2) for each subpopulation: (a) the relative size of the subpopulation, (b) the DNA index (DI), and (c) the fractions of cells in the cell cycle phases G_1,

METHODS IN CELL BIOLOGY, VOL. 33

S, and $G_2 + M$. The problems that must be solved thus concerns sample acquisition and storage, standardization, staining, flow cytometry (FCM), and statistical and analysis (deconvolution).

II.   Basic Principles of the Methods

The integrated set of methods (Christensen *et al.*, 1978; Vindeløv *et al.*, 1983a–d) developed to solve the problems just described are outlined in Fig. 1. The FCM DNA analysis is performed on a suspension of nuclei. This has two major advantages. DNA-unspecific fluorescence from the

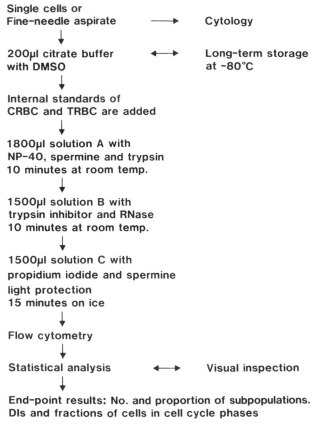

FIG. 1.   Overview of methods and results.

cytoplasm is avoided, and fine-needle aspirates that contain a variable fraction of bare nuclei are well suited as starting material. The possibility of simultaneous analysis of cytoplasmic or plasma membrane-associated structures is thus sacrificed for easy and accurate determination of nuclear DNA in tissues and solid tumors. The analysis is performed on unfixed material. This is essential to avoid a potentially selective cell loss caused by centrifugation steps, and keeps the requirement of cells at a minimum. Clumping and staining artifacts caused by a fixative are avoided. Samples can be long-term stored by freezing in a citrate buffer with dimethyl sulfoxide (DMSO) (Vindeløv et al., 1983a).

The key to accurate and reproducible DI determination, as well as statistical analysis of multiple overlapping populations, is adequate internal standardization. This is achieved by adding a mixture of chicken (CRBC) and trout erythrocytes (TRBC) to the sample before staining (Vindeløv et al., 1983c). The peaks produced by the standards provide two points of reference at DI values of ∼ 0.30 (CRBC) and 0.80 (TRBC), and allow deconvolution of the histogram and DI determination independent of zero-point shift.

The preparation consists of three steps (Vindeløv et al., 1983b). Clean nuclei are obtained in the first step (solution A) by the combined action of the nonionic detergent Nonidet P-40 (NP-40) and trypsin. In addition, trypsinization increases the fluorescence of nuclei with dense chromatin such as granulocytes, presumably by splitting some chromosomal proteins. Spermine is essential for the stability of the unfixed nuclei during trypsinization. RNase treatment in the second step (solution B) prevents dye binding to double-stranded RNA. Trypsin inhibitor is added, because trypsin activity after addition of propidium iodide (PI) results in unstable nuclei. In the final step (solution C) PI is added and the spermine concentration further increased for optimal stability. The PI binds to double-stranded nucleic acid by intercalation (Crissman et al., 1975; Krishan, 1975).

III. Applications

Flow cytometric DNA analysis has been established as a useful research tool for a number of years. A review of applications in our laboratory may be found in Vindeløv and Christensen (1989). The methods have been used for monitoring the stability of cell lines, for sensitivity testing, and for studying the action of anticancer drugs in vitro. Likewise the methods have been used to monitor the cell cycle perturbations produced by radiation, chemotherapy, and hormonal substances in murine and human tumors in experimental animals, and in malignant turmors in patients. Since the same methods can be applied at these different levels of complexity, the results are directly comparable and may serve as a link between experimental and clinical results.

DNA analysis is currently assessed as a possible prognostic parameter in neoplastic disease (Merkel *et al.*, 1987). Final evaluation awaits the results of large prospective series of uniformly staged and treated patients, examined with high-resolution DNA analysis.

IV. Materials

Citrate Buffer

Sucrose (BDH), 85.50 g (250 mM)
Trisodium citrate · H$_2$O (Merck), 11.76 g (40 mM)
Dissolve in distilled water (dH$_2$O), ~ 800 ml
Add DMSO (Merck) 50 ml
Add dH$_2$O to a total volume of 1000 ml
Adjust to pH 7.6

Stock Solution

Trisodium citrate · 2H$_2$O (Merck), 1000 mg (3.4 mM)
NP-40 (Shell), 1000 μl (0.1% v/v)
Spermine tetrahydrochloride (Serva, 35300), 522 mg (1.5 mM)
Tris (Sigma 7–9, T-1378) 61 mg (0.5 mM)
Add dH$_2$O to a total volume of 1000 ml

Solution A

Stock solution, 1000 ml
Trypsin (Sigma, T-0134), 30 mg
Adjust to pH 7.6

Solution B

Stock solution, 1000 ml
Trypsin inhibitor (Sigma, T-9253), 500 mg
RNase A (Sigma, R-4875), 100 mg
Adjust to pH 7.6

Solution C

> Stock solution, 1000 ml
> PI (Fluka), 416 mg
> Spermine tetrahydrochloride (Serva, 35300), 1160 mg
> Adjust to pH 7.6

The citrate buffer is stored at 4°C. The staining solutions are stored in aliquots of 5 ml in capped plastic tubes at −80°C. The tubes with solution C are wrapped in aluminum foil for light protection of the PI. Before use the solutions are thawed in a water bath at 37°C, but not heated to 37°C. Solutions A and B are used at room temperature (RT). Solution C is kept in an ice bath.

Internal Standards

Chicken blood can be obtained by heart puncture and collected in a tube containing 50 IU of heparin per milliters of blood. The blood is subsequently diluted by citrate buffer. Rainbow trout blood can be obtained by caudal vein puncture after anesthesia with MS 222 (Sandoz, Basel, Switzerland) and should be mixed with citrate buffer immediately. The concentrations of CRBC and TRBC are determined by counting in a hemocytometer. The red cell concentrations are adjusted by dilution with citrate buffer to CRBC = 145×10^4 cells/ml, and TRBC = 255×10^4 cells/ml. These suspensions are then mixed in equal volumes to obtain a final RBC concentration of 2×10^6 cells/ml and a ratio of CRBC/TRBC = 4 : 7, which will produce peaks of equal height in the histogram. The mixture of standards can be long-term stored at −80°C in aliquots of (for example) 100 μl (sufficient for three to five samples), after freezing as described in Section V,B.

V. Methods

A. Sample Acquisition

Fine-needle aspiration is used for initial mechanical disaggregation of normal tissues, lymphomas, and solid tumors. It is a gentle and rapid procedure, causing less debris than cutting with knives or scissors. It is therefore used, even when a surgical biopsy specimen is available. The tumor is secured by pinning it with injection needles to a plate of Styrofoam. In the case of very small specimens the tissue is wrapped in dialysis tube and secured during biopsy by pinning this to the plate as before. Biopsy samples as small as $2 \times 2 \times 2$ mm can be successfully aspirated in this way. A 0.5×25 mm (25 gauge $\times$ 1) needle on a 20-ml disposable syringe fitted on a one-hand-operated handle (Cameco, Täby, Sweden) is adequate. Longer needles may be required for *in vivo* aspiration, and thinner needles may give a better cell yield in fibrous tumors, such as some breast cancers. Suction is applied only when the point of the needle is within the tissue. The needle is moved back and forth in different directions within the tumor. The aspirate should stay within the needle during aspiration. The needle is flushed with citrate buffer (200 μl), and the cell yield is checked by counting in the hemocytometer. Aspiration is repeated until 10^6 cells have been obtained. The cells are stored by freezing (see Section V, B), or stained and analyzed on the same day. Two additional aspirates are spread on slides for cytologic examination.

B.   Storage

Samples and internal standards not stained and analyzed on the same day as obtained, are stored at $-80°C$ after freezing (Vindeløv *et al.*, 1983a). Samples have been kept in this way for as long as 5 years, without change in the DNA histograms. Samples can be frozen either as aspirates in citrate buffer or as surgical biopsy specimens in a dry tube. It is essential that each sample is frozen and thawed only once. The cell suspension in citrate buffer is frozen in polypropylene tubes (38×12.5 mm test tubes with screw cap, Nunc, Roskilde, Denmark) immersed in a mixture of dry ice and ethanol ($-80°C$). Before use the samples are thawed in a water bath at 37°C. The sample should only be thawed and not heated to 37°C.

C.   Standardization and Staining

Staining with PI (Vindeløv *et al.*, 1983b) is performed after addition of the internal standards (Vindeløv *et al.*, 1983c) to 200 μl of sample cell suspension in citrate buffer. The standards are added in an amount equal to 20% of the cells in the sample. With the concentrations chosen, 20% of the sample cell count in 0.1 μl (the volume of the hemocytometer used in our laboratory) is equal to the number of microliters of standard to be added to 200 μl of sample. Staining is performed by stepwise addition of the staining solutions. Solution A (1800 μl) is added to 200 μl of tumor aspirate in citrate buffer and the tube is inverted to mix the contents gently. After 10 minutes at RT, during which the tube is inverted two or three times, 1500 μl of solution B is added. The solutions are again mixed by inversion of the tube, and after 10 minutes at RT 1500 μl ice-cold solution

C is added. The solutions are mixed and the sample filtered through a 25-μm nylon mesh into tubes wrapped in aluminum foil for light protection of the PI. The samples are kept in an ice bath until analysis, which should take place between 15 minutes and 3 hours after addition of solution C (Vindeløv *et al.*, 1983b). If cells are scarce the volumes of staining solution can be halved to increase the cell concentration.

There are few critical aspects of this simple procedure. Pure, analytic-grade reagents should be used for the solutions. Accurate weighing of the reagents and accurate pipetting of the solutions is important. The samples should be handled gently throughout, since agitation will increase debris and clumping.

D. Flow Cytometric Analysis

The FACS 3 flow cytometer (Becton Dickinson, Mountain View, CA) is used in our laboratory. The cell concentrations are optimized for this instrument. Other flow systems with different sample flow rates may require different cell concentrations for optimal results. A 50-μm nozzle tip is used and cells are analyzed at a rate of 100–200 cells/second. If cells are scarce and the sample important, the rate of measurement is reduced as low as 10–20 cells/second. The point is that the sample flow rate should be kept low enough to ensure a thin stream of sample, with the cells intersecting the laser beam in the same path. After fitting an elliptical lens the rate of measurement has been doubled without any loss in resolution. Stock solution is used as sheath liquid to prevent changes in dye binding caused by differences in ionic strength (Vindeløv *et al.*, 1983b). The blue 488-nm line of the argon ion laser is used for excitation. Two interference filters, LWP 580 and SWP 660 (Optisk Laboratorium, Lyngby, Denmark), are used for the emission.

E. Statistical Analysis

The endpoint results mentioned in Section I are determined by statistical analysis of the DNA distribution (Christensen *et al.*, 1978). Visual inspection of the histogram plays a role in determining the quality of the analysis and the number of subpopulations present. Figure 2A shows an example of a DNA histogram. Statistical analysis or deconvolution described briefly, involves the following steps:

1. Normal distributions are fitted to the peaks of the standards. Debris is subtracted by fitting a truncated exponential between the CRBC and TRBC peaks, and the zero-point is corrected, based on the means of the standards.

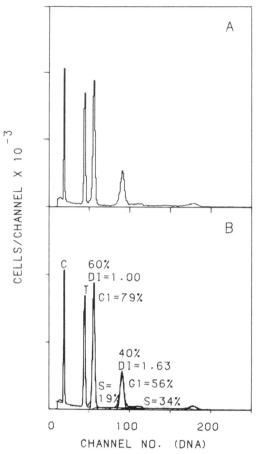

FIG. 2.   (A) A DNA distribution from breast cancer. Fine-needle aspiration of surgical biopsy stored by freezing. Visual inspection indicates a good-quality histogram with two subpopulations. (B) The histogram after statistical analysis. The peaks produced by the CRBC (C) and TRBC (T) are indicated. The first subpopulation is strictly diploid (DI = 1.00) and constitutes 60% of the cells. The second subpopulation is hypertriploid (DI = 1.63) and constitutes 40% of the cells. The percentages of cells in the cell cycle phases are indicated on the figure. The CV values of the G_1 peaks are 2.3% and 2.2%.

2. A model describing the G_1 and $G_2 + M$ peaks with normal distributions, and the S phase with an exponential function of a polynomial of a given degree, is fitted to the histogram by maximum likelihood. For histograms with more than one subpopulation, a mixture of this density is fitted (Fig. 2).

3. The DI values of the G_1 peaks are calculated by comparison with known values for normal cells (Vindeløv *et al.*, 1983d), and the areas under the curves are estimates of the fractions of cells in the cell cycle phases G_1, S, and $G_2 + M$.

Figure 2B shows the deconvoluted histogram and the estimated values. Several important features may be pointed out.

1. The very accurate zero-point determination obtained by the standards.
2. A point of reference (TRBC) close to the diploid peak makes DI determination more accurate.
3. A slight nonlinearity of the relationship of fluorescence to DNA has been determined experimentally, and corrected for in the computer program. This allows prediction of the $G_2 + M$–mean location and thereby deconvolution of heterogeneous tumors with overlapping populations, as shown in Fig. 2.
4. Histograms with a treatment-induced perturbation of the cell cycle, can be deconvoluted by increasing the degree of the polynomial used for fitting the S phase. An example is shown in Fig. 3. The S phase distribution is estimated as an additional endpoint result.

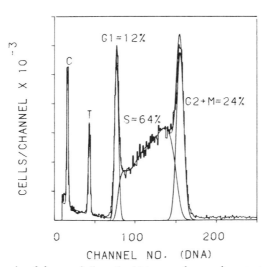

FIG. 3 An example of deconvolution of a histogram from a drug-perturbed cell population. The small-cell lung cancer cell line NCI-N592 was exposed to the alkylating agent BCNU. The S phase was fitted by an eighth-degree polynomial. The standards and the percentages of cells in the cell cycle phases are indicated on the figure.

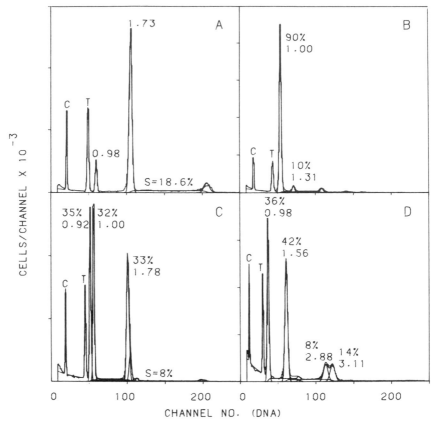

FIG. 4   (A) A DNA histogram from small-cell carcinoma of the lung. Fine-needle aspiration *in vivo* of a lymph node metastasis. A single aneuploid (DI = 1.73) population is present.
The peak with DI = 0.98 could represent normal cells and possibly some tumor cells. The CV
values of the G_1 peaks are 1.9% and 1.7%. (B) A distribution from carcinoma of the oral
cavity. Fine-needle aspiration of a small surgical biopsy. Ninety percent of the cells are strictly
diploid. A small subpopulation with DI = 1.31 is present. The S phases are confounded and
cannot be estimated reliably. The CV of the diploid G_1 peak is 2.2%. (C) A distribution from
a non-Hodgkin's lymphoma. Fine-needle aspiration *in vivo*. Three subpopulations of nearly
the same size are present. The separation of the diploid (DI = 1.00) and the hypodiploid
(DI = 0.92) peaks illustrate the need for a high resolution. The CV values range from 1.6% to
1.8%. The S phases of the diploid and hypodiploid subpopulations are confounded. An
average value can be calculated. (D) A distribution from breast cancer. Fine-needle aspiration of surgical biopsy. Four subpopulations are present. Their sizes and DI values are
indicated on the figure. Only one or two of the S phases can be estimated. The CV values vary
from 2.1 to 2.8.

VI.   Results

The methods were developed with emphasis on obtaining optimal results in a wide range of cells and tissues, in particular solid tumors. The staining method was developed in 1978 as a modification of an initial attempt (Vindeløv, 1977). In the past 10 years we have analyzed ~17,000 samples from clinical and experimental studies. Satisfactory results have been obtained in all types of cells and tissues examined with the exception of sperm. The samples analyzed include (1) normal tissues: human lymphocytes, granulocytes, and spleen, mouse lymphocytes, bone marrow, spleen, liver, kidney, and thymus; (2) human neoplasms: lung cancer, breast cancer, lymphoma, leukemia, bladder cancer, and cancer of the oral cavity; (3) human tumors in nude mice: breast cancer, lung cancer, melanoma, and colon cancer; (4) mouse ascites tumors: JB-1, L1210, Ehrlich, and P388. Some examples are shown in Fig. 4. Coefficients of variation of ~2% are obtained routinely. The theoretical and practical problems of resolving minor DNA differences as well as long-term reproducibility and comparability of results obtained by the methods have been examined in some detail (Vindeløv et al., 1983a,d).

ACKNOWLEDGMENTS

This work was supported by grants from The Danish Cancer Society, The Danish Medical Research Council, and The Lundbeck Foundation.

REFERENCES

Christensen, I. J., Hartmann, N. R., Keiding, N., Larsen, J. K., Noer, H., and Vindeløv, L. L. (1978). In "Pulse Cytophotometry" (D. Lutz, ed.), Part III, pp. 71–78. European Press Medicon, Ghent.

Crissman, H. A., Mullaney, P. F., and Steinkamp, J. A. (1975). Methods Cell Biol. 9, 179–246.

Krishan, A. (1975). J. Cell Biol. 66, 188–193.

Merkel, D. E., Dressler, L. G., and McGuire, W. L. (1987). J. Clin. Oncol. 8, 1690–1703.

Vindeløv, L. L. (1977). Virchows Arch. Abt. B Cell Pathol. 24, 227–242.

Vindeløv, L. L., and Christensen, I. J. (1989). Eur. J. Haematol. Suppl. 48, 42:69–76.

Vindeløv, L. L., Christensen, I. J., Keiding, N., Spang-Thomsen, M., and Nissen, N. I. (1983a). Cytometry 3, 317–322.

Vindeløv, L. L., Christensen, I. J., and Nissen, N. I. (1983b). Cytometry 3, 323–327.

Vindeløv, L. L., Christensen, I. J., and Nissen, N. I. (1983c). Cytometry 3, 328–331.

Vindeløv, L. L., Christensen, I. J., Jensen. G., and Nissen, N. I. (1983d). Cytometry 3, 332–339.

Chapter 15

DNA Analysis from Paraffin-Embedded Blocks

DAVID W. HEDLEY

Departments of Medicine and Pathology
Princess Margaret Hospital
Toronto, Ontario M4X 1K9
Canada

I.   Introduction and Applications

This method is concerned with evaluating DNA flow cytometry (FCM) as a guide to prognosis in cancer. Clinical management of individual patients requires some idea of their prognosis, for example the likelihood of recurrence following surgical removal of the primary tumor, and the probable life expectancy of patients with recurrent disease. At the present time the major prognostic indicators are the extent of disease (i.e., clinicopathological stage), site of origin, and histological subtype. Assessment of cellular differentiation at the light microscope (LM) level can give additional information, but the grading systems used are very subjective and show poor reproducibility when assayed by different observers.

Flow cytometric measurement of cellular DNA content can give information about tumor ploidy and proliferative activity, both of which are objectively determined and potentially important indicators of biological aggression. In order to define the precise role of this technique in routine clinical practice, large numbers of patients with similar types of cancer need to be followed up to meaningful endpoints. Normally fresh, unfixed tumor tissue is used for DNA FCM, and a considerable length of time is therefore required to accrue sufficient samples and allow for adequate follow-up. For some rare but important tumors, such as cancers of children, even major institutions perform only a few biopsies each year.

For many decades it has been routine practice to fix tumor biopsy specimens in formaldehyde, and then to dehydrate them using ethanol and embed in paraffin wax in order to facilitate section cutting using a microtome. Paraffin blocks are remarkably durable, and because of an occasional need to reexamine a particular biopsy sample, they tend to be kept indefinitely. A few years ago we showed that they could be used as an alternative to fresh material as a starting point for DNA FCM (Hedley *et al.*, 1983, 1985). The major advantage is that one can take a series of blocks from comparable patients treated many years previously, and determine whether DNA histograms obtained retrospectively were predictive of clinical outcome.

The method for DNA analysis using paraffin blocks involves the cutting of thick microtome sections, dewaxing with xylene, rehydrating through graded ethanols, and finally producing of a cell or nuclear suspension by enzymatic digestion. This can then be stained and measured as for fresh unfixed material. Apart from being more cumbersome than techniques that use fresh tissue, the paraffin blocks method tends to give somewhat more cell debris and higher CV values. For these reasons it is better suited to performing large retrospective studies to establish the clinical utility of DNA FCM, rather than for routine purposes where fresh tissue is available.

II.   Methods

A.   Selection of Blocks

Apart from needing full clinical details such as the patient's outcome, parallel thin sections should be examined by LM to ensure that an adequate proportion of the material consists of cancer rather than stromal tissue. Ideally at least 10% of the total cell population should be malignant, to prevent small DNA aneuploid peaks from being lost against the background of normal host cells.

B.   Section Cutting

Sections are normally cut using a microtome. Partially sectioned nuclei are a major source of debris, which can be minimized by using 50-μm or thicker sections. The number of sections required for DNA FCM varies between one and four, depending on the cellularity of the block.

C.   Dewaxing

Thick sections usually curl up tightly as they come off the microtome, and do not require unrolling. Place them in 10-ml glass centrifuge tubes (caution: xylene dissolves most plastics), and add 3 ml xylene. Let stand for 10 minutes at room temperature (RT), then aspirate the xylene using a Pastcur pipct and add a second 3 ml of xylene for 10 minutes and reaspirate. We currently strongly recommend the use of Histoclear (National Diagnostics, Somerville, NJ) as a substitute for xylene, because it is nontoxic and biodegradable.

D. Rehydration

After removing xylene or Histoclear, add 3 ml 100% ethanol and allow to stand for 10 minutes. Then aspirate and replace, sequentially, with 95%, 70% and 50% ethanol for 10 minutes each. During this process the sections become soft and friable. The extent to which this takes place varies between blocks, and care is needed not to aspirate tissue along with the ethanol. Centrifugation may be required to pellet material in suspension. Finally, wash twice using distilled water, again using centrifugation if the sample is clearly breaking up.

E. Enzyme Treatment

Prepare a 0.5% solution of pepsin in 0.9% saline. Note that inferior grades contain substantial amounts of other enzymes as contaminants. The recommended preparation is Sigma catalog no. P7012, activity 2500–3200 units/mg protein. Adjust pH to 1.5 by adding 2 N HCl. Add 1 ml of the pepsin solution and place the tubes in a 37°C water bath for 30 minutes, with intermittent vortexing. Toward the end of this period the sample often becomes turbid. Examination in a hemocytometer using phase-contrast microscopy should reveal either intact cells or bare nuclei resembling those seen in the parallel thin section.

F. Cell Counting

Before staining, the concentration of cells or nuclei should be adjusted to that appropriate for DNA FCM. In our experience, cell counts using a hemocytometer tend to underestimate the number of nuclei present. Alternatively, they can be made using a Coultcr counter, with the volume threshold set at 8 μm to eliminate debris.

G. Staining

Centrifuge and resuspend at a final concentration of 10^5–10^6 cells/ml in a buffered medium such as Hanks' balanced salt solution (HBSS). Originally we used RPMI-1640 tissue culture medium, but this is overstringent. Stain DNA with 1 μg/ml 4', 6'-diamidino-2-phenylindole dihydrochloride (DAPI) (Boehringer Mannheim), and keep at RT for 30 minutes after addition of the dye. Immediately before running add RNase (type 1A, Sigma), final concentration 0.1% (w/v), and filter. [Note that if using an argon ion-based flow cytometer, propidium iodide (PI) may be used instead of DAPI as the DNA fluorochrome (see later and Schutte *et al.*, 1985).]

H.   Flow Cytometry

Once the nuclear suspension has been prepared and stained from paraffin-embedded tissue, the sample is run exactly as one derived from unfixed material. In mercury arc-based flow cytometers, use a UG1 or similar ultraviolet (UV) excitation filter. Alternatively, tune an ion laser into the UV range. Fluorescence is measured in the range 480–530 nm.

III.   Comparison with Fresh Tissue

Comparison of DNA histograms obtained when the same tumor issue is examined fresh and following paraffin embedding shows that they are remarkably similar (Fig. 1). In particular, there is an excellent correlation between DNA indices obtained using the two techniques, as shown in Fig. 2. Although S-phase percentage estimates are generally in agreement, the correlation has been found to be to much less close than is the case with DNA index (DI) (Kute *et al.*, 1988). Admixture with diploid host cells is a major source of error in S-phase percentage measurements using tumor biopsy samples, and microtome sections cut from paraffin blocks probably contain a greater proportion of normal stroma (which can be trimmed away when fresh tissue is available). DNA histograms from paraffin blocks tend to show wider CV values and contain more cell debris than is seen with fresh material, which further reduces the accuracy of S-phase percentage estimates.

Finally, it should be noted that although it is customary to use a biological standard such as chick red blood cells (CRBC) when running fresh material, no entirely satisfactory standard has been developed for use with paraffin blocks. The reason for this is that formalin fixation reduces, and enzyme digestion increases the intensity of DNA staining. Unfortunately these effects differ between cell types, so that, for example, CRBC and human peripheral blood lymphocytes (PBL) do not give the same ratio of peak channel numbers following fixation and digestion as they do when run fresh. By using one of the methods discussed here (Section IV, C), normal-appearing tissue could be isolated from the block being studied for process-

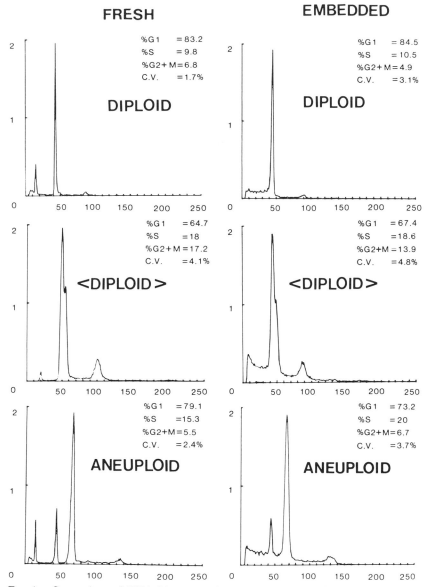

Fig. 1. Comparison of DNA histograms from the same tumor tissue examined either fresh or following fixation and paraffin embedding. Note that the paraffin blocks method gives a close approximation to the results using fresh tissue, but that the histograms contain more debris to the left of the diploid G_1 peak, and the CV values are slightly greater. From Hedley *et al.*, *Journal of Histochemistry and Cytochemistry*, 1983, vol. 31, pp. 1333–1335. With permission.

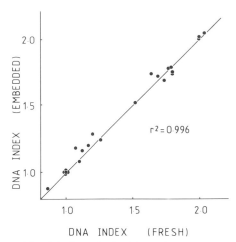

FIG. 2. Comparison of DNA indices obtained from the same tissue examined either fresh or following fixation and paraffin embedding. The single hypodiploid tumor was identified as such by the use of an internal biological standard added to the unfixed sample. From Hedley *et al.*, *Journal of Histochemistry Cytochemistry*, 1983, vol. 31, pp. 1333–1335. With permission.

ing and used as an internal standard, but a detailed comparison with the same tissue run fresh has not been published to date. Inability to define the normal diploid G_1 peak means that hypodiploid tumor populations cannot be identified with certainty, and most institutions therefore adopt the convention of treating the G_1 peak with lowest DNA content as normal diploid when calculating DI. This means that uncommon but perhaps significant hypodiploid tumors are erroneously assigned a DI > 1.0

IV.   Modifications of Original Method

A large number of variations of the original 1983 method have been described; some are intended simply to suit in-house practice, whereas others are more generally useful in certain situations (for review see Hedley, 1989).

A.   Alternative DNA Fluorochromes

Of the fluorochromes evaluated, in our hands DAPI gave the lowest CV values, but it requires UV excitation. When using an argon ion laser it is often more convenient to use one of the strong visible-light wavelengths, in which case PI at 50 μg/ml may be substituted. Its use is described in detail by Schutte *et al.* (1985). Some laboratories have found that it gives results as good as or better than DAPI.

B. Enzymatic Treatment

Prolonged digestion with 0.5% pepsin gives increasingly broad G_1 peaks and excessive cell debris (unpublished observations), presumably because of increasing DNA fragmentation. Although a detailed evaluation of methods of enzyme treatment has yet to be published, it has been suggested that trypsin or Protease (pronase; type XXIV, Sigma) may yield less cellular debris than pepsin. Schutte *et al.* (1985) describe the use of 0.25% trypsin (Difco) in citrate buffer [3 mM trisodium citrate, 0.1% (v/v) Nonidet P-40 (NP-40), 1.6 mM spermine tetrachloride, 0.5 mM Tris, pH 7.6]. Rehydrated sections were incubated overnight at 37°C. The use of Protease was first described by van Driel-Kulker *et al.* (1985). Rehydrated sections were placed in 2 ml of 0.05% Protease (9.7 U/mg) in PBS, and incubated at 37°C for 30 minutes with intermittent vortexing. Although developed to prepare cells for image analysis, this technique has been used with FCM.

C. Selecting Areas of Interest

Microtome sections from blocks prepared specifically for microscopy often contain large amounts of nonmalignant cells or can show areas of morphological tumor cell heterogeneity that one would like to study in isolation. Parallel stained sections can be used to map out areas of interest in the block, which can then be cut out using a scalpel or a 4-mm skin biopsy punch. Alternatively, rehydrated thick microtome sections can be placed on glass slides, and areas of interest isolated by scraping away irrelevant areas using a scalpel and dissecting microscope (Oud *et al.*, 1986).

V. Common Problems

The common problems are excessive cell debris and broad peaks, and often these coexist. Assuming that the flow cytometer is properly aligned and giving satisfactory results with fresh tissue, poor-quality DNA histograms obtained from paraffin blocks are the result of either unsatisfactory starting material or bad technique. The manner in which the original specimen was handled is very important. Small specimens rapidly placed in formalin are much better fixed than large pieces of tissue, or when there is delay in fixation (as occurs, for example, with autopsy material). Neutral buffered formalin or formalin–acetone–acetic acid generally gives satisfactory results, while the use of more specialized fixatives such as Bouin's fluid or the mercury-based fixatives often results in uninterpretable DNA histograms. Provided that it has been adequately fixed, the age of the block does not appear to affect results of subsequent FCM. It has been shown that material from some pathology departments gives generally poorer results than is found with other institutions. Faced with the problem of excessive debris or broad peaks, an FCM laboratory setting up to examine paraffin blocks would be well advised to obtain thick sections from a block that another facility has shown to produce good results; paraffin-embedded material is eminently transportable. Failure to produce good results then indicates a problem of technique.

Poor results not attributable to starting material or machine performance require a thorough review of technique. The published method should be followed in detail; with particular emphasis on the grade and activity of the pepsin used. If the problem persists, then it is most likely that the cells are unusually susceptible to the pepsin solution used, with excessive DNA degradation taking place. Possible solutions to this problem would be to try a weaker pepsin mixture, or to substitute trypsin or Protease using one of the variant methods discussed earlier. It has been suggested in meeting abstracts that either of these enzymes yields less debris than pepsin.

VI.   Multiparametric Analysis Using Paraffin Blocks

Although a detailed description of the techniques is beyond the scope of this chapter, monoclonal antibodies (mAb) have been used to identify oncogene products or proteins associated with proliferation in nuclei obtained from paraffin blocks. Obviously these proteins will be digested to a variable degree during enzyme treatment, but despite this some interesting results have been obtained, and the scope for two-parameter DNA versus antibody staining is considerable. The reader is referred to Bauer et al. (1986) and Watson et al. (1985) for further technical information. (See also Bauer, Chapter 24, this volume.)

REFERENCES

Bauer, K. D., Clevenger, C. V., Endow, E. K., Murad, T., Epstein, A. L., and Scarpelli, D. G. (1986). *Cancer Res.* **46,** 2428–2434.

Hedley, D. W. (1989). *Cytometry* **10,** 229–241.

Hedley, D. W., Friedlander, M. L., Taylor, I. W., Rugg, C. A., and Musgrove, E. A. (1983). *J. Histochem. Cytochem.* **31,** 1333–1335.

Hedley, D. W., Friedlander, M. L., and Taylor, I. W. (1985). *Cytometry* **6,** 327–333.

Kute, T. E., Gregory, B., Galleshaw, J., Hopkins, M., Buss, D., and Case, D. (1988). *Cytometry* **9,** 494–498.

Oud, P. S., Hanselaar, T. G. J. M., Reubsaet-Veldhuizen, J. A. M., Meijer, J. W. R., Genmink, A. H., Pahlplatz, M. M. M., Beck, H. L. M., and Vooijs, G. P. (1986). *Cytometry* **7,** 595–600.

Schutte, B., Reynders, M. M. J., Bosman, F. T., and Blijham, G. H. (1985). *Cytometry* **6,** 26–30.

Van Driel-Kulker, A. M. J., Mesker, W. E., van Velzen, U., Tanke, H. J., Feichtinger, J., and Ploem, J. S. (1985). *Cytometry* **6,** 268–272.

Watson, J. V., Sikora, K. E., and Evan, G. L. (1985). *J. Immunol. Methods* **83,** 179–192.

Chapter 16

Detection of M and Early-G_1 Phase Cells by Scattering Signals Combined with Identification of G_1, S, and G_2 Phase Cells

ELIO GEIDO and WALTER GIARETTI

IST, Istituto Nazionale per la Ricerca sul Cancro
Laboratorio di Biofisica
16132 Genova, Italy

MICHAEL NÜSSE

GSF, Gesellschaft für Strahlen- und Umweltforschung
Institut für Biophysikalische Strahlenforschung
D-6000 Frankfurt-am-Main
Federal Republic of Germany

I. Introduction

Cell structural properties measured by flow cytometry (FCM) have been used to discriminate subpopulations of white blood cells (Hoffman *et al.*, 1980), to resolve subpopulations in tumors (Dolbeare *et al.*, 1983), to separate mitotic cells from interphase cells (Darzynkiewicz *et al.*, 1977, 1987; Benson *et al.*, 1984; Larsen *et al.*, 1986; Zucker *et al.*, 1988), and to evaluate chromatin conformational changes in isolated nuclei (Papa *et al.*, 1987). The molecular basis for the observed alterations in cytoplasm

149

and chromatin conformation and structure is unknown but could involve the nuclear matrix and the cytomatrix, as reviewed by Pienta *et al.*, (1989).

A new FCM method (Giaretti *et al.*, 1989; Nuesse *et al.*, 1989) combining light-scattering measurements and detection of bromodeoxyuridine (BrdUrd) incorporation via fluorescent antibodies and quantitation of cellular DNA content by propidium iodide (PI) allows identification of additional compartments in the cell cycle. Thus, while cell staining with anti-BrdUrd antibodies and PI reveals the G_1, S, and $G_2 + M$ phases of the cell cycle, differences in light scattering allow separation of G_2 phase cells from M-phase cells and subdivision of G_1 phase into two compartments with G_{1A} representing postmitotic cells that mature to G_{1B} cells ready to initiate DNA synthesis.

The method involves fixation of cells in 70% ethanol, extraction of histones with 0.1 N HCl at 0°C, thermal denaturation of DNA at 80°–95°C for variable durations in relation to cell type. Mitotic cells show much lower 90° scatter than G_2-phase cells, and G_1 postmitotic cells (G_{1A}) show lower 90° scatter than G_1 cells about to enter the S phase (G_{1B}). In addition, depending on the conditions of thermal denaturation, M-phase cells can show lower PI fluorescence emission compared to G_2 cells. The thermal denaturation appears to enhance the differences in chromatin structure of cells in the various phases of the cell cycle to the extent that cells could be separated on the basis of the 90° scatter, which is mainly dependent on reflective and refractive components in the nucleus. Light scattering is correlated with chromatin condensation, as judged by microscopic evaluation of cells sorted on the basis of light scatter. The method has the advantage over the parental BrdUrd/DNA bivariate analysis (Dolbeare *et al.*, 1983, 1985; Beisker *et al.*, 1987) in allowing the G_2 and M phases of the cell cycle to be separated, and the G_1 phase to be analyzed in more detail.

II. Application

Fixation was with 70% ethanol, as commonly used by others (Dolbeare *et al.*, 1983, 1985; Beisker *et al.*, 1987). Different cell types grown either as monolayers or as suspension culture were giving similar results (see Section III). Thermal denaturation conditions, as used in our protocol, were analyzed by a systematic investigation (see Section IV). Only two different instruments have been used so far (see Section V).

III.   Materials

A.   Cell Lines

1. A strain of hyperdiploid Ehrlich ascites tumor cells. (EAT, clone F5, DNA index, DI = 1.22) was growing in suspension cultures. The cells were maintained at 37°C under 6% CO_2 in a special medium (A2 medium supplemented with 20% horse serum (Nüsse, 1981). A somewhat continuously growing culture was obtained by daily dilution of the cells in fresh medium. Cell cycle data of these cells were published earlier (Nüsse, 1981).

2. V79 and Chinese hamster embryo cells (obtained from Dr. S. Cram) were maintained at 37°C under 6% CO_2 in α-MEM (GIBCO, Grand Island, NY) supplemented with 15% fetal calf serum (FCS). The cells were subcultured three times a week.

3. MCF-7 cells (courtesy of Dr. A. Nicolin) were maintained in Dulbecco's minimum essential medium (DMEM) (Flow Laboratories, Ayrshire, UK) supplemented with 10% FCS, L-glutamine, and nonessential amino acids (GIBCO). The cells were subcultured weekly in Falcon plastic T flasks at 2.7×10^3 cells/cm^2 seeding density. Cell cycle data of these cells were published earlier (Bruno et al., 1988).

4. PB-3 and PB-1 cells, purchased from Dr. J. F. Conscience (Ball et al., 1983; Conscience and Fischer, 1985; Giaretti et al., 1990), were maintained in RPMI-1640 medium (GIBCO) supplemented with 10% FCS, L-glutamine, nonessential amino acids (GIBCO). In the case of PB-3 cells, 20% of serum-free medium conditioned by the WEHI-3 myelomonocytic leukemia cell line as source of interleukin 3 (IL-3) was added.

B.   Enrichment of Mitotic Cells and BrdUrd Pulse-Labeling

Colcemid was added at a concentration of 0.2 μg/ml for several time intervals. Control cells with no Colcemid were run in parallel. Mitotic indices (MI) were estimated under a light microscope using conventional staining techniques. The cells were pulse-labeled with BrdUrd. In all experiments 10 μM BrdUrd was added to the cultures for 15 minutes at 37°C; no additional substances reported to increase incorporation of BrdUrd (Beisker et al., 1987) were used. The cells were trypsinized after the BrdUrd pulse, washed with PBS if necessary, and fixed in cold 70% ethanol.

C.   Monoclonal Anti-BrdUrd Antibodies

Monoclonal anti-BrdUrd antibodies obtained from the following sources have been used:
F. Dolbeare, Lawrence Livermore National Laboratory, Livermore, CA (IU-1 and IU-4; Vanderlaan and Thomas, 1985; Vanderlaan *et al.*, 1986); Becton Dickinson (Mountain View, CA); Partec (Arlesheim, Switzerland); Eurodiagnostics (Apeldoorn, The Netherlands).

D.   Staining Procedure

For staining, $2-3 \times 10^6$ fixed cells were sedimented and incubated in 2 ml phosphate-buffered saline (PBS) containing RNase A (1 mg/ml) at 37°C for 20 minutes. The cells were then sedimented and incubated in ice-cold 0.1 N HCl for 10 minutes, washed once in cold distilled water (dH_2O), resuspended in 2 ml dH_2O, and heated for different time intervals (usually between 20 and 40 minutes) at 95°C. The cells were rapidly cooled in ice water; next, 5 ml of PBS containing 0.5% Tween-20 (PBST) was added and the cells were sedimented. The cells were then resuspended for 30 minutes at room temperature in 0.4 ml of anti-BrdUrd antibody diluted between 1 : 100 and 1 : 1000 in PBST containing 0.5% bovine serum albumin (BSA). After washing with PBST the cells were stained in 0.4 ml FITC-conjugated goat anti-mouse IgG antibody (Sigma) diluted 1 : 50 in PBST containing 0.5% BSA. After 20 minutes the cells were washed again and resuspended in PBS containing 20 $\mu g/ml$ PI.

IV.   Critical Aspects of the Procedure

Cell loss due to their aggregation during heat treatment and a rather large CV of the G_1 peak were observed occasionally. These two effects depended on both the cell type under study and the temperature and duration of the thermal treatment. The thermal denaturation, which is necessary to produce single-stranded DNA in order to make the incorporated BrdUrd accessible to the anti-BrdUrd antibodies, appears to affect both cell structure and fluorescence emission. The effects varied depending on the phase of the cell cycle. Usually 95°C and 20 minutes treatment resulted in a good separation of M versus G_2 and G_{1A} versus G_{1B} cells. No separation was observed after a heat treatment with a temperature <80°C.

Prolongation of thermal treatment (as long as 50 minutes) induced much lower PI fluorescence emission of M cells without affecting extensively their 90°-scattering signals. In conclusion, the preparation conditions for different cell types should be optimized by carefully varying temperature and duration of the heat treatment. The other steps in our staining procedure, such as 0.1 HCl treatment, or the use of different anti-BrdU antibodies, did not influence the results.

V.   Instruments

This procedure was tested using two different instruments.

A FACS 440 dual-laser flow sorter (Becton Dickinson) was located in the Laboratory of Biophysics (IST), Genoa, Italy. With this instrument we have performed list-mode analysis and cell sorting. The four parameters, red (PI) and green (FITC) fluorescence as well as forward (FSC) and 90° scatter (90° SC) were measured simultaneously, the data being stored in real time as four parameters per cell in list mode. Any combination of two of the four parameters was studied after the measurements. Bivariate BrdUrd/DNA or scatter/DNA distributions were displayed as "contour plots" (lowest contour line between 2% and 5% of maximum). Red fluorescence and scatter were always displayed on a linear scale, green fluorescence on a log or linear scale. At least 20,000 cells were measured for each histogram. Mitotic and interphase nuclei were sorted on slides based on differences in their 90° SC signals and PI fluorescence. The sorted cells were then analyzed under a fluorescence microscope to identify mitotic and interphase cells by differences in the nuclear morphology.

A Cytofluorograf 30 L (Ortho Diagnostic), in the Institute for Biophysical Radiation Research (GSF), Frankfurt, FRG, was equipped with a data acquisition system to store and analyze two-parametric histograms (no list mode).

For both instruments, excitation of PI and FITC was provided by the 488-nm line (500 mW) of an argon ion laser (model 2025, Spectra Physics). The emitted fluorescence was collected at two wavelength ranges, namely between 510 and 560 nm (bandpass filter) for FITC antibody staining and at wavelengths >610 nm for PI fluorescence.

VI.   Results

Figure 1A shows a typical BrdUrd (FITC)/DNA (PI) content bivariate plot for exponentially growing EAT cells obtained with a FACS 440 using list mode analysis (Giaretti *et al.*, 1989). Three separated cell populations are observed: cells that have incorporated BrdUrd and show high FITC fluorescence (greater than channel 20 on *y* axis), which represents the

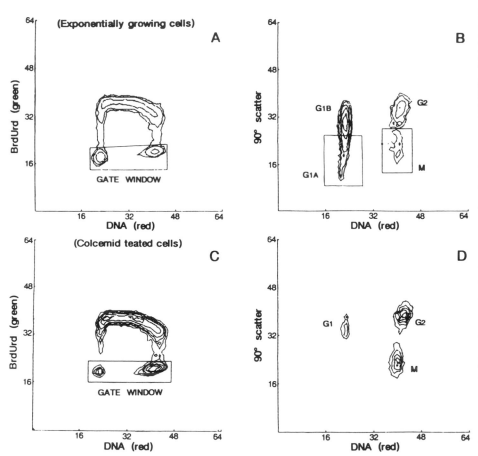

FIG. 1.   Bivariate contour plots obtained from list-mode FCM analysis of EAT cells subjected to BrdUrd pulse-labeling (refer to text for detailed explanation). In (B) can be clearly observed the presence of early-G_1 cells (G_{1A}) separated from late-G_1 cells (G_{1B}) and G_2 cells separated from M cells. In (D) the accumulation of M cells by colcemid and the disappearance of G_{1A} cells can be clearly observed.

S-phase cells; and two groups of unlabeled cells that show different DNA double strand content (equivalent to 18 and 36 channels on the x or DNA axis), which represent G$_1$ and G$_2$ + M phase cells, respectively.

By electronic gating on the BrdUrd/DNA parameters, it was possible to analyze further any group of cells for the forward and 90° parameters. With the gating set as shown in Fig. 1A and 1C, only unlabeled cells are selected. The 90° scattering by these cells is shown in Fig. 1B and 1D. The G$_2$ and M cells exhibit markedly different scattering properties. The scatter from G$_2$-phase cells is almost double that of M-phase cells. The identity of these populations was confirmed by determination of MI and by cell sorting. More than 95% of cells sorted from the group labeled M exhibited mitotic figures, whereas none of the G$_2$-phase cells showed a mitotic figure. Additionally, and depending mainly on the duration of heat treatment, G$_2$ and M-phase cells can display different PI fluorescence intensities although they have the same DNA content. M-phase cells show less PI fluorescence compared to G$_2$-phase cells. Under optimal conditions, M-phase cells are found between G$_1$- and G$_2$-phase cells (see for details Nuesse et al., 1989).

Interestingly, the 90° scatter of unlabeled G$_1$-phase cells also showed two populations though not as well resolved as the G$_2$ and M populations. As seen in Fig. 1C, Colcemid treatment, which blocks cells in mitosis, resulted in the disappearance of G$_1$-phase cells with low 90° scatter (G$_{1A}$). The reappearance of this population after release of the block was also observed (not shown). Discussion of potential applications of this technique is presented in several reports (Nuesse et al., 1989; Giaretti et al., 1989).

References

Ball, P. E., Conroy, M. C., Heusser, C. H., Davis, J. M., and Conscience, J. F. (1983). *Differentiation* **24**, 74.

Beisker, W., Dolbeare, F., and Gray, J. W. (1987). *Cytometry* **8**, 235–239.

Benson, M. C., McDougal, D. C., and Coffey, D. S. (1984). *Cytometry* **5**, 515–522.

Bruno, S., Di Vinci, A., Geido, E., and Giaretti, W. (1988). *Breast Cancer Res. Treat.* **11**, 221–229.

Conscience, J. F., and Fischer, F. (1985). *Differentiation* **28**, 291.

Darzynkiewicz, Z., Traganos, F., Sharpless, T., and Melamed, M. R. (1977). *J. Histochem. Cytochem.* **25**, 875–880.

Darzynkiewicz, Z., Traganos, F., Carter, S. P., and Higgins, P. J. (1987). *Exp. Cell Res.* **172**, 168–179.

Dolbeare, F., Gratzner, H., Pallavicini, M. G., and Gray, J. W. (1983). *Proc. Natl. Acad. Sci. U. S. A.* **80**, 5573–5577.

Dolbeare, F., Beisker, W., Pallavicini, M. G., Vanderlaan, M., and Gray, J. W. (1985). *Cytometry* **6**, 521–530.

Giaretti, W., Nuesse, M., Bruno, S., Di Vinci, A., and Geido, E. (1989). *Exp. Cell Res.* **182,** 290–295.

Giaretti; W., DiVinci, A., Geido, E., Marsano, B., Minks, M., and Bruno, S. (1990). *Cell Tissue Kinet.* **23,** in press.

Hoffman, R. A., Kung, P. C., and Hansen, P. (1980). *Proc. Natl. Acad. Sci. U. S. A.* **77,** 4914.

Larsen, J. K., Munch-Peterson, B., Christansen, J., and Jorgensen, K. (1986). *Cytometry* **7,** 54–63.

Nuesse, M. (1981). *Cytometry* **2,** 70–79.

Nuesse, M., Juelch, M., Geido, E., Bruno, S., Di Vinci, A., Giaretti, W., and Ruoss, K. (1989). *Cytometry* **10,** 312–319.

Papa, S., Capitani, S., Matteucci, A., Vitale, M., Santi, P., Martelli, A. M., Maraldi, N. M., and Manzoli, F. A. (1987). *Cytometry* **8,** 595–601.

Pienta, K. J., Partin, A. W., and Coffey, D. S. (1989). *Cancer Res.* **49,** 2525–2532.

Vanderlaan, M., and Thomas, C. B. (1985). *Cytometry* **6,** 501–515.

Vanderlaan, M., Watkins, B., Thomas, C., Dolbeare, F., and Stanker L. (1986). *Cytometry* **7,** 499–507.

Zucker, R. M., Elstein, K. H., Easterling, R. E., and Massaro, E. J. (1988). *Cytometry* **9,** 226–231.

Chapter 17

Controls, Standards, and Histogram Interpretation in DNA Flow Cytometry

LYNN G. DRESSLER

Division of Molecular and Cellular Diagnostics
University of New Mexico Cancer Center
University of New Mexico School of Medicine
Albuquerque, New Mexico 87131

I. Introduction

Controls and standards as well as defined criteria to interpret DNA histograms are essential components of DNA flow cytometry (FCM) methodology. One of these controls is cytologic or histologic examination of the specimen. This control is necessary to evaluate the quality of the specimen and to assess the percentage of tumor versus nontumor cells present. Another control is microscopic review of the fluorescently stained cells or nuclei. This control provides information regarding specific uptake of dye into the nucleus, integrity of cells or nuclei (broken vs intact), and the degree of cell aggregation (clumping). Microscopic screening of histologic and fluorescent samples thus allows one to evaluate the quality of the specimen and may also be helpful in the interpretation of difficult DNA histograms.

Essential to the determination of relative DNA content is a DNA reference standard. Both internal and external standards have been used as normal DNA content reference standards from a variety of sources: chicken red blood cells (CRBC), trout red blood cells (TRBC), human peripheral blood lymphocytes (hPBL), and nonmalignant tissue from the same organ as the specimen. Other standards include commercially available fluorescent reference beads or fixed CRBCs to aid in instrument alignment and setting up gates or thresholds on the flow cytometer.

157

Evaluation of DNA histograms involves the incorporation of defined parameters to allow consistency and quality control of data interpretation. A defined set of evaluation criteria are required that include limits and definitions of the coefficient of variation (CV), the mean value of DNA content of the measured cell population, the DNA index (DI), specific definitions for DNA content (ploidy)—that is, normal (DNA diploid) versus abnormal (DNA aneuploid) and subgroups of abnormal (hypodiploid, hyperdiploid, near diploid, near tetraploid, tetraploid, hypertetraploid, and multiploid). Criteria for evaluability of cell cycle analysis should also be defined.

In this chapter, technical details for controls, standards, and criteria used to evaluate and interpret DNA FCM data that have proved efficient and reproducible will be presented. The following guidelines should prove useful to investigators setting up DNA FCM methods in their laboratories and hopefully will stimulate suggestions, improvements, and eventually standardization.

II.   Application

Application and incorporation of quality control parameters for DNA content analysis into daily procedures have been variously described by several investigators; however, their routine use in research or clinical laboratories is often inadequate. Because controls, standards, and interpretation criteria may vary with the state of the sample, the following discussion will be divided into two sections: (1) techniques applicable to fresh or frozen tissue and (2) techniques applicable to formalin-fixed, paraffin-embedded tissue.

A.   Fresh or Frozen Tissue

1.   CYTOLOGIC CONTROLS

Cytologic controls are practical in fresh and frozen tissue specimens, in which an aliquot of the sample is obtained following tissue dissociation or, in the case of body fluids, upon receipt of the specimen. Cytologic controls can be obtained as smears or sometimes more easily as cytocentrifuged preparations stained with Giemsa or hematoxylin and eosin. In addition, evaluation of the fluorescent cells or nuclei, stained with DNA-specific fluorochromes, such as propidium iodide (PI) or 4′,6′-diamidino-2-phenylindole (DAPI), can be performed provided the fluorescence microscope being used has appropriate filter combinations to view the emitted fluorescence of the dyes used.

2. DNA REFERENCE STANDARDS

Several different external and internal DNA standards have been described in the literature (Vindelov *et al*; 1983; Iverson and Laerum, 1987; Jakobsen, 1983; see also Chapters 10–15 of this volume). An external standard is one that is prepared in a separate tube from the sample and run in parallel with the sample. An internal standard is one that is actually added to the sample tube prior to fluorochrome staining. Ideally, a sample should be run in duplicate, where one sample tube contains the internal standard and the other does not. This allows for quality control of staining variability. However, because a large quantity of available sample is needed, running duplicate samples is not always feasible or practical. Most investigators have chosen to use either an internal standard or an external standard or in some cases a combination of both. There are currently two different strategies for using DNA reference standards: one is to use a hypodiploid reference(s) to identify where the human diploid G_0/G_1 peak should be located and the other is to use a normal diploid reference to superimpose with the expected normal diploid peak in the histogram.

Vindeløv and colleagues (Vindeløv *et al.*, 1983) described a system using the ratio of two hypodiploid references, CRBC and TRBC, as internal controls, to help identify the position of the human diploid G_0/G_1 population. Nuclei of CRBC and TRBC have different DNA contents, which are both less than a normal human lymphocyte (33% and 80%, respectively). Aliquots of both TRBC and CRBC are added to the tumor specimen prior to staining, and the ratio of the peak positions of the trout G_0/G_1 to the chicken G_0/G_1 are used to normalize the histogram and identify the human diploid G_0/G_1 position.

Normal diploid controls, such as unstimulated hPBL or nonmalignant tissue from the same organ as the tumor sample are also used as DNA reference standards. Human PBL are the most widely used DNA reference standard and are generally used as an external standard. Human PBL and CRBC are also used as an external standard for instrument alignment, setting high voltages and gains, setting gates and threshold for light scatter, size or fluorescence intensity on the flow cytometer (see Section V for details).

Chicken red blood cells alone, either fixed or unfixed, can be used as an external standard for instrument alignment and to set size or fluorescence thresholds. The author uses unfixed PI-stained CRBC in phosphate-buffered saline (PBS) solution to set the threshold for fluorescence intensity on the flow cytometer. Events having fluorescence intensity less than that of the CRBC nuclei are considered debris and are gated out of the final histogram. This allows collection of events more likely to represent tumor events and less likely to represent debris. Collection of ungated data also allows the user to estimate the degree of debris in the histogram. The peak position of CRBC, however, can be somewhat variable in the low-channel regions, and because of this, CRBC by themselves are generally not recommended to identify the DNA diploid G_0/G_1 peak.

In our laboratory, unstimulated PBL are used as an external DNA reference standard. The hPBL are first run on the flow cytometer and adjustments in gain and/or high voltage are made to position the peak height of the G_0/G_1 population at channel 60 (on a 256 scale). This channel number allows for a tetraploid tumor (4N G_0/G_1 and 8N G_2M) to be on scale. We routinely collect 5000 events for the hPBL standard. We then run the tumor sample and collect 25,000 events. To confirm the position of the diploid G_0/G_1 peak, an aliquot of hPBL nuclei is added to the sample tube to a final concentration of 20%. The "mix" of DNA reference standard and sample are then rerun on the flow cytometer and the exact number of events is collected in the mix as was collected for the sample (Dressler and Bartow, 1989). In the case of an aneuploid tumor, the diploid peak height will increase and the aneuploid peak height will correspondingly decrease in the mixed relative to the unmixed sample (Fig. 1). In the case of a diploid tumor, one should obtain a superimposition of the G_0/G_1 peaks and a proportional decrease in the height of the G_2M peak.

B.   Formalin-Fixed, Paraffin-Embedded Tissue

1.   HISTOLOGIC CONTROLS

Samples received fixed and embedded in paraffin wax can be easily evaluated by histologic review of the 5-μm sections immediately preceding and following the thick sections cut for DNA flow assay. Review of these "top" and "bottom" hematoxylin and eosin (H&E)-stained sections allows for assessment of the amount of tissue area composed of tumor cells and ensures that the selection of sections by FCM was appropriate; that is, the sections contained representative proportions of tumor cells.

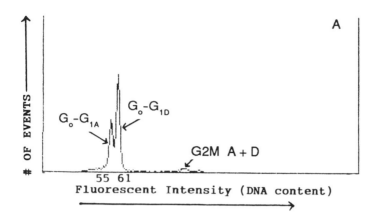

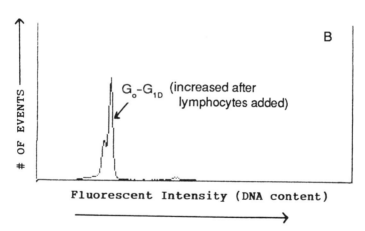

Fig. 1. Use of external control to identify human DNA diploid G_0/G_1 peak. (A) Histogram of a breast tumor obtained fresh from surgery. Two distinct peaks are observed at channels 55 and 61, respectively. To confirm which of these two peaks represents the normal DNA diploid population, an aliquot of unstimulated normal human peripheral blood lymphocytes (hPBL) was mixed with the tumor sample (to a final concentration of 20%). (B) The peak corresponding to channel 61 is seen to have increased after the addition of hPBL and confirms this peak as representing the normal DNA diploid G_0/G_1 (G_0-G_{1D}) population. The DNA aneuploid tumor was assigned an interpretation of "DNA hypodiploid" and a DI of 0.90 was calculated.

2. DNA REFERENCE STANDARDS

An optimal normal DNA reference standard for paraffin block samples would be nonmalignant epithelial tissue from the same patient and organ that was fixed and processed in parallel with the tumor sample; ideally nonmalignant tissue from the same paraffin block as the sample. If nonmalignant tissue is to be used as a DNA reference, it must have a highly cellular epithelial component to allow enough cells to be obtained for assay. An alternative DNA reference standard, which may be more practical in certain tissue types, is an uninvolved lymph node from the cancer dissection. For biopsy specimens or when lymph nodes or corresponding nontumor tissue is not available, a normal lymph node or tonsil block from another patient, whose tissue was processed in the same laboratory, at about the same time, and by the same methods as the tumor specimen, can be used.

The use of control tissue in the paraffin block procedure, however, does not provide a DNA reference standard as consistent as the trout/chick ratio or hPBL used for fresh or frozen tissue. Differences in fixation and/or processing as well as differences in chromatin structure of the control tissue as compared to the tumor tissue, can affect the binding of dyes, specifically the intercalating dyes, resulting in different fluorescence intensity of diploid nuclei (see Fig. 2). In our laboratory, we have experienced this difficulty in <6% of all paraffin block cases. Listed here are several suggestions for minimizing difficulty with the paraffin block technique.

1. If a lymph node is submitted as control, it should be uniformly well fixed.
2. If fixation of the lymph node or any other block is inconsistent or poor, an alternate block should be selected.
3. Prescreening of corresponding histologic slides stained with hematoxylin and eosin prior to assay is very helpful in assessing the quality of fixation and tumor composition in the selected block.
4. In cases where blocks contain tumor only in specific areas, the technician can dissect out the indicated areas of the block to maximize recovery of tumor tissue. This selection technique is also useful for obtaining both normal and neoplastic tissue from the same block and for excluding areas of necrosis or host inflammatory cells from the tumor specimen.
5. CRBC nuclei can be quite useful in the paraffin-embedded procedure to help determine whether a shift in fluorescence is instrument related or sample related.

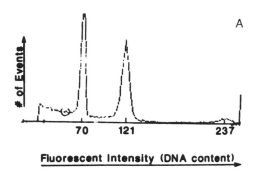

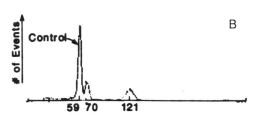

FIG. 2. Problems with controls in the paraffin block procedure. (A) Histogram obtained from a breast tumor that was formalin-fixed and paraffin-embedded. Two distinct peaks are seen at channels 70 and 121. (B) Uninvolved lymph node tissue from the same patient was mixed with the tumor sample and the "mix" was rerun on the flow cytometer. As can be seen in the histogram, a distinct peak is observed at channel 59, corresponding to the "control" lymph node. Obviously, this lymph node specimen could not be used to confirm the position of the normal DNA diploid peak. Differences in fluorescence intensity in the paraffin block technique have been observed by others; however, in our experience this difficulty is observed in <6% of all cases run and thus is not a major drawback in the technique.

C. DNA Histogram Interpretation

Numerous papers have been published using DNA FCM measurements to characterize a variety of solid tumors. However, standard quality control parameters are not universally in place in the clinical and research laboratories at the present time. Before standardization can occur, definitions for ploidy status, criteria for evaluability, and modeling parameters for S phase need to be understood and well described. The following guidelines have proved efficient and practical and provide consistency and quality control in DNA histogram interpretation.

1. DNA PLOIDY STATUS

In order to distinguish results obtained by DNA flow cytometry from
data obtained by cytogenetic techniques, it was recommended by the
Convention on Nomenclature in 1984, that the terms "normal" and "abnormal stemline" be used instead of diploidy and aneuploidy, respectively
(Hiddeman *et al.*, 1984). The term "DNA aneuploid" was agreed on as being
a synonym for "abnormal stemline." In our laboratory, the terms "DNA
diploid" and "DNA aneuploid" are used for normal and abnormal stemlines, respectively.

DNA ploidy status of tumors is defined based on the amount of DNA
relative to normal. The DNA index (DI) is a value given to express the
amount of DNA content relative to normal and is calculated by the
following equation.

$$DI = \frac{\text{mean or modal channel no. of DNA aneuploid } G_0/G_1 \text{ peak}}{\text{mean or modal channel no. of DNA diploid } G_0/G_1 \text{ peak}}$$

A DNA diploid population (G_0/G_1) is given a DI of 1.00 by definition.
DNA aneuploid tumors can have less than the 2N amount of DNA and are
termed "DNA hypodiploid" or, more frequently, have more than the 2N
amount of DNA and are termed "DNA hyperdiploid." Those tumors that
have a 4N amount of DNA are termed "DNA tetraploid," while those that
are "near" 4N are sometimes called "DNA near tetraploid," and those
greater than 4N are "DNA hypertetraploid." Tumors may also have more
than one abnormal population, in which case they are referred to as
multiple aneuploid or DNA multiploid (Dressler and Bartow, 1989).

There are situations however, when the interpretation of DNA diploid
versus DNA aneuploid is not straightforward. For example, a fresh tumor
may show a single peak with a slight left shoulder. Addition of control
lymphocytes causes a single peak to be observed, shifted to the left by one
channel, and no shoulder is visible. Because two distinct peaks cannot be
resolved, it cannot be determined by FCM whether or not there are indeed
two distinct populations. Some investigators have interpreted this type of
histogram as "diploidlike," "near-diploid," or "questionable ploidy."
Others have specified a percentage of error allowable, within the bounds of
which a tumor is to be interpreted as DNA diploid without modifying
terms (i.e., DI $1.00 \pm 10\%$ = diploid.)

It should be emphasized that only a fraction of nuclear DNA is stainable
(accessible to dyes) and that the extent of DNA accessibility varies depending on the dye used, chromatin structure, and staining procedure (Darzynkiewicz *et al.*, 1984). The terms "DNA ploidy" or "DNA index" are

thus rather operational, and actual DNA content of the studied cells may be different from that established using "normal" cells as the DNA reference standard.

2.   CRITERIA FOR EVALUATION

The use of certain criteria help the investigator to evaluate and interpret a DNA histogram. It should be understood that these criteria may not be applicable to all tumor types, and individual sets of criteria may be necessary for disease groups and specific studies, especially if the specimen has been obtained from a patient undergoing treatment. The following discussion of interpretation criteria is based on the author's experience with a variety of solid tumors with an emphasis on breast cancer. For a more detailed discussion please refer to Dressler and Bartow (1989).

One of the most important criteria is coefficient of variation (CV) of the mean DNA content of the G_1 cell subpopulation. Its value corresponds to the ability to resolve two G_1 peaks in mixed-cell populations and helps to assess quality of sample preparation and instrument alignment. The larger the CV, the wider the peak and the greater the chance of missing an abnormal population that lies close to 2N, or the ability to resolve two abnormal populations. In addition, the wider the CV, the less accuracy will be achieved in trying to estimate the S-phase region of the histogram Therefore, criteria for CV are often set to allow for distinction between which histograms are evaluable and which are not. It is important to define how the CV is determined and it is essential to report CV values and ranges in all published studies. Histograms with CV values of < 5% are generally acceptable to most investigators (Dressler *et al.*, 1987).

Another important criterion for evaluating histograms, one that can be very controversial, is the definition of a DNA aneuploid peak. Many investigators do not report their definitions. Those investigators who have reported definitions for DNA aneuploid generally only consider an abnormal stemline present when two distinct G_0/G_1 populations are observed (Bauer *et al.*, 1986; Dressler and Bartow, 1989). In our laboratory, a peak is defined as being DNA aneuploid if at least 10% of the total events collected in the histogram are found in the G_0/G_1 peak and a corresponding G_2M can be identified. There are examples, however, of discrete, symmetrical nondiploid G_0/G_1 peaks being observed that comprise < 10% of the total events collected, that also have a corresponding G_2M. These cases are usually termed "questionable ploidy," and another sample is requested or the assay is repeated on a separate aliquot of the same sample to confirm ploidy interpretation.

An exception to this definition of DNA aneuploidy, in our laboratory, is the case of a tetraploid tumor, having a 4N amount of DNA. Because one cannot distinguish DNA diploid G_2M cells from DNA tetraploid G_0/G_1 cells in the 4N peak, a DNA tetraploid tumor is defined as having at least 15–20% of the total events collected in the 4N peak. Other investigators have defined a DNA tetraploid tumor as having 10% of cells in the 4N peak on the histogram (Baildam *et al.*, 1987). Still others do not include DNA tetraploid tumors as being among the DNA aneuploid group (Ewers *et al.*, 1984), but rather refer to any even multiple of 2N as being euploid (diploid, tetraploid, octaploid). Microscopic examination of the stained specimen at the time of FCM analysis to quantify the proportion of aggregated cells or nuclei is essential in interpreting tetraploidy in DNA histograms. In addition, some form of electronic doublet discrimination is also essential to more precisely define DNA tetraploidy versus clumps.

Overall, we use four classifications to describe the DNA histogram: DNA diploid, DNA aneuploid, questionable ploidy, and uninterpretable. "Questionable ploidy" includes (1) inability to resolve a shoulder; (2) a small discrete peak falling short of 10% of the total collected events; (3) a wide CV; or (4) a single G_0/G_1 peak, with a broadened G_2M (4N) region. "Uninterpretable" refers to a sample that is (1) all debris, (2) one that does not yield a sufficient number of cells for analysis, or (3) one in which very poor resolution was obtained. In these instances, we either try to repeat the assay if there is sufficient material available, or request another specimen (e.g., another paraffin block) to be submitted for assay. Specific criteria for evaluability that have been useful in our laboratory to provide consistency in interpretation are in Section III.

III. Methods

Cytospin Protocol for Cytologic Evaluation

1. Following tissue dissociation, place ~100,000 cells in suspension (50–100 μl) in a 12×75 mm tube.
2. Add an equal volume of 95% ethanol to fix samples.
3. Gently vortex the fixed sample and add 150–200 μl to the cytospin funnel.
4. Centrifuge for 5 minutes at 350 g.
5. Let air dry and follow routine staining procedure for cytologic specimens (hematoxylin and eosin, Giemsa, Wright, etc.).

Histologic Protocol

1. Cut one 5 μm section immediately preceding and directly following thick sections. Stain according to routine method.
2. Evaluate section for percentage of tumor versus nontumor. At least 15% of tissue section should be composed of tumor cells for quality assurance.
3. Evaluate for necrosis, hemorrhage, and degree of fibrosis.
4. If tumor is localized to one area, or both tumor and nontumor are present in separate areas on the same block, mark block with razor blade to outline specific areas of interest.
5. When microtoming, each "outlined" part of section can be placed in separate tubes and processed independently.

Fluorochrome Control

1. Immediately before running sample on flow cytometer, measure out 20 μl of sample and apply to glass slide as a wet preparation.
2. Add a small coverslip and seal edges with nail polish to prevent sample from drying out.
3. View under fluorescence microscope and record percentage of cells/ nuclei staining, intensity of stain (0–3; 3 highest), integrity of cells and nuclei (percentage broken or intact), degree of noncellular material staining (0–3; 3 highest), percentage of cells/nuclei in aggregates versus single-cell dispersion, and size of particles (small, medium, large).

DNA Reference Standards

To obtain a large number of reference aliquots, the author routinely uses unstimulated hPBL, obtained as whole blood from the local blood bank prepared as follows:

1. Dilute blood 1:1 with PBS containing 1% heparin (sodium salts; 1000 units/ml) in a sterile glass bottle.
2. Layer over Ficoll–Paque (four parts diluted blood to one part Ficoll–Paque), and centrifuge for 30 minutes at 750 g.
3. Carefully pipet off lymphocyte layer and wash cells in sterile Hanks' balanced salt solution (HBSS).
4. Centrifuge at 250 g for 5 minutes. Resuspend in 1 ml HBSS and count cells.
5. Resuspend cells in appropriate volume of freezing medium (Stone et al., 1985) to allow a concentration of 5×10^6 cells/ml to be frozen in 1.0-ml aliquots.

Freezing Medium

RPMI (or MEM, Medium 199) + 20% serum, 100 ml
NaCl, 1.765 g
Dimethyl sulfoxide (DMSO), 14.940 ml

1. Resuspend cells in cryopreservation medium. Concentration should not exceed 5×10^6 cells/ml for ease of FCM preparation.
2. Place vials containing cells in $-70°C$ freezer.
3. Keep sample at $-70°C$ until day of assay.

IV. General Criteria for Interpretation of DNA Histograms

A. Histologic Evaluation

1. Five-micron sections (top and bottom) or cytospin stained with hematoxylin and eosin must contain $\geq 15\%$ tumor cells; percentage of lymphocytes, degree of necrosis, fibrosis, or presence of hemorrhage should be recorded.
2. Control should be confirmed as being uninvolved with tumor.

B. Evaluation of DNA Fluorescent Stained Cells

The following characteristics are noted for each sample:
1. Percentage of cells/nuclei in a single-cell suspension versus in aggregates
2. Integrity of cells/nuclei (i.e., percentage broken vs intact)
3. Intensity of stain (1–3; where 3 is brightest)
4. Size and shape of particles
5. Percentage of debris in the sample

C. Ploidy Status

1. DNA DIPLOID

A single-peak tumor with G_0/G_1 CV $\leq 5.0\%$, and a corresponding G_2M peak is identified, whose diploid position is confirmed by mix of control with tumor specimen.

2. ANEUPLOID

This describes a histogram showing more than one discrete G_0/G_1 peak, where a corresponding G_2M peak can be identified and the percentage of events in the abnormal G_0/G_1 peak is $\geq 10\%$ of the total number of events collected. The CV of the DNA aneuploid G_0/G_1 peak is $\leq 5.00\%$ for S-phase analysis.

In hypodiploidy a discrete, abnormal G_0/G_1 peak is observed, where mix must identify one peak as normal. The abnormal G_0/G_1 peak should contain $\geq 10\%$ of all events collected.

In tetraploidy the 4N peak should contain 15–20% of all events collected and a corresponding G_2M population is identified (8N). Presence of cell aggregates (clumping) should be rigorously estimated. Extensive cell aggregation mimics presence of cells or nuclei with DI = 2.00. Doublet discrimination on the flow cytometer is also helpful. (It should be recognized that this value is somewhat arbitrary in that normal tissues including liver, bladder, thyroid, and testes contain significant fractions of tetraploid cells.)

3. PLOIDY STATUS UNINTERPRETABLE

Several different situations can result in uncertainty about ploidy status. Among them are the following:

1. An insufficient number of tumor cells is found on stained section or cytospin (i.e., <15%).
2. A poor-quality histogram, caused by excess debris or poor resolution, is obtained.

4. QUESTIONABLE PLOIDY

Conditions in this category include:

1. A single peak is obtained for the tumor, whose position cannot be confirmed in "diploid area" when control is added. (It should be noted that DNA diploid tumors can demonstrate small shifts in fluorescence relative to external diploid reference standards due to chromatin structure-related variations in DNA stainability independent of ploidy.)
2. An unresolvable shoulder.
3. A small discrete peak falling short of 10% of the total events collected.
4. Single-peak tumors with a wide CV (>5.00%).
5. A single G_0G_1 peak with a broadened G_2M region.

V. Daily Instrument Setup Procedures

1. Align instrument with fluorescent beads (or glutaraldehyde-fixed CRBC). Measure and record integrated red fluorescence (IRFL) and forward-angle light scatter (FALS). Coefficient of variation on both should be <2.00%.

2. Run external control (PBL nuclei for fresh or frozen samples). Adjust high voltage and/or gain to position G_0/G_1 peak in channel 60. On a linear scale (0–255) this will allow a tetraploid tumor (4N) and corresponding G_2M (8N) to be on scale. Ensure that CV is <3.5% (this is usually the case when the beads have been used for proper alignment in step 1).

3. Set threshold intensity (for fresh or frozen samples) with unfixed, stained CRBC nuclei. (CRBC should be found at approximately channel 20.)

4. Run external diploid control (hPBL for fresh or frozen samples) and collect 5000 events. For control and samples, record channel numbers for G_0/G_1 and G_2M peak(s), corresponding CV values, high-voltage setting, and ratio of G_2M peak position to the G_0/G_1 peak position (for both DNA diploid and DNA aneuploid populations). (This ratio gives a control for the stoichiometric relationship of the stain system and should be fairly consistent throughout the run using the same staining system.)

5. Following stabilization of sample flow rate, run the tumor specimen and collect 25,000–50,000 events, if possible. During collection of first 10,000 events, carefully monitor the following:
   a. Shoulders on peaks (debris vs off-center alignment)[1]
   b. Potential G_0/G_1 peaks observed in channels higher than 120 (G_2M events will be off scale)[2]

---

[1] If shoulders are observed, we stop data acquisition and then reacquire to ensure that we are not observing artifact due to baseline drift or off-center alignment. External standards or beads are run again to ensure proper alignment, if necessary. Acquiring a two-parameter histogram of fluorescence versus time in list mode can also assess possible drifting and hydraulic stability.

[2] If the specimen shows a potential aneuploid G_0/G_1 peak whose position is greater than channel 120, only 10,000 events are initially collected at this voltage. The voltage and/or gain are then lowered to adjust the position of the abnormal peak to channel 120 or lower and 25,000 events are collected at this lowered voltage. This adjustment allows the corresponding G_2M to be on scale and makes possible cell cycle analysis.

Acknowledgment

I thank Dr. Kenneth Bauer of Northwestern University, Chicago for his review of this chapter.

References

Baildam, A. D., Zaloudik, T., Howell, A. Barnes, D. M., Turnbull, L., Swindell, R., Moore, H., and Sellwood, R. A. (1987). *Br. J. Cancer* **55,** 553–559.

Bauer, K. D., Merkel, D. E., Winter, J. N., Harder, R. J., Hauck, W. W., Wallemart, C. B., Williams T. J., and Variakojis, D. (1986). *Cancer Res.* **46,** 3173–3178.

Darzynkiewicz, Z., Traganos, F., Kapuscinski, J., Staiano-Coico, L., and Melamed, M. R. (1984). *Cytometry* **5,** 355–363.

Dressler, L. G., and Bartow, S. A. (1989). *Semin. Diagn. Pathol.* **6,** in press.

Dressler, L. G., Seamer, L., Owens, M. A., Clark, G. N., and McGuire W. L., (1987). *Cancer Res.* **47,** 5294–5302.

Ewers, S. B., Langstorm, E., Baldetorp. B., and Killander, D. (1984). *Cytometry* **5,** 408–419.

Hiddeman, W., Schumann, J., Andreeff, M., Barlogie, B., Herman, C. J., Leif, R. C., Mayall, B. H., Murphy, R. F., and Sandberg, A. A. (1984). *Cytometry* **5,** 445–446

Iverson, O. E., and Laerum, O. D., (1987). *Cytometry* **8,** 190–196.

Jakobsen, A. (1983), *Cytometry* **4,** 161–165.

Stone, K. R., Craig, B. R., Palmer, J. O., Rivkin, S. E., and McDivit, R. (1985). *Cytometry* **6,** 357–361.

Vindeløv, L. L., Christensen, I. J., and Nissen, N. I. (1983). *Cytometry* **3,** 323–327.

Chapter 18

Cell Division Analysis Using Bromodeoxyuridine-Induced Suppression of Hoechst 33258 Fluorescence

RALPH M. BÖHMER

Ludwig Institute for Cancer Research
Melbourne Tumour Biology Branch
Melbourne, Victoria 3050
Australia

I. Introduction

When offered in sufficient concentration to cells, bromodeoxyuridine (BrdUrd) competes with thymidine for incorporation into newly synthesized DNA. If such BrdUrd-substituted cells are stained with the thymidine specific fluorescent dyes Hoechst 33258 (HO258) or Hoechst 33342 (HO342), their fluorescence is partly suppressed and no longer proportional to DNA content (Latt, 1973, 1977; Latt *et al.*, 1977). At a certain level of BrdUrd incorporation, combined with a certain dye concentration, this fluorescence suppression (quenching) is of such an extent that, after the beginning of BrdUrd incorporation, the fluorescence of a proliferating cell stays exactly constant in spite of the increase in DNA content during S phase, until, at division, the fluorescence is halved. This effect can be exploited to determine all cycle phase durations in exponentially proliferating cell cultures, the kinetics of division after stimulating quiescent cells, and the fractions of non proliferating cells. All these data can be obtained by single-parameter flow cytometry (FCM) using one of the Hoechst (HO) dyes (Böhmer, 1979, 1982; Beck, 1981a,b; Nüsse, 1981; Rabinovitch, 1983).

After BrdUrd incorporation, the HO fluorescence is no longer proportional to cellular DNA content, so that the single-parameter staining

METHODS IN CELL BIOLOGY, VOL. 33

with HO does not yield information about the cell's actual location within the cell cycle. This additional information can be obtained by staining the cells simultaneously with a DNA-specific dye that is not subject to the BrdUrd-induced quenching. The most useful dyes for this purpose are ethidium bromide (EB) and propidium iodide (PI) (Latt, 1977; Böhmer and Ellwart, 1981a,b; Nogushi *et al.*, 1981). Bivariate FCM analysis based on staining with HO and EB can provide complete information about the cell cycle progression within the time interval between the beginning of BrdUrd incorporation and cell harvest (Böhmer and Ellwart, 1981a, b; Noguchi *et al.*, 1981; Ellwart *et al.*, 1982; Böhmer, 1982; Kubbies and Rabinovitch, 1983).

II. Application

The BrdUrd-HO-EB technique has proved itself a most efficient tool for studying the time course of cell division following mitogenic stimulation of quiescent cells. The biochemical events that are involved in the control of proliferation appear to be located in the G_0 and G_1 phases of the cell cycle, whereas the completion of S phase and mitosis appears to proceed by constant schedules that are affected only by cytotoxic conditions. For the study of growth control mechanisms, the kinetics of entry into mitosis are therefore providing the same information as the kinetics of entry into S phase. Entry into S phase can be studied using incorporation of tritiated thymidine followed by autoradiography, or, flow cytometrically, by labeling the cells with BrdUrd-specific antibodies after BrdUrd incorporation. Compared with these techniques, the one-step staining with HO–EB is simple and fast, facilitating the extensive screening of variables that might affect the stimulation process. Apart from the practical advantages, the use of mitosis as a cell-kinetic marker is superior in terms of precision because mitosis is a discrete event, while the accumulation of tritiated thymidine or BrdUrd at the beginning of the S phase occurs gradually.

When the HO–EB staining is applied to cell populations that were already proliferating (with cells in all phases of the cycle) at the beginning of BrdUrd treatment, the histogram sequences contain information about the duration of all cell cycle phases, including the rate of DNA synthesis in all parts of the S phase. Unfortunately, the quantitative analysis of these variate data requires complex computer software, which is not yet commercially available. The quantitative evaluation of single-parameter histograms (HO staining alone) for the analysis of exponentially proliferating populations during BrdUrd exposure has been described by several authors (Beck, 1981a,b; Nüsse, 1981; Böhmer, 1979, 1982). An equivalent alternative for bivariate HO–EB histograms is based on the use of BrdUrd-specific antibodies (see Dolbeare *et al.*, Chapter 21, this volume). Although the latter technique requires evaluation programs of similar complexity, it has gained a greater popularity than the HO–EB technique, because such programs have been made available, and possibly also because the relatively difficult antibody labeling presents a more adequate challenge to the skilled researcher. Therefore, this article will deal only with the simple analysis of cell division kinetics following mitogenic stimulation of quiescent cell populations. Modifications and other applications of the technique are described in Poot *et al.*, Chapter 19, and Crissman and Steinkamp, Chapter 20 this volume.

III. Materials

A. Stock Solution of BrdUrd

Dissolve 100 mg of 5-bromodeoxyuridine together with 30 mg of deoxy-cytidine (dC) in 10 ml distilled water (dH_2O) at room temperature (RT) (concentration 10 mg/ml BrdUrd and 3 mg/ml dC). Pass through steriliz-ing filter. Prepare aliquots and store in freezer at −20°C. Do not expose BrdUrd solution to more light than necessary for handling. The stock solution seems to deteriorate after a few months in the freezer and there-fore should not be stored for extended periods of time. When thawing an aliquot of stock solution, inspect for the presence of any precipitate. Warm up to 37°C and shake well to dissolve it.

B. Dye Solution for Cell Staining

Dissolve 20 mg of HO258 together with 100 mg of EB in 10 ml of H_2O (concentration 2 mg/ml HO258 and 10 mg/ml EB). Prepare aliquots and store in freezer. Deterioration of this stock solution has not been observed. To make the final dye solution, mix 250 ml of pure ethanol with either 750 ml of phosphate-buffered saline (PBS; without calcium and magne-sium) or 750 ml of Tris-HCl buffer (pH ~7.3). Add 1 ml of the dye stock solution to make a final dye concentration of 2 μg/ml HO258 and 10 μg/ml EB. These concentrations have to be exact. This solution can be kept for several months in a brown bottle at 4°C. A 1-ml dispenser on the bottle is most useful.

IV.   BrdUrd Incorporation and Cell Handling

A.   BrdUrd Exposure

Supplement the growth medium with 10 μg/ml BrdUrd (plus 3 μg/ml dC). This BrdUrd concentration was found to be appropriate for cell types from various species and tissue origins. Few batches of fetal calf serum (FCS) have been found that, when used at the usual concentration of 10% in the growth medium, reduced BrdUrd incorporation levels. This could be countered by increasing the BrdUrd concentration in the medium to 15–20 μg/ml. The speed of cell cycle progression in the first round of BrdUrd incorporation is not affected at the incorporation levels required for this technique.

Depending on the experiment, either add BrdUrd stock solution to the medium in which the cells already are, or make up the BrdUrd-supplemented medium first, then wash the cells and incubate in the new medium.

B.   Cell Harvest

1.   SUSPENSION CULTURES

If sequential harvests from one suspension culture are to be made, resuspend the sedimented cells well (make sure no cells stick to the flasks) and take out an appropriate aliquot of the suspension. Approximately 10^5 cells are enough for more than one good histogram. Avoid cell-kinetic perturbation of the remaining cell culture by keeping the time outside the incubator to a mininum (minimize light exposure and drop in temperature). Some types of suspension cultures show extensive cell aggregation (e.g., phytohemagglutinin-treated lymphocytes) or are affected by intercellular contacts. In such cases, it may be favorable to prepare aliquots of the suspension at the time of stimulation and harvest a whole culture well at each time point.

2.   MONOLAYER CULTURES

Remove all growth medium carefully and add just enough trypsin–EDTA solution to cover the monolayer. Some types of cells require quick prerinse with only EDTA, but make sure this does not prematurely cause cell detachment. The speed of cell detachment during trypsin treatment varies with the growth conditions and with the cell cycle status of the cells.

Therefore, do not rely on predetermined trypsinization times, but always watch the detachment process in the microscope. When the detachment process is completed, add PBS (without calcium and magnesium), supplemented with 10% serum to stop the action of trypsin.

C. Cell Staining

Centrifuge the suspensions, decant the supernatants, and drain the tubes by placing them upside down on a tissue. Add 1 ml of dye solution, and shake the tubes vigorously and immediately to minimize cell aggregation. Minimum staining time before FCM analysis is 30 minutes. If the cells are to be analyzed within a few hours after staining, they should be left at RT (in the dark). When stored at 0–4°C, the cell samples do not deteriorate within 1 month, so that the probes from several experiments can be collected and analyzed together at a convenient time. This is the advantage of the recommended staining solution over an alternative one using the dyes HO342 and EB–PI in PBS with 0.1% Triton X-100. Although the latter staining yields slightly sharper histograms as early as 5 minutes after staining, the samples deteriorate rapidly and cannot be stored for later analysis.

V. Flow Cytometry

A. Instrumentation

Any flow cytometer that is equipped with an ultraviolet (UV) light source for dye excitation, can measure fluorescence emission in two different spectral ranges, and allows bivariate data analysis, can be used. Illumination may be provided by the UV lines of an argon or krypton laser or the light of a mercury arc lamp, filtered with a short-pass or bandpass filter transmitting only below 400 nm. The blue fluorescence of the HO dye is selected using either the combination of a 400-nm long-pass filter and a 480-nm short-pass filter, or a 440-nm bandpass filter. The red fluorescence of EB–PI is selected by a 610-nm long-pass or 610-nm bandpass filter. Care must be taken to use the appropriate dichroic mirrors (beam splitters) in all positions of the optical configuration with respect to their wavelengths of transmission and deflection. Errors in the optical configuration were identified as the most common problem for novice users of this technique.

A data handling system is required that can display and store bivariate histograms (scattergrams). The system must further provide the possibility to define several regions in the bivariate histograms and to determine the number of particles recorded in these regions. Rectangular regions parallel to the axes are sufficient in many cases, but sometimes the histograms require regions that are tilted. Elliptical regions that can be set in any orientation and shape are ideal for a precise quantitation of cluster contents. Some flow cytometers offer the possibility to distort electronically (compensate) the histogram shape and orientation, based on the subtraction of fluorescence values. This can be used to make some histograms suitable for less flexible region setting programs.

B. How to Run the Probes?

The staining solution contains 25% ethanol, and this appears to be the cause of two problems that have to be observed. During the first 30 seconds after starting to run a probe, the red and blue fluorescence is shifting continuously, and the recording must not be started before the clusters have stabilized in one position. This shift is presumably due to dilution of the ethanol concentration by the sheath fluid remaining in the tubing after backflush between the probes. The other problem is a blurring of the histograms at higher running speeds. To achieve sharp histograms, the probes have to be run very slowly. The magnitude of this effect is different for different flow cytometers and depends on the characteristics of the flow chamber and the excitation optics. It can be somewhat reduced by slightly defocusing the exciting light beam. Generally, the alignment of the flow cytometer should not be optimized with fluorescent plastic beads but with cells in the dye solution at cell densities similar to those in the probes to be analyzed.

C. Numerical Standard

The total number of cells in the samples can be determined by the ratio of cells to reference particles added to the suspensions in known amount. The criterion for the choice of reference particles is their location in the histograms. The cluster of reference particles (including the clusters of doublets, triplets, etc.) must not overlap with the clusters of cells. Chicken erythrocytes (CRBC) can be used as reference particles for human cells, as long as the erythrocytes are free of doublets. They should be fixed and prestained in the standard HO–EB dye solution with 25% ethanol, kept at a density of $\sim 10^6$ cells/ml and added 1 part in 10 parts to the cell suspensions. The frequency of erythrocyte doublets becomes significant after longer storage periods. Then fresh erythrocytes must be prepared. Ultraviolet-excitable fluorescent plastic beads (e.g., Fluoresbrite, carboxy, 4.54 μm, Polysciences) are suitable for all kinds of cells. Their blue fluorescence is higher than the one of the cells, and their red fluorescence much lower. Care must be taken with the various electronic thresholds of the flow cytometer, to ensure that neither beads nor cells are lost under or above a threshold. Beads should be added to the cell suspension immediately before analysis, since they tend to stick to the tubes.

VI.   Data Analysis and Results

The analysis described here applies only to the division kinetics of growth-stimulated cultures where all cells were at 2C DNA content at the time of BrdUrd addition. In this case, each dividing cell has incorporated BrdUrd during the whole S phase, which leads to distinct clusters of cells before and after division, without intermediate fluorescence values that would require a more complex histogram evaluation, as mentioned earlier.

The histogram evaluation is demonstrated in Fig. 1, which shows examples from a series of histograms obtained after stimulating quiescent normal human skin fibroblasts with different concentrations of epidermal growth factor (EGF). The first histogram (A) was recorded 26 hours after exposing the cells to a concentration of EGF (10 ng/ml) that induces their maximal mitogenic stimulation. The majority of cells has not yet divided and the cells are distributed over the G_1, S, and G_2/M phases of the cycle. This distribution is made visible by the DNA-proportional EB fluorescence (ordinate), which is not subject to BrdUrd-induced fluorescence quenching. All these cells (designated ND on the abscissa) have the same HO fluorescence. A small proportion of cells have already divided (designated D on the abscissa), and this has halved both fluorescence values.

The fraction of cells that have divided is determined from the cluster contents D and ND as

$$f_D = \frac{(D/2)}{(D/2 + ND)}$$

These values are plotted in Fig. 1B as a function of time after EGF stimulation for various concentrations of the mitogen.

The other two histograms (Fig. 1A; panels II and III) are derived from cultures stimulated with suboptimal concentrations of EGF (0.2 and

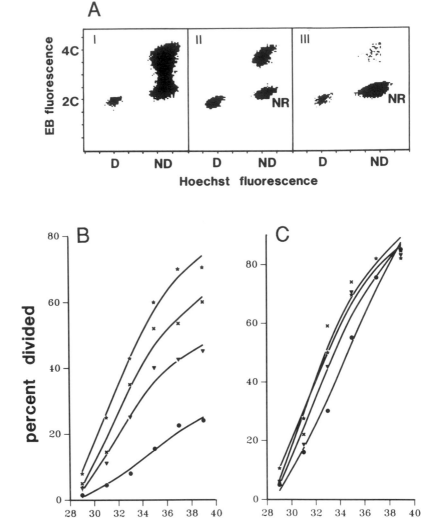

FIG. 1. (A) Bivariate histograms obtained at different times after mitogenic stimulation with different concentrations of epidermal growth factor (EGF). (I) 26 hours, 10 ng/ml EGF; (II) 35 hours, 0.2 ng/ml EGF; (III) 35 hours, 0.03 ng/ml EGF. (B) Time course of cell division following stimulation with various concentrations of EGF (ng/ml): *, 10; **X**, 0.12; ▼, 0.06; ●, 0.015. Proportions of divided cells are normalized on the entire cell population (see text). (C) As in (B), but with the proportion of divided cells normalized on only the population of responding cells (see text). The curves in (B) and (C) are fitted to the data points with an approximation algorithm using a cubic spline.

0.03 ng/ml, respectively) and harvested after 35 hours. Both histograms show a lack of cells in S phase. Part of the cells that have not divided (ND) are in G_2/M (i.e., shortly before mitosis), and part are at 2C DNA content. From the lack of cells in S phase one can conclude that the flux from G_1 into S had ceased several hours before the time when the cells were harvested. Therefore, these cells remaining at 2C DNA content can be considered as cells that did not respond to the mitogenic stimulus (designated NR, nonresponders). To investigate the kinetics of division of the responding cell population, the nonresponders have to be subtracted from the population of nondivided cells; that is, the percentages of divided cells have to be determined by normalizing on the responding population instead of the whole population. The fraction of nonresponders is determined from the histograms as

$$f_{NR} = \frac{NR}{[ND + (D/2)]}$$

Since f_{NR} is slightly variable from culture to culture for any one mitogen concentration, it should be averaged from histograms made at times when a discrete cluster NR becomes quantifiable as a result of the lack of cells in S phase. This mean value $\overline{f_{NR}}$ can then be used to renormalize the fractions of divided cells at all time points.

$$f_D = \frac{(D/2)}{[(D/2) + ND)]} \times (1 - \overline{f_{NR}})$$

The division kinetics of the responding population is plotted in Fig. 1C. One can see that the slopes of all curves are identical and that there is also no significant time shift. This result implies that the only effect of suboptimal concentrations of the mitogen EGF is a reduced proportion of responders in the population. These data, contrary to prevailing concepts of probabilistic growth control, indicate that the reduced mitogen concentration does not result in reduced rates of entry into the cell cycle within the responding cell population.

Total cell numbers in a probe can be determined either by Coulter counter or by a numerical standard in the histogram, as described earlier. Since the proportion f_D of divided cells is known from the histogram, the absolute cell number in the probe can be compensated for cell division:

$$N_{corr} = N - 0.5 f_D N$$

These division-corrected cell numbers must remain constant over the period of investigation unless cell loss occurs, which can thus be quantitated by comparison with the cell numbers at the start of the experiment.

VII.   Critical Aspects of the Method

A.   Cell Aggregation

The occurrence of cell aggregation represents a problem for the cell division analysis. This is because the aggregate of two divided cells with 2C DNA content gives the same signal as one cell in the G_2/M phase. This leads to an overestimation of the nondivided cell fraction. The degree of cell aggregation varies with type of cell culture and the culture conditions, and it was found difficult to dissociate the aggregates completely using the standard staining technique formerly described. The degree of cell aggregation can be judged by the occurrence of cell triplets and quadruplets in the histograms. Solution: preparation of naked nuclei by standard enzymatic techiques prior to staining with HO–EB, or doublet discrimination by pulse-width analysis, as available in some types of flow cytometers.

B.   Second-Round Cell Division

In some cases of mitogenic stimulation experiments, a proportion of cells goes through a second cell cycle before all stimulated cells have undergone their first division. The data evaluation principles as described before require that the numbers of cells after mitosis be divided by two, and, accordingly, the numbers of cells that have undergone division twice have to be divided by four. In the second cell cycle the further fluorescence suppression by BrdUrd is only weak despite the continued presence of BrdUrd. Therefore, mitotic cells in the second round appear at the same fluorescence as mitotic cells in the first round of division, and, accordingly, those cells that have divided twice cannot be distinguished from those divided once, and their numbers will be divided by two instead of four. Beginning at the time when second-round mitosis occurs, this leads to a gradual underestimation of the proportions of nondivided or nonresponding cells. The solutions include the following.

1.   Design experiments in such a way that analysis can be discontinued at the time when second-round mitosis occurs. Since cells in second S phase can be recognized in the bivariate histograms (Böhmer and Ellwart, 1981a), it is possible to estimate when second cell division becomes a significant factor.

2.   A variation of the BrdUrd–HO method has been proposed (Kubbies and Rabinovitch, 1983) that uses higher BrdUrd incorporation levels and lower concentrations of the HO dye to achieve a small additional fluores-

cence suppression the second S phase. This method achieves a distinction between first and second division, but the distinction was found to be only marginal, and an extremely good histogram resolution (machine performance) is required for quantitative analysis. For standard routine applications, this technique appears less recommendable, because the histograms look complicated and a misinterpretation of the various cell clusters is possible.

C.   Selective Cell Loss

Like most cell-kinetic techniques, the FCM analysis is based on the quantitation of relative cell numbers, in our case the relative amounts of divided and nondivided cells. These relative numbers become meaningless when nonrandom cell loss occurs. To give a hypothetical example: under certain conditions, only a small proportion of a quiescent population responded to the mitogenic stimulus, but all the nonresponsive cells died, disintegrated, and consequently did not appear in the histogram. Then the histograms will show 100% stimulated cells, which is obviously not what really happened. Therefore, the occurrence of cell loss needs to be monitored carefully. This can be done using the division-corrected total cell counts as described earlier, since, after compensation for cell division, a decrease in cell numbers reflects the cell loss. The cell division data can be corrected for such cell loss when some conditions are fulfilled.

The loss must be known to occur from one particular group of cells only; for example, either only the divided or only the nondivided cells are lost. It will normally be very difficult to establish that such an assumption is valid.

The culture-to-culture variation in absolute cell numbers at the start of the experiment must be very small. This can be achieved by observing the following points:

1. Accurate pipetting at the stage of cell seeding.
2. An even distribution of cells on the culture substrate must be achieved, because local cell crowding will lead to a variation in culture proliferation after seeding. In most types of small culture vessels, cell crowding occurs because of vibrations in the incubator. Therefore, it is advisable to use an incubator with convection fan switched off for the first hours after seeding until the cells have adhered.
3. When serum deprivation is used to achieve quiescence, the cultures should be washed once before incubation in the quiescing medium, since variations in the amounts of residual serum may lead to variations in the degree of proliferation before the onset of quiescence.

The following most common causes for cell loss have been identified:

1. Incomplete cell harvest from monolayers. This can occur because the speed of cell detachment varies with the growth conditions and the cell cycle status. Solution: monitor the cell detachment process strictly in every flask.

2. Detachment of part of the population from the monolayer during the period of investigation. In most cases where this effect was encountered, the detaching cells were dead and, though collected in the harvesting process, could not be stained to appear in the DNA histograms. In some types of adherent cell cultures this detachment seems to occur as a regular element of the growth control process, and such cultures require a sophisticated approach to cell-kinetic studies.

3. Cell adherence to the sedimentation area in suspension cultures. Solution: before collecting an aliquot of suspension, ensure by vigorous pipetting that all cells are in suspension.

4. Cell death in suspension cultures: many growth conditions that cause cell cycle perturbance also cause cell death, and it was found that dead cells, though seemingly intact based on microscopic inspection, took up much less dye after fixation than those cells that were alive at the time of fixation. Thus, they were not recorded in the DNA histograms. This effect was found for all standard methods of DNA-specific staining, and may be due to DNA degradation following death in culture. A general solution for this problem is not available.

Wherever cell loss occurs, the analysis of FCM data must be based on both relative and absolute cell counts, and in many cases meaningful cell-kinetic information can only be obtained with the help of model calculations incorporating both absolute and relative cell numbers as well as assumptions about the cell cycle stage in which cells are lost.

REFERENCES

Beck, H. P. (1981a). *Cytometry* **2,** 170.
Beck, H. P. (1981b). *Cell Tissue Kinet.* **14,** 163.
Böhmer, R. M. (1979). *Cell Tissue Kinet.* **12,** 101.
Böhmer, R. M. (1982). *Prog. Histochem. Cytochem.* **14** (4).
Böhmer, R. M., and Ellwart, J. (1981a). *Cytometry* **2,** 31.
Böhmer, R. M., and Ellwart, J. (1981b). *Cell Tissue Kinet.* **14,** 653.
Ellwart, J., Böhmer, R. M., and Dörmer, P. (1982). *Exp. Cell Res.* **139,** 111.
Kubbies, M., Rabinovitch, P. S. (1983). *Cytometry* **3,** 276–281.
Latt, S. A. (1973). *Proc. Natl. Acad. Sci. U. S. A.* **70,** 3395.
Latt, S. A. (1977). *J. Histochem. Cytochem.* **25,** 913.
Latt, S. A., George, Y. S., and Gray, J. W. (1977). *J. Histochem. Cytochem.* **25,** 927.
Noguchi, P. D., Johnson, J. B., and Browne, W. (1981). *Cytometry* **1,** 390.
Nüsse, M. (1981). *Cytometry* **2,** 70.
Rabinovitch, P. S. (1983) *Proc. Natl. Acad. Sci. USA* **80,** 2951–2955.

Chapter 19

Cell Cycle Analysis Using Continuous Bromodeoxyuridine Labeling and Hoechst 33258–Ethidium Bromide Bivariate Flow Cytometry

MARTIN POOT AND HOLGER HOEHN

Department of Human Genetics
University of Würzburg Biocenter
Am Hubland, 87 Würzburg
Federal Republic of Germany

MANFRED KUBBIES

Boehringer Mannheim Research Center
8122 Penzberg
Federal Republic of Germany

ANGELIKA GROSSMANN

Department of Comparative Medicine
University of Washington
Seattle, Washington 98195

YUHCHYAU CHEN AND PETER S. RABINOVITCH

Department of Pathology
School of Medicine
University of Washington
Seattle, Washington 98195

METHODS IN CELL BIOLOGY, VOL. 33

I.   Introduction

A limitation shared by current cell cycle techniques is that continued growth beyond a single cell cycle cannot be discriminated. In conventional, univariate DNA flow cytometry (FCM) this is because the G_1 product of mitosis is indistinguishable from the G_0 cell that has never entered the S phase. The newer bivariate techniques using acridine orange (Darzynkiewicz *et al.*, 1976) or 5′-bromodeoxyuridine (BrdUrd) antibody staining (Gray and Mayall, 1985) distinguish replicating from resting cells, but so far they cannot enumerate the times a given cell has replicated within a given observation period.

Since the mid-1980s, our laboratories have designed modifications of the BrdUrd–Hoechst dye (BrdUrd–HO) quenching technique that was originally developed by Latt for mitotic cells (Latt, 1973), and applied to interphase cells and FCM by Latt (1977), Böhmer (1979), Nuesse (1981), Beck (1981), and others. These modifications consisted of careful titrations of BrdUrd and HO concentrations in relation to cell types and cell densities, and in the definition of staining conditions that would yield highly reproducible bivariate flow cytograms of a variety of human and murine cell cultures, both primary and permanent. The unique feature of these cytograms is that they permit the tracing of dividing cells throughout multiple cell cycles. Moreover, a number of kinetic parameters that characterize replicative heterogeneity and cell cycle traverse can be derived by analyzing temporally spaced samples of a growing cell culture.

II.   Application

Applications to date have ranged over several cell types, examining questions in both cell biology and clinical medicine. The method was initially applied to human diploid fibroblast (hDF) and lymphocyte cultures (Rabinovitch, 1983; Kubbies and Rabinovitch, 1983). With hDF cultures, it was shown that their *in vitro* growth rate is regulated both by a noncycling cell fraction and by the probability by which resting cells are recruited into the cell cycle (Rabinovitch, 1983). The mitogen response of human peripheral blood lymphocytes (hPBL) was studied as a function of cell culture conditions and donor age (Kubbies *et al.*, 1985a,b); contrary to previous reports, striking differences that relate to donor age were found only between prepubertal and postpubertal donors (Schindler *et al.*, 1988). Currently, a number of highly specific cell cycle lesions are being defined in human genetic disorders whose common denominator is a disturbance of cellular proliferation. For example, the cell-kinetic lesion in Fanconi's anemia (FA) consists of accumulations of cells within the G_2 compartments of the first and second cell cycles after activation (Kubbies et al., 1985c; Schindler et al., 1985, 1987a). Strongly elevated G_2-phase accumulations are also displayed by ataxia telangiectasia (AT) cells in response to X irradiation. In both conditions, the quantitative assessment of cell blockage in the G_2-phase compartment via the BrdUrd–HO technique has proved useful as a clinical test (Schindler et al., 1985, 1987b). Characteristic cell cycle alterations have also been found in Bloom's syndrome (Poot et al., 1989). In addition to the definition of specific cell cycle defects, the continuous BrdUrd-labeling method permits a highly sensitive in vitro monitoring of growth factor effects (Kubbies et al., 1987a), cytotoxic agents (Vogel et al., 1986; Poot et al., 1988a), and cellular differentiation (Seychab et al., 1989; Poot et al., 1990).

III.   Materials

A.   Cell Cultures

The diverse cell types that are amenable to the assay are illustrated by the following examples. Human peripheral blood mononuclear cells were obtained by Ficoll–Hypaque centrifugation from anticoagulated venous blood. Murine splenic B lymphocytes are separated through a Percoll gradient from finely minced spleens. The source of DF-like cells were skin biopsy samples or second-trimester amniotic fluids. Mouse primary and spontaneously transformed mesenchymal cells were from explants of lung and kidney. NIH-3T3 cells, the murine lymphoid CTLL line, and human Epstein–Barr virus (EBV)-transformed cell lines were obtained from public cell banks. With the exception of PBL and murine splenic B-lymphocytes (which are naturally quiescent), all other cell types need to be rendered quiescent prior to exposure to mitogens and BrdUrd in order to examine the kinetics of mitogen stimulation in the manners shown subsequently. Table I lists the commonly used cell types, their culture conditions, and the protocols used to induce proliferative quiescence. Also shown in Table I are the conditions under which growth was induced in a particular culture (i.e., plating density, types of mitogens, BrdUrd concentration). These conditions must be carefully observed. It is recommended that all cultures be grown in commercial tissue culture-grade plastic ware; the cultures must be shielded from light by aluminum foil wrappings; incubation is at 37.5°C in humidified incubators gassed with 5% CO_2 and air.

TABLE I

CELL CULTURE AND STAINING CONDITIONS EMPLOYED FOR BIVARIATE BrdUrd–HO–EB FLOW CYTOMETRIC ANALYSIS OF SIX PROTOTYPE CELL TYPES[a]

				Cell type		
Condition	hPBL	hDF	EBV-LCL	NIH-3T3	M-BCL	M-CTLL
Cell culture media	RPMI-1640, 16% FBS, 20 μM αTG	MEM, 10% FBS	RPMI-1640, 10% FBS	MEM/HAM-F10, 50:50, 10% FBS	RPMI-1640, 16% FBS	RPMI-1640, 10% FBS, 20 U/ml IL-2
Quiescence	Natural	48 Hours 0.1% FBS	4 Days unfed	24 hours 0.5% FBS	Natural	10% FBS, 0.2 U/ml IL-2
Plating density	10^4 cells/ml	2.5×10^3 cells/cm^2	5×10^4 cells/ml	2.5×10^3 cells/cm^2	10^4 cells/ml	10^4 cells/ml
Mitogen	PHA, CD3	10% FBS	10% FBS	10% FBS	LPS + anti-μ	IL-2
BrdUrd	100 μM	65 μM	100 μM	100 μM	100 μM	65 μM
dCyt	—	65 μM	100 μM	100 μM	—	—
HO258	1.2 μg/ml	1.2 μg/ml	1.2 μg/ml	1.2 μg/ml	1.2 μg/ml	1.2 μg/ml
EB	1.5 μg/ml	2.0 μg/ml	2.0 μg/ml	2.0 μg/ml	1.5 μg/ml	1.5 μg/ml
Staining density	4×10^5 cells/ml	5×10^5 cells/ml	10×10^5 cells/ml	5×10^5 cells/ml	4×10^5 cells/ml	4×10^5 cells/ml

[a] hPBL, Human peripheral blood lymphocytes; hDF, human diploid fibroblastlike cells; EBV-LCL, human Epstein–Barr Virus-transformed lymphoblastoid cell lines; NIH-3T3 cells, murine 3T3 cells (NIH strain); M-BCL, murine splenic B lymphocytes; M-CTLL, murine interleukin 2-dependent T-lymphocyte cell line; FBS, fetal bovine serum; αTG, α-thioglycerol; IL-2, interleukin 2; PHA, phytohemagglutinin; LPS, lipopolysaccaride; anti-μ, anti-B-cell differentiation antigen.

B. Cell Culture Harvest

At the desired times after exposure to mitogens in the presence of BrdUrd (usually beginning at 20–24 hours and ending between 72 and 96 hours), adherent cell cultures are harvested by standard trypsinization. Suspended cells are pipetted briefly, transferred to 15-ml centrifuge tubes, pelleted at 400 g for 10 minutes and resuspended in 10 ml culture medium containing 10% fetal bovine serum (FBS) and 10% dimethyl sulfoxide DMSO. The samples are then stored at $-20°C$ to $-40°C$ in the dark until analysis. Storage times of as long as 2 years have not noticeably affected the quality of the FCM analysis.

IV. Procedures

A. Staining Solution

The staining solution consists of (final concentration):

> 100 mM Tris, pH 7.4
> 154 mM NaCl
> 1 mM CaCl$_2$
> 0.5 mM MgCl$_2$
> 0.1% Nonidet P-40 (NP-40)
> 0.2% bovine serum albumin (BSA)
> 1.2 μg/ml Hoechst 33258 (HO258)

Batches of 100 ml of this solution are prepared weekly from 100× concentrated stock solutions in deionized water. This working solution is stored in dark glass bottles at 4°C and can be filtered immediately prior to use if necessary.

B. Staining Procedure

The staining procedure consists of two parts. First, the thawed cells are pelleted at 400 g for 5 minutes, after which they are resuspended in ice-cold staining solution, and incubated during 15 minutes at 4°C in the dark. The cell concentration during staining is very important; it may range from 4–8 × 10^5 cells/ml depending on cell type (see Table I). Second, from a 100× concentrated stock solution, ethidium bromide (EB) is added to a final concentration of 1.5 or 2.0 μg/ml, depending on cell type (see Table I). After a further 15 minutes in the dark at 4°C, samples are ready for analysis by FCM; typical flow rates are 200–500 cells/seconds.

C.   Critical Aspects

1. A very critical aspect of the continuous BrdUrd-labeling assay is the cell culture methodology. Cells must be in good condition. Otherwise, cellular debris with liberated proteases and nucleases will cause cell clumping and confusing bivariate analysis patterns. Poorly controlled cell culture conditions and use of suboptimal-quality mitogens (such as growth factors, serum, PHA, etc.) are the most likely reasons for poor proliferation.

2. At least as critical as optimal cell culture is the avoidance of short-wavelength light from the time of first exposure to the halogenated base analog. Careful attention must be paid to this requirement throughout the entire harvesting, freezing, and staining steps. These procedures must be carried out in the dark with minimal lighting by a red darkroom lamp. BrdUrd-substituted DNA is extremely sensitive to short-wavelength light, and any such exposure will lead to suboptimal dye binding, chromatin damage, and nuclear decay.

3. A number of factors bear on the degree of the desired quenching of the HO258 fluorochrome (Kubbies and Rabinovitch, 1983); (1) the density at which cells are grown and exposed to the base analog is critical. Plating densities of 2.5×10^3 cells/cm^2 for adherent cultures, and of 2×10^5 cells/ml for suspension cultures should not be exceeded (see Table I). Higher seeding densities, particularly in the case of fast-growing cells, will lead to rapid depletion of BrdUrd. (2) Presumably because of variable amounts of nucleotide precursors, different batches of sera may affect the efficacy of HO dye quenching. (3) The final HO and EB dye concentrations given in Table I must be carefully observed; we also recommend that fluorochromes from different commercial sources be tested side by side in order to screen out inferior dye batches. (4) As one would expect from the AT-base pair affinity of the HO dye, both the AT/GC base pair ratio and the interspersion pattern will affect the quenching efficiency of BrdUrd-substituted chromatin. AT-rich genomes produce a greater quenching effect than GC-rich genomes (Kubbies and Friedl, 1985).

4. BrdUrd is cytotoxic. The continuous exposure to halogenated nucleoside analogs may entail a number of adverse effects on cellular functions (e.g., metabolic changes due to alterationss in the balance of nucleotide pools, direct DNA damage, alterations in DNA–protein interaction). Directly or indirectly, these effects could impair proliferation. Our experience shows that proliferation is not measurably affected at BrdUrd concentrations in the order of 65–100 μM in two frequently used cell types: hPBL and NIH-3T3 cells. Provided that they do not belong to a subline deficient in thymidine kinase, NIH-3T3 cells were found to tolerate BrdUrd concentrations up to 300 μM before they show signs of growth inhibition. On

the other hand, human lymphoblastoid cell lines and hDF are sensitive to BrdUrd. Fibroblasts experience a 10–15% reduction of 72-hour cell counts when exposed to 10 μM BrdUrd, but no further reduction occurs in the interval between 10 and 65 μM. The BrdUrd sensitivity of hDF and of EBV-transformed lymphoblastoid cells is expressed as arrest in the G_1 compartments of the second and third cell cycles after activation, but not in the G_0 to S-phase transition. Using hypoxic (5% v/v) rather than atmospheric oxygen cell culture conditions, these arrests can be minimized. Incorporation of BrdUrd apparently sensitizes cultured fibroblasts toward the growth-inhibitory effects of ambient and elevated oxygen (Poot *et al.*, 1988b).

V.   Controls, Standards

Given the complex type of information yielded, and the high level of resolution required by BrdUrd–HO FCM, several standards have to be included to avoid artifacts of various types. First, regarding a possible cytotoxic effect of BrdUrd incorporation, the growth of cells under analysis has to be tested with a concentration series of the halogenated pyrimidine analog. Analysis can best be done with conventional univariate DNA FCM using DAPI (4, 6-diamidino-2-phenylindole) as DNA-staining fluorophore concomitant with cell counting. In case accumulation of cells in the G_2 phase of the cell cycle or a severe BrdUrd-dependent reduction in cell growth is encountered, lowering the BrdUrd concentration and/or the addition of an equimolar amount of deoxycytidine (dCyt) can be considered. Second, the quality control of the culture medium, the batch of FBS, and possible mitogens to be used can be performed by analyses with BrdUrd–HO FCM using a cell type of known growth characteristics. Third, the staining procedure should be tested with a cell type of known AT content in its DNA in order to assure optimal staining and FCM resolution. Fourth, the flow cytometer in use has to be carefully adjusted and optimized. In our experience chicken red blood cells (CRBC) are well suited to this purpose, but also other cell types, such as NIH-3T3 or nonstimulated hPBL can be used. Instrument adjustment can also be performed with a sample of the quiescent cells frozen immediately after initiation of a given experiment. This allows checking of the procedure for rendering cells quiescent. The cluster of cells in G_0/G_1 should be focused such that a coefficient of variation (CV) of 3–5% is achieved with respect to the HO258 and to the EB axis.

VI. Instruments

Most of the published work with BrdUrd–HO FCM has been performed on an ICP-22 epiillumination system, but also the laser-powered Ortho Cytofluorograph 50H has been used. The ICP-22 instrument (PHYWE, Goettingen, FRG) uses for excitation an Osram HBO 100 W/2 mercury arc lamp combined with an FT 450 dichroic mirror (Ditrich Optics), a UG1 and a BG38 glass filter (Schott, Mainz, FRG). Hoechst fluorescence is collected with a FT 450 dichroic mirror (Ditrich Optics) and a K45 glass filter (Schott), and fluorescence from EB is selected with a K65 glass filter (Schott). The Ortho Cytofluorograph 50H and other cell sorters use 351- to 364-nm excitation from an argon ion laser. The filter combinations on such instruments are similar to those employed with the ICP-22.

VII. Results

Figure 1 illustrates the response of hPBL (obtained from a 52-year-old donor) to polyclonal activation by PHA. After initiation of the culture, cell aliquots were harvested at the times indicated. The most prominent signal cluster in the 35-hour panel represents G_0/G_1 cells. As activated cells enter the S phase, their EB fluorescence increases; because of the quenching of HO258 fluorescence by BrdUrd-substituted chromatin, there is a concomitant shift of replicating cells to lower fluorescence intensities on the HO axis. At 45 hours, the first cells have divided and arrive in the G_1 compartment of the second cell cycle (G_1'). During the second round of replication in the presence of BrdUrd, S-phase cells increase both their EB and HO258 fluorescence, which results in a distribution that is a "mirror image" of the first cycle distribution. Further divisions result in signal distributions that run parallel to those of the second cycle; because of overlapping distributions, cells beyond the G_1 phase of the fourth cell cycle (G_1''') can no longer be distinguished. In the sequence shown in Fig. 1, recruitment of cells from G_0 still occurs at 80 hours after mitogen stimulation (note the presence of first-cycle fluorescence signals). At 80 hours and 96 hours, signals are seen that indicate nuclear lysis or degradation in the third- and fourth-cycle populations (the signals connecting the G_1' and G_2'' clusters to the origin of the cytogram).

Figure 2 illustrates the activation and cell cycle progression of serum-deprived hDF-like cells by exposure to 10% serum. DNA synthesis starts between 18 and 22 hours (not shown), and reaches its maximum at

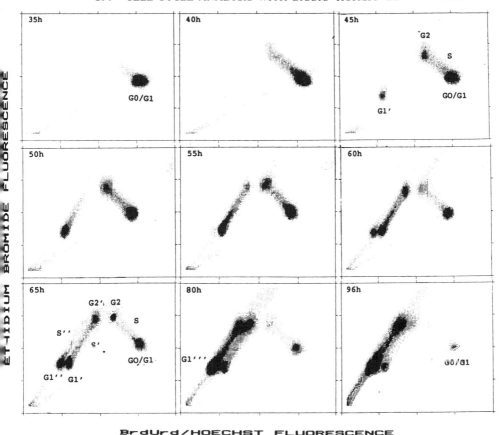

ETHIDIUM BROMIDE FLUORESCENCE

BrdUrd/HOECHST FLUORESCENCE

FIG. 1. Bivariate scattergrams of human peripheral blood lymphocytes stimulated with PHA and cultured for various time intervals.

30 hours after stimulation. Most of the cells have arrived in the second cell cycle by 42 hours; a portion of these enters the third cell cycle (54-hour to 90-hour panels). However, there is a preponderance of G_1 over S- and G_2-phase cells at these later times. This indicates that hDF-like cells are subject to increasing cell cycle arrest as they enter the G_1 phases of their second and third cell cycles.

In Fig. 3, a sequence of cytograms obtained with mouse NIH-3T3 cells is shown. DNA synthesis starts as early as 10 hours after release from quiescence (not shown), and cells are well into the first cycle, with some already in the G_1' compartment, after 18 hours of serum stimulation. At

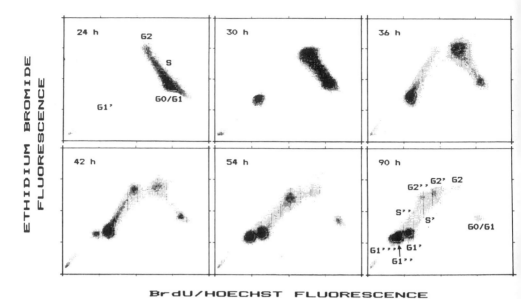

FIG. 2.   Bivariate scattergrams of human primary diploid fibroblasts grown in MEM with 10% FBS for various time intervals.

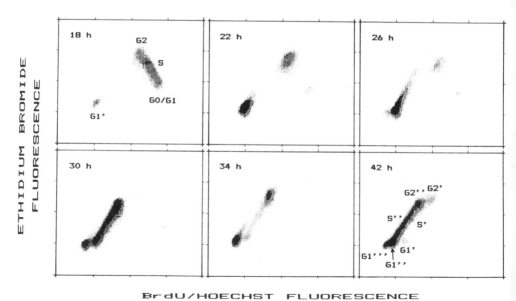

FIG. 3.   Bivariate scattergrams of mouse NIH-3T3 cells grown in DMEM/Ham's F-10 with 10% FBS for various time intervals.

22 hours, most cells are in the G_2 and in the G_1' compartment, and very few reside in G_0/G_1. Four hours later, the majority of cells is in the second and third cycle, whereas a few are still retained in the first cycle. The striking degree of synchrony with which cells traverse the various cell cycle compartments is easily evident by visual examination. At 30 hours of serum stimulation, almost all cells are in the second and third cell cycle, and only a minute fraction of cells is still in the G_0/G_1 and in the G_2 compartment. The latter two compartments are devoid of cells after 34 hours, and cells continue to traverse the second and third cycle. The G_1 of the fourth cycle (G_1''') is reached at 42 hours, whereas the second cycle is at this time point almost entirely empty. Beyond 42 hours of culture, analysis becomes difficult because of the less satisfactory separation between the third and the fourth cycle. NIH-3T3 cells typically show no perceptible arrest in any cell cycle compartment during these culture conditions.

Figures 4A and B illustrate the procedure by which the distribution of cells within individual cell cycles is determined. The cell distributions in first, second, and third cycle are electronically framed, rotated, and projected onto a single axis. The resulting univariate fluorescence distributions can then be processed by conventional curve fitting (bottom row of panels in Fig. 4A and B) in order to determine the distribution of cells in the G_1, S, and G_2/M phases of the cell cycle. These distributions at the time of analysis ("real data") are listed at the bottom of each panel in the sequence (from left to right): G_1, S, G_2/M. By taking into account the number of times a given population has divided since the beginning of the experiment, these "real" data are converted to "original" data, which better reflect the behavior of cells in the starting population. This is done by dividing the "real" number of cells in the second cycle by two, and those within the third cycle by four, and so on. By this simple mathematical procedure, one obtains a quantitative and accurate measure of the proliferative history of a cell population. Note that Fig. 4B, which is a 72-hour harvest cytogram from a patient with FA, shows much greater accumulations of cells in the G_2/M phases of the first and second cell cycles than does Fig. 4A, which is from a healthy lymphocyte donor.

Figure 5 illustrates the conversion of the results of sequential, temporally spaced cell cycle distribution analysis into a kinetic analysis. This conversion is realized by plotting the cell cycle distribution data of each time point (obtained by the procedure explained in the discussion of Fig. 4) on a semilogarithmic scale. Such plots of the proportions of cells in the first, second, and third cell cycle versus time can be fitted to a number of cell cycle models. The curves shown in Fig. 5 are the results of computer fits to a modified version of the Smith–Martin transition probability model (Smith and Martin, 1973; Rabinovitch, 1983; Kubbies et al., 1985a, b).

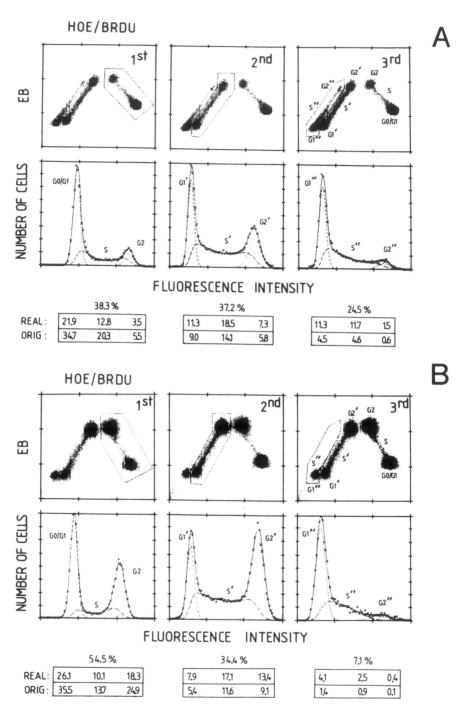

FIG. 4. Example of computer analysis of bivariate scattergrams. The top row indicates electronic framing of each cell cycle component of the scattergram; the second row shows univariate cell distributions obtained via projection of the framed data onto a single axis parallel to each respective cell cycle. (A) Peripheral blood lymphocytes from a normal donor 72 hours after PHA stimulation. (B) Peripheral blood lymphocytes from a donor with Fanconi's anemia 72 hours after PHA stimulation.

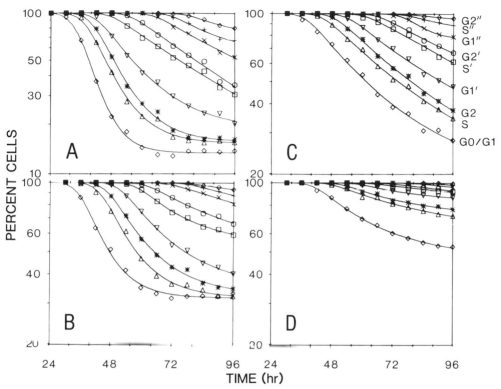

FIG. 5. Exit kinetics of peripheral blood lymphocytes from four donors of different ages. (A) Newborn; (B) 15 years of age; (C) 35 years of age; (D) 75 years of age. Ordinate: percentage of cells that have not exited beyond a particular cell cycle compartment. Abscissa: time (hours) after PHA stimulation. The respective cell cycle compartments are indicated to the right of (c).

Each such line thus represents a distribution curve for the times of transition from one compartment to the next. The following cell-kinetic information can be derived from the mathematical analysis of these curves: duration of the initial lag phase (i.e., the time between induction and first entrance into the S phase), minimum duration of cell cycle phase (given by the x-axis intercept of two successive exit curves), compartment-specific transition probabilities (slopes of the exit curves), and compartment-specific cell cycle arrest (difference of the extrapolated plateau-phase levels between successive exit curves). The examples of exit curves shown in Fig. 5 illustrate the changes of these kinetic parameters as a function of donor age for PHA-activated 96-hour cultures of hPBL. Other applications of this comprehensive cell cycle analysis are presented in several reviews (Rabinovitch *et al.*, 1988; Kubbies *et al.*, 1989).

ACKNOWLEDGMENTS

We are indebted to Miss J. Koehler (Würzburg) for the preparation of the figures. The authors are supported by Deutsche Forschungsgemeinschaft grant HO 849-2-1 (M. P., M. K., H. H.) and by NIH grant AG 01751 (A. G., Y. C., P. S. R.).

REFERENCES

Beck, H. P. (1981). Cytometry 2, 146–158.
Böhmer, R. M. (1979). Cell Tissue Kinet. 12, 101–114.
Darzynkiewicz, Z., Traganos, F., Sharpless, T., and Melamed, M. R. (1976). Proc. Natl. Acad. Sci (U. S. A.) 73, 2881–1884.
Gray, J. W., and Mayall, R. H., eds. (1985). Cytometry 6, (6).
Kubbies, M., and Friedl, R. (1985). Histochemistry 83, 133–137.
Kubbies, M., and Rabinovitch, P. S. (1983). Cytometry 3, 276–281.
Kubbies, M., Schindler, D., Hoehn, H., and Rabinovitch, P. S. (1985a). Cell Tissue Kinet. 18, 551–562.
Kubbies, M., Schindler, D., Hoehn, H., and Rabinovitch, P. S. (1985b). J. Cell. Physiol. 125, 229–234.
Kubbies, M., Schindler, D., Hoehn, H., Schinzel, A., and Rabinovitch, P. S. (1985c). Am. J. Hum. Genet. 37, 1022–1030.
Kubbies, M., Hoehn, H., and Rabinovitch, P. S. (1987a). Expl. Cell Biol. 55, 225–236.
Kubbies, M., Schindler, D., Hoehn, H., Friedl, R., and Rabinovitch, P. S. (1987b). In "Clinical Cytometry and Histometry" (G. Burger et al., eds.), pp. 243–245. Academic Press, Orlando, Florida.
Kubbies, M., Hoehn, H., Schindler, D., Chen, Y. C., and Rabinovitch, P. S. (1989). In "Flow Cytometry" (A. Yen, ed.), CRC Press, Boca Raton, Florida, pp. 1–26.
Latt, S. A. (1973). Proc. Natl. Acad. Sci. U. S. A. 70, 3395–3402.
Latt, S. A. (1977). J. Histochem. Cytochem. 25, 913–922.
Nuesse, M. (1981). Cytometry 2, 70–77.
Poot, M., Schindler, D., Kubbies, M., Hoehn, H., and Rabinovitch, P. S. (1988a). Cytometry 9, 332–339.
Poot, M., Esterbauer, H., Rabinovitch, P. S., and Hoehn, H. (1988b). J. Cell Physiol. 137, 421–429.
Poot, M., Nicotera, T., Rüdiger, H. W., and Hoehn, H. (1989). Free Rad. Res. Commun. 7, 179–187.
Poot, M., Rizk-Rabin, M., Hoehn, H., and Parlovitch, J. H. (1990). J. Cell Physiol. 143, 279–286.
Rabinovitch, P. S. (1983). Proc. Natl. Acad. Sci. U. S. A. 80, 2951–2960.
Rabinovitch, P. S., Kubbies, M., Chen, Y. C., Schindler, D., and Hoehn, H. (1988). Exp. Cell Res. 174, 309–318.
Schindler, D., Kubbies, M., Hoehn, H., Schinzel, A., and Rabinovitch, P. S. (1985). Lancet II, 937.
Schindler, D., Seyschab, H., Poot, M., Hoehn, H., Schinzel, A., Fryns, I. P., Tommerup, N., and Rabinovitch, P. S. (1987a). Lancet II, 1398.
Schindler, D., Kubbies, M., Hoehn, H., Schinzel, A., Rabinovitch, P. S. (1987b). Am. J. Pediatr. Hematol. Oncol. 9, 172–176.
Schindler, D., Kubbies, M., Priest, R. E., Hoehn, H., and Rabinovitch, P. S. (1988). Mech. Ageing Dev. 44, 253–263.
Seyschab, H., Friedl, R., Schindler, D., Hoehn, H., Rabinovitch, P. S., and Chen, U. (1989). Eur. J. Immunol. 19, 1605–1612.
Smith, J. A., and Martin, L. (1973). Proc. Nat. Acad. Sci U.S.A. 70, 1263–1267.
Vogel, D. G., Rabinovitch, P. S., and Mottet, N. K. (1986). Cell Tissue Kinet. 19, 227–336.

Chapter 20

Detection of Bromodeoxyuridine-Labeled Cells by Differential Fluorescence Analysis of DNA Fluorochromes

HARRY A. CRISSMAN AND JOHN A. STEINKAMP

Cell Biology Group
Los Alamos National Laboratory
Los Alamos, New Mexico 87545

I. Introduction

Studies on DNA synthesis, cell cycle traverse, and cell proliferation have been significantly advanced through the use of radiolabeled DNA precursors such as tritiated thymidine. Recently, the nonradioactive base analog, 5-bromodeoxyuridine (BrdUrd), has been substituted for tritiated thymidine, since BrdUrd is safer to handle and sensitive methods for its detection are now currently available. Also, under appropriate conditions this base analog is substituted stoichiometrically for thymidine in newly replicated DNA in a time- and concentration-related manner (Bick and Davidson, 1974). One widely used technique for cytological detection of BrdUrd-substituted DNA employs an immunofluorescent assay of BrdUrd developed by Gratzner (1982). This technique requires the partial denaturation of cellular DNA by heat or acid treatment to expose the incorporated BrdUrd to the antibody. Dolbeare *et al.* (1983) modified this procedure by including the fluorescent counterstain propidium iodide (PI) to measure total DNA content. Using two-color, flow cytometric (FCM) analysis, cells containing incorporated BrdUrd are readily detected and their cell cycle position is easily assessed (see Dolbeare *et al.*, Chapter 21, this volume).

199

Another cytochemical method for detecting BrdUrd-substituted DNA was first demonstrated by Latt (1973) using the A-T base-binding fluorochrome Hoechst 33258, (HO) which is quenched when bound to A–BrdUrd regions in double-stranded DNA. However, attempts to use the BrdUrd–HO fluorescence-quenching technique in FCM have failed to provide resolution and sensitivity comparable to the immunofluorescent assay. This was due to the lack of a sensitive method for quantitatively measuring the reduction in HO fluorescence when bound to BrdUrd. In those studies, BrdUrd-labeling periods of 6 hours (Ellwart and Dohmer, 1985) or longer (Bohmer, 1979, 1981; Bohmer and Ellwart, 1981; Kubbies and Rabinovitch, 1983) were required for cells to incorporate sufficient amounts of the base analog for detecting HO quenching.

In this chapter, we describe a sensitive two-color FCM method for detecting BrdUrd-labeled DNA in cells treated for ≤ 30 minutes (Crissman and Steinkamp, 1987). The technique uses two nonintercalating, DNA-specific fluorochromes: Hoechst 33342 (HO) and GC-binding mithramycin (MI) (Crissman and Tobey, 1974; Crissman et al., (1979), a dye whose fluorescence in the presence of BrdUrd remains stoichiometric to DNA content, under the conditions employed. Using dual-wavelength excitation, the blue (HO) and green-yellow (MI) fluorescence emissions are measured, and a differential amplifier (Steinkamp and Stewart, 1986) subtracts the blue fluorescence from the green-yellow fluorescence signal amplitude on a cell-by-cell basis; the resulting difference signal is then amplified. Cells in S phase exhibit a significant BrdUrd–HO quenching and produce a greater differential fluorescence signal compared with that of cells in the G_1 and $G_2 + M$ phase, which, except for minor differences in HO and MI stainability, show relatively small (near zero) fluorescence differences. The technique is simple and rapid, and requires only one-step staining. It is mild and therefore minimizes cell loss and loss of other important cellular markers such as DNA and/or chromatin, RNA, and proteins, including cellular antigens.

II.   Applications

Essentially, this technique represents a very sensitive approach for measuring the small decrease in HO fluorescent intensity when the dye is bound to BrdUrd incorporated into DNA during short pulse-labeling periods. Quantitative FCM measurement of the HO quenching is accomplished by electronic cell-by-cell subtraction of the HO fluorescent signal from the signal of a second dye that is essentially unaffected by BrdUrd. In this section we briefly provide the basic principle and theory for the method.

The technique employs a selective combination of two DNA-specific fluorochromes that have different spectral properties and that do not compete for the same binding sites on DNA. One of the dyes is HO which binds to AT regions and is quenched by BrdUrd. In the initial studies, the GC-binding MI was used with HO since, under appropriate conditions, MI is not affected by BrdUrd, and neither dye–dye interference nor energy transfer between the two dyes significantly affects FCM BrdUrd analysis. In the absence of BrdUrd, both dyes bind stoichiometrically to DNA content, and thus, during FCM analysis, the integrated areas of the fluorescent signals of both HO and MI should be equal when the instrument gain settings are adjusted so that the G_1 peak of the HO–DNA content and the G_1 peak of the MI–DNA content histograms are in the same channel number. When the two fluorescent signals are subtracted electronically, on a cell-by-cell basis, the values obtained for every cell will be near zero, theoretically. Under idealized conditions a bivariate profile plotted for the MI minus HO fluorescent signal differences versus MI–DNA content would show a straight line at the zero value running across the entire cell cycle. Using this scheme in analysis of a cell population that had been pulse-labeled with BrdUrd (i.e., only S-phase cells are labeled), then zero values for the MI-HO fluorescent signal differences would be obtained only for the G_1 and $G_2 + M$ cells, since, as explained before, in these subpopulations the fluorescent intensity of both dyes remains proportional to DNA content. However, in S-phase cells that contain BrdUrd, the HO fluorescent signal is reduced in proportion to the quantity of BrdUrd-substituted DNA, while the MI fluorescent intensity in these same cells still remains proportional to the DNA content. Flow cytometric analysis of S-phase cells on a cell-by-cell basis will thus yield MI minus HO difference values that sensitively reveal the amount of HO quenching in each cell and that proportionally reflect cellular BrdUrd content.

III.   Materials

Solutions of BrdUrd (5-bromodeoxyuridine, Sigma Chemical Co.) are freshly prepared in distilled water (dH_2O) at 1.0 mg/ml. This solution should be sterilized if BrdUrd labeling is performed on cells that are to be maintained in culture for long time periods, such as during pulse–chase or continuous labeling experiments. For pulse-labeling and direct ethanol fixation, sterilization of BrdUrd is not required. The BrdUrd solution should be protected from direct light.

Saline GM solution and 95% ethanol are prepared and used for ethanol fixation (70% final concentration) of cells as described in Chapter 10 (Crissman and Tobey, this volume).

Hoechst 33342 (Calbiochem, San Diego, CA) is dissolved in dH$_2$O at 1.0 mg/ml and stored in the refrigerator for at least 1 month in a foil-wrapped container. The properties of the HO dye have been presented in Chapter 9 (Crissman et al., this volume).

Mithramycin (MI) (Pfizer Co., Groton CT), 2.5 mg/vial, is dissolved in phosphate-buffered saline (PBS) at 1.0 mg/ml and stored in the refrigerator for at least 1 month. DNA-binding properties and other essential features of MI are discussed in Chapter 10 (Crissman and Tobey, this volume).

Solutions of MgCl$_2$ (250 mM) are prepared in dH$_2$O and may be stored at room temperature (RT) for at least 1 month.

IV.   Staining Procedure

Fixed-cell samples are centrifuged and the ethanol fixative removed by aspiration. The cells are resuspended in the dye solution containing HO and MI at concentrations of 0.5 μg/ml and 5.0 μg/ml in PBS supplemented with 5.0 mM MgCl$_2$. The cell density should be ~7.5 × 10^5 cells/ml stain solution. Cell samples can be analyzed in the stain solution from 30 minutes to 4 hours following staining at RT.

V.   Critical Aspects

Routinely CHO cells are labeled in culture with 30 μM BrdUrd for 1 hour at 37°C. The cultures are not exposed to strong light over this period, since the BrdUrd is light-sensitive and can deteriorate. In addition, cells incorporating BrdUrd into their DNA can suffer chromatin damage and lose viability if exposed to strong light and thus fail to incorporate sufficient amounts of BrdUrd for FCM detection.

The length of the BrdUrd-labeling period is generally varied depending on the cell cycle time and the rate of DNA synthesis. Rapidly growing

L1210 cells (i.e., $T_G = 12$ hours), for example, may require only a 30-minute pulse, but some slow-growing cells may require pulses of 1–2 hours before they have incorporated sufficient quantities of BrdUrd. In some cell types the pulse duration as well as the BrdUrd concentration must be determined empirically. Excessively high concentrations of BrdUrd can be toxic to cells. Preliminary growth studies and FCM analyses of BrdUrd-treated and untreated cell populations may be necessary to determine optimal conditions for BrdUrd labeling.

VI. Controls and Standards

Initially, staining experiments should be performed on untreated cells to determine the lowest concentrations of HO and MI that can be combined to obtain DNA content histograms of good quality with low coefficient of variation (CV) values. Ideally, at the dye concentrations selected for two-color staining, both the HO–DNA content and the MI–DNA content histograms will yield the same percentage of cells in the various phases of the cell cycle. Also, the CV values in both of the DNA histograms will be as low as possible and very close to the same numerical value. Under these conditions the dye–dye interactions will be minimal and the subtraction of the HO fluorescent signal from the MI signal, on a cell-by-cell basis, will be as close to the ideal value of zero as possible. Using the optimized HO and MI dye concentrations for staining and analysis of BrdUrd pulse-labeled cell populations will ensure the most sensitive and accurate FCM measurement of the BrdUrd content.

VII. Instrument

The FCM instrument used is described in detail in Chapter 29 (Crissman et al., this volume); however, only two lasers were used, one operating in the ultraviolet (UV) (333.6–363.8 nm) and one tuned to 457.9 nm. The lasers were separated by 250 μm to provide sequential excitation and analysis of each fluorochrome (i.e., HO and MI, respectively). The HO fluorescence was measured over the 400–495 nm range while the MI fluorescence was measured above 495 nm. The electronic gains were adjusted so that the G_1 peaks of the FCM-generated HO– and MI–DNA content histograms are initially in the same channel number. The fluorescence signals are then subtracted electronically (i.e., MI minus HO) on a cell-by-cell basis (Steinkamp and Stewart, 1986), and these difference measurements reflect quantitatively the quenching of HO fluorescence, which is directly proportional to cellular BrdUrd content in cells that had synthesized a portion of their DNA over the pulse-labeling period. The arbitrary adjustment of the G_1 peak positions in the same channel sets the zero value for the subtraction process for all cells across the cell cycle. The assignment of the zero value is valid, since the G_1 cells would not be expected to contain BrdUrd and thus not exhibit HO quenching, so the MI minus HO value should be zero.

VIII.   Results

Untreated Chinese hamster (line CHO) cells examined by this technique show an equal affinity for both dyes as seen by the linear relationship in staining throughout the cell cycle in Fig. 1A. However, slight differences in stainability between the two dyes are detected with increased sensitivity and amplified in the MI–HO fluorescence difference versus DNA content (MI) profile (Fig. 1B). In contrast, a population of CHO cells treated for 30 minutes with 30 μM BrdUrd shows a significant quenching of HO fluorescence in S phase as seen in Fig. 1C. The G_1 and $G_2 + M$ subpopulations are essentially unaffected. Comparison of the MI–DNA distributions in Fig. 1A and C indicates that MI fluorescence was not affected by the BrdUrd-substituted DNA nor by the quenching of the HO fluorescence. Profiles obtained for untreated and BrdUrd-treated CHO populations stained with MI, in the absence of HO (data not shown), were the same as shown in Fig. 1A and C. The G_1 peak positions in the single-parameter HO–DNA profile are also the same in Fig. 1A and C. However, the magnitude of the MI–HO fluorescence differences, reflecting BrdUrd–HO quenching, shown in Fig. 1D, is most significant for cells in S phase that have incorporated BrdUrd in the 30-minute pulse-labeling period. The percentage of BrdUrd containing S-phase cells detected by this technique (i.e., relative fraction of cells in the boxed region in Fig. 1D) was 35%. This value is in good agreement with the FCM computer-fit S-phase calculation of 37% and also with a 33% thymidine-labeling index derived by standard autoradiography. The bivariate contour distribution for the BrdUrd-treated population (Fig. 1D) is also similar to distributions obtained for BrdUrd-treated CHO cells using the BrdUrd–antibody technique as shown previously by Dolbeare et al. (1983).

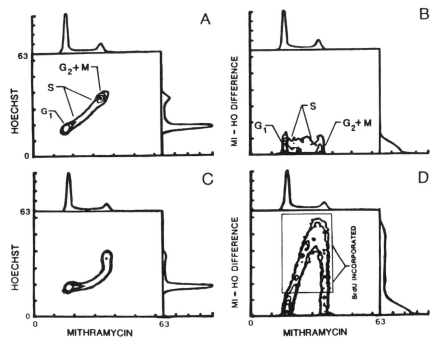

Fig. 1. DNA content (HO and MI) and MI–HO signal difference frequency distribution histograms and the corresponding bivariate contour diagrams for untreated CHO cells (A, B) and for CHO cells treated in culture with 30 μM BrdUrd for 30 minutes (C, D). The x and y axes are linear relative units. The MI–HO difference signal amplitudes from G_1, S, and $G_2 + M$ phase control (untreated) cells are shown in the bivariate diagram (B). The difference signals from cells with zero or slightly negative difference values, due to stainability, were accumulated along the x axis of this display, since the negative outputs from the signal difference amplifier cannot be visualized in the bivariate diagram unless a positive offset voltage is added to the amplifier output (see Crissman and Steinkamp, 1987). In (D), the cells that have incorporated BrdUrd are easily visualized, whereas the data counts from most G_1 and $G_2 + M$ phase cells remain along the x axis. The difference amplifier gain was initially adjusted and fixed to give a maximum expansion of the fluorescence difference signal range from cells that had incorporated BrdUrd. The untreated cell population was analyzed at this same instrument gain setting.

ACKNOWLEDGMENTS

This work was supported by the Los Alamos National Flow Cytometry and Sorting Research Resource funded by the Division of Research Resources of NIH (grant P41-RR01315) and the Department of Energy.

REFERENCES

Bick, M., and Davidson, R. L. (1974). *Proc. Natl. Acad. Sci. U. S. A.* **71**, 2082–2086.

Bohmer, R. M. (1979). *Cell Tissue Kinet.* **12**, 101–110.

Bohmer, R. M. (1981). *Cytometry,* **2**, 31–34.

Bohmer, R. M., and Ellwart, J. (1981). *Cell Tissue Kinet.* **14**, 653–658.

Crissman, H. A., and Steinkamp, J. A. (1987). *Exp. Cell Res.* **173**, 256–261.

Crissman, H. A., and Tobey, R. A. (1974). *Science* **184**, 1297–1298.

Crissman, H. A., Stevenson, A. P., Kissane, R. J., and Tobey, R. A. (1979). *In* "Flow Cytometry and Sorting" (M. R. Melamed, P. F. Mullaney, and M. L. Mendelsohn, eds.), pp. 243–261. Wiley, New York.

Dolbeare, F., Gratzner, H., Pallavacini, M., and Gray, J. W. (1983). *Proc. Natl. Acad. Sci. U. S. A.* **80**, 5573–5577.

Ellwart, J., and Dohmer, P. (1985). *Cytometry* **6**, 513–520.

Gratzner, H. (1982). *Science* **218**, 474–475.

Kubbies, M., and Rabinovitch, P. S. (1983). *Cytometry* **3**, 276–281.

Latt, S. A. (1973). *Proc. Natl. Acad. Sci. U. S. A.* **70**, 3395–3399.

Steinkamp, J. A., and Stewart, C. C. (1986). *Cytometry* **7**, 566–574.

Chapter 21

Using Monoclonal Antibodies in Bromodeoxyuridine–DNA Analysis

FRANK DOLBEARE, WEN-LIN KUO, WOLFGANG BEISKER,
MARTIN VANDERLAAN, AND JOE W. GRAY

Biomedical Sciences Division
Lawrence Livermore National Laboratory
Livermore, California 94550

I. Introduction

The utility of bromodeoxyuridine (BrdUrd) as a marker for cell cycle traverse studies has been substantially increased by the introduction of monoclonal antibodies (mAb) against BrdUrd incorporated into cellular DNA (Gratzner, 1982; Raza *et al.*, 1984; Vanderlaan and Thomas, 1985, Vanderlaan *et al.*, 1986). These antibodies are useful as immunological reagents to stain cells containing BrdUrd fluorescently so that intensity of fluorescence is proportional to the amount of incorporated BrdUrd. The intensity of fluorescence is great enough to permit easy microscopic or flow cytometric (FCM) analysis of BrdUrd incorporation. The cytokinetic utility of the BrdUrd labeling has been further increased by the technique of simultaneous measurement of DNA content and the amount of incorporated BrdUrd (BrdUrd–DNA analysis; Dolbeare *et al.*, 1983). The BrdUrd–DNA assay is based on a procedure for simultaneously staining cells with dyes that fluoresce at different wavelengths (e.g., fluorescein and propidium iodide, PI). The procedure requires that the DNA is partially denatured to expose incorporated BrdUrd to a specific antibody. Denaturation is necessary because antibodies developed so far bind only to BrdUrd (or other halogenated pyrimidines) in single-stranded DNA. The remaining undenatured DNA is then stained with PI. Green fluorescence

METHODS IN CELL BIOLOGY, VOL. 33

from the fluorescein-conjugated antibody is a measure of BrdUrd incorporation. Red fluorescence from the PI is a measure of DNA.

Several methods have been used for denaturation of DNA. These include use of high-molarity HCl (1.5–4 M) (Dolbeare *et al.*, 1983), elevated temperature in 50% formamide (Dolbeare *et al.*, 1985) or in low ionic strength solution (Beisker *et al.*, 1987a), and 0.08 M NaOH (Gratzner, 1982). Restriction endonucleases in combination with exonuclease III, while not causing physical denaturation of the DNA, have been used to expose halopyrimidines in DNA for immunochemical staining (Dolbeare and Gray, 1988). The protocol described here uses thermal denaturation of the DNA at low ionic strength.

II. Application

The primary application of the technique is in the quantification of phase fractions, phase durations, doubling time, labeling index, and growth fractions. One may use multiple sample points (Dolbeare *et al.*, 1983) or single sample points (Begg *et al.*, 1985) to obtain labeling index and potential doubling times. The method can also be used to measure DNA repair (Beisker *et al.*, 1989).

The method has been applied to the analysis of dispersed solid tumors, bone marrow, spleen, liver, bladder epithelial, neural cells, dermal cells, cultured mammalian cells, and cell nuclei.

III. Materials

1. Antihalopyrimidine antibody: We describe here procedures using mAb against BrdUrd and iododeoxyuridine (IdUrd). A list of commercially available anti-BrdUrd antibodies is shown in Table I. Animal antisera derived against one of the halopyrimidines may also be used. The final sensitivity of the measurement depends on the purity, affinity, and specificity of the particular antibody. High-affinity antibodies permit quantification of low levels of BrdUrd incorporation, for example, <0.1% substitution (Beisker *et al.*, 1987b). Generally, anti-BrdUrd antibodies are more efficient at stoichiometric binding of BrdUrd when BrdUrd substitution is <100% (where steric factors limit binding) and >1% substitution (below which diva-

TABLE I

COMMERCIAL SOURCES OF ANTIBROMODEOXYURIDINE ANTIBODIES[a]

Anti-BrdUrd	Sources
IgG purified	Serotec Ltd.
76-7 (IgG$_1$)	AMAC, Inc.
	Biodesign INC.
	Immunotech S.A.
B44 (purified IgG$_1$)	Becton Dickinson Immunocytometry Systems
B44-FITC	Becton Dickinson
Bu20 (supt)	Accurate Chemical and Scientific Corp.
	Dako Corp.
	Dimension Laboratories, Inc.
Bu5.1 (IgG$_{2a}$ purified)	Paesel GmbH
Bu5-1 (purified-FITC)	Paesel GmbH
Bu1-75 (rat ascites)	Accurate Chemical and Scientific Corp.
	Sera-Lab Ltd.
IU-4 (purified)	Caltag Laboratories, Inc.
Br 3	Caltag Laboratories, Inc.
ABDM	Bio Cell Consulting
BMC9318 (IgG$_1$)	Boehringer Mannheim Biochemicals

[a] Most of the sources are listed in *Linscott's Directory of Immunological and Biological Reagents*, Fifth Edition (Linscott's Directory, Mill Valley, CA). Complete addresses of all of the above sources and their international distributors are given in the directory.

lent binding is limited). The higher affinity antibodies also permit working at greater antibody dilutions.

2. Antibody-diluting buffer: Standard phosphate-buffered saline (PBS) commonly described for tissue culture use containing 2× SSC (SSC = 0.15 M NaCl + 0.015 M sodium citrate), 0.5% Tween-20 (Sigma Chemical Co., St. Louis, MO), and blocking protein. The blocking protein may be 1% bovine serum albumin (BSA), 1% gelatin, or 2% dry defatted milk protein (Carnation).

3. Ribonuclease A stock solution: RNase A (Sigma) at 0.5 mg/ml in PBS. This solution should be stored refrigerated with 0.1 mg/ml sodium azide.

4. Paraformaldehyde solution (EM grade, Polyscience, Inc., Warrington, PA): Stock solution is 0.25% or 1% paraformaldehyde in PBS, pH 7.2.

5. 0.1 M HCl plus 0.5% Triton X-100: 5 g of Triton X-100 (New England Nuclear) in 1 liter of 0.1 M HCl.

6. Wash buffer: 5 g of Tween-20 (Sigma) in 1 liter of PBS.

7. Goat anti-mouse IgG–fluorescein conjugate: This antibody may be obtained from a number of sources. We have used goat anti-mouse IgG–FITC from Becton Dickinson, Cappell Labs, Sigma Chemical Co., and U. S. Biochemicals with good results. Other sources of this antibody are listed in Linscott's Directory of Immunological and Biological Reagents (see Table I). Depending on the particular utility (e.g., membrane labeling in addition to BrdUrd–DNA analysis), one may require either a blue-fluorescing, aminomethylcoumarin acetic (AMCA), or a red-fluorescing, Texas red, Princeton red, or phycoerythrin conjugate.

8. PI (Sigma): The working solution is 10 μg/ml in PBS, pH 7.2. A stock solution of 1 mg/ml PI in 70% ethanol stored in the refrigerator is stable for at least a year.

IV.   Staining Procedure

1. Use an aliquot of cells that has been previously labeled with BrdUrd [or chlorodeoxyuridine (CldUrd) or IdUrd] and fixed in cold 50% ethanol or methanol–acetic acid (3 : 1 v/v). Centrifuge at 500 g for 1 minute. (Note: avoid overcentrifugation, which can lead to serious cell clumping.) Pour off supernatant. Suspend cells by gentle vortexing.

2. Add 1.5 ml of RNase stock solution and incubate for 10 minutes at 37°C.

3. Centrifuge at 500 g for 1 minute, pour off, vortex pellet, and suspend cells in 3 ml of 0.25–1.0% paraformaldehyde solution for 30 minutes at room temperature (RT).

4. Centrifuge, decant, and vortex pellet. Wash with 3 ml PBS; centrifuge, decant, and vortex pellet.

5. Suspend cells in 1.5 ml of 0.1 M HCl–Triton X-100 for 10 minutes on ice.

6. Add 5 ml of PBS and centrifuge for 2 minutes at 500 g. Drain pellets well before vortexing.

7. Suspend cells in 1.5 ml of distilled water (dH$_2$O) and place in water bath at 95°C for 10 minutes. This is the DNA-denaturing step and often must be adjusted depending on the cell type to maximize denaturation while avoiding extensive cell loss.

8. Remove samples from hot-water bath and place in an ice–water mixture until the suspensions are cold. Then add 3 ml of PBS, mix,

and centrifuge at 500 *g* for 2 minutes. Drain, vortex, pellet. At this point some clumping may be observed. Disperse clumps by pipetting or syringing through a 25-gauge needle. Failure to disperse the clumps completely may prevent access to BrdUrd-labeled cells. These cells may appear later as apparently unlabeled, resulting in a measurement error.

9. Suspend cells in 100 μl of diluted anti-BrdUrd antibody for 30 minutes at RT. Staining is better at 25°C than at 4°C for the short incubation time.

10. Add 5 ml of wash buffer, centrifuge at 500 *g* for 2 minutes. Drain well, and vortex pellet.

11. Add 100 μl of diluted second antibody (goat anti-mouse IgG–FITC conjugate) for 20 minutes at RT. This step may be omitted if a direct conjugate anti-BrdUrd is used.

12. Add 5 ml of wash buffer, centrifuge, drain well, and vortex pellet.

13. Suspend cells in 1.5 ml of PI working solution.

14. Analyze with flow cytometer with 488-nm excitation with a 514-nm bandpass filter to pass green fluorescence (FITC = incorporated BrdUrd) and a 600-nm long-pass filter to pass red PI fluorescence (= DNA content). If AMCA is used as the fluorophore for anti-BrdUrd, then excitation is with 363 nm and a 450-nm bandpass filter is used.

V.   Critical Aspects of the Procedure

1. Generally, antibody binding to BrdUrd requires that the DNA is partially denatured. Therefore, the level of bound antibody and fluorescence quantifying the incorporated BrdUrd depends on the amount of denaturation. Extended denaturation will increase the amount of bound antibody but will also reduce the amount of double-stranded DNA and thus reduce the amount of bound PI. Complete denaturation of the DNA would mean that the DNA distribution would be very poor with a loss of resolution of the G_1, or $G_2 + M$ phases of the cell cycle. Essentially a smeared distribution would result. Also, increased cell loss may result from the extended denaturation time. Increasing the denaturation time beyond the 10 minutes, however, may also help to resolve the $G_2 + M$ into G_2 and M peaks on the bivariate distribution because of the difference in denaturation of DNA in mitotic and G_2-phase cells (Nüsse *et al.*, 1989; see also Geido *et al.*, Chapter 16, this volume).

2. Paraformaldehyde fixation will lower the sensitivity of the staining reaction, probably by reducing the denaturability of the DNA. Even with 1% paraformaldehyde fixation, however, quantitation of a 10 nM BrdUrd 30-minute pulse is attainable. The paraformaldehyde treatment also helps to prevent cell loss, especially of lymphoid cells, during the staining procedure. We have found also that the fixation step improves the quality of the DNA histogram with lower coefficients of variation (CV) for the G_1 peak.

3. Nonspecific fluorescence is due primarily to nonspecifically bound antibody (anti-BrdUrd or fluorescein-conjugated second antibody). Additional washes after antibody treatment or incorporation of 1–5% blocking protein in the wash buffer can reduce nonspecific binding of antibody. Since most nonspecific antibody binding is in the cytoplasm and on the cytoplasmic membrane, using nuclei rather than whole cells can also greatly reduce nonspecific fluorescence.

4. Cell loss generally results from clumping and cell adherence to the walls of the test tube being used to process the cells. Lymphoid cell loss is generally greater from this procedure but may be reduced greatly by using lower centrifuge speeds during the pelleting of cells. Use 400–800 g maximal for 3 or 4 minutes. Clumping is also higher when cells have been fixed in methanol–acetic acid than when fixed in 50–70% ethanol. Frequently large clumps appear after the thermal denaturation step. Clumps may be disaggregated by a combination of mild vortexing and syringing the suspension gently through a 25-gauge needle. Cell adherence to the centrifuge tube can be decreased either by siliconizing the tubes or by using microfuge tubes to reduce tube surface area.

5. When staining cells that have double halopyrimidine labels (e.g., with BrdUrd pulse and continuous IdUrd label), one should be concerned with the differences in specificities and affinities of the specific antibodies. If very specific antibodies are used, there may not be a problem. As an example of halogen specificity, IU-4 has affinity for the halopyrimidines in the following order: IdUrd > BrdUrd > CldUrd. On the other hand, Br-3 has the following specificity: BrdUrd = CldUrd > > IdUrd. If some cross-reactivity occurs with one of the antibodies (e.g., IU-4 will react with IdUrd and at lower affinity with BrdUrd), then add that antibody at a much lower concentration after the specific antibody Br-3 has incubated with the cells for 20–30 minutes. Then continue the incubation for an additional 30 minutes. In this way the Br-3 will saturate BrdUrd sites but not react with the IdUrd sites. Adding the IU-4 then will preferentially bind only to the exposed IdUrd sites. Using a lower concentration of this antibody will prevent displacement of the Br-3 from the BrdUrd sites.

6. Using correct optical filters during the FCM analysis will prevent the crosstalk that often plagues investigators. Propidium iodide exhibits a

broad band of fluorescence ranging from 530 to 700 nm. If a 550-nm short-pass filter coupled with a 500-nm long-pass filter is used for the green fluorescence, then some PI fluorescence will be observed in the green fluorescence channels, causing a skewing of the BrdUrd histogram. Using excess PI concentration will also aggravate this problem. We recommend a 5–10 μg/ml final concentration of PI.

VI.   Controls, Standards

The following standards should be run.

1. A negative control should be a sample with anti-BrdUrd omitted, but with goat anti-mouse IgG–fluorescein conjugate added. If the anti-BrdUrd is a direct conjugate, then the measure of nonspecific binding can be accomplished by using cells that have not been pulsed with BrdUrd.
2. An autofluorescence standard can be included that has been through the protocol without added antibodies or PI.
3. To determine whether crosstalk is occurring (i.e., whether any PI fluorescence is being detected by the fluorescein detector), add PI but no antibodies after the thermal denaturation step. If the green fluorescence is above background, then crosstalk is present. Using a very broad-band green filter (i.e., with a range of 520–550 nm or higher) allows some PI fluorescence through into the fluorescein detector. Also, use of too much PI will allow some PI fluorescence into the fluorescein detector.

VII.   Instruments

Any flow cytometer equipped with a single argon ion laser with two photomultiplier tubes is adequate. The instrument should also be equipped with a log amplifier to accommodate the large range of fluorescence signal generated by the anti-BrdUrd fluorescence. Both fluorescein and PI can be excited at 488 nm. Use a 514-nm bandpass filter for the fluorescein fluorescence (BrdUrd content) and a 600-nm long-pass filter for the PI fluorescence (DNA content). The incorporation of a doublet eliminator will prevent accumulation of doublet G_1 and early S-phase cells in windows that should contain either G_2 or late S-phase cells.

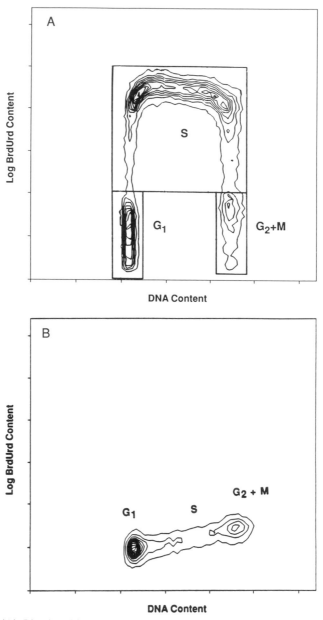

FIG. 1. (A) Bivariate histogram showing DNA content (x axis, PI fluorescence) and BrdUrd content (y axis, log fluorescein fluorescence) following a 30-minute pulse of 1 μM BrdUrd. (B) Bivariate histogram showing DNA content and BrdUrd content of CHO not receiving BrdUrd pulse.

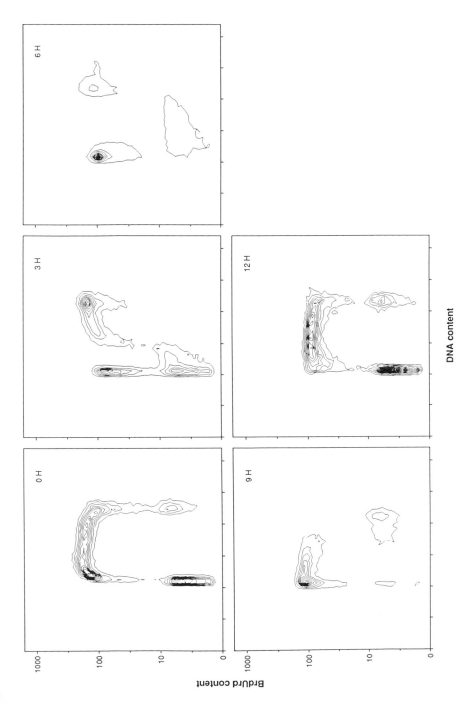

DNA content

FIG. 2. A series of bivariate DNA-BrdUrd distributions taken after a 30-minutes pulse of 10 μM BrdUrd followed by a thymidine chase and subsequent sampling at 3-hour intervals.

215

VIII. Results

Figure 1A shows a bivariate DNA–BrdUrd contour histogram generated by CHO cells stained according to the protocol just given. This is the
kind of histogram generated after a 30-minute pulse of 1 μM BrdUrd when
the cells are fixed immediately after the pulse. G_1 and G_2 + M populations
should have only background green fluorescence. S-Phase cells have
green fluorescence and produce the horseshoe-shaped pattern with mid-S-
phase cells having the highest fluorescence. The fraction of cells with
S-phase fluorescence divided by the total population will give the labeling
index. Cells that have not been incubated with BrdUrd but stained by the
same protocol will exhibit only background fluorescence. Figure 1B shows
a similar bivariate distribution for CHO cells not pulsed with BrdUrd.

The cell-kinetic applicability of the method is demonstrated by Fig. 2,
where a single BrdUrd pulse was given at $t = 0$ followed by a thymidine
pulse–chase after 30 minutes. Samples were taken at 30 minutes (0 hours)
and at $t = 3, 6, 9$, and 12 hours, fixed in 50% ethanol, and stained by the
protocol just given. S-Phase cells that incorporated BrdUrd show the
typical green fluorescence after 30 minutes. This same cohort of S-phase
cells progress through the cell cycle, with the fluorescence appearing in
G_2 + M and daughter G_1 cells at 3 and 6 hours, and progressing further
into G_1 and back into Sphase at 9 hours, with most of the label reappearing
in S phase at approximately one cell cycle time (i.e., 12 hours after the
initial pulse of BrdUrd).

References

Begg, A. C., McNally, N. J., Shrieve, D. C., and Karcher, H. A. (1985). *Cytometry* **6,**
620–626.

Beisker, W., Dolbeare, F., and Gray, J. W. (1987a). *Cytometry* **8,** 235–239.

Beisker, W., Hittelman, W. N., and Eisert, W. G. (1987b). *Cytometry Suppl.* **1,** 84.

Dolbeare, F., and Gray, J. W. (1988). *Cytometry* **9,** 631–635.

Dolbeare, F., Gratzner, H., Pallavicini, M., and Gray, J. W. (1983). *Proc. Natl. Acad. Sci.
U. S. A.* **80,** 5573–5577.

Dolbeare, F., Beisker, W., Pallavicini, M., Vanderlaan, M., and Gray, J. W. (1985).
Cytometry **6,** 521–530.

Gratzner, H. G. (1982). *Science* **218,** 475–476.

Nüsse, M., Julch, M., Geido, E., Bruno, S., Di Vinci, A., Giaretti, W., and Russo, K.
(1989). *Cytometry* **10,** 312–319.

Raza, A. G., Preisler, H. D., Myers, G. I., and Bankert, R. (1984). *N. Engl. J. Med.* **310,**
991–991.

Vanderlaan, M., and Thomas, C. B. (1985). *Cytometry* **6,** 501–505.

Vanderlaan, M., Watkins, B., Thomas, C., Dolbeare, F., and Stanker, L. (1986). *Cytometry*
7, 466–507.

Chapter 22

Identification of Proliferating Cells by Ki-67 Antibody

HEINZ BAISCH

Institut für Biophysik und Strahlenbiologie
Universität Hamburg
D-2000 Hamburg 20
Federal Republic of Germany

JOHANNES GERDES

Division of Molecular Immunology
Forschungsinstitut Borstel
D-2061 Borstel
Federal Republic of Germany

I. Introduction

The monoclonal antibody Ki-67 was found by chance in studies aimed at the production of monoclonal antibodies (mAb) to nuclear antigens specific to Hodgkin and Sternberg Reed cells (Gerdes *et al.*, 1983).

Ki-67 recognizes a human nuclear antigen that is present in proliferating cells but is absent in resting cells. This was shown in lymphocytes stimulated with phytohemagglutinin A (PHA) (Gerdes *et al.*, 1984b). While untreated lymphocytes were negative, the stimulated cells began to express Ki-67 antigen 24–36 hours after addition of PHA. The transition of lymphocytes from G_0 resting stage to G_1 cycling stage was measured using the differential DNA–RNA staining with acridine orange (Darzynkiewicz, Chapter 27, this volume). Cells in all cycle phases including G_1 were

METHODS IN CELL BIOLOGY, VOL. 33

positive for Ki-67, while G_0 cells consistently did not express this antigen. Thus with the help of Ki-67 a rapid determination of the growth fraction (GF) of a given cell population has become possible. The association of Ki-67 with proliferation was additionally documented by showing its positive reaction with proliferating HL-60 cells and its strong suppression after induction of HL-60 cells to differentiation by phorbol ester TPA (Gerdes *et al.*, 1983).

When the size of the tumor GF as measured with Ki-67 was compared to histopathological tumor grading, a significant positive correlation between mean values of GF and histologically defined grades of malignancies was found in various tumor types. Considering single values, however, there was a remarkable scatter of the sizes of GF in all histologically defined grades (Gerdes *et al.*, 1984a, 1986, 1987a,c; Vollmer *et al.*, 1986; Lelle *et al.*, 1987; Lokhorst *et al.*, 1987; Giangaspero *et al.*, 1987). Thus, as further substantiated by a retrospective clinical study (Gerdes *et al.*, 1987b), the determination of tumor GF with Ki-67 may be a more objective aid to estimate the outcome of an individual tumor case and may be of help in outlining individual treatment protocols.

The determination of the GF with the help of Ki-67 may also be used in combination with other antibodies or dyes. Flow cytometry (FCM) is particularly useful to measure cells immunostained with Ki-67 and having DNA simultaneously counterstained with propidium iodide (PI) (Baisch and Gerdes, 1987; Palutke *et al.*, 1987). This approach enables one to estimate Ki-67 expression in relation to cell position in the cell cycle, on a cell-by-cell basis. However, cells that cease proliferation as a result of nutritional deprivation may still express Ki-67, as shown for U937 cells growing at plateau at high cell densities (Baisch and Gerdes, 1987). Ki-67 has also been used together with antibromodeoxyuridine (anti-BrdUrd) antibodies (Larsen, 1987; see Larsen, Chapter 23, this volume) and antibodies against cytoplasmic and membrane-located antigens (Campana *et al.*, 1988).

Although the antibody Ki-67 was discovered in 1983, thus far we know very little about the antigen that is detected by this antibody. There is evidence that Ki-67 antigen is a nonhistone protein that is highly susceptible to protease treatment. This sensitivity may be one of the reasons why this antigen still awaits further characterization. Despite uncertainty regarding the nature of the detected antigen, because the association of Ki-67 with cell proliferation has been documented in numerous studies, these antibodies have become a routine tool in studies of the proliferative capacity of cells and tissues.

II.   Application

The Ki-67 antibody can be used for cell-kinetic investigations of cell cultures, stimulated lymphocytes, and solid tumors as well as normal tissue. This is true for human material and some other species. For stimulation of lymphocytes by PHA or other inducers, the method is fast and more convenient than autoradiography (Palutke *et al.*, 1987). Moreover, it provides additional information, when double-staining Ki-67 plus PI is applied.

For solid tumors the method offers a fast and convenient estimation of proliferation capacity. Preparation of cell suspensions for FCM may be a problem, and it must be determined whether the results are representative for the whole tumor. We recommend performing a cryosection Ki-67 staining in parallel. From these slides the spatial distribution of Ki-67-positive (i.e., proliferating) cells within the tumor can be determined.

III.   Materials and Methods

Since fixation and preparation of cell suspensions may yield different results for various cells and tissues, we describe two different methods: a procedure without fixation, and a method for preparing cell suspensions from solid tumors. Several methods should be tested for every cell type to select the optimal one. Section III, E outlines the Ki-67 staining of cryosections we recommend its application in parallel to FCM measurements for solid tumors.

A.   Acetone Fixation of Cell Suspensions

1. Spin 10^6 trypsinized cells or cells in suspension (e.g., mononuclear cells after Ficoll–Hypaque separation) at 400 g for 5 minutes, and decant.
2. Add 2 ml NaCl 0.15 M at 4°C, resuspend. Drop cell suspension into 8 ml ice-cold pure acetone while gently shaking.
3. Store at −20°C or less for at least 24 hours, and as long as several months.
4. Centrifuge, decant, and add 0.1% RNase in Tris buffer. Incubate 20 minutes at 37°C, spin, and decant.

5. Add 100 μl Ki-67 [1 : 10 dilution with phosphate-buffered saline (PBS) containing 1% bovine serum albumin (BSA)], mix, and incubate 30 minutes at room temperature (RT; $18° - 24°C$) gently shaking, add PBS, spin, and decant.

6. Add 100 μl fluorescein-conjugated goat anti-mouse antibody (1 : 40 dilution with PBS–BSA), mix, and incubate 30 minutes at RT, add PBS, spin, and decant.

7. Add 0.5 ml PI (2 μg/ml in PBS), stain 20 minutes at RT in the dark. Measure cell fluorescence by FCM.

B.   Paraformaldehyde Fixation

1. Spin 10^6 cells (400 g, 5 minutes), decant, and resuspend pellet in the remainder (50–100 ml) of fluid.

2. Add 1 ml 0.5% paraformaldehyde in PBS. Incubate 5 minutes at 4°C, spin, decant, and resuspend.

3. Add 1 ml 0.1% Triton X-100 in PBS. Mix, incubate 5 minutes at 4°C, spin, decant, and resuspend.

4. Incubate with Ki-67 and anti-mouse FITC antibody as described in Section III, A (steps 5 and 6).

5. Add 1 ml ice-cold 70% ethanol while shaking. Incubate 10 minutes at 4°C, spin, decant, and resuspend.

6. Add 1 ml 0.112% sodium citrate in PBS. Incubate 20 minutes at 4°C, spin, decant, and resuspend.

7. Add 0.5 ml PI in PBS (2 μl/ml), stain 20 minutes at RT in the dark, and analyze cells by FCM.

The whole procedure can also be done in culture plates (Palutke et al., 1987).

C.   Analysis of Unfixed Cells

This procedure is described by Larsen (1987).

1. Prepare cell suspension with 2×10^6 cells/ml (see Section III,A, step 1) and resuspend in appropriate volume of buffer C (0.25 M sucrose, 40 mM Na citrate); divide into 50-μl samples containing 10^5 cells. For storage at $-80°C$ 5% dimethyl sulfoxide DMSO has to be added to buffer C.

2. Add 200 μl PI solution [20 μg/ml PI, 0.1% Nonidet P-40 (NP-40), 0.2 mg/ml RNase (Sigma IA), 1% BSA, 0.5 mM EDTA, in PBS with Ca and Mg; adjust to pH 7.2]. Incubate for 15 minutes at RT.

3. Add 25 μl Ki-67 antibody (undiluted); incubate 60 minutes at RT.

4. Add 25 μl FITC-conjugated anti-mouse IgG antibody (1 : 5 in buffer C); incubate 60 minutes at RT.
5. Analyze in FCM instrument. Any instrument capable of exiting in the blue (e.g., 488 nm) and sufficiently sensitive to measure low fluorescence intensities can be used.

D. Enzymatic Cell Isolation from Solid Tumors

1. Cut 0.5 g tumor tissue into 1-mm cubes with scalpels in a Petri dish.
2. Add PBS, decant after pieces have subsided.
3. Add 10 ml 37°C solution D [10 ml PBS containing 0.1 mg/ml $CaCl_2$ and 0.1 mg/ml $MgCl_2$ + 0.8 ml DNase (2.5 mg DNase/ml PBS) + 0.8 ml 5% Difco Bacto Trypsin]. Incubate 10 minutes at 37°C gently agitating, let tissue fragments sediment, and decant.
4. Add 10 ml 37°C solution D; incubate 45 minutes at 37°C with gentle agitation.
5. Shake vigorously 10 times, add 0.8 ml DNase (2.5 mg DNase/ml PBS), and pass through a 53-μm filter.
6. Spin cell suspension (200 g, 5 minutes), decant, and proceed as described. Section III,A, step 2; Section III,B, step 2; or Section III,C, step 1.

E. Immunostaining of Cryostat Sections with Ki-67

We recommend performing either the three-step immunoperoxidase technique as described by Stein et al. (1980) or the alkaline phosphatase anti-alkaline phosphatase (APAAP) method as described by Cordell et al. (1984). Tissue preparation, tissue sectioning, and fixation are identical for both methods:

1. Cut small tissue samples ($\sim$ 8 × 8 × 8 mm), insert into plastic tubes, and cover the tissue with sterile PBS or saline. Close the tube and insert it into liquid nitrogen (3–5 minutes) to freeze the tissue rapidly. Store frozen tissues at −70°C. (Tissues may be stored for several years without altering the Ki-67 antigenicity.)
2. Cut 4–6 μm cryostat sections and let them dry in air overnight.
3. Fix section for 15 minutes in acetone at RT and subsequently for 15 minutes in chloroform at RT.

Immunoperoxidase Staining

For immunoperoxidase staining the slides are then incubated with mAb Ki-67 for 30 minutes at RT. After three brief washes in Tris-buffered saline

(TBS), the sections are incubated with peroxidase-conjugated antimouse serum for 30 minutes (e.g., DAKO, Copenhagen; rabbit anti mouse-PO at a dilution of 1 : 20), and, after a further washing, with peroxidase-conjugated antirabbit serum (e.g., Jackson goat antirabbit-PO at a dilution of 1 : 100). Subsequently, peroxidase reaction is carried out according to the principles described by Graham and Karnowsky (1966). Finally slides are counterstained with hemalum and mounted.

APAAP Method

Slides are incubated with Ki-67 for 30 minutes and subsequently incubated with unlabeled rabbit antimouse serum for 30 minutes (e.g., DAKO rabbit antimouse-Ab at a dilution of 1 : 20). Subsequently, the sections are treated with APAAP complexes (e.g., DIANOVA, Hamburg, FRG, APAAP at dilution 1 : 50) for 30 minutes. The incubation with rabbit antimouse serum and APAAP is repeated once, and thereafter the alkaline phosphatase-developing reaction is performed with the modified new fuchsin method as described by Stein et al. (1985). Finally, slides are counterstained with hemalum and mounted.

IV. Critical Aspects

Cell fixation may be a problem when different cell types are used. It is suggested that the acetone and the paraformaldehyde fixation as well as the method without fixation should be tried. It is important to have the acetone of high purity (e.g., Merck, Acetone p.a., Catalog no. 14, Darmstadt, FRG). For some cells a milder fixation may be required, namely: Replace Step 3 (Section III, A) with the following: drop 2 ml of cell suspension in 2 ml ice-cold 50% acetone, then add 6 ml ice-cold absolute acetone while gently shaking.

The isolation procedure of cells from solid tumors may be modified depending on the type of tissue studied. Trypsin can be replaced by collagenase or a combination of hydrolytic enzymes (Ensley et al., 1987; see also Cerra et al., Chapter 1, this volume). In some situations it may be easier to stain isolated bare nuclei in suspension, since the Ki-67 epitope is located in the nucleus. However, caution is required with use of proteolytic enzymes; it is known, for instance, that pepsin destroys the antigen.

After proper staining the Ki-67 fluorescence is restricted to the nucleus. Since cytoplasmic versus nucleus stainability cannot be assessed in most FCM instruments, Ki-67 localization has to be determined by microscope.

Both cytoplasmic fluorescence and unspecific staining have to be as low as possible. The unspecific staining is determined by taking into account the FCM measurement of autofluorescence and fluorescence of control samples treated only with FITC-conjugated anti-mouse IgG antibody. The unspecific flourescence should not be higher than twice the intensity of autofluorescence. A reduction of unspecific fluorescence can be achieved by preincubation of cells in normal goat serum (diluted 1 : 5–1 : 10 in PBS) before Ki-67 incubation (Section III,A, Step 3). The dilutions of Ki-67 and FITC-conjugated antimouse antibodies as given in this protocol may be altered depending on the batch of particular antibodies used, for optimal results.

V.   Controls and Standards

To ensure a proper staining procedure, a negative and a positive control should be run before measuring the samples of a particular experiment. Peripheral blood lymphocytes (PBL) after Ficoll–Hypaque separation can be used as negative control, and stimulated (PHA or concanavalin A) lymphocytes are positive for Ki-67. The latter also contain ~ 10–20% negative cells in addition to the 80–90% positive cells, and therefore can be used as negative and positive control in a single measurement. Ki-67 is also expressed in numerous cell lines such as U937 or HL-60. Such lines (if growing exponentially) can also serve as positive controls. The most convenient way is to fix and store a great number of lymphocytes and HL-60 cells (for example) as negative and positive controls, respectively. In addition, for every cell type autofluorescence and nonspecific fluorescence (without first antibody) have to be measured.

For preparation from solid tumors we recommend cutting cryosections from the same piece of tumor that has been used for FCM measurement. Staining as described in Section III,E will provide a control for FCM and in addition show the localization of Ki-67 positive cells within the tumor.

VI.   Examples of Results

Figure 1 is a histogram of Ki-67-stained U937 cells, 3 days after split (Fig. 1A), and the same cells stained with only FITC-conjugated anti-mouse IgG antibody (Fig. 1B), both stained using the method given in

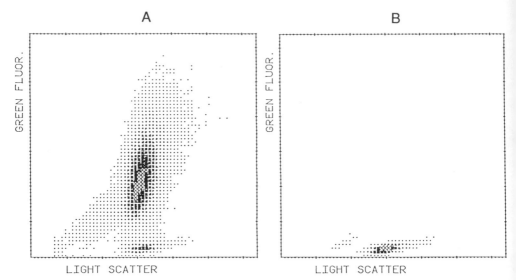

Fig. 1. U937 cells stained with monoclonal Ki-67 and FITC-conjugated antimouse antibody (A) and control without first antibody (B), according to the method in Section III,A. All scales are linear.

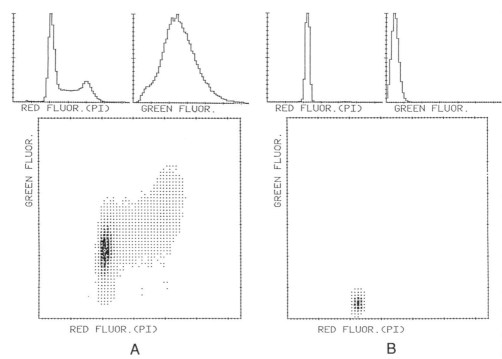

Fig. 2. Ki-67-stained HL-60 cells (A) and human peripheral blood lymphocytes (B), using the method in Section III, B. All scales are linear.

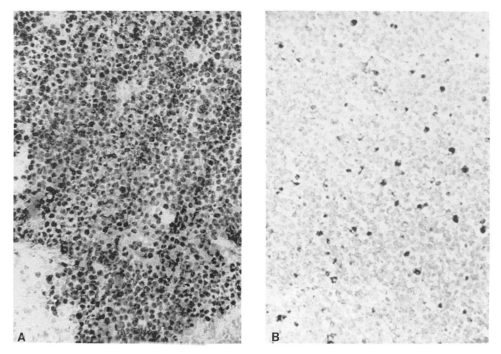

FIG. 3.   Ki-67 peroxidase-stained cryosections of (A) fast-growing and (B) slowly growing human malignant lymphomas.

Section III, A. Figure 2 shows HL-60 cells (Fig. 2A) and lymphocytes from PBL (Fig. 2B) after incubation with Ki-67, stained using the method given in Section III,B. Figure 3 show cryosections of fast-growing tumor (Fig. 3A) and slowly growing tumor (Fig. 3B) stained with Ki-67 according to the method given in Section III,E. The black nuclei represent positive Ki-67 staining, whereas the grayish color represents the hemalum counterstaining.

REFERENCES

Baisch, H., and Gerdes, J. (1987). *Cell Tissue Kinet.* **20**, 387–31.
Campana, D., Coustan-Smith, E., and Janossy, G. (1988). *J. Immunol. Methods* **107**, 79–88.
Cordell, J. L., Falini, B., Erber, W. N., Ghosh, A. K., Abdulaziz, Z., Macdonald, S., Pulford, K. A. F., Stein H., and Mason D. Y. (1984). *J. Histochem. Cytochem.* **32**, 219–229.
Ensley, J. F., Maciorowski, Z., Pietraszkiewicz, H., Klemic, G., KuKuruga, M., Sapareto, S., Corbett, T., and Crissman, J. (1987). *Cytometry* **8**, 479–487.

Gatter, K. C., Dunnill, M. S., Gerdes, J., Stein, H., and Mason, D. Y. (1986). *J. Clin. Pathol.* **39**, 590–593.

Gerdes, J. (1985). *Int. J. Cancer* **35**, 169–171.

Gerdes, J., Schwab, U., Lemke, H., and Stein, H. (1983). *Int. J. Cancer* **31**, 13–20.

Gerdes, J., Dallenbach, F., Lennert, K., Lemke, H., and Stein, H. (1984a). *Hematol. Oncol.* **2**, 365–371.

Gerdes, J., Lemke, H., Baisch, H., Wacker, H. H., Schwab, W., and Stein, H. (1984b). *J. Immunol.* **133**, 1710–1715.

Gerdes, J., Lelle, R. J., Pickartz, H., and Heidenreich, W. (1986). *J. Clin. Pathol.* **39**, 977–980.

Gerdes, J., Pickartz, H., Brotherton, J., Hammerstein, J., Weitzel, H., and Stein, H. (1987a). *Am. J. Pathol.* **129**, 486–492.

Gerdes, J., Stein, H., Pileri, S., Rivano, M. T., Gobbi, M., Ralfkiaer, E., Nielsen, K. M., Pallesen, G., Bartels, H., Palestro, G., and Delsol, G. (1987b). *Lancet* **2**, 448–449.

Gerdes, J., van Baarlem. J., Pileri, S., Schwarting, R., Unnik, J. A. M., and Stein, H. (1987c). *Am. J. Pathol.* **128**, 330–334.

Giangaspero, F., Doglioni, C., Rivano, M. T., Pileri, S., Gerdes, J., and Stein, H. (1987). *Acta Neuropathol.* **74**, 179–182.

Graham, R. C., Jr., and Karnovsky, M. J. (1966). *J. Histochem. Cytochem.* **14**, 291–302.

Larsen, J. (1987). *Cytometry Suppl.* **1**, 350.

Lelle, R. J., Heidenreich, W., Stauch, G., and Gerdes, J. (1987). *Cancer* **59**, 83–88.

Lokhorst, H. M., Boom, S. E., Bast, B. J. E. G., Peters, P. J., Tedder, T. F., Gerdes, H., Petersen, E., and Ballieux, R. E. (1987). *J. Clin. Invest.* **79**, 1401–1411.

Palutke, M., KuKuruga, D., and Tabaczka, P. (1987). *J. Immunol. Methods* **105**, 97–105.

Stein, H., Bonk, A., Tolksdorf, G., Lennert, K., Rodt, H., and Gerdes, J. (1980). *J. Histochem. Cytochem.* **28**, 746–760.

Stein, H., Gatter, K. G., Asbahr, H., and Mason, D. Y. (1985). *Lab. Invest.* **52**, 676–683.

Vollmer, E., Roessner, A., Gerdes, J., Mellin, W., Stein, H., Chong-Schachel, S., and Grundmann, E. (1986). *J. Cancer Res. Clin. Oncol.* **112**, 281–282.

Chapter 23

Washless Double Staining of a Nuclear Antigen (Ki-67 or Bromodeoxyuridine) and DNA in Unfixed Nuclei

JØRGEN K. LARSEN

The Finsen Laboratory
Rigshospitalet, University Hospital
DK-2100 Copenhagen, Denmark

I. Introduction

This article presents a new methodological approach to the double staining of a nuclear antigen and DNA. In the first step of this procedure the cells are lysed with detergent into a suspension of pure nuclei. This is done in order to optimize the exposure of DNA and the specific nuclear antigen to the staining reagents without denaturing the antigen, and to minimize the amount of extranuclear material that might contribute nonspecifically to the measured nuclear fluorescence. If an FITC-conjugated, specific antibody is available, the method may be simplified to two steps. If not, an antibody sandwich is built up in the single nucleus by addition of a specific monoclonal nonconjugated antibody and a secondary, FITC-conjugated antibody, in a second and a third step, sequentially. Thus, the success of the three-step method is based on the following assumptions:

1. The specific nuclear antigen does not dissolve from the nuclear matrix as a result of the lysis procedure.
2. The specific antibody distributes freely in and out of the nucleus, and to the antigenic site in the nucleus.
3. Localization of the FITC-conjugated secondary antibody to nuclear-bound primary antibody is not prevented by a limited surplus of primary antibody in solution.

METHODS IN CELL BIOLOGY, VOL. 33

4. The measurement of increased green fluorescence, corresponding to increased expression of specific antigen in the individual nucleus, is not prevented by a limited surplus of FITC-conjugated antibody in the solution.

5. The antibody sandwich is stable during the necessary period of measurement.

For the special purpose of staining bromodeoxyuridine (BrdUrd) with an antibody recognizing BrdUrd in single-stranded DNA, two additional steps have to be inserted in the procedure before the application of specific antibody, in order to denature the DNA by HCl and subsequently to restore the pH.

II. Applications

So far, applications have been limited to the hematopoietic system and to cell cultures. The method has primarily been used for flow cytometric (FCM) analysis of the Ki-67–DNA and BrdUrd–DNA distribution in phytohemagglutinin (PHA)-stimulated lymphocytes (Larsen et al., 1987). The detailed staining procedure for this application is given in Section III. Results, including the comparison with measurements on fixed cells, are reported elsewhere (Larsen et al., in press). The method has furthermore been used for analysis of the BrdUrd–DNA distribution in monolayer cultures of butyrate-treated bladder cancer cell lines (Larsen et al., 1989) and of the p62–c-*myc* oncoprotein–DNA distribution in normal and transformed murine mast cells (Giaretti et al., 1990).

The washless staining of unfixed nuclei makes it possible in a simple way to extend the FCM DNA analysis (cell cycle distribution, DNA ploidy of stemlines) with a secondary cell-kinetic parameter (DNA synthesis, cycling–noncycling distribution, cell cycle-associated oncoprotein expression). As a probe for a secondary cell-kinetic parameter, the Ki-67 antibody recognizes a nuclear antigen naturally expressed in cycling cells (see Baisch and Gerdes, Chapter 22, this volume) and therefore might be more practical for clinical investigations than the anti-BrdUrd antibody. This is because its use is not dependent on complicated vital or supravital labeling and subsequent denaturation of the labeled molecule. However, for the use of Ki-67–DNA analysis on biopsy specimens from heterogenous solid tumors, the present method needs modification, because complete lysis of the cytoplasm and permeabilization of the nucleus are hard to accomplish for all present cell populations simultaneously.

In comparison with established methods using fixed cells (Ki-67: Baisch and Gerdes, 1987; BrdUrd: Dolbeare *et al.*, 1983), the present method for washless staining of unfixed nuclei has the following characteristics:

1. Cell suspensions can be frozen and stored.
2. The staining procedure is simple and might be adapted for automatic processing.
3. Staining time is generally <1 hour for Ki-67–DNA and <2 hours for BrdUrd–DNA.
4. Small samples (consisting of ≤100,000 cells) can be analyzed.
5. The loss of nuclei is negligible.
6. DNA is measured with high precision (CV = 2–4%).
7. There are few false signals of DNA hyperploidy (minimum aggregation).
8. The fraction of the antigen-positive subpopulation is adequately and reproducibly measured and is similar to that measured in fixed samples (nuclei of cycling cells are Ki-67-positive; nuclei of continuously labeled cells are BrdUrd-positive).

III.   Staining Procedures

Monodisperse cell suspensions (harvested from cultures of lymphocytes, leukemia cell lines, trypsinized monolayer cell cultures, etc.) are centrifuged, resuspended in ice-cold freezing buffer [sucrose 250 mM, dimethyl sulfoxide (DMSO) 5% v/v, Na citrate 40 mM, pH 7.6] to a minimum concentration of 10^5 cells in 50 μl, and stored at −80°C. Cell suspensions from trypsinized monolayer cultures are washed once in phosphate-buffered saline (PBS) with trypsin inhibitor or serum before storage.

A.   Ki-67–DNA Staining (Direct)

Samples of frozen, unfixed cells are thawed, and aliquots of ~10^5 cells are stained in a series of steps, during which the sample tubes, protected against intensive light, are slowly agitated in an ice bath mounted on a mixer (~100 rpm). No washings are applied. The reagents are added stepwise on top of each other to a final sample volume of 250–300 μl. The samples are not filtered before FCM.

1. For 15 minutes add 200 μl lysis–DNA-staining solution [calcium- and magnesium-free Dulbecco's PBS, Nonidet P-40 (NP-40; BDH) 0.5%

v/v, propidium iodide (PI; Sigma) 20 μg/ml, RNase (Sigma, R-5503) 0.2 mg/ml, EDTA 0.5 mM, pH 7.2].

2. For at least 30 minutes, add 25 μl FITC-conjugated monoclonal Ki-67 antibody [DAKO F-788, diluted 10× in PBS with 1% bovine serum albumin (BSA; Behring) to ~10 μg/ml mouse IgG] or equivalent amount of irrelevant, isotype control antibody (e.g., FITC-conjugated antithyroglobulin antibody (DAKO)].

B. Ki-67–DNA Staining (Indirect)

1. For the first 15 minutes, add 200 μl lysis–DNA-staining solution.
2. For the subsequent 15 minutes, add 25 μl monoclonal Ki-67 antibody (DAKO-PC, M-722, diluted 10× in PBS with 1% BSA to ~10 μg/ml mouse IgG) or equivalent amount of irrelevant, isotype control antibody [e.g., anti-Von Willebrand factor antibody (DAKO M-616)].
3. For at least 15 minutes, add 25 μl FITC-conjugated rabbit antimouse antibody [DAKO F-313, F(ab')$_2$ fragment, diluted 10× in PBS with 5% normal rabbit serum (DAKO X-902)].

C. BrdUrd–DNA Staining

For the BrdUrd–DNA analysis, cell cultures are incubated with BrdUrd (5-bromo-2'-deoxyuridine) at a final concentration of 10–50 μM for a period of 4–96 hours. Cells should be protected from intensive light exposure during the incubation with BrdUrd as well as during the staining procedure. For elimination of nonincorporated BrdUrd, the thawed cells are washed once in PBS with 1% BSA before staining. Cells are resuspended to a concentration of ~10^5 cells in 50 μl PBS with 1% BSA. During all steps, except step 2 below (HCl, room temperature, RT), sample tubes are slowly agitated in an ice bath.

1. For 15 minutes, add 100 μl lysis–DNA-staining solution.
2. Add 25 μl 1 N HCl, and agitate for 30 minutes at RT.
3. Add 75 μl 1 M TRIS (TRIZMA base, Sigma T-1503), pass the sample carefully through a pipet (Gilson Pipetman, 200-μl tip), and agitate for 5 minutes (return to ice bath).
4. For 15 minutes add 25 μl anti-BrdUrd antibody (DAKO-BrdUrd, M-744, clone Bu20a, diluted 10× in PBS with 1% BSA; alternatively ABDM antibody, Bio Cell Consulting, Grelligen, Switzerland, diluted 20×; or anti-BrdUrd antibody, Becton Dickinson 7580, diluted 10×) or equivalent amount of irrelevant, isotype control antibody, for example anti-Von Willebrand factor antibody.

5. For at least 15 minutes, add 25 μl FITC-conjugated rabbit antimouse antibody, diluted 10× in PBS with 5% normal rabbit serum.

IV. Critical Aspects of the Procedure; Controls and Standards

In any series of measurements, the functioning of the reagents, the staining technique, and the flow cytometer should be controlled by analysis of simultaneously stained samples with known proportions of antigen-positive and negative cells (as recommended by Baisch and Gerdes, Chapter 22, this volume). Repeated measurement of the control samples makes it possible to observe changes in level of green fluorescence detection that might occur during measurement of a long series of samples. Chicken (CRBC) and trout erythrocytes (TRBC), having ~35% and 80% of the human diploid G_0/G_1 DNA content and being Ki-67-negative and BrdUrd-negative, may be useful as internal standards when added to the sample before staining (Vindeløv et al., 1983b). Stained samples may be kept in the refrigerator and measured the following day.

With respect to the tolerance in composition of the lysis–DNA-staining solution, the Ki-67 staining of nuclei of PHA-stimulated lymphocytes and HL-60 leukemia cells is adequate within the limits of 0.1 – 1% NP-40, 0.5 – 10 mM EDTA, 0–1% BSA, 1 : 10–1 : 1 concentration of PBS, and pH 7.2–8.0. However, the staining of Ki-67 is impaired by treatment with 1% formaldehyde, 0.15 N HCl (the dose used for BrdUrd staining; Section III, C), 0.1% citric acid, 0.01% dithiothreitol (DTT), or trypsin (as in solution A of Vindeløv et al., 1983a). The staining can be speeded up by increasing the lysis efficiency (more detergent, less BSA, lower ionic strength, higher pH) and/or the force of sample agitation. Centrifugation of the unfixed nuclear suspension results in more-or-less selective aggregation and loss of nuclei.

To ensure operation within the range of stoichiometric detection of antigen expression, it is important to maintain optimal proportion between the amount of antibody and the number of nuclei, allowing saturation of the antigen also in the samples with highest fractions of antigen-positive nuclei. At the same time, the dose of FITC-conjugated antibody should be kept below the critical level, at which background fluorescence from free FITC-conjugated antibody submerges the green-fluorescence pulses from antigen-negative nuclei. With respect to instrumentation, measurement of the nuclear FITC fluorescence should be optimized as for PI fluorescence, by using a relatively narrow core stream and a critically limited observation area for the fluorescence-collecting optics.

With the direct Ki-67–DNA staining technique, a lack of balance between dose of antibody and number of nuclei, as might be indicated by an unexpectedly high count rate on the flow cytometer, can be corrected by addition of supplementary volumes of lysis–DNA-staining solution and FITC-conjugated Ki-67 antibody.

With the indirect method, it is important that the samples to be compared contain an equal number of cells, because supplementary dosage of primary antibody is not possible, after the secondary antibody already has been added. For each particular antigen, one must find the optimal dosage by running samples with various mixtures of antigen-positive and -negative cells, at various combinations of doses of primary and secondary antibodies.

In case of doubt, and always when new cell lines or biopsy material are to be studied, it is recommended that the quality of the nuclear suspension under the fluorescence microscope be controlled, and it should be taken into account that even nuclei looking very pure are not necessarily permeable enough to give optimal access for the antibody to the antigenic sites.

V.   Results

Figure 1 shows examples of the estimates of the Ki-67–DNA distribution (Fig. 1a–c), and the BrdUrd–DNA distribution (Fig. 1d–f). Examples of the DNA distribution are shown in Fig. 2. These estimates can be obtained with washless staining of unfixed nuclei of normal human peripheral blood lymphocytes (PBL), isolated as the mononuclear fraction by density centrifugation of peripheral blood. The nuclei of unstimulated PBL are confined to a single peak with low FITC fluorescence and low PI fluorescence (G_0 phase). After stimulation with PHA for 48 and 96 hours, an increasing fraction of the PBL nuclei occurs with significantly increased FITC fluorescence and distributed throughout the G_1, S, and G_2M phases according to their PI fluorescence. The nuclei of all cells recruited into the cell cycle are thus recognized as Ki-67-positive, in concordance with the results obtained with alternative methods (Baisch and Gerdes, 1987; Palutke et al., 1987; Sasaki et al., 1987; Drach et al., 1989; see also Baisch and Gerdes, Chapter 22, this volume). The BrdUrd–DNA distribution of PBL incubated with 50 μM BrdUrd from 24 hours after addition of PHA, at which time the cells start entering the cell cycle, shows at 48 hours (Fig. 1e)

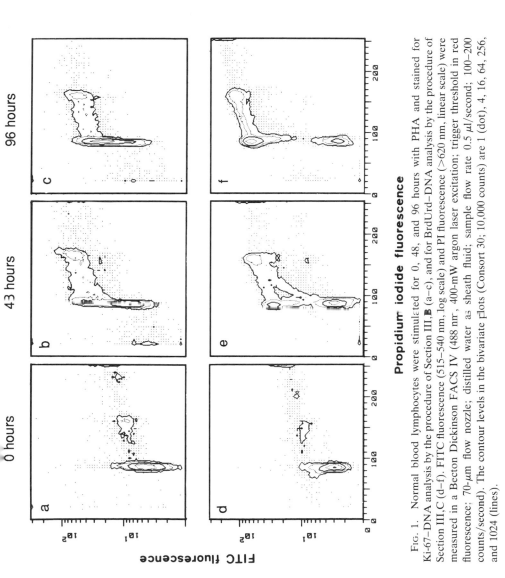

Propidium iodide fluorescence

FIG. 1. Normal blood lymphocytes were stimulated for 0, 48, and 96 hours with PHA and stained for Ki-67–DNA analysis by the procedure of Section III,B (a–c), and for BrdUrd–DNA analysis by the procedure of Section III,C (d–f). FITC fluorescence (515–540 nm, log scale) and PI fluorescence (>620 nm, linear scale) were measured in a Becton Dickinson FACS IV (488 nm, 400-mW argon laser excitation; trigger threshold in red fluorescence; 70-μm flow nozzle; distilled water as sheath fluid; sample flow rate 0.5 μl/second; 100–200 counts/second). The contour levels in the bivariate plots (Consort 30; 10,000 counts) are 1 (dot), 4, 16, 64, 256, and 1024 (lines).

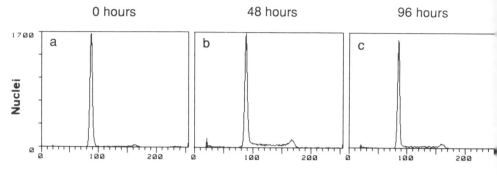

Propidium iodide fluorescence

FIG. 2. Univariate PI fluorescence distributions from the Ki-67–DNA. Stained samples shown in Figs. 1a–c (Section III,B).

two different lanes of S-phase nuclei with differing FITC fluorescence, indicating cells in the first and second S-phase transit, whereas at 96 hours (Fig. 1f) no more cells are entering S from G_0.

ACKNOWLEDGMENTS

This work was supported by the Danish Medical Research Council, the Danish Cancer Society, the Simon Spies Foundation, and the Mogens and Jenny Vissing Foundation.

REFERENCES

Baisch, H., and Gerdes, J. (1987). *Cell Tissue Kinet.* **20,** 387–391.
Dolbeare, F., Gratzner, H., Pallavicini, M. G., and Gray, J. W. (1983). *Proc. Natl. Acad. Sci. U.S.A.* **80,** 5573–5577.
Drach, J., Gattringer, C., Glassl, H., Schwarting, R., Stein, H., and Huber, H. (1989). *Cytometry* **10,** 743–749.
Giaretti, W., Di Vinci, A., Geido, E., Marsano, B., Minks, M., and Bruno, S. (1990). *Cell Tissue Kinet.,* in press.
Larsen, J. K., Mortensen, B. T., Christiansen, J., Christensen, I. J., Lykkesfeldt, A., Thorpe, S. M., Rose, C., and Mouridsen, H. T. (1987). *Cytometry Suppl.* **1** (Abstr. 350).
Larsen, J. K., Christensen, I. J., Christiansen, J., Mortensen, B. T., Goclawska, A., and Kieler, J. (1989). *Cell Tissue Kinet.* **22,** 198 Abstr.
Palutke, M., KuKuruga, D., and Tabaczka, P. (1987). *J. Immunol. Methods* **105,** 97–105.
Sasaki, K., Murakami, T., Kawasaki, M., and Takahashi, M. (1987). *J. Cell. Physiol.* **133,** 579–584.
Vindeløv, L. L., Christensen, I. J., and Nissen, N. I. (1983a). *Cytometry* **3,** 323–327.
Vindeløv, L. L., Christensen, I. J., and Nissen, N. I. (1983b). *Cytometry* **3,** 328–331.

Chapter 24

Analysis of Proliferation-Associated Antigens

KENNETH D. BAUER

Department of Pathology
Northwestern University School of Medicine
Chicago, Illinois 60611

I. Introduction

Recent developments have made a number of monoclonal antibody (mAb) probes that recognize intracellular antigens associated with cell growth or proliferation available to researchers. Many of these proteins are localized to the cell nucleus, an area of the cell that, until recently, was largely overlooked for immunofluorescence (IF) measurements by flow cytometry (FCM). Major considerations for the previous inattention to the cell nucleus include the fact that, while it is easy to conceptualize mAb saturating epitopes (required for protein quantification by IF) on the outside of the cell, the high density of intracellular proteins as well as regions of hydrophobicity makes this same possibility more complex when applied to nuclear proteins. Second, a considerably higher background staining has been observed for years by many investigators when IF methods are applied to the analysis of intracellular antigens.

In order to achieve appropriate conditions for introduction of the mAb of interest into the cell, one must first fix and permeabilize the cell. This has been accomplished to date using one of two basic strategies: (1) alcohol fixation or (2) brief incubation with paraformaldehyde at low concentration followed by short detergent (e.g., Triton X-100) treatment. The alcohol fixatives ethanol and methanol have historically been the preferred

METHODS IN CELL BIOLOGY, VOL. 33

fixatives for simultaneous DNA and cell surface IF analysis (Braylan *et al.*, 1982). Other data (Jacobberger *et al.*, 1986) indicate the utility of cold-alcohol fixation for the analysis of DNA and nuclear proteins. Alcohol fixatives have been shown in some cases, however, to cause the loss and/or redistribution of nuclear proteins, possibly due in part to hypotonicity and dehydrating effects (Clevenger *et al.*, 1985; Epstein and Clevenger, 1985).

An alternative fixation strategy for intracellular IF and DNA analyses involves brief sequential incubations with paraformaldehyde or formaldehyde followed by a detergent such as Triton X-100 (Clevenger, *et al.*, 1985; Mann *et al.*, 1987). The aldehyde fixatives formaldehyde and paraformaldehyde have been reported to form crosslinks between protein end groups that are largely reversible in aqueous solution (Pearse, 1980). While excellent for IF, these fixatives when used with intercalating DNA fluorochromes such as propidium iodide (PI), yield comparatively poor-quality DNA staining as indicated by a broad coefficient of variation (CV) of the mean DNA content of the G_1/G_0 peak. When fixation with these aldehydes is followed by brief detergent treatment, enhanced IF intensity has been documented relative to paraformaldehyde only, suggesting that the detergent provides improved accessibility of epitopes on chromatin proteins (Clevenger *et al.*, 1985). This fixation–permeabilization strategy also results in superior DNA staining in terms of low CV and increased fluorescence intensity (Clevenger *et al.*, 1985). Despite the usefulness of this fixation method for the analysis of many intracellular proteins, epitopes on many proteins can be sequestered by cross-linking fixatives, such that they are no longer recognizable by the mAb. In such cases, alcohol fixatives represent a reasonable alternative.

Recent reports have documented the analysis of a number of intracellular proliferation-associated proteins by FCM. Among those analyzed to date are the oncoproteins p53 and c-*myc* (Darzynkiewicz *et al.*, 1986; Rabbitts *et al.*, 1985); Ki-67 antigen, a nucleolus-associated protein that appears to discriminate G_0 from cycling cells in a number of cell systems (Baisch and Gerdes, 1987); cyclin or PCNA, a proliferation-associated protein that is preferentially expressed in late G_1 phase and S phase relative to other cell cycle phases (Kurki *et al.*, 1986); and p105, an interchromatin granule-associated protein that appears to be differentially expressed in G_0 versus G_1 and G_2 versus M phases of the cell cycle (Clevenger *et al.*, 1987a, b). Surprisingly, little has been done to date to examine possible differences in expression of such antigens in proliferative disorders or neoplasia. Given the loss or alteration of growth control mechanisms associated with cancer, such analysis could provide useful insight into the biology of cancer and could have diagnostic and/or prognostic utility.

II.   Application

Analysis of the type just described appears applicable to a wide spectrum of cells, including cells isolated from tumor tissue. General proteases such as trypsin and pepsin, which are often required to dissociate such tissues, however, can destroy protein(s) of interest, such that the interpretation of IF results becomes very difficult (Lincoln and Bauer, 1989). To date, nearly all analysis of proliferation-associated proteins has been performed on whole cells, although some reports suggest the feasibility of such measurements using isolated nuclei obtained from paraffin-embedded pathology material (Watson, 1986; Anastasi et al., 1987; Bauer et al., 1986b).

Two recent studies by Kastan and colleagues indicate that the analysis of proliferation-associated antigens using indirect IF with FITC may be coupled with direct IF analysis of leukocyte surface antigens (Kasten et al. 1989a, b). With the exception of these studies, the analysis of proliferation-associated proteins to date has been restricted almost exclusively to one-color IF + DNA content analysis, along with light scatter measurements. This reflects the fact that FITC has been the fluorochrome of choice for intracellular IF analysis. Whereas FITC (MW 389) is a small fluorochrome, other fluorochromes conjugatable to antibodies are either of large molecular weight (e.g., R-phycoerythrin, MW 240,000), which can result in steric inhibition of antibody binding to nuclear proteins, or overlap spectrally with DNA fluorochromes such as PI (e.g., Texas red). Engelhard et al., (1990) have introduced a three-color dual-laser method that allows for the simultaneous analysis of IF (using an FITC-conjugated antibody), DNA content (using 4', 6-diamidino-2-phenylindole, DAPI), and total cellular protein (using sulforhodamine 101). With the recent development of low molecular weight red-exciting dyes that can be conjugated to antibodies (Waggoner, 1986), dual-color IF analysis of intracellular proteins appears likely in the near future.

III.   Materials

A.   Formaldehyde and Triton X-100 Solutions for Cell Fixation and Permeabilization

Formaldehyde solution (10%) is diluted in phosphate-buffered saline (PBS), pH 7.2, to a final concentration of 0.5%. A highly purified solution is recommended and is available from Polysciences, Inc. (Warrington, PA;

formaldehyde, 10%, ultrapure, EM grade, catalog no. 4018. The stock solution is stable for 12 months at room temperature (RT). Diluted formaldehyde kept at 4°C is stable for months. The pH of this solution, however, should be checked occasionally to assure that it has not acidified.

Triton X-100 (0.1%) (Sigma Chemical Co., St. Louis, MO) is prepared in PBA (Triton–PBA : PBS + 0.1% Na azide + 0.1% bovine serum albumin, BSA) and stored at 4°C.

B. Alternate Fixation Using Methanol

Absolute methanol is chilled to between −70° and −80°C.

C. Primary Antibody Solution for Indirect IF

The primary antibody is suspended in PBA. To optimize quantitative IF measurements by FCM, additional factors that require careful consideration include titration of the mAb of interest relative to an isotype control at the same concentration in an effort both to optimize the saturation of epitopes and to assure the specificity of the IF staining. Optimal staining for intracellular antigens often occurs in the 1–4 $\mu g/ml$ range. The use of a carrier protein (e.g., 0.1–0.5% BSA or 1–4% goat serum) in both the primary and secondary antibody solutions minimizes nonspecific staining by the antibody of interest.

The number of commercially available sources of mAb recognizing proliferation-associated antigens is increasing rapidly. Table I provides a partial listing of such reagents and vendors.

D. Secondary Antibody Solution for Indirect IF

To enhance the specificity of antibody staining with lower background, fluorescein isothiocyanate (FITC)–goat anti-mouse immunoglobulin $F(ab')_2$ fragments as opposed to whole antibody are recommended for staining. This should be titered to saturation. Second, the use of antibodies that have been adsorbed against human Ig to minimize nonspecific staining can be of considerable value in terms of reducing nonspecific antibody adherence.

E. DNA-Staining Solutions

RNase A is required to eliminate RNA binding by PI. Since many RNase preparations available are relatively crude preparations containing

TABLE I

SOURCE OF COMMERCIALLY AVAILABLE MONOCLONAL ANTIBODIES TO
NUCLEAR PROTEINS

Monoclonal antibody	Vendor[a]
Anti Ki-67 antigen	DAKOPATTS, Santa Barbara, CA, catalog no. M722 AMAC, Inc., Westbrook, ME, catalog no. 0256
Anti-proliferating cell nuclear antigen (PCNA)	American Biotech Plantation, FL, catalog no. ABT 151P Boehringer Mannheim Biochemicals, Indianapolis IN, catalog no. 12 02 693
Anti-c-*fos* oncoprotein	Oncogene Science, Inc., Manhasset, NY, catalog no. OP17
Anti-c-*myc* oncoprotein	Cambridge Research Biochemicals, Valley Stream, NY, catalog no. OM-11-906
Anti-c-*myb* oncoprotein	Microbiological Associates, Bethesda, MD, catalog no. Myb 132, myb 133

[a] Since this list was compiled additional sources for these reagents may
be available.

other enzymatic contaminants including DNase and proteases, a purified
RNase preparation is recommended. This can be accomplished by either
boiling a crude preparation for 5 minutes or purchasing a chromatographi-
cally-purified RNase A (e.g., Worthington Biochemical Corp., Freehold,
NJ, catalog code RASE). Stock RNase (3600 units/ml) solution is diluted
in PBS and stored at −20°C.

Propidium iodide is maintained as a stock solution at 1 mg/ml in distilled
water (dH$_2$O). A purified PI is recommended for DNA staining (e.g.,
Calbiochem, San Diego, CA, catalog no. 537059). Propidium iodide and
other DNA fluorochromes are suspected carcinogens and should be handled and discarded appropriately. Particular caution is recommended when
weighing out pure dyes in powder form to avoid inhalation.

IV.   Staining Procedures

A.   Cell Fixation with Paraformaldehyde

1. Fix $\sim 1 - 2 \times 10^6$ cells in 1 ml of 0.5% formaldehyde solution (dilute stock solution 1 : 20 in PBA) for 30 minutes on ice.
2. Centrifuge cells 10 minutes at 200 g.
3. Add 0.1% Triton–PBA solution (10^6 cells/ml). Incubate for 3 minutes on ice.

B.   Alternate Fixation Using Methanol

1. Centrifuge 2×10^6 cells at 1000 rpm for 10 minutes. Pour off supernatant, resuspend cell pellet.
2. Rinse cells twice in 1 ml HEPES–Hanks' or PBS, carefully resuspending cell pellet in each case.
3. Add 100 μl PBS. Add 900 μl $-70°$C MeOH dropwise while vortexing tube gently.
4. Incubate 30 minutes at $-20°$C.
5. Centrifuge, pour off supernatant, resuspend pellet.
6. Repeat step 2.
7. Store cells in HEPES–Hanks' or PBS, or proceed directly to Triton X-100 incubation for indirect-IF staining procedure.

C.   Indirect Immunofluorescence Staining

1. Wash once in PBA (*optional*).
2. Resuspend in remaining supernatant and add 30 μl of primary mAb at appropriate concentration (and isotype control at the same concentration in a parallel tube). Dilute the mAb in 970 μl of PBA per 10^6 cells. Incubate 1 hour at 4°C.
3. Wash in 0.1% Triton–PBA, then in 1 ml of PBA, carefully aspirating down to cell pellet. Centrifuge.
4. Add 50 μl FITC-conjugated goat anti-mouse immunoglobulin antibody (properly diluted, e.g. 1 : 20, in PBA) to resuspended cell pellet. Incubate 30 minutes at 4°C.
5. Centrifuge. Wash in 0.1% Triton–PBA.
6. Repeat step 5.

D.   Propidium Iodide Staining

1. RNase-treat 180 units/ml (dilute frozen RNase stock 1 : 20 in PBA) 20 minutes at 37°C, centrifuge 200 g for 10 minutes.
2. Pour off supernatant and resuspend in PI (50 μg/ml) for at least 1 hour in the dark at 4°C. (This concentration represents a 1 : 20 dilution of the PI stock solution in dH$_2$O.)

E.   Synthetic Peptide Inhibition

This method is outlined by Kastan *et al.* (1989b).

1. Previous to addition of primary antibody, add 40 μl FBS (final concentration 4%) to each tube.
2. Add 10μl of a 1 μg/5 μl solution of appropriate peptide to each tube (it is also valuable to add an irrelevant synthetic peptide as a control).

V.   Critical Aspects

A.   Antigen Sensitivity to Proteolysis

Trypsin is a proteolytic enzyme that hydrolyzes peptide bonds whose carbonyl group is provided by lysine or arginine. Pepsin is well recognized to have a broad range of nonspecific proteolytic activities. Among other cleavage sites, pepsin has been shown to cleave preferentially bonds of leucine, phenylalanine, methionine, and tryptophan to another hydrophobic residue. Finally, even proteolytic enzymes that in principle are highly specific for a particular protein (e.g., collagenase) are often sold in a form contaminated with other enzymatic components that can destroy or alter protein(s) of interest. Thus, for quantification of a particular protein using indirect IF methods described, it is important to address whether freshly dissociated cells obtained from the tissues without proteolytic enzymatic treatment show comparable levels of the protein of interest as compared to cells obtained in the absence of proteolytic treatment.

Useful for such analyses are *in vitro* cell lines grown in suspension containing the antigen of interest, which can be divided into two groups (control vs enzyme-treated) and then analyzed for IF by FCM and/or by immunoblot analysis. An example of this approach is provided in Lincoln and Bauer (1989).

B. Antigen Verification Using Sodium Dodecyl Sulfate (SDS)–Gel Electrophoresis and Immunoblot Analysis

Recent years have seen the rapid introduction of novel mAb probes with purported specificity to proliferation-associated antigens including oncoproteins. While such probes provide potentially powerful new FCM-based quantitative IF assays, it is important to realize that a mAb recognizes an epitope that consists of an oligopeptide on the order of six to seven amino acid residues as opposed to the whole protein of interest. Thus, given the thousands of proteins within a cell, a critical test involves independent verification of the specificity of the mAb. This can be easily accomplished using SDS–polyacrylamide gel electrophoresis (PAGE), followed by immunoblot analysis using the same mAb in combination with avidin–biotin immunoperoxidase staining to evaluate whether one or more protein bands are recognized.

In addition to this application, another use of SDS–PAGE and immunoblot analysis for antigen verification stems from possible artifacts in quantitative IF analysis of nuclear proteins resulting from chromatin structure differences. Specifically, the accessibility of epitopes recognized by the mAb of interest can vary in relation to proliferation- or differentiation-dependent changes in chromatin structure. To evaluate the possibility of chromatin effects, parallel analyses are recommended in which IF levels by FCM are compared with immunobiochemical detection by SDS–PAGE followed by immunoblot analysis on cell lysates (Clevenger *et al.*, 1987a,b).

C. Verification of Proliferation-Associated Expression of Intracellular Antigens

In assessing the proliferation-associated expression of a nuclear antigen, it is also important to recognize that a substantial increase (up to 4-fold) in total cell mass and protein content occurs as a consequence of normal cell growth during the transition from G_0 to G_1 (Darzynkiewicz *et al.*, 1967). As a result, increments in IF between G_0 and G_1 may simply reflect a general protein increase rather than the true proliferation-associated protein. Both PAGE and immunoblot densitometry can be used to verify proliferation-associated expression of an antigen of interest by loading a *constant* level of protein per well on the electrophoresis apparatus from these cell cycle subpopulations. Thus, the modulation of a specific protein can be independently evaluated on the basis of constant total protein content.

A possible alternative to the approach just described is provided by a three-color FCM method, in which a specific antigen is examined simultaneously with DNA and total protein contents (Engelhard *et al.*, 1990). This approach allows for the direct assessment of the relationship between modulations in a specific protein and total protein content.

One limitation in the analysis of antigen expression in relation to cell proliferation is the fact that conventional DNA FCM using dyes such as PI cannot resolve G_0 versus G_1 populations. Thus, differential expression of an antigen between these kinetically distinct subpopulations cannot be assessed. To circumvent this problem, the authors (Bauer *et al.*, 1986a) have documented the feasibility of restaining sorted cells previously stained for IF with the acidic acridine orange method of Darzynkiewicz *et al.* (1975; see also Darzynkiewicz, Chapters 27 and 32, this volume), which identifies G_0 versus G_1 populations in many cell systems. Thus, by sorting cells of varying IF, one can directly relate the IF to the above cell cycle compartments which cannot be resolved when IF analysis is coupled with conventional DNA staining.

D. Synthetic Peptide Blocking

An alternate strategy for assessing the specificity of a mAb involves blocking with a synthetic peptide representing the epitope recognized by the mAb. When using such a peptide in excess prior to the addition of the primary mAb, IF staining should be reduced to control levels if the mAb is specific to this epitope. It is important to recognize, however, that this type of blocking does *not* assure that the mAb is specific to a single protein (see discussion in Section V, A above). One report (Kastan *et al*, 1989b) documents the utility of synthetic peptide blocking for examining expression of c-*myb*, c-*myc*, and c-*fos* in normal hematopoietic cells.

VI. Controls and Standards

Current standards for instrument alignment most commonly utilize nonbiological particles with relatively stable and homogeneous fluorescence intensity and light-scattering characteristics. A second type of standard that appears crucial is a staining standard. Cells known to contain the antigen of interest can serve as a useful staining standard for quantitative IF measurements. By parallel IF staining of test cells, this type of biological standard can be used to measure relative levels of the antigen of interest. While *in vitro* cells can be quite useful for this purpose, it is important to recognize that protein modulation in *in vitro* systems is frequent in relation to culture growth phase (e.g., exponential vs plateau) and/or passage. Thus, a large number of cells from a *single* culture placed in aliquots into vials and frozen in a cryogenic freezer can provide a consistent biological standard analyzable from run to run for this purpose. A useful variation on this general strategy involves the use of two (or more) such standards varying in antigen content to address critically the specificity and/or the linearity of quantitative IF measurements.

VII.   Instruments

Simultaneous analysis of DNA content and proliferation-associated antigens is generally performed using the combination of PI as the DNA fluorochrome and FITC-conjugated antibody for IF analysis. Both of these dyes are excited by the 488-nm laser line on argon ion lasers. Thus, analysis to date has been restricted almost exclusively to laser-based FCM systems. All commercial laser-based instruments should allow for this analysis. One filter combination available for this purpose is as follows: 488-nm laser interference filter, 560-nm short-pass dichroic, 530-nm short-pass + 525-nm bandpass filter (green FITC IF), 640-nm long-pass filter (red PI fluorescence).

VIII.   Results

Two examples are provided that illustrate FCM analysis of cell proliferation-associated antigens. Figure 1 illustrates dual-parameter anti-p34 green fluorescence versus PI red fluorescence contour plots for human pokeweed mitogen (PWM)-stimulated lymphocytes. Results from 0-hour cultures and 72-hour PWM-stimulated cultures are illustrated on the upper and lower panels, respectively. Note that whereas the 0 hour cultures are homogeneous in both red (DNA) and green (anti-p34) fluorescence, the 72-hour cultures illustrate two discrete populations of cells in the G_1/G_0 DNA region (red fluorescence channels 17–25). The dimmer population corresponds to G_0-like cells whereas the brighter population (i.e., $\sim$ 3-fold

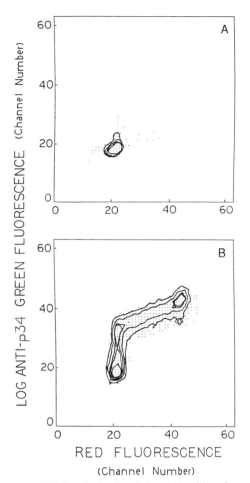

FIG. 1. Dual-parameter FCM analysis of peripheral blood lymphocytes (PBL) fixed with 0.5% formaldehyde–0.1% Triton X-100 and stained with anti-p34 mAb (indirect IF) and PI. (A) 0 hour PBL. (B) 72 hours PBL cultures stimulated with pokeweed mitogen (5 μg/ml) in culture medium. For experimental details, see Bauer (1986a). Anti-p34 mAb recognizes a heterochromatin protein and was produced by Dr. Alan Epstein, University of Southern California. Isocell contour levels of 15, 45, 135, and 225 cells are shown.

increase in p34) corresponds to G_1-like cells. See Bauer *et al.* (1986a) for details.

Figure 2 illustrates dual-parameter FCM results for Raji cells stained for p105, an interchromatin-associated protein using anti-p105 indirect IF (green fluorescence) and DNA content (red fluorescence). Zero hour

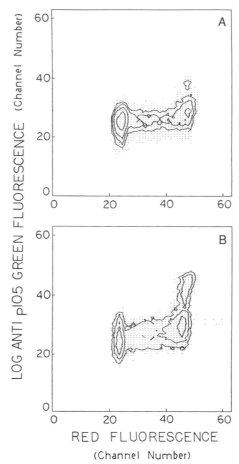

FIG. 2. Dual-parameter FCM profiles of Raji (Burkitt's lymphoma) cell line fixed and permeabilized with 0.5% formaldehyde–0.1% Triton X-100 and stained with anti-p105 (indirect IF) and PI. (A) Exponentially growing Raji cells. (B) Raji cells incubated in the presence of colchicine (1 μg/ml) in culture medium for 8 hours. Experimental details are provided in Clevenger *et al.* (1987b). Anti-p105 recognizes mAb an interchromatin-associated protein and was produced by Dr. Alan Epstein, University of Southern California. Isocell contour levels of 10, 30, 90, and 270 cells are shown.

cultures are illustrated in the upper panel, whereas the lower panel shows results following 8 hours incubation with colchicine (1 μg/ml). Note that the G_2M DNA region (red-fluorescence channels 43–51) illustrates two distinct populations of varying p105 content. The brighter population corresponds to mitotic cells whereas the dimmer population corresponds to cells in G_2 phase. The bright mitotic population constitutes only a small proportion of the cells in 0 hour cultures but is increased markedly following 8 hours of colchicine treatment. See Clevenger *et al.* (1987b) for details. Other examples of the analysis of proliferation-associated antigens are referenced throughout this manuscript.

REFERENCES

Anastasi, J. A., Bauer, K. D., and Variakojis, D. V. (1987). *Am. J. Pathol.* **128,** 573.

Baisch, J., and Gerdes, J. (1987). *Cell Tissue Kinet.* **20,** 287.

Bauer, K. D., Clevenger, C. V., Williams, T. J. and Epstein, A. L. (1986a). *J. Histochem. Cytochem.* **34,** 2428.

Bauer, K. D., Clevenger, C. V., Endow, R. K. Murad, T. M., Epstein, A. L., and Scarpelli, D. G. (1986b). *Cancer Res.* **46,** 2428.

Braylan, R. C., Benson, N. A., Nourse, V. and Kruth, H. S. (1982). *Cytometry* **2,** 337.

Clevenger, C. V., Bauer, K. D., and Epstein, A. L. (1985). *Cytometry* **6,** 208.

Clevenger, C. V., Epstein, A. L., and Bauer, K. D. (1987a). *J. Cell. Physiol.* **130,** 336.

Clevenger, C. V., Epstein, A. L., and Bauer, K. D. (1987b). *Cytometry* **8,** 280.

Darzynkiewicz, Z., Dokov, V., and Pienkowski, M. (1967). *Nature (London)* **214,** 1265.

Darzynkiewicz, Z., Traganos, F., Sharpless, T., and Melamed, M. R. (1975). *Exp. Cell Res.* **90,** 411.

Darzynkiewicz, Z., Staiano-Coico, L., Kunicka, J., DeLeo, A., and Old, L. J. (1986). *Leuk. Res.* **10,** 1383.

Engelhard, H. H., Krupka, J. L., and Bauer, K. D. (1990). *Cytometry* (In Press).

Epstein A. L., and Clevenger C. V. (1985). *In* "Recent Advances In Non-histone Protein Research" (I. Bekhor, ed.), Vol. 1, pp. 117–137. CRC Press, Boca Raton, Florida.

Jacobberger, J. W., Fogelman, D., and Lehman, J. M. (1986). *Cytometry* **7,** 356.

Kastan, M. B., Slamon, D. J., and Civin, C. I. (1989a). *Blood* **73,** 1444.

Kastan, M. B., Stone, K. D., and Civin, C. I. (1989b). *Blood* **74,** 1517.

Kurki, P., Vanderlaan M., Dolbeare F., Gray J., and Tan, E. M. (1986). *Exp. Cell Res.* **166,** 209.

Lincoln, S. T., and Bauer, K. D. (1989). *Cytometry* **10,** 456.

Mann, G. J., Dyne, M., and Musgrove, E. A. (1987). *Cytometry* **8,** 509.

Pearse, A. G. E. (1980). *In* "Histochemistry—Theoretical and Applied" Vol. 1, p. 97.

Rabbitts, P. H., Watson, J. V., Lamond, A., Forster, A., Stinson, M. A., Evan, G., Fischer, W., Atherton, E., Sheppard, R., and Rabbitts, T. H. (1985). *EMBO J.* **4,** 2009.

Waggoner, A. S. (1986). *In* "Applications of Fluorescence in the Biomedical Sciences" (Taylor, D. L., Waggoner, A. S., Murphy, R. F., Lanni, F., and Birge, R. R. (eds.), pp. 3–28. Liss, New York.

Watson, J. V. (1986). *Cytometry* **7,** 400.

Chapter 25

The Stathmokinetic Experiment: A Single-Parameter and Multiparameter Flow Cytometric Analysis

FRANK TRAGANOS

*Cancer Research Institute of the
New York Medical College at Valhalla
Valhalla New York 10595*

MAREK KIMMEL[1]

*Investigative Cytology Laboratory
Memorial Sloan-Kettering Cancer Center
New York, New York 10021*

I. Introduction

The stathmokinetic or "metaphase-arrest" technique, first described by Puck and Steffen (1963), was introduced as a procedure for estimating cell production or birthrates. By determining the percentage of mitotic cells in a culture at various times following the addition of a chemical agent that arrests cells during division, it is possible to obtain a plot of the increase in mitosis as a function of time. If the growth fraction (GF) is known, the slope of the mitotic accumulation curve will, under certain circumstances (see later), provide an estimate of the cell cycle duration or cell doubling time (Aherne *et al.*, 1977; Wright and Appleton, 1980).

This stathmokinetic approach has been widely used over the past several decades to analyze the kinetics of cell growth both *in vitro* and *in vivo*. Wright and Appleton (1980) reviewed various stathmokinetic techniques and discussed their advantages and limitations. Several reviews detailing

[1] Present address: Department of Statistics, Rice University, Houston, Texas 77251.

METHODS IN CELL BIOLOGY, VOL. 33

flow cytometric (FCM) analysis of stathmokinetic experiments are available (Darzynkiewicz, 1984; Darzynkiewicz et al., 1986; Gray et al., 1986), as are mathematical analyses and interpretation of data obtained from such experiments (Kimmel and Traganos, 1985, 1986; Kimmel et al., 1983; Macdonald, 1981; Sharpless and Schlesinger, 1982).

II. Theoretical and Practical Considerations

Before describing the use of FCM to analyze stathmokinetic experiments, it is necessary to discuss the theoretical and practical problems involved in this approach.

A.   Asynchronous, Exponential Growth

Direct analysis of a stathmokinetic experiment presumes that the entire cell population of interest is growing asynchronously and exponentially. While this is generally true of immortal cell culture lines, other *in vitro* cell systems (e.g., mitogen-stimulated lymphocytes) and most *in vivo* cell populations do not fulfill this criteria. Thus, when cell populations contain quiescent cells (G_0, G_{1Q}), differentiating cells (G_{1D}), cells in transition between cycling and quiescence (G_{1T}, S_T, G_{2T}), or dying cells, the analysis of a stathmokinetic experiment becomes complex (Kimmel et al., 1986) or impossible (e.g., when these cell cycle compartments cannot be precisely determined). In the present instance, the simplest model, in which all cells are presumed to be growing exponentially, will be discussed.

B.   Choice of Stathmokinetic Agents

The choice of a stathmokinetic agent is another important variable. The perfect stathmokinetic agent blocks cells in mitosis, is nontoxic, and does not perturb cell progression through the other phases of the cell cycle. Several classes of agents have been used for this purpose, but all must be used carefully.

The classical mitotic inhibitor is colchicine, a plant alkaloid that binds to tubulin, preventing its polymerization and thus the formation of the mitotic spindle (Wilson et al., 1976). Colcemid (N-desacetyl-N-methyl colchicine) is a less toxic derivative also used as a mitotic inhibitor (Stubblefield et al., 1967). Another family of "spindle poisons" are the *Vinca* alkaloids vincristine, vinblastine, and the vinblastine analog vindesine. *Vinca* alkaloids should be used with caution, however, since at high concentrations they have been found to affect RNA and DNA synthesis as well as amino acid transport into cells (Creasy, 1975). In our hands, however, vinblastine has been found to be the mitotic inhibitor of preference for normal and tumor cells of lymphocytic origin. Other potentially useful stathmokinetic agents include the ansa macrolid maytansine (Wolpert-DeFillipes *et al.*, 1975), the bis-dioxopiperazines ICRF-159 (razoxane) and its (+)isomer ICRF-187 (Bakowski, 1976) and nocodazole (DeBrabander *et al.*, 1976). Maytansine shares the same mechanism of action as the *Vinca* alkaloids, while nocodazole inhibits microtubule formation. The bis-dioxopiperazines inhibit cell division but not subsequent DNA synthesis, which results in the formation of multinucleated cells (Hallowes *et al.*, 1974). The use of the podophyllotoxins VM-26 and VP-16 as stathmokinetic agents should be avoided, since they appear to have their primary effect on S- and G_2-phase cells (Krishan *et al.*, 1975).

C. Optimal Concentration of the Stathmokinetic Agent

Regardless of the choice of stathmokinetic agent, a dose–response curve must be performed with any agent for each cell system studied to determine the optimal dose that is low enough to minimize toxicity yet adequate to prevent "leakage" of cells through the mitotic block (Darzynkiewicz *et al.*, 1986).

D. Duration of Stathmokinesis

Even under optimal experimental conditions, cells arrested in mitosis die after a period of time, restricting the duration of the experiment. Depending again on cell type, mitotic cell death (and disintegration) can be observed as early as 3.5 hours (Clarke, 1971; Morris, 1967) or as late as 12 hours (Frei *et al.*, 1964; Smith *et al.*, 1974) after addition of the agent. Generally, disintegration of mitotic cells after short periods of incubation with inhibitors only occurs *in vivo*, so that *in vitro* incubation times of 6–12 hours are routine. Nevertheless, for slow-growing populations, few cells will have entered mitosis by the time the first cells blocked in mitosis begin to die. While correction procedures can be utilized to estimate mitotic cell loss (Puck *et al.*, 1964; Aherne and Camplejohn, 1972), the kinetics of rapidly growing cell populations can be determined more precisely. To obtain maximum data on cell cycle kinetics, one would like to have the ability to collect stathmokinetic data for at least half the normal doubling time of the cell system being studied.

E. Specificity

Often, a delay is detected between the addition of a stathmokinetic agent and the accumulation of cells in mitosis. The delay may indicate that there is a lag between the addition of an agent and its effect on mitosis due to the time required to penetrate the cell membrane, or, alternatively, the agent may have induced a transient delay in progression from G_2 phase to mitosis (Fitzgerald and Brehaut, 1970; Darzynkiewicz et al., 1984). More subtle effects of stathmokinetic agents such as the slowing down of cell transit from G_1 to S, through S phase, or from S to G_2 phase are more difficult to detect. However, even these minor effects can be uncovered with careful analysis (see later).

F. Cell Concentration

While it is possible to derive kinetic information from synchronized cell populations, uncontrolled, partial synchronization of exponentially growing cells is to be avoided. Partial synchronization can occur when cells are split from a highly dense (or confluent) culture. Ideally, to ensure exponential growth, cells to be analyzed should be fed and the culture diluted appropriately for several cell cycles prior to the addition of the stathmokinetic agent.

The starting cell concentration should be within 30–50% of the concentration of plateau phase cultures, since too low a cell concentration can produce a "lag" in cell growth and a suboptimal number of cells harvested at each time point. Obviously, too high a cell concentration means the culture is no longer in exponential growth.

G. Sampling

If a large, single culture is to be sampled numerous times during the stathmokinetic experiment (generally the method of choice for suspension cultures), changes in temperature and pH should be kept to a minimum to avoid minor perturbations (e.g., cell synchronization in G_1 phase).

When adherent cell cultures are to be used, multiple plates (dishes) originating from a single culture or pooled cultures, should be set up 12–24 hours before addition of the stathmokinetic agent. In any event, adequate time should be provided for cells to become adherent and exponential growth to resume, since such agents may affect cell adherence. Generally, prewarmed, "conditioned" medium is useful in minimizing the time necessary for cell adherence. Conditioned medium can be prepared by removing medium from 1- to 2-day-old exponentially growing cultures and eliminating dead cells by centrifugation. Cells from individual dishes can be trypsinized at appropriate intervals (30–60 minutes) after addition of the stathmokinetic agent. Since cells arrested in mitosis tend to be only loosely attached or floating, both the supernatant and the trypsinized cell fraction need to be collected and pooled prior to fixation.

III. Single-Parameter versus Multiparameter Approach

A. Single-Parameter Measurement

Rapid, unbiased, and accurate measurements of the cell cycle distribution of large cell populations can be accomplished by FCM. The simplest and most straightforward approach to the FCM analysis of stathmokinesis involves staining the cells with a dye specific for cellular DNA content and measuring the accumulation of cells in the $G_2 + M$ population over time. (Note that G_2-phase and mitotic cells contain identical amounts of DNA and cannot be independently quantified by this approach.) This single-parameter approach has correlated well with the mitotic accumulation curve obtained by visual counting of mitotic figures for a variety of cell lines (Barfod and Barfod, 1980). Dosik *et al.* (1981) extended the technique so that, in addition to measuring cell accumulation in $G_2 + M$, they could also analyze cell numbers in G_1 and S phase during the course of the stathmokinetic experiment. This provided data on the rate of cell exit from G_1 and the transit of cells through S phase. This approach provides a powerful method for the analysis of cell kinetics and can be applied to studies of drug effects on cells. Macdonald (1981) developed a mathematical model for analysis of data obtained in this way.

B. Multiparameter Approach

A second, multiparameter FCM approach provides precise enumeration of cells in G_1, G_{1A}, S, G_2, and M based not only on DNA content (as before), but also on changes in chromatin structure as reflected by the sensitivity of DNA to denaturation (Darzynkiewicz *et al.*, 1977a,b). The staining procedure, utilizing the metachromatic dye acridine orange (AO), which is described in detail in this volume (Darzynkiewicz, Chapter 32), when coupled with the stathmokinetic approach, can provide a great deal of kinetic information.

Several other multiparameter FCM approaches have been described that allow for the discrimination of mitotic cells from G_2 cells. Larsen *et al.* (1986) demonstrated that mitotic cells could be identified using several fluorochromes [e.g., ethidium bromide (EB), propidium iodide (PI), mithramycin (MI)] following nuclei isolation and formaldehyde fixation. In other studies that utilize partial DNA denaturation to allow staining of incorporated 5-bromodeoxyuridine (BrdUrd) for the identification of S-phase cells, it was noted that a mitotic cell population having lower stainability with PI than G_2 cells could be resolved (Moran *et al.*, 1985). This population would have been difficult to identify in circumstances in which S-phase cells failed to incorporate BrdUrd because its PI fluorescence overlapped with S-phase cells (Trinkle *et al.*, 1988). However, Zucker *et al.* (1988) and Epstein *et al.* (1988) discovered that the $G_2 + M$ population identified by PI or EB staining, respectively, could be subdivided into G_2- and M-phase cells based on 90° (right-angle) light scatter. Nüsse *et al.* (1989) combined the BrdUrd–PI staining technique with measurement of the additional parameters of forward and/or right-angle light scatter. In this way, analysis of G_1, S, G_2, and mitotic cells can be obtained even when noncycling or slowly cycling S-phase cells are present (see Geido *et al.*, Chapter 16, this volume).

IV. Experimental Procedures for Flow Cytometric Analysis

A. Cell System

Two cell systems were used in this study: the murine leukemia L1210 and Friend erythroleukemia cell lines. These cell lines grow in suspension and have similar kinetics. Both have doubling times of ~12 hours when in exponential growth. For stathmokinetic experiments these cells should be grown within the concentration range of 2–6×10^5 cells/ml.

B. Stathmokinetic Agent

Previous studies (Darzynkiewicz *et al.*, 1986; Traganos *et al.*, 1987) had shown that both cultures were sensitive to low concentrations of vinblastine. In each instance, vinblastine was added at a concentration of 0.05 μg/ml, which was sufficient to block cells in mitosis but not so toxic as to affect cell progression through other cell cycle phases or cause a significant loss of mitotic cells over the course of 6–7 hours in culture.

C. Fixation

Various fixation procedures for use with DNA stains and for some of the multiparameter approaches noted earlier are reviewed in the appropriate chapters in this volume. Working with fixed samples is advantageous, since staining and analysis can be performed at the investigator's convenience. While fixation is not necessary for many DNA-staining protocols that use detergents to permeabilize cells, care must be exercised when using detergent-based staining protocols to ensure that mitotic cells are not lost; many of these techniques result in isolation of nuclei and mitotic cells have no nuclear membrane.

V. An Example of Single-Parameter Data

The DNA histograms displayed in Fig. 1 are representative of a typical stathmokinetic experiment, although only selected time points are included. The cells were stained with the ultraviolet (UV)-excited, DNA-specific dye DAPI (4', 6-diamidino-2-phenylindole) (Darzynkiewicz *et al.*, 1984) and excited with the light from a mercury arc lamp in an ICP 22A flow cytometer.

A variety of software programs exist for determining the cell cycle phase distribution of exponentially growing cells (Baisch *et al.*, 1979), although few deal successfully with perturbed distributions, especially when G_1-phase cells do not represent the major "peak" in the distribution. It should also be noted that displays as in Fig. 1, which are normalized to peak height rather than cell number, can result in distorted views of the distributions; for example, although the peak heights of G_1 cells at 0 and 2 hours are equivalent, there are 9% fewer cells in the latter.

The stathmokinetic information available from the data presented in Fig. 1 will be discussed as a specific subset of the multiparameter analysis, since the data are handled in exactly the same manner.

VI. Single-Parameter and Multiparameter Analysis of a Stathmokinetic Experiment

The first FCM techniques that allowed for the discrimination of mitotic cells as a separate population from G_2 cells were based on the differential denaturation of DNA *in situ* by heat (Darzynkiewicz *et al.*, 1977a) or acid

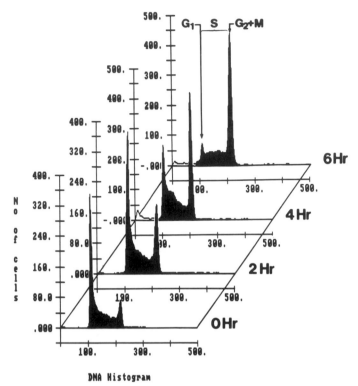

DNA Histogram

Fig. 1. Single-parameter flow DNA distributions of L1210 cells exposed to a stathmokinetic agent. Exponentially growing L1210 cells were exposed to 0.05 μg/ml vinblastine sulfate for 0, 2, 4, and 6 hours. Cells were stained with the DNA fluorochrome DAPI (see text). Flow cytometric measurement of the cell cycle distributions demonstrates the accumulation of cells in the $G_2 + M$ cell cycle phase with time of exposure to the stathmokinetic agent.

(Darzynkiewicz et al., 1977b) followed by staining with the metachromatic fluorochrome AO (see Darzynkiewicz, Chapter 32, this volume, for details). In addition to the mitotic cells forming a separate population, interphase cells could, in addition to being divided into G_1, S, and G_2 cells, be further subdivided into G_{1A} and G_{1B} cells (see later). Thus, a total of five distinct cell cycle phases could be identified as opposed to the three available from a single-parameter DNA distribution.

When the green (double-stranded DNA) and red (single-stranded DNA) AO luminescence values are transformed to total (red + green) luminescence equivalent to DNA content and α_t (red/total luminescence)

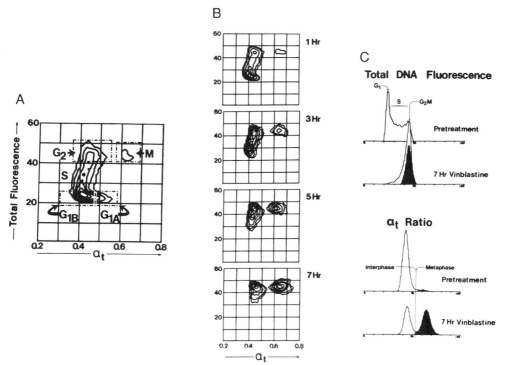

FIG. 2. Differential stainability of FL cells with acridine orange (AO) following partial denaturation of DNA *in situ* provides identification of five cell cycle compartments for the analysis of stathmokinesis. Following exposure to acid conditions (pH 1.3), staining of DNA with AO results in green fluorescence when bound to double-stranded DNA and red luminescence when bound to single-stranded DNA. (A) Transformation of the data to total fluorescence (red + green) and the ratio α_t (red/total fluorescence) results in a distribution depicted in the contour map. Mitotic cells (M) form a separate population from G_2-phase cells based on their α_t ratio, while postmitotic G_{1A} cells can be identified as a portion of the G_1 population with α_t values in excess of S-phase cells. (B) Addition of the stathmokinetic agent vinblastine results in accumulation of cells in mitosis. (C) Analysis of the cell cycle distribution at various times during stathmokinesis can be made along the total-fluorescence axis (top) with the aid of the α_t ratio (bottom). Thus, M cells (darkened area) can be identified as a subpopulation within the $G_2 + M$ peak of the total fluorescence distribution.

representing the degree of DNA denaturation *in situ*, the characteristic distribution observed in Fig. 2A is obtained for exponentially growing cells.

The panel of four contour maps in Fig. 2B illustrates the progressive changes in the distribution observed following addition of vinblastine. The

G_{1A} compartment empties, followed by G_{1B}, early-, and then mid-S-phase cells. When individual single-parameter distributions of the data are examined (Fig. 2C), the "DNA" histograms can be seen to change, as in single-parameter experiments. The α_t distributions, however, allow for the identification of mitotic cells, which can then be discriminated from G_2 cells as the experiment progresses.

In the following discussion, Sections VI,A–C describe the analysis of stathmokinetic data that can be obtained from either single or multiparameter analysis. Sections IV,D and E deal with populations that can only be identified by multiparameter approaches such as the one previously described.

A.   The $G_2 + M$ Accumulation Curve

The action of the stathmokinetic agent causes an accumulation of cells in mitosis over time with a concomitant decrease in G_1- and S-phase cells. In a single-parameter analysis, mitotic cells cannot be distinguished from G_2-phase cells. As a result, it is necessary to treat them as a single population. The $G_2 + M$ accumulation curve yields information about the doubling time. The following procedure can be employed:

1. For each time point $t_i > 0$, the "collection function" is computed, equal to $1 + f_i$ where f_i is the fraction of $G_2 + M$ cells at time t_i.
2. The natural logarithm of the collection function is plotted versus time [i.e., $\ln(1 + f_i)$ against t_i for all t_i]. (Equivalently, $1 + f_i$ may be plotted on semilog coordinates.)
3. The plot is interpolated by a straight line.
4. The slope (λ) of this line is computed.
5. The quantity $T_d = \ln(2)/\lambda = 0.693/\lambda$ is an estimate of the doubling time of the cell population.

It should be noted that the doubling time is not exactly equal to the generation (interdivision) time. The equality holds if the GF is 100%, growth is exponential, no cell death occurs, and all cell cycle phases have nonrandom durations.

B. G_1 Exit Kinetics

Since cell reentry into G_1 is prevented by the stathmokinetic agent, the rate of emptying of the G_1 compartment can be measured. The curve representing cell exit from G_1 during stathmokinesis has two distinct slopes, an exponential tail preceded by a concave shoulder (Fig. 3). The

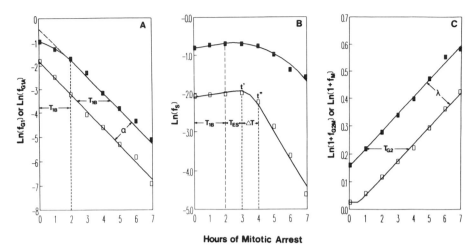

Hours of Mitotic Arrest

FIG. 3.   Example of analysis of stathmokinetic data based on the G_{1A} to G_{1B} transition model. Experimental data on FL cells (symbols) from Fig. 2 are interpolated by a mathematical model (continuous lines). In the model, the durations of the S, G_2, and M phases are assumed to be constant, while G_1 consists of an exponentially distributed G_{1A} and a constant G_{1B} subphase. (A) The exit data from G_1 (■) and G_{1A} (□) in semilog scale. The slope (α) of the straight-line interpolation of the G_{1A} curve or of the tail of the G_1 curve equals 0.7 hr^{-1}. From this, the mean duration of G_{1A} can be determined to be $T_{1A} = 1/\alpha = 1.4$ hour. The point at which the concave shoulder of the G_1 curve ends gives an estimate of the duration of G_{1B}, $T_{1B} = 2$ hours. (A similar estimate is provided by measuring the horizontal distance between the tails of the G_{1A} and G_1 curves.) (B) Transit data through S phase (■) and through an early-S window (□) in semilog scale. The ascending portion of the S transit curve ends at $t = T_{1B} = 2$ hours, followed by the concave part that extends beyond the last experimental point at $t_7 = 7$ hours, providing $T_{1B} + T_S \geq t_7 = 7$ hours, hence $T_S \geq 5$ hours. The descending, straight-line portion of the S transit is not apparent on the graph. The early-S window curve exhibits all three phases (ascending, concave, and straight-line descending). Their boundaries are approximately at $t' = 3$ hours and $t'' = 4$ hours, indicating that the window starts at about $T_{ES} = 1$ hour into S (since $t' = T_{1B} + T_{ES}$) and extends for about $\Delta T = 1$ hour ($t'' = T_{1B} + T_{ES} + \Delta T$). (C) The accumulation of cells in $G_2 + M$ and M, plotted as "collection curves"—that is, $\ln(1 + f_{G2M})$ and $\ln(1 + f_M)$, respectively. Following a lag of ~ 30 minutes duration in the M collection curve, both curves are interpolated by straight lines with identical slopes $\lambda = 0.062$ hr^{-1}, which corresponds to the doubling time $T_d = \ln(2)/\lambda = 0.69/0.062 = 11.1$ hours. The horizontal distance between the two curves provides the estimate of G_2 transit time, $T_{G2} = 2.4$ hours. The total interdivision time estimated from the foregoing data is $T \cong T_{1A} + T_{1B} + T_S + T_{G2} + T_M = 1.4 + 2 + 5 + 2.4 + 0.5 \cong 11.3$ hours, which is not very different from T_d.

biphasic nature of this curve is consistent with the idea that there are two parts to G_1 phase: an indeterminate phase (G_{1A}) responsible for the stochastic nature of cell exit from G_1, and a part of G_1 (G_{1B}) through which cells pass in a fixed length of time. This simple probabilistic model of the cell cycle (Smith and Martin, 1973) has been used to explain why cell exit from G_1 phase is not synchronous even when cells are synchronized, for instance, in the preceding mitotic phase of the cell cycle.

If this model is assumed, the duration of G_1 is different for individual cells within G_1 phase. Therefore, G_1, as a whole, can be characterized only by its "mean duration" or "mean transit time." This mean is the sum of the nonrandom duration of the G_{1B} subphase and the mean duration of the exponentially distributed, random G_{1A} subphase. The procedure for calculating these durations is as follows:

1. The natural logarithms of the cell fractions measured in G_1 (the G_1 exit curve) are plotted against time. (Alternatively, the G_1 exit curve can be plotted on semilog paper.)
2. The tail of the curve is approximated by a straight line.
3. The slope of this line is computed.
4. The time at which the concave portion of the exit curve ends and the straight-line portion begins (T_{1B}) is found.
5. The mean duration of G_1 is computed as $T_1 = T_{1A} + T_{1B}$, where $T_{1A} = 1/\alpha$. (T_{1A} is the mean duration of G_{1A} and T_{1B} is the duration of G_{1B}.)

C. S-Phase Transit

The number of cells in S phase as a function of time after the beginning of stathmokinesis can be determined based on DNA content. In fact, it is possible, and often useful, to examine "windows" in various portions of S phase (e.g., early-, mid-, or late-S phase). If these windows are of constant size (constant number of channels), then, at the onset of the experiment, fewer cells would be present in a "late-S" window than in an "early-S" window, reflecting the exponential age distribution of the population. The number of cells in S phase (or windows in S) should increase slightly with time, which again reflects the exponential cell age distribution of the population. This is followed by a second, concave phase and by a third phase characterized, ideally, by an exponential tail. The slope of this tail in semilog coordinates should be similar to the straight-line portion of the G_1 exit curve. The "deflection" point between phases two and three would be later in windows located later in S. The combined length of the ascending and concave portions of the curve provides a measure of the combined duration of the deterministic (G_{1B}) portion of G_1 phase plus any

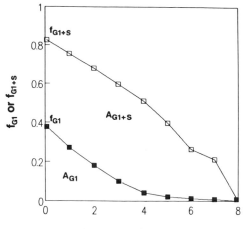

Hours of Mitotic Arrest

FIG. 4.   Nonparametric estimation of the mean duration of G_1 and S phase. G_1 ($\blacksquare$) and $G_1 + S$ ($\square$) exit data, both in linear scale. The estimate of the mean duration of G_1 is calculated as $E(T_1) = (A_{G1} + f_{G1}/\lambda)/2 = (0.82 + 0.38/0.062)/2 = 3.5$ hours, where $\lambda = 0.062$ hr^{-1} is estimated from the collection curves in Fig. 3C, $f_{G1} = 0.38$ is the fraction of G_1 cells at the beginning of stathmokinesis, and $A_{G1} = 0.824$ is the area under the G_1 exit curve computed from the graph above based on piecewise linear approximation of the data. $E(T_1) = 3.5$ hours is close to the previous, parametric estimate $T_1 = T_{1A} + T_{1B} = 3.4$ hours. The estimate of the mean joint duration of G_1 and S, $E(T_1 + T_s) = (A_{G1+s} + f_{G1+s}/\lambda)/2 = (3.83 + 0.83/0.062)/2 = 8.6$ hours, with λ, f_{G1+s}, and $(A_{G1+s} + f_{G1+s}/\lambda)/2 = (3.83 + 0.83/0.062)/2 = 8.6$ hours, with λ, f_{G1+s}, and A_{G1+s} computed as before. Subtraction $E(T_s) = E(T_1 + T_s) - E(T_1) = 8.6 - 3.5 = 5.1$ hours provides the estimate of S transit time, which is close to that obtained by the parametric approach as described in Fig. 3B.

portion of S phase that precedes the lower threshold of the S-phase window(s). If, for instance, two narrow windows were placed in very early- and very late-S phase, the difference in time at which the deflection took place between the respective curves would provide a rough estimate of the minimum duration of S phase. Alternatively, if the "window" is identical with the entire S phase, the combined duration ($T_{1B} + T_S$) of G_{1B} and S is defined by the deflection point. If T_{1B} is known from the analysis of the G_1 exit curve, T_S can be found by subtraction.

D.   Mitotic Accumulation Curve

As noted already, the mitotic accumulation curve often demonstrates a small lag immediately after addition of the stathmokinetic agent. Nevertheless, it is essential that the remainder of the curve be exponential over, preferably, at least half of the cell cycle period. A deviation from exponential indicates that the mitotic block is "leaky," the cells are not in asynchronous or exponential growth, or there is selective death in some portion of the cycle. Similarly, as in the case of the $G_2 + M$ accumulation curve discussed previously, the mitotic accumulation curve is calculated as the natural logarithm of $(1 + f_i)$, where f_i now represents the fraction of cells in mitosis at time t_i. As such, the full spectrum of values will vary between 0 and $\ln(2)$.

In a multiparameter analysis, both the M and $G_2 + M$ accumulation curves are generally available. The slopes of the straight-line (exponential) portions of each should be identical (and equal to λ) if there are no perturbations in either cell cycle phase. Both slopes provide an estimate of the cell population doubling time ($T_d = \ln(2)/\lambda$). This property of M and $G_2 + M$ accumulation curves is true for a very general class of cell cycle models, including arbitrary probabilistic distributions of any cell phase, providing the population is in exponential growth (Kimmel, 1985). The distance along the time axis between the two curves provides an estimate of the duration of G_2 phase (T_{G2}).

E. G_{1A} Exit Kinetics

The final piece of information unique to some multiparameter approaches, including the present AO technique, is the independent analysis of G_{1A} exit kinetics.

The G_{1A} compartment has been identified by its characteristic G_1 total fluorescence (DNA content) but increased α_t ratio (Fig. 2). G_{1A} cells are believed to have a somewhat more condensed chromatin structure than do G_1 cells capable of initiating DNA synthesis (i.e., G_{1B} cells) (Darzynkiewicz et al., 1981). These G_{1A} cells are, in principle, similar to the G_{1A} cells determined by other staining protocols (e.g., based on RNA or protein content) (Darzynkiewicz and Traganos, 1982). Their kinetic characteristics are thought to be responsible for the "exponential tail" of the G_1 exit curve. When measured as an independent population, the slope of the G_{1A} exit curve parallels the straight-line portion of the G_1 exit curve of the same experiment. The exponential character of the G_{1A} slope is evident from the outset of stathmokinesis. The slope (α) of the G_{1A} exit curve is equal to the inverse of the mean transit time (i.e., $T_{1A} = 1/\alpha$, or $\alpha = 1/T_{1A}$) within the Smith and Martin (1973) framework of the cell cycle. The distance along the time axis between the G_1 and G_{1A} exit curves (T_{1B}) provides an estimate of the length of the deterministic portion (G_{1B}) of the G_1 phase. This value should be similar to the T_{1B} value obtained from the G_1 exit curve.

VII. Nonparametric Analysis of Exit Curves

There are instances in which the "tails" of semilog plots of G_1 and G_{1A} exit curves are *not* parallel and/or cannot be approximated by straight lines. In such cases, the Smith and Martin (1973) model must be abandoned and the estimation of the mean transit time in G_1 (or G_{1A}) should not be carried out as previously described. An alternative method has been proposed in such situations (Kimmel, 1985) based on probabilistic distributions of transit times. In this approach, the mean transit time in G_1 is computed from the expression $T_1 = (A + f_{G1}/\lambda)/2$, where A is the area under the graph of the G_1 exit curve (plotted in linear coordinates), f_{G1} is the fraction of cells in G_1 at time $t = 0$ (the beginning of the stathmokinetic experiment), and λ is the slope of the $G_2 + M$ or M accumulation curve (see example in Fig. 4). This method is quite exact provided that the experiment is of sufficient duration for the nearly complete emptying of the G_1 compartment. If this is not achieved, the area A will be underestimated, as will T_1. A strictly analogous procedure is, in principle, applicable to the analysis of the joint exit curves of $G_1 + S$. The outcome is the summary transit time through these phases $(T_1 + T_S)$. However, emptying of both G_1 and S phase (a prerequisite for this calculation) is normally difficult to achieve, especially for slowly growing cells (see Section II, D).

VIII. General Approach to the Analysis of Cell Cycle Perturbations

As we have seen, the stathmokinetic experiment can provide detailed information about cell proliferation and cell cycle kinetics. However, the technique is most useful in detecting and quantifying alterations in cell proliferation induced either by the addition of perturbing agents (e.g., drugs, growth factors, irradiation) or by alterations in culture conditions (e.g., serum deprivation, temperature or pH changes). To illustrate the potential of the method, the discussion that follows has been limited to the area of drug-induced perturbations and divided into a general, idealized discussion of drug action and a specific discussion, supported by actual data, of CI-921, a derivative of the chemotherapeutic agent amsacrine (Traganos *et al.*, 1987). In the first instance, drug action will be presumed to be limited to a specific cell cycle phase, while the discussion of CI-921 will demonstrate that, in actuality, even drugs that are presumed to have cell cycle phase specificity affect other cell cycle phases in sometimes subtle ways.

Before proceeding, it should be noted that the results of stathmokinetic experiments are most easily interpreted when drug concentrations utilized in such experiments are cytostatic and not cytotoxic, since drug-induced cell death requires significantly more sophisticated models for analysis (Kimmel and Traganos, 1985). In addition, while it is difficult to detect, one should be aware of the possibility of synergism between the perturbing and stathmokinetic agents.

A.   Drug Effects on G_1-, S-, and G_2-Phase Cells

If a drug's action was limited to "blocking" cell transit through G_2 phase (G_2 block), then such an agent would cause a change in the M but not the $G_2 + M$ accumulation curve. (Note that this type of drug effect would not be detectable in single-parameter analysis of stathmokinetic experiments.) A deflection (slowdown) in the M accumulation curve, which could be manifested as a change (decrease) in slope, should occur at a time point prior to T_{G2}, the time difference between the straight-line portions of the two accumulation curves. This must be true because if the block was in late-S phase, the $G_2 + M$ accumulation curve would also be affected. Whether the M accumulation curve merely changes slope or plateaus indicates that there is either a slowdown in transit or a leaky block in G_2 or the block is permanent. A change to a negative slope would indicate that the stathmokinetic agent or the drug, or the combination of the two, caused the destruction of cells in mitosis.

The term "point of action" of a drug is often used to describe the length of time (and thus, the cell cycle phase) prior to mitosis that the drug effect is manifested. Thus, if a drug caused a deflection (change in slope) of the M accumulation curve after 90 minutes, then the point of action of the drug would be 90 minutes prior to mitosis; if G_2 was 2 hours long, the block would be in G_2 phase. In making these calculations, the closer together the time points of the stathmokinetic experiment, the more accurate the estimate of the point of drug action.

A drug that blocks cells in S phase would cause a change in slope of both the M and $G_2 + M$ accumulation curves and affect the descending portion of the S-phase transit curve. The later in S phase the block, the sooner one would observe the deflection in the M and $G_2 + M$ accumulation curves. The $G_2 + M$ accumulation curve would be affected first, since the block is located closer to G_2 than M. However, the point of action of the drug should be calculated from the M accumulation curve and is equal to T_{G2} plus that portion of S located beyond the block. Again, a slowdown or leakage through the block in S phase would result in a change in slope rather than a plateau in either the M or $G_2 + M$ accumulation curves.

Analysis of the S-phase transit curve depends on the fraction of S phase that is sampled and the location of that fraction (window) in S. If the S-phase window includes all of S phase, a block of S-phase transit (no matter where in S phase) will affect the slope of the descending portion of the curve. Presumably, G_1 cells continue to enter S phase although only those S-phase cells located beyond the block when the drug was added would exit S phase. If, for instance, a block occurred in late S phase, the S-phase transit curve would appear to continue to rise (often obscuring the deflection point) until the curve plateaued when no more G_1-phase cells were available to enter S phase. Note that the slope of the descending portion of the S-phase transit curve is normally similar to that of the G_{1A} exit curve or the straight-line portion of the G_1 exit curve for unperturbed populations. A slowdown in S-phase traverse, or a leaky block, rather than causing a plateau, would lead to a change in slope of the descending portion of the curve.

Pinpointing the action of an S-phase-blocking agent is made easier if several "windows" in S phase are sampled. Thus, if an agent blocked cells in late-S phase, transit curves for windows located in early- or mid-S phase would be more-or-less unaffected while the late-S-phase transit curve would be altered as described before. It then follows that the earlier in S phase the agent acts, the more windows (i.e., those located beyond the point of action of the drug) will be affected.

Drug-induced changes in cell exit from G_1 phase effect all the curves already described but may not be evident from M and $G_2 + M$ accumulation curves unless a large proportion of the cell cycle duration is covered by the stathmokinetic experiment. When information on both the G_1 and G_{1A} exit kinetics is available, the following effects can be discerned: (1) In cultures treated with a G_{1A}-arresting drug, the G_{1A} exit curve will change slope or plateau while the G_1 exit curve will only be affected after a delay equivalent to the duration of G_{1B}. (2) If the number of cells remains constant in both these compartments, one can conclude that cells are arrested in both G_{1A} and G_{1B}. (3) Arrest of cells in G_{1B} will be manifested as a change in the G_1 exit curve without any change in slope of the G_{1A} exit kinetics. As before, a slowdown in or leakage through these blocks results in a change in slope of the exit curve rather than a plateau.

The effect of a G_1 block on the S-phase transit curve(s) depends on the location of the window(s). Generally, a block anywhere in G_1 will affect the deflection point on any S-phase transit curve, since fewer G_1-phase cells will eventually enter S phase (i.e., only those beyond the block), shortening the time before the percentage of cells in the window begins to decrease. A change (slowdown) in exit from G_1 will change (decrease) the slope of the ascending portion of the S-phase transit curve. In addition, such a slowdown would move the deflection point of the S-phase transit curve to the right (later in time). As would be expected, the transit curves of windows located in early S phase would be the first to be affected by a change in G_1 exit kinetics.

Changes in the slope of the $G_2 + M$ accumulation curve would occur at a time equivalent to the length of S phase and that portion of G_1 phase beyond the block. The M accumulation curve would be the last to be affected, since the point of action would be equal to $T_{G2} + T_S$ plus some portion of G_1.

B.   Effects of CI-921 on Cell Cycle Kinetics

As an example of a practical application of the stathmokinetic technique, the following provides a brief discussion of the action of the drug CI-921 on exponentially growing FL cells.

CI-921 binds to DNA by intercalation, is slightly more lipophilic, and has a higher association constant for DNA than the parent compound amsacrine (Baguley et al., 1984). Preliminary studies (Baguley et al., 1984) suggested that, as with most intercalators that are considered S-phase-active agents, cells become blocked in G_2 phase at subcytotoxic concentrations. Later studies (Traganos et al., 1987) have provided a more detailed description of the drug's effect based on the stathmokinetic experiment.

The slope of the mitotic accumulation curve in the absence of drug provides an estimation of the growth rate ($0.062 \, h^{-1}$) which corresponds to a doubling time of 11.1 hours. The addition of 50 nM CI-921 immediately affected the accumulation of cells in mitosis (Fig. 5), causing a plateau within 30 minutes of its addition. Therefore, the point of action of CI-921 could be said to be in G_2, 30 minutes prior to mitosis.

Careful examination of the drug-treated M accumulation curve indicates a slight negative slope of the curve. Since the block in M accumulation occurred at 30 minutes, the mitotic cells continue to "age" with increasing time of stathmokinesis. The decrease suggests a minimal loss (disintegration) of mitotic cells over time.

The distance between the M and $G_2 + M$ accumulation curves provides an estimate of T_{G2} that, in this experiment, was approximately ~ 2.4 hours (Fig. 5). The action of CI-921 on the S–G_2 transition appears to occur after the first hour of treatment. Between the first and second hour of treatment, there was a slowdown in entrance into G_2 phase; presumably no cells were exiting G_2 at this time as witnessed by the M accumulation curve for the drug-treated culture. Changes in the slope of the drug-treated $G_2 + M$ accumulation curve, while indicating that the cells are not permanently

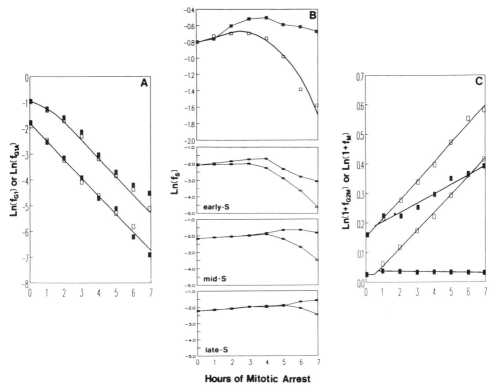

Hours of Mitotic Arrest

FIG. 5. Stathmokinetic data for FL cells perturbed by CI-921 (■) compared to unperturbed data as in Fig. 3 (□). (A) G_1 and G_{1A} exit data in semilog coordinates. The continuous curves were generated by a mathematical model, as in Fig. 3A. There appear to be no major differences between drug-treated and control exit kinetics. (B) S-phase transit data (upper panel) and early-S, mid-S and late-S transit data (three smaller panels) in semilog coordinates. Unperturbed S-phase transit data were interpolated by a mathematical model as in Fig. 3B; perturbed S data and all S-window data were interpolated by piecewise linear curves. The data from the drug-treated culture indicate increased accumulation in S phase (details in text). (C) $G_2 + M$ and M collection curves. Unperturbed data approximated by straight lines as in Fig. 3C. Drug-treated data approximated by least-squares straight lines, based on all data points except $t = 0$. The slope of the perturbed $G_2 + M$ collection function is $\lambda' = 0.031$ hr^{-1}, corresponding to a doubling time $T_d' = 22.4$ hours. The difference between this value and that for the unperturbed population ($T_d = 11.1$ hours) may be considered a rough summary measure of drug action. The slightly negative slope of the perturbed M collection function suggests a deterioration in mitotic arrest. A delay in drug action (~ 30 minutes) seems to be present in both the $G_2 + M$ and M curves (details in the text).

blocked in S phase, are difficult to interpret precisely. The following represents one possible interpretation of these data.

CI-921 caused a 1 hour delay in transit through early-S phase (Fig. 5). The delay increased to ~ 3 hours in mid-S phase. However, cells in a late-S window appeared still to be accumulating by 7 hours and had not yet reached the "deflection" point. This "calculus" of delays should be considered with caution. It suggests that at 50 nM, CI-921 causes a "pure" delay in S phase of > 3 hours. However, the G_2M accumulation curve shows a delay of no more than 2 hours (until the fifth hour of the experiment when the delay increases significantly). The point is that the delays in the mid- and late-S phase windows appear at ~ 5 hours after addition of the drug. These effects are not inconsistent with the "distributed effects" scenario in which the drug initially affects early-S phase cells and these effects propagate toward mid- and late-S phase.

During the initial 2.25 hours, the percentages of G_1 cells in both control and drug-treated cultures decreased by 60–65% (Fig. 5), after which the G_1 compartment emptied at a faster rate. Only at the last time point was there any variation in the two curves. However, the measured difference of $\sim 1.5\%$ in the two populations is not much greater than the error in the measurement. The similarity between the concave and straight-line portions of the two G_1 exit curves suggest that neither the G_{1A} nor the G_{1B} compartment was affected by the drug. Independent analysis of the G_{1A} exit curve confirms that over the first 7 hours of continuous exposure, exit from G_{1A} was not affected by 50 nM CI-921.

As seen from the example of CI-921, cytostatic agents may act in a "distributed" manner, affecting cells in various subcompartments of the cell cycle to varying degrees. In some cases, this may make the graphical analysis outlined earlier difficult, even if a subdivision of the S phase into several disjointed windows is employed. There is, however, an alternative to the graphical analysis described previously. Mathematical modeling can be used to extract information on drug action from the detailed, channel-by-channel FCM data. Fitting the mathematical model to the data, it is possible to construct a "drug action curve" depicting the distribution of drug action over the cell cycle. This approach requires computer calculations and, while they are too complex to discuss in the present forum, examples can be found in Kimmel and Traganos (1985, 1986).

In summary, the technique of stathmokinesis provides a useful tool to characterize drug action *in vitro*. Of course, as the foregoing example demonstrates, care must be exercised in interpreting drug effects in this system. Clearly, agents such as CI-921 are toxic for S-phase cells, but at the cytostatic concentrations used in stathmokinetic experiments, drug effects become more complex. Rather than observing an S-phase cell kill, an effect on S-phase transit was observed that ultimately resulted in a G_2 accumulation. Nevertheless, the approach just described, when applied with proper care, provides a wealth of information that would be impossible to obtain in such detail by any other means.

ACKNOWLEDGMENTS

Support for this work was provided by Public Health Service grants CA 23296 and CA 28704.

REFERENCES

Aherne, W. A., and Camplejohn, R. S. (1972). *Exp. Cell Res.* **74**, 496–501.
Aherne, W. A., Camplejohn, R. S., and Wright, N. A. (1977). "An Introduction to Cell Population Kinetics." Arnold, London.
Baguley, B. C., Denny, W. A., Atwell, G. J., Finlay, G. J., Rewcastle, G. W., Twidgen, S. J., and Wilson, W. R. (1984). *Cancer Res.* **44**, 3245–3251.
Baisch, H., Beck, H. P., Christensen, I. J., Hartmann, N. R., Fried, J., Dean, P. N., Gray, J. W., Jett, J. H., Johnston, D. A., White, A. R., Nicolini, C., Zeitz, S., and Watson, J. V. (1979). *In* "Flow Cytometry IV" (O. D. Laerum, T. Lindmo, and E. Thorad, eds.), pp. 152–158. Universitetsforlaget, Oslo.
Bakowski, M. T. (1976). *Cancer Treat. Rev.* **3**, 95–107.
Barfod, I. J., and Barfod, N. M. (1980). *Cell Tissue Kinet.* **13**, 1–8.
Clarke, R. M. A. (1971). *Cell Tissue Kinet.* **4**, 263–272.
Creasy, W. A. (1975). *In* "Antineoplastic and Immunosuppressive Agents, Part II" (A. C. Sartorelli and D. G. Johns, eds.), pp. 670–694. Springer-Verlag, Berlin.
Darzynkiewicz, Z. (1984). *In* "Growth, Cancer and the Cell Cycle" (P. Skehan and S. Friedman, eds.), pp. 249–278. Humana Press, Clifton, New Jersey.
Darzynkiewicz, Z., and Traganos, F. (1982). *In* "Genetic Expression in the Cell Cycle" (G. M. Padilla and K. S. McCarty, eds.), pp. 103–128. Academic Press, New York.
Darzynkiewicz, Z., Traganos, F., Sharpless, T., and Melamed, M. R. (1977a). *J. Histochem. Cytochem.* **25**, 875–880.
Darzynkiewicz, Z., Traganos, F., Sharpless, T., and Melamed, M. R. (1977b). *Exp. Cell Res.* **110**, 201–214.
Darzynkiewicz, Z., Traganos, F., and Melamed, M. R. (1981). *Cytometry* **1**, 98–108.
Darzynkiewicz, Z., Williamson, B., Carswell, E. A., and Old, L. J. (1984). *Cancer Res.* **44**, 83–90.
Darzynkiewicz, Z., Traganos, F., and Kimmel, M. (1986). *In* "Techniques in Cell Cycle Analysis" (J. W. Gray and Z. Darzynkiewicz, eds.), pp. 291–336. Humana Press, Clifton, New Jersey.
DeBrabander, M. J., Van De Veire, R. M. L., Aerts, F. E. M., Borgers, M., and Janssen P. A. J. (1976). *Cancer Res.* **36**, 905–916.
Dosik, G. M., Barlogie, B., White, A. R., Gohde, W., and Drewinko, B. (1981). *Cell Tissue Kinet.* **14**, 121–134.
Epstein, R. J., Watson, J. V., and Smith, P. J. (1988). *Cytometry* **9**, 349–358.
Fitzgerald, P. H., and Brehaut, L. A. (1970). *Exp. Cell Res.* **59**, 27–35.
Frei, E., Whang, J., Scoggins, R. B., Van Scott, E. J., Rall, D. P., and Ben, M. (1964). *Cancer Res.* **24**, 1918–1928.

270   FRANK TRAGANOS AND MAREK KIMMEL

Gray, J. W., Dolbeare, F., Pallavicini, M. G., and Vanderlaan, M. (1986). In "Techniques in Cell Cycle Analysis" (J. W. Gray and Z. Darzynkiewicz, eds.), pp. 93–138. Humana Press, Clifton, New Jersey.

Hallowes, R. C., West, D. G., and Helmann, K. (1974). Nature (London) 247, 487–490.

Kimmel, M. (1985). Math. Biosci, 74, 111–123.

Kimmel, M., and Traganos, F. (1985). Cell Tissue Kinet. 18, 91–110.

Kimmel, M., and Traganos, F. (1986). Math. Biosci. 80, 187–208.

Kimmel, M., Traganos, F., and Darzynkiewicz, Z. (1983). Cytometry 4, 191–201.

Kimmel, M., Darzynkiewicz, Z., and Staiano-Coico, L. (1986). Cell Tissue Kinet. 19, 289–304.

Krishan, A., Paika, K., and Frei, E., III (1975). J. Cell Biol. 66, 521–530.

Larsen, J. K., Munch-Petersen, B., Christiansen, J., and Jorgensen, K. (1986). Cytometry 7, 54–63.

Macdonald, P. D. M. (1981). In "Biomathematics and Cell Kinetics" (M. Rottenberg, ed.), pp. 125–142. Elsevier, Amsterdam.

Morani, R., Darzynkiewicz, Z., Staiano-Coico, L., and Melamed, M. R. (1985). J. Histochem. Cytochem. 33, 821–827.

Morris, W. T. (1967). Exp. Cell Res. 48, 209–217.

Nüsse, M., Julch, M., Geido, E., Bruno, S., DiVinci, A., Giaretti, W., and Ruoss, K. (1989). Cytometry 10, 312–319.

Puck, T. T., and Steffen, J. (1963). Biophys. J. 3, 379–397.

Puck, T. T., Sanders, P., and Petersen, D. (1964). Biophys. J. 4, 441–455.

Sharpless, T. K., and Schlesinger, F. H. (1982). Cytometry 3, 196–200.

Smith, J. A., and Martin, L. (1973). Proc. Natl. Acad. Sci. U.S.A. 70, 1263–1267.

Smith, R. S., Thomas, D. B., and Riches, A. C. (1974). Cell Tissue Kinet. 7, 529–535.

Stubblefield, E., Klevecz, R. R., and Deaven, L. (1967). J. Cell. Physiol. 69, 345–354.

Traganos, F., Bueti, C., Darzynkiewicz, Z., and Melamed, M. R. (1987). Cancer Res. 47, 424–432.

Trinkle, L. S., Swope, V. B., Abdel-Malek, Z. A., and Nordlund, J. J. (1988). Cytometry 9, 432–435.

Wilson, L., Anderson, K. A., and Chin D. (1976). Cold Spring Harbor Conf. Cell Prolif. Cell Motil. 1051–1064.

Wolpert-DeFillippes, M. K., Adamson, R. H., Cysyk, R. L., and Johns, D. G. (1975). Biochem. Pharmacol. 24, 751–754.

Wright, N. A., and Appleton, D. R. (1980). Cell Tissue Kinet. 13, 643–663.

Zucker, R. M., Elstein, K. H., Easterling, R. E., and Massaro, E. J. (1988). Cytometry 9, 226–231.

Chapter 26

Detection of Intracellular Virus and Viral Products

JUDITH LAFFIN AND JOHN M. LEHMAN

Department of Microbiology and Immunology
Albany Medical College
Albany, New York 12208

I. Introduction

Flow cytometric (FCM) analysis of viral infection can detect both viral and cellular events qualitatively and quantitatively. Our group and others have used the FCM technology to gain an understanding of the events that are necessary in permissive and nonpermissive infections with simian virus 40 (SV40). This virus belongs to the papovavirus group, has a closed circular double-stranded DNA genome of $\sim 3 \times 10^6$ Dal (5200 bp), which codes for three coat proteins (VP_1, VP_2, VP_3) and three nonstructural proteins, large tumor (T) antigen (94 kDa), small t antigen (17 kDa) and the agnoprotein (Brady and Salzman, 1986). The large T antigen has received considerable attention because this protein is known to initiate viral DNA replication in permissive cells and may have multiple functions in the transformation of nonpermissive cells. The development of monoclonal antibodies (mAb) to different epitopes of T antigen has permitted FCM analysis of the appearance of this viral protein. In addition, simultaneous measurements with other discriminating factors (DNA, RNA, and/or other proteins) could be made and this allowed quantity, distribution, and population dynamics of T antigen to be correlated to the appearance of these macromolecules (Lehman and Defendi, 1970; Lehman *et al.*, 1979).

Simian virus 40 is capable of a lytic infection (permissive) in monkey kidney cells and a transforming infection (nonpermissive) in numerous

271

other cells (mouse, rat, human, etc.). To date, FCM studies have provided information in defining the relationship of the initiation of host cell DNA synthesis with the appearance and quantity of T antigen during the lytic cycle. The appearance, quantity, and correlation of T antigen (epitope PAb101) has been reported for permissive cells (Lehman *et al.*, 1988) and nonpermissive human cells (Laffin *et al.*, 1989). Presently under study is transformation by wild-type and temperature-sensitive SV40 in human, mouse, and chinese hamster embryo (CHE) cells. These investigations are providing information on the differential expression of T-antigen epitopes in the evolving transformed cell and their association with other known oncogenic indicators.

The following procedure outlines the staining protocol for SV40 T antigen in cells cultured in monolayers. The mAb, PAb101, was employed; however, other antibody preparations once concentrated, purified, and characterized can be utilized. Further, this technology has been applied to other viral model systems to obtain information on various parameters of pathogenesis at the cellular level, which might then be applied to the clinical disease. Preliminary results indicated that the technique could be of value both as a rapid diagnostic test and as a method to study aspects of the human infection (Elmendorf *et al.*, 1988). Some problems and solutions related to this method are also pertinent to the detection of other than virus intracellular antigens, and are discussed in Chapter 24 of this volume, by Bauer.

II.   Materials and Methods

Procedures are summarized as follows:

1. Fixation
2. Reaction with primary antibody
3. Washes to remove primary antibody
4. Reaction with FITC-labeled secondary antibody
5. Washes to remove unbound secondary antibody
6. Incubation with RNase and propidium iodide (PI) (DNA staining)
7. Filtration of samples (prior to run on flow cytometer)

Following fixation of cells, which requires ~ 30 minutes for 12 samples, the staining procedure is relatively simple but long. It may take from 6 to 8 hours to complete, but only a small fraction of the effort is labor-intensive. Thus, several runs of 12 samples can be processed in succession. Furthermore, the last incubation with RNase and PI can be combined and stained overnight at 4°C. Filtration is performed prior to flow analysis and can be accomplished in 10 minutes (12 samples). Time for data analysis varies:

A. Reagents for Fixation and Staining

1. PHOSPHATE-BUFFERED SALINE WITHOUT CA^{2+} AND MG^{2+} (PBS)

NaCl, 8.0 g
KCl, 0.2 g
Na_2PO_4, 1.2 g
KH_2PO_4, 0.2 g

Dissolve in 1 liter of distilled water (dH_2O), adjust pH to 7.4, sterilize, and store at room temperature (RT).

2. TRYPSIN–EDTA

2.5% Trypsin (GIBCO, Grand Island, NY), 10.0 ml
EDTA (tetrasodium ethylenediaminetetraacetate), 0.1 g
PBS, 90.0 ml

Filter, sterilize, and store at 4°C.

3. WASH SOLUTION (WS)

Normal goat serum,* 100.0 ml
PBS, 900.0 ml
Triton X-100, 20.0 μl
Sodium azide, 1.0 g

*Heat-inactivate serum for 60 minutes at 56°C. Store at 4°C.

4. PROPIDIUM IODIDE (PI)

PI (Calbiochem–Behring, La Jolla, CA), 1.0 mg
PBS, 100.0 ml
Triton X-100, 20.0 μl
Sodium azide, 0.1 g

Store at 4°C in the dark.

5. RNase

RNase A (Sigma, St. Louis, MO), 100.0 mg
PBS, 100.0 ml
Sodium azide, 0.1 g

Store at 4°C.

6. Antibodies

a. Primary Antibody. A mAb or a polyclonal antiserum with high specificity is diluted in WS as determined by titration (Section III,D,1).

b. Secondary Antibody. Commercially prepared affinity-purified goat anti-mouse IgG (ab')$_2$ fragments purchased from Boehringer—Mannheim (Indianapolis, IN) as a fluorescein conjugate (fluorescein isothiocyanate, FITC) is diluted in WS.

B. Protocol

The following example demonstrates the processing of monolayers of SV40-infected human diploid fibroblasts (HDF) in 60-mm dishes.

Fixation Procedure

1. Remove culture media and rinse once with PBS.
2. Add 1.0 ml trypsin–EDTA (warmed to 37°C) to each plate. Remove trypsin–EDTA after 15–30 seconds.
3. Place at 37°C for ~ 5 minutes or until cells detach. (Caution: Do not leave in trypsin–EDTA any longer than necessary.)
4. Add 1.0 ml of WS and resuspend cells using a 1000 μl pipetman.
5. Transfer to a 1.5-ml microfuge tube. Centrifuge in a variable-speed microfuge (Fisher model 59A) at 5000 rpm for 15 seconds.
6. Remove supernatant; resuspend cells in 1.0 ml of cold PBS by vortexing gently.
7. Count cells and adjust concentration to 1×10^6 cells/ml.
8. Centrifuge as in step 5 above and remove supernatant.
9. Add 100 μl of cold PBS to cell pellet and thoroughly resuspend. Immediately add 900 μl of methanol at −20°C and fix for a minimum of 15 minutes at −20°C.
10. Samples can be immediately stained or stored at −20°C for future use.

Staining with Primary Antibody

1. Centrifuge as in step 5 of above fixation procedure, remove fixative, and rinse once with 1.0 ml cold PBS.
2. Add 0.5 ml of primary antibody to each tube, then vortex gently.
3. Incubate in a 37°C water bath for 2.0 hours.

Washes to Remove Unbound Antibody

1. Centrifuge (fixation procedure, step 5), and then remove supernatant.
2. Add 0.5 ml of WS and vortex gently.
3. Place on ice for 15 minutes.
4. Repeat wash (steps 1–3 above).

Staining with FITC-Labeled Antibody

1. Add 0.5 ml of secondary antibody to each tube, then vortex gently.
2. Incubate in a 37°C water bath for 2.0 hours.

Washes to Remove Excess Antibody

Repeat above washing protocol (steps 1–4).

DNA Staining

1. Add 0.5 ml of RNase solution and gently vortex.
2. Incubate at 37°C for 30 minutes.
3. Add 0.5 ml of PI solution to each tube and vortex (final volume 1.0 ml).
4. Place at 4°C in the dark for 30 minutes. (As an alternative, both the RNase and PI solutions can be added to the samples after the final wash, mixed, and placed in dark at 4°C overnight.)

FILTRATION OF SAMPLE

Filter each sample prior to processing on the flow cytometer. This is accomplished by passing the sample through nylon mesh (50 μm) to remove cell clumps. The filtration should be repeated if samples have to be rerun at another time. Some preparations have been reanalyzed after a week at 4°C with little loss in signal, but generally samples should be processed within 48 hours.

Note that the antibody incubation times are optimal for the SV40 cell culture system. It may be the starting point for other antigen–antibody measurements. For example, shorter incubations were adequate for assaying cytomegalovirus (CMV)-infected cells for early antigen (Elmendorf *et al.*, 1988).

III.   Critical Parameters

A.   Criteria for Detection of Viral Antigen

1.   SAMPLE PROCESSING

Sample choice is directed by the question being addressed but the initial processing of the cells may be handled by one of the following methods.

1. Experimental cell culture models, such as the SV40–HDF system, require removal of the cell monolayers with an enzyme (i.e., trypsin) without damaging the cells (Jacobberger *et al.*, 1986; Lehman *et al.*, 1988; Laffin *et al.*, 1989).
2. Clinical samples including blood, bladder washings, and lung lavages must be prepared by a purification and concentration procedure (Elmendorf *et al.*, 1988). Cells can then be fixed and stained (Section II,B).
3. Archival specimens (paraffin sections) may be used if the viral antigen is nuclear and stable. Techniques for the removal of paraffin and staining for DNA and a nuclear antigen have been described (Anastasi *et al.*, 1987). Since cells are fixed, the sample may be processed immediately.
4. Solid tissue must first be processed into a single-cell suspension using a method that neither damages the cells nor destroys the antigen. Cells are then fixed and stained (Section II,B).

2.   BACKGROUND

The amount of background fluorescence detected may vary with cell type. In studies with HDF, background fluorescence was high in the untransformed cell. When the cells were transformed by the SV40 T antigen, there was a lower background signal as a result of reduced autofluorescence and smaller cell size. This made it imperative to establish standards of known negative and positive populations as controls for background fluorescence.

3. DISCRIMINATING FACTORS

In addition to assaying for the antigen in question, it is desirable to use a second parameter relevant to the study to assist in identification of the positive population. For example at early postinfection time points of SV40-infected HDF, between 5 and 10% of the total cell population was positive for the PAb 101 epitope. When distribution of T antigen was examined in these cells using a second marker (DNA) to help discriminate the positive population, the histogram revealed that the majority of positive cells were in the G_2 peak (40–50%).

4. CELL SIZE

The size of the cell may influence the fluorescence measurements in several ways. First, it is more difficult to remove unbound antibody and PI from larger cells, and second, antigen concentrated in the nucleus of a small cell could have a brighter signal (peak value) than the same quantity of antigen diffused in the cytoplasm of a larger cell. The second difficulty can be eliminated by acquiring antigen signal using area (pulse integration) mode. Comparison of cell size can be monitored during flow analysis with forward-angle light scatter. Additionally, the nuclear/cytoplasmic ratio may be evaluated and the location of the antigen identified by examination under the fluorescent microscope.

5. CELL NUMBER

Cell number should always be kept constant. Ideal sample size is between 1.0 and 1.5×10^6 cells/ml. This allows for cell loss but retains the ability to collect between 10^4 and 10^5 cells per histogram. Lower cell counts are acceptable, but flow rate is adversely affected unless sample volume is adjusted.

B. Technical Difficulties in Fixation

Two factors that influence the selection of a fixative are (1) that the antigen is not lost or changed during processing and storage, and (2) that the background fluorescence is not increased. A number of fixation techniques are available for analysis of intracellular antigens using FCM. To stain nuclear antigen and DNA, a method of lysis by detergent or osmotic pressure can be used, but results in the loss of cytoplasm. A second method utilizes detergent to permeablize the cell, followed by fixation with formaldehyde. A third method, alcohol fixation (methanol is used in our laboratory), both permeabilizes and fixes the cells (Jacobberger, *et al.*, 1986). Methanol-fixed samples can be stored at −20°C for > 1 year with neither significant loss in quantity or quality of antigen nor change in background fluorescence, but repeated warming of a fixed sample appeared to be deterimental. Many potential problems can be identified by a preliminary evaluation under the fluorescent microscope. To this end, the cells should be grown on coverslips and then fixed and stained with the same reagents used for FCM.

In subsequent steps, fixation may result in a nonselective loss of cells. This is a consequence of cells adhering to surfaces and to each other. To maximize cell recovery, contact with new surfaces should be reduced by dispensing reagents into the sample tube, and then resuspending by gentle vortexing. The pellet is dispersed easily when care is taken not to compact the cells. Aggregation is also reduced by the presence of serum and Triton X-100 in the staining solutions.

It is important to establish that the populations analyzed by FCM are true representation of the original culture and that fixation did not result in modification of these populations. Changes in both the internal structure and the surface of an infected cell necessitate special attention during fixation and staining to prevent this type of alteration. During the lytic cycle of SV40 in monkey kidney (CV1) cells, a concomitant increase in DNA content and cell mass with eventual cell lysis is observed. At late time points postinfection, there may be a selective loss of these large fragile cells by either lysis or removal during filtration. Additionally, the clumping caused by cell debris can bias the results. Careful monitoring of the sample condition at the time of collection allows adjustments to be made (i.e., shorter time in trypsin–EDTA).

C.   Antibodies

One of the principal advantages of the FCM analysis of a population is the quantitation of viral and/or host proteins on a per-cell basis. This demands that the antibody be specific and in saturation. Constant cell number per sample is required when the total population has a high level of epitope expression. A small percentage of very positive cells may not deplete the antibody, but if the sample contains a high proportion of cells with numerous epitopes, the results suggest an erroneous lower value of the antigen per cell (average FITC signal). This is a consequence of insufficient content of primary or secondary antibodies, with the result that not all epitopes are labeled.

Specificity and affinity of the antibody must be established. This was accomplished by employing a well-characterized antibody or determining reactivity of an undefined antibody using Western blot analysis and/or immunoprecipitation. Again, examination of known positive and negative cells on coverslips may help determine whether staining was specific.

D. Controls

The sensitivity of this method, necessitates including controls that will identify the problems in each of three areas: (1) antibody, both primary and secondary, (2) DNA content, and (3) sample. Some aspects have been covered in previous papers (Jacobberger et al., 1986; Laffin et al., 1989).

1. ANTIBODY CONTROLS

Primary antibodies are generally mAb prepared from either cell culture or mouse ascites. After a 50% cut by ammonium sulfate, the antibodies are desalted and then further purified on a protein A–sepharose column or stored at −20°C in the presence of 0.1% sodium azide. Care should be taken not to contaminate the supply with proteases and DNases, which would lower the antibody titer and destroy DNA staining. The second antibody is a commercial anti–mouse IgG F(ab')$_2$, which proved to be very consistent. When needed, a dilution of each is made in WS and kept at 4°C.

Each primary antibody is titrated using a standard positive and negative cell for that epitope (e.g., A58 and B1 were positive and negative, respectively, for T antigen epitope reactive with PAb101. See Fig. 1). The FITC-labeled secondary antibody is diluted 1 : 40, 1 : 80, and 1 : 60 and then tested with each of the primary antibody dilutions. The highest concentration that provides a > 90% specific staining and does not alter other parameters should be selected. Examples of dilutions used were a 1 : 40 from a 50% cut of PAb101 supernatant and a 1 : 80 for the secondary. As a check, the positive and negative controls for that epitope are stained with the selected antibody concentration and processed with each experiment. Periodic microscopic examination of the antibody control samples is performed to confirm staining distribution.

2. DNA CONTENT

Included in each experiment described in this chapter were A58 and B1 cells (see Fig. 1) used as standards for DNA content. These were checked for their x and y coordinates at the beginning and end of the instrument run and compared to previous assays. In addition, a methanol-fixed mouse

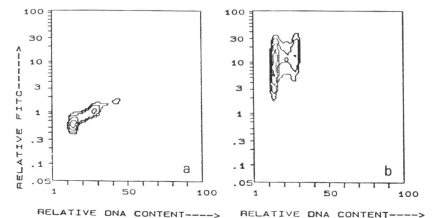

FIG. 1.   Multiparameter contour plots of chinese hamster embryo (CHE) cells stained with propidium iodide (PI) for their DNA content and by indirect immunofluorescence (FITC) for the presence of the early viral protein, SV40 large T antigen (PAb101 epitope). (a) the B1 (T antigen-negative) CHE line used as a standard to establish the level of background fluorescence. The G_1 of the B1 standard had a DNA content with a peak channel of 20 and contained 79% (CV of 8.0%) of the total population. (b) The average FITC value for the entire population after background subtraction (see text) was 0.09 compared to 11.58 of the T antigen-positive CHE line A58 (each histogram contains 10,000 cells). Of the cells in the A58 standard, 99% were positive for T antigen, while the B1 standard contain < 1% positive cells when gated as indicated in the text. These values were routinely monitored at the beginning and during each flow analysis to permit comparison between experiments and establish specificity of staining. The histograms are presented as collected, but smoothed for illustrative purposes. All analyses were done on raw data after background subtraction.

lymphocytes control, treated with the RNase solution and stained with PI, was analyzed as a DNA content indicator and to confirm the red alignment. Each experiment also contained cells with a predicted DNA content for that cell type, stained in the same manner as the experimentals. Samples with questionable DNA content should be karyotyped.

3.   SAMPLE

Autofluorescence and nonspecific binding are determined by staining both infected and uninfected cells as follows: (1) one set with primary antibody and PI and no secondary, (2) one set with secondary and PI and no primary, (3) one set with PI only, and (4) one set with unrelated antibody of the same isotype as the primary, secondary, and PI. These samples are stained as described, substituting WS for the omitted reagent. The data obtained then allowed the selection of the proper control. An example of a normal cell with a high background is the uninfected HDF used in the human transformation experiments (Laffin *et al.*, 1989). These cells were then included as a control in each experiment with HDF.

IV. Flow Analysis

A. Instrumentation

Fluorescence measurements described in this chapter were performed with a Cytofluorograf IIs model H.H., and a 2151 data analysis system (Ortho Diagnostics, Inc., Westwood, MA), using a 5-W argon ion laser (Coherent, Palo Alto, CA) at a wavelength of 488 nm and 50 mW. Similar results have been obtained using a FACS IV (Becton-Dickinson, Sunnyvale, CA) and an EPICS V (Coulter Electronics, Hialeah, FL). The laser was aligned immediately before use with fluorescein-labeled microspheres (Polysciences no. 9847). For alignment, a yellow filter (40 nm width) was used for red fluorescence. The coefficient of variation (CV) was typically 1.0% for low-angle light scatter, and 0.8% for both green and red. A58 and B1 cells (controls described in Section III,D,1) were used to detect instrument variation. For analysis, no filter was used on light scatter, but a barrier filter (535-nm bandpass, 40-nm width) was used for green and a barrier filter (640-nm long-pass) used for red. Data were collected and stored in a two-parameter cytogram as follows. First a gated population using light scatter versus red fluorescence (DNA) eliminated noncellular material. Cells passing through this gate were then analyzed using red area versus red peak for selection of single cells. Collected cells were then displayed in a third dual cytogram of red area versus green area. These data were temporarily maintained in the Ortho data analysis system and then transferred to an IBM PC-AT for analysis and storage. Between 10,000 and 50,000 cells were collected on each control and sample.

B. Data Analysis

Direct evaluation of the data presented in this chapter was performed from the flow cytometer computer display. Estimation of the percentage positive and the relative DNA distribution may be obtained when all controls are determined adequate. This lengthens the time between samples and increases the chance for variation. Each sample takes an average of 2–5 minutes to process. So when possible, analysis using flow computer programs was delayed until after all samples were measured.

For storage and further analysis, the preliminary information was expanded by transfer of data to an IBM PC–AT. The broad range of FITC staining in many SV40 experiments necessitated collection of green fluorescence in log mode, which was then converted on the IBM to linear as follows: $y' = 10[(y - 40)/30]$. Red fluorescence had been collected in linear with the control diploid cells (B1 and A58) used to establish consistent location (channel number). For analysis, the cell number was correlated to both the x axis (red, DNA) and the y axis (green, T antigen). The mean y value was defined as the average FITC value after fluorescence compensation, which was necessary because of the overlap of the red and green signals. This can be accomplished either by a program available from the manufacturer of the flow cytometer or a computer subtraction program (Lehman *et al.*, 1988; Laffin *et al.*, 1989). The mean y value was used as the level of expression of the viral protein. The negative control was established by gating a known non-antigen-expressing population of the same cell type so that <1% of this population was above the gate. For example, in SV40–HDF studies, uninfected HDF were used as the negative control. The percentage of positive was defined as the percentage of cells in a sample above this negative population. This allowed for correlation of cell cycle compartment with quantity and percentage of cells expressing the viral protein.

V.   Applications

Multiparameter FCM can be used in basic studies of virus–host interactions as well as clinical investigations. Viral transformation with both DNA and RNA tumor viruses results in cell cycle changes. Recently the use of T antigen mAb as probes has helped to identify and characterize structually and functionally distinct forms of the protein. The techniques (ELISA, immunoprecipitation, RIA) employed in these studies averaged the quantity of antibody-bound epitope for the whole population. Flow analysis can monitor the appearance of these epitopes, both quantitatively and qualitatively on a per-cell basis during cell cycle progression (Lehman *et al.*, 1988; Laffin *et al.*, 1989). This permits possible identification of cells that differentially express these epitopes (see Fig. 2), which would then allow for isolation and characterization of these populations.

There are many medically important viral infections that are difficult to identify in the clinical laboratory. Cytomegalovirus (CMV) infection is the cause of significant morbidity and mortality in patients with human immunodeficiency virus (HIV) as well as immunocompromised transplant

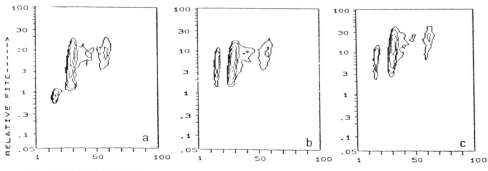

FIG. 2. A comparison of PAb101 and PAb430 epitope staining in SV40-infected HDF cells 12 weeks postinfection. Three aliquots of the same fixed-cell preparation were stained with saturating concentrations of (a) PAb101, which recognized a carboxy epitope of T antigen, (b) PAb430, which recognized an amino epitope, and (c) PAb101 PAb430 combined. All conditions and analysis were as described in Fig. 1 and text, with each histogram representing 10,000 cells. The peak channels for the relative DNA content of the G_1, G_2/G_1 tetraploid, and the G_2 tetraploid populations were at 15, 30, and 60, respectively. (a) Using PAb101 the average FITC value for the entire population was 4.96, as compared to (b) PAb430, with a value of 5.66. The major difference between PAb101 and PAb430 staining patterns was the G_1 population, where >80% of the cells were negative for PAb101, but positive for PAb430. (c) The combined staining of PAb101 and PAb430 resulted in a doubling of the average FITC value, indicating that competition between epitopes PAb101 and PAb430 of T antigen in HDF cells is minimal.

patients. Identification of virus infection, which may take as long as 8 weeks, is confirmed by the combination of viral replication in tissue culture, a rise in antibody titer, and the appearance of inclusion bodies by histopathology. Since rapid identification would allow for more effective antiviral chemotherapy, a number of new techniques using DNA probes and mAb specific to CMV early proteins have been developed but are labor-intensive. The antibodies to CMV are commercially available and have proved to be excellent for FCM analysis. In the cell culture model system, early CMV antigen was detected as early as 30 minutes postinfection, earlier than standard methods (Elmendorf et al., 1988). Additionally, the antigen can be measured and the number of cells infected can be determined. This may be important in assessing the disease state and the efficacy of drug treatment.

Possibly the best application in viral diagnosis would be the ability to detect a small population of infected cells and follow that population in such diseases as HIV or CMV. Human immunodeficiency virus is ideal for this type of assay; there are mAb available to various viral antigens, and with the identification of a second marker (T helper cell, macrophage, etc.), small numbers of infected cells could be identified (McSherry *et al.*, 1990). In addition, a three-color immunofluorescence technique (Rabinovitch *et al.*, 1986) is available for the simultaneous measurement of two mAb and DNA content using standard 488-nm excitation. Fluorescein isothiocyanate (green) is used for an internal viral antigen, phycoerythrin (orange) for a surface marker, and 7–amino actinomycin D (red) for the DNA. This would increase sensitivity and permit detection of the rare positive cell.

ACKNOWLEDGMENTS

The authors gratefully acknowledge the contributions of Mr. David Fogleman and Dr. James Jacobberger in development of the staining protocol, and Ms. Jo Ann D'Annibale in typing the manuscript. These studies were supported by grant CA41608 from the National Cancer Institute.

REFERENCES

Anastasi, J., Bauer, K. D., and Variakojis, D. (1987). *Am. J. Pathol.* **128,** 573–282.

Brady, J. N., and Salzman, N. P. (1986). *In* "The Papovaviridae: The Polyomaviruses" (N. P. Salzman, ed.), Vol. 2, pp. 1–26. Plenum, New York.

Elmendorf, S., McSharry, J., Laffin, J., Fogleman, D., and Lehman, J. M. (1988). *Cytometry* **9,** 254–260.

Jacobberger, J. W., Fogleman, D., and Lehman, J. M. (1986). *Cytometry* **7,** 356–364.

Laffin, J., Fogleman, D., and Lehman, J. M. (1989). *Cytometry* **10,** 205–213.

Lehman, J. M., and Defendi, V. (1970). *J. Virol.* **6,** 738–749.

Lehman, J. M., Horan, P. K., and Cram, L. S. (1979). *In* "Flow Cytometry and Cell Sorting" (M. R. Melamed, P. F. Lidmo, and M. L. Mendelsohn, ed s.), pp. 409–420. Wiley-Liss, New York.

Lehman, J. M., Laffin, J., Jacobberger, J. W., and Fogleman, D. (1988). *Cytometry* **9,** 52–59.

Lehman, J. M., and Jacobberger, J. W. (1990). *In* "Flow Cytometry and Cell Sorting" (M. R. Melamed, P. F. Lindmo, and M. L. Mendelsohn, eds.), pp. 623–631. Wiley-Liss, New York. (in press).

McSherry, J. J., Costantino, R., Robbiano, E., Echols, R., Stevens, R., and Lehman, J. M. (1990). *J. Clin. Microbiology* **28,** 724–733.

Rabinovitch, P. S., Torres, R. M., and Engel, D. (1986). *J. Immunol.* **136,** 2769–2775.

Chapter 27

Differential Staining of DNA and RNA in Intact Cells and Isolated Cell Nuclei with Acridine Orange

ZBIGNIEW DARZYNKIEWICZ

Cancer Research Institute of the
New York Medical College at Valhalla
Valhalla, New York 10595

I. Introduction

The fluorochrome acridine orange (AO) can differentially stain double-versus single-stranded nucleic acids. This dye intercalates into double-stranded nucleic acids and, in the intercalated form when excited in blue light, fluoresces green with maximum emission at 530 nm. Interactions of AO with single-stranded nucleic acids results in condensation (transition to solid state) and subsequent agglomeration (precipitation) of the product. The luminescence (phosphorescence) of these condensed products is in red wavelength, with maximum emission at 640 nm (Kapuscinski *et al.*, 1982; review, Darzynkiewicz and Kapuscinski, 1990). The shift in the emission spectrum when the dye binds to different substrates received the name metachromasia, and AO thus is a metachromatic fluorochrome.

To obtain differential staining of DNA versus RNA it is necessary to denature selectively any double-stranded (ds) RNA, ensuring that all cellular RNA is single-stranded (ss), while DNA still remains in its native, double-helical conformation. Cell treatment with AO in the presence of chelating agents (EDTA, citrate) results in such selective denaturation of RNA (Darzynkiewicz *et al.*, 1975, 1976; Traganos *et al.*, 1977). That is, the

METHODS IN CELL BIOLOGY, VOL. 33

chelating agents, by breaking the protein–RNA interactions, disrupt the structure of ribosomes making rRNA more sensitive to denaturation. Acridine orange subsequently denatures dsRNA. In this staining reaction AO thus has two distinct functions: it (i) selectively denatures dsRNA and (ii) differentially stains RNA, after its conversion to ss form, versus DNA. Higher affinity of AO to ssRNA than to ssDNA is responsible for the specificity of the denaturing reaction (Darzynkiewicz and Kapuscinski, 1990). Selective denaturation of RNA (while DNA remains undenatured) can be achieved only within a very narrow concentration range of free dye at a given ionic milieu. Too high an AO concentration leads to denaturation of DNA (both RNA and then DNA stain red), and too low produces incomplete RNA denaturation (while all DNA stains green, a portion of RNA also stains green).

There are several modifications of the technique of differential staining of RNA and DNA with AO (Darzynkiewicz et al., 1975, 1976, Traganos et al., 1977). Of these, the method presented in this chapter has been applied most extensively and appears to offer the best differential resolution of stainability between DNA and RNA. The technique consists of two steps. In the first step the cells are permeabilized by detergent (Triton X-100) in the presence of acid (0.08 N HCl) and serum proteins. The serum proteins and low pH prevent cell lysis by detergent (Darzynkiewicz et al., 1976), while nucleic acids remain insoluble in these unfixed cells. Treatment with HCl dissociates (and partially extracts) histones and other nuclear proteins from DNA, unmasking the latter for interactions with AO. This leads to an ~ 3-fold increase in stainability of DNA, and decreases intercellular variability in DNA stainability due to differences in chromatin structure (Darzynkiewicz et al., 1984). Detergent disperses cell aggregates, which results in lowering the proportion of cell doublets in suspension. Subsequent staining of these permeabilized cells with AO in the second step, at higher pH and in the presence of EDTA, results in green fluorescence that is specific for DNA (F_{530}), and RNA-specific red luminescence (F_{600}). Both steps should be done at 0–4°C to prevent DNA denaturation (which may occur in the first step) and to suppress the activity of endogenous nucleases.

Because in an average cell rRNA and tRNA consist of ~ 75% and 15% of total RNA, respectively, the measured red luminescence of the whole cells, which is proportional to total-cell RNA (Bauer and Dethlefsen, 1980), represents predominantly these two species of RNA. The green fluorescence is stoichiometrically related to DNA content (Coulson et al., 1977).

In isolated cell nuclei nearly all RNA is a precursor of rRNA and is localized in nucleoli.

II. Application

The method can be applied to a variety of cell types. Especially good specificity of nucleic acids' staining can be obtained in cells of lymphocyte or monocyte lineage (nonstimulated or stimulated lymphocytes, leukemias, lymphomas, myelomas). Most proliferating tissue culture lines of normal or tumor origin also exhibit good differential stainability of DNA and RNA (reviews, Darzynkiewicz, 1988; Darzynkiewicz and Traganos, 1990; Darzynkiewicz et al., 1980). Exceptions are cells that contain large quantities of glycosaminoglycans or proteoglycans such as heparin (mast cells), hyaluronic acid (fibroblasts), or keratan sulfate (differentiated epidermal cells). Interactions of AO with these polymers results in yellow or red luminescence, thus decreasing the specificity of RNA and DNA detection. In these cases, however, the method can be applied to isolated cell nuclei where, regardless of the cell type, only nucleic acids stain with AO.

There are other methods to estimate RNA content in cells or cell nuclei. Cell staining with propidium following exhaustive digestion with DNase selectively stains dsRNA (Frankfurt, 1980). In contrast to the AO-based technique, however, this method does not provide any information about the cell cycle distribution and thus does not allow a comparison of RNA content in cell populations at the same phase of the cell cycle. Simultaneous staining of DNA and RNA with a combination of Hoechst 33342 and pyronin Y (PY) (Shapiro, 1981) offers higher sensitivity of detection of small amounts of RNA, but requires a double-laser excitation source. Furthermore, PY stains fluorescently only dsRNA and the staining is not always stoichiometric. Namely, PY appears to discriminate between rRNA conformation in polysomes and single ribosomes, and RNA stainability with PY drops when polysomes dissociate (Darzynkiewicz et al., 1987; Traganos et al., 1988).

The main application of the method stems from the observation that cellular RNA content is closely associated with cell proliferation and differentiation. Correlated measurements of RNA and DNA content made it possible to discriminate several distinct compartments of the cell cycle characterized by different kinetic properties, in addition to the traditional four main phases of the cycle (Darzynkiewicz, 1984; Darzynkiewicz et al., 1980). Based on RNA content it was also possible to distinguish noncycling cells from their cycling counterparts in a variety of cell types. RNA measurements were also introduced into the clinic, for characterization of leukemias, lymphomas, myelomas, and other proliferative diseases (review, Darzynkiewicz, 1988). RNA content was found, in these studies, to be a strong prognostic parameter in all investigated malignancies.

III. Materials

A. Stock Solution of Acridine Orange

Dissolve AO in distilled water (dH_2O) to obtain 1 mg/ml concentration. It is essential to have AO of high purity. Polysciences, Inc. (Warrington, PA) offers AO of the highest purity ("Electro Pure," catalog no. 4539). The stock solution may be kept in the dark (in dark or foil-wrapped bottles) at 4°C for several months without deterioration.

B. First Step: Solution A

Triton X-100, 0.1% (v/v)
HCl, 0.08 N (final concentration)
NaCl, 0.15 N (final concentration)

Solution A may be prepared by adding 0.1 ml of Triton X-100 (Sigma Chemical Co.), 8 ml of 1.0 N HCl, and 0.877 g NaCl, and dH_2O to a final volume of 100 ml. This solution may be stored at 4°C for several months.

C. Second Step: Solution B

Acridine orange, 6 μg/ml (20 μM)
EDTA–Na, 1 mM
NaCl, 0.15 M
Phosphate–citric acid buffer, pH 6.0

This solution may be made as follows:
1. Prepare 100 ml of buffer by mixing 37 ml of 0.1 M citric acid with 63 ml of 0.2 M Na_2HPO_4.
2. Add 0.877 g NaCl, stir until dissolved.
3. Add 34 mg of EDTA disodium salt. Equivalent amounts of tetrasodium EDTA may be also used. EDTA in acid form may also be used, but it requires a longer time to dissolve. Stir until dissolved.
4. Add 0.6 ml of the stock solution of AO (1 mg/ml).

Solution B is also stable and can be stored for several months at 4°C in dark. Solutions A and B may be kept in automatic-dispensing pipet bottles set at 0.4 ml for solution A and 1.2 ml for solution B—the latter in a dark bottle.

D.   Nuclear Isolation Solution

Prepare a solution containing 10 mM Tris buffer (pH 7.6), 1 mM sodium citrate, 2 mM MgCl$_2$, and 0.1% (v/v) nonionic detergent Nonidet (NP-40).

IV.   Staining Procedure

A.   Staining and Unfixed Cells

1. Transfer a 0.2-ml aliquot of the original cell suspension (not more than 2×10^5 cells) to a small tube (e.g., 5 ml volume). Chill on ice.
2. Add gently 0.4 ml of ice-cold solution A. Wait 15 seconds, keeping on ice.
3. Add gently 1.2 ml of ice-cold solution B. Measure cell luminescence during the next 3–15 minutes (equilibrium time).

The sample should be kept on ice prior to and during the measurement.

Vortexing or syringing cells when immersed in solution A, especially in the absence of any serum or proteins in the original cell suspension, results in disintegration of their membrane and isolation of cell nuclei. RNA content of these isolated nuclei can then be measured. However, visual inspection of the nuclei under phase-contrast or ultraviolet (UV) microscopy is essential to estimate the efficiency of the isolation, which can be controlled by selecting optimal time and speed of vortexing, or number of syringings.

B.   Staining of Fixed Cells

1. Fix cells in suspension in 70% ethanol, at 0°C.
2. Centrifuge cells, remove all ethanol, rinse once, and resuspend in Hanks' buffered saline (HBSS), at a cell density $<2 \times 10^6$ per 1 ml.
3. Withdraw 0.2 ml of cell suspension and stain them, using solutions A and B, as described already for fresh, unfixed cells. In the case of fixed cells the presence of Triton X-100 in solution A is not necessary, although it does not interfere with their staining.
4. To assess the contribution of RNA to the detected luminescence, after removal of ethanol and suspension in HBSS, the cells can be incubated with RNase (10^3 units per 1 ml) for 30 minutes at 37°C, prior to staining with AO.

The advantage of staining fixed cells is the stability of the staining reaction, due perhaps to inactivation of the endogenous nucleases. Also, more reliable estimates of the specificity of staining, due to the possibility of using exogenous RNase or DNase, can be made. The disadvantage is, however, lower resolution of DNA measurements [higher coefficient of variation (CV) of the mean green fluorescence of $G_{0/1}$ cells] and increased cell aggregation.

C.   Isolation and Staining of Unfixed Nuclei from Solid Tumors

1. The tissues may be kept frozen (below −40°C) prior to nuclear isolation. Place the freshly resected (or frozen and then thawed) tissue in the nuclear isolation solution and trim to remove the necrotic, fatty, and other undesirable portions.

2. Transfer the trimmed tissue to a new aliquot of the isolation solution and mince finely with scalpel or scissors. Mix small tissue fragments by vortexing and/or pipetting with a Pasteur pipet or syringe with a large-gauge needle. Observe the release of nuclei by sampling the suspension and viewing it under the phase-contrast microscope. If release of nuclei is inadequate, transfer the minced tissue suspended in the isolation solution, into homogenizer with a glass or Teflon pestle. Homogenize by pressing the pestle several times; check the efficiency of nuclear isolation. The nuclei should be clean, unbroken, and lacking cytoplasmic tags.

    Collect the nuclear suspension from above the remaining tissue fragments (allowed to sediment for ∼ 1 minute) with a Pasteur pipet and filter through 40–60 μm pore nylon mesh. Dilute the suspension with the isolation buffer, if necessary, to have no more than 2×10^6 nuclei per 1 ml of the isolation solution in the final suspension. All isolation steps should be done on ice.

3. Withdraw a 0.2-ml aliquot of nuclei suspension and stain identically as described for whole cells; that is, mix with 0.4 ml of solution A and then with 1.2 ml of solution B.

4. Control samples can be incubated with RNase. To this end, 1 ml of final nuclear suspension is treated with 10^3 units of RNase A for 20 minutes at 24°C, and 0.2-ml aliquots of this suspension are then processed in exactly the same way as nuclei that were not subjected to treatment with RNase.

Note that RNA leaks slowly from the isolated nuclei into the buffer. It is advisable therefore to stain the nuclei as soon as possible after isolating them. After the staining, RNA complexed with AO is less soluble.

V.   Critical Aspects of the Procedure

A.   Preservation of Intact Cells

To ensure that unfixed cells are permeabilized but do not disintegrate, the presence of serum (proteins) or serum albumin in the first step is required. Serum can be provided by making the original cell suspension in tissue culture medium containing 10–20% (v/v) serum [or in buffered saline solution containing 10% serum or 1% (w/w) albumin]. The cells thus can be taken from tissue culture without prior centrifugation, washing, etc., and directly stained, as described in the procedure. Do not vigorously shake, pipet, or vortex cell suspensions after addition of the detergent (unless trying to isolate the nuclei), because the cells are fragile and can easily be broken.

B.   Critical Concentration of Acridine Orange

Differential staining of DNA versus RNA requires proper concentration of free (unbound) AO in the final staining solution and at the time of the actual act of measurement under equilibrium conditions (~ 20 mM). The following problems relating to this requirement may occur.

1. When the cell number (density) in the original suspension exceeds 2×10^6 cells per 1 ml (or even less when cells are highly hyperdiploid and/or rich in RNA), the amount of bound AO is proportionally high and therefore free dye concentration may be significantly decreased. The RNA denaturation is then incomplete and part of RNA stains green.

Solution: Dilute the original cell suspension, or if higher cell densities have to be used (e.g., to have high cell flow rates), use a higher AO concentration in solution B.

2. With some instruments (most cell sorters) in which cell measurements take place outside the nozzle (i.e., in air), a significant diffusion of dye from the sample to the sheath stream takes place after the stream leaves the nozzle, prior to its intersection with the laser beam. This breaks the equilibrium and lowers the actual AO concentration in the sample at the time of measurement. Dye diffusion is also a problem in some home-made instruments that have flow channels characterized by extremely narrow and long sample streams.

Solution: Increased AO concentration (up to 60 μM) in solution B and increased flow rates compensate for the diffusion effect. Wherever possible, channels with altered geometry (wider sample stream and/or shorter distance between the nozzle and intersection with the laser beam) should be used. Optimal dye concentration for a particular instrument can be established by preparing a series of solution B with different AO concentrations (e.g., from 20 to 60 μM), and testing at which concentration the cells in $G_{0/1}$ cell cluster have the same green fluorescence (the lowest CV of the green fluorescence mean value, corresponding to a lack of correlation between green and red luminescence of the $G_{0/1}$ cells). Thus, as is evident in Figs. 1 or 2, the $G_{0/1}$ cell cluster ought to be horizontal (or vertical if axes are reversed) but never skewed (diagonal).

3. Staining cells on microscopic slides, mounted under coverslips in solution B, in equilibrium with AO—to be viewed by UV microscopy, for example as a control to flow cytometry (FCM) in order to observe localization of the reaction—requires a higher concentration of AO compared with that applied in FCM. This is because AO is absorbed on glass or plastic surfaces, which lowers the free-AO concentration in the solution in which the cells are mounted. Alternatively, the slides should be prewashed with solution B prior to their use, to saturate the binding sites with AO.

C. Overlap of the "Green" and "Red" Emission Spectra of AO; Sensitivity of RNA Detection

One of the limitations of the AO technique is relatively low sensitivity of RNA detection. This is primarily due to the emission spectrum overlap: the green fluorescence of AO intercalated to DNA has a long "tail" toward a higher wavelength, which is significant even in red as far as 620–670 nm. Therefore, RNA measurements in cells (or cell nuclei) characterized by a high DNA/RNA ratio lack sensitivity, being obscured by the high component due to AO bound to DNA. The following strategies improve the measurements:

1. Long-pass filters transmitting above 640 or 650 nm, rather than 610 or 620 nm, should be used for measurements of red luminescence, representing AO bound to RNA. This significantly reduces the DNA-associated spectral component.

2. It is difficult to achieve optimal excitation of AO interacting with both DNA and RNA simultaneously. Namely, excitation of AO bound to DNA by intercalation is maximal at 490 nm. On the other hand, maximum excitation of AO in the condensed complexes with RNA is at 455 nm; hence for the highest sensitivity of RNA measurements the excitation should be at 455–460 nm. Unfortunately, at that low wavelength the excitation of AO bound to DNA is inadequate. The compromise excitation wavelength may thus be the 478-nm line of the argon ion laser. In light of all of these considerations, the optimal excitation of both RNA- and DNA-bound AO, and thus the best simultaneous RNA and DNA stain-

ability can be achieved in flow cytometers that have a mercury lamp as a source of illumination, using blue excitation filters that transmit a relatively wide band of excitation light (i.e., between 450 and 490 nm).

VI.   Controls, Standards

To assay specificity of staining, parallel samples prefixed in ethanol and suspended in buffered saline containing Mg^{2+} may be treated with either RNase or DNase as follows.

1. 10^3 units per 1 ml of RNase A (e.g., high-purity preparation of RNase provided by Worthington Biochemical Corp., Freehold, NJ; code name "RASE") for 30 minutes at 37°C, or
2. 1 mg/ml of DNase I (Worthington) for 30 minutes at 37°C

Following these incubations, the cells should be stained with AO, passing through solutions A and B, as described before. The percentage loss of red or green luminescence as a result of treatment with RNase or DNase is an indication of the specificity of staining of RNA or DNA, respectively.

In the case of unfixed cells, cell suspensions already treated with solutions A and B may be subsequently treated with 10^3 units of RNase A and incubated for 30 minutes at 24°C prior to fluorescence measurements.

Nonstimulated, peripheral blood lymphocytes appear to be the best standard for RNA and DNA. The RNase-treated and untreated nonstimulated lymphocytes should be measured to establish the extent of RNase-specific red luminescence. Thus, for instance, if the mean red luminescence of untreated and RNase-treated lymphocytes is 50 and 20 channels, respectively, then the RNase-specific component equals $50 - 20$, or the equivalent of 30 channels. (Note that nonstimulated lymphocytes have very low RNA content and therefore have a proportionally high nonspecific component in RNA staining, mostly because of spectrum overlap from AO bound to DNA.) In the next step, RNA content of the measured cells is compared with that of the lymphocytes. To this end, the other cells have to be measured under conditions identical to those used for lymphocytes, and their RNase-specific component of mean red luminescence should be expressed as a multiplicity of the lymphocytes' RNA. Thus, for instance, if the other cells' mean red luminescence equals 120 channels before and 20 channels after RNase treatment, their RNase-specific component is equivalent to 100 channels. Their RNA index is therefore 100:30 (i.e., 3.3) when expressed as a multiplicity of lymphocytes' RNA.

Because lymphocytes are not uniform, and because B and T lymphocytes differ somewhat with respect to RNA content, a more accurate standard would be purified populations of B or T lymphocytes. If the measured cells have a much larger RNA content than lymphocytes, or if they are aneuploid, and therefore can be distinguished from lymphocytes based on differences in red or green luminescence, it is convenient to use lymphocytes as an internal standard.

Note that because the extent of the spectrum overlap depends on the excitation wavelength and the emission filters used, it can vary from instrument to instrument, depending on minor differences in the filters' specification. Therefore, the proportion of the RNase-specific luminescence to total red luminescence varies as well. This variation does not allow one to express RNA content as a ratio of the mean intensity of red luminescence of the measured cell population to that of the lymphocytes (without an adjustment for the RNase specificity), because such an index cannot be compared between different laboratories or flow cytometers.

The lymphocytes may also serve as a standard of DNA content, to estimate the DNA index from the mean (modal) intensity of the green fluorescence of the $G_{0/1}$ population.

VII.   Instruments

Acridine orange, having maximal absorption at ~ 455–490 nm, is optimally excited in blue light. The 488-nm line of the argon ion laser is the most commonly used. In the instruments illuminated by a mercury lamp, blue excitation filters can be used (BG 12, bandpass interference filters transmitting light between 460 and 500 nm). The emission filters and dichroic mirror combination should discriminate green fluorescence (measured at $530 + 15$ nm) and red luminescence (measured preferably above 640 or 650 nm).

As discussed before, the geometry of the flow channel affects the staining reaction ("the diffusion problem"). The author has experience with the following instruments, which can be categorized into three groups:

1. FC-4800, FC-4801 (Biophysics); FC-200, FC-201, and ICP-22 (all Ortho Instruments Co.), FACS Analyzer, FACScan (Becton Dickinson), Profile and Elite (Coulter Electronics), and PARTEC. Good differential stainability of RNA–DNA with AO was observed using all of the listed instruments. They have no "diffusion problem" and do not require an increased AO concentration compared to this protocol. Mercury lamp

illumination in ICP-22, FACS Analyzer, and PARTEC instruments make them especially well designed for measurement of cells with low RNA content.

2. FACS II, FACS III, FACStar (Becton Dickinson), EPICS IV and C (Coulter Electronics), and 30H Cytofluorograph (Ortho) require higher $(25-50 \ \mu M)$ concentration of AO in solution B to compensate for the dye diffusion.

3. Due to extensive dye diffusion it is difficult to obtain good stainability using the Ortho 50H Cytofluorograph with the original sorting channel. Substitution of this channel with the nonsorting one (as in the 30H Cytofluorograph) improves the staining.

VIII. Results

Figure 1 illustrates stainability of HL-60 leukemic cells undergoing myeloid differentiation in the presence of dimethyl sulfoxide (DMSO). As is evident, differentiation is associated with a significant loss of RNA and cell arrest in the G_1 compartment. Evidence of proper staining is the horizontal, nonslanted position of the G_1 cluster. If the AO concentration is too low, the G_1 cluster deviates toward the diagonal, showing a correlation between green and red luminescence.

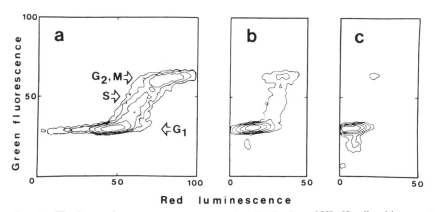

FIG. 1. The isometric contour maps representing distribution of HL-60 cells with respect to their green and red luminescence values after staining with AO. (a) Exponentially growing cells; (b) cells growing in the presence of 1.5% (v/v) DMSO for 3 days; (c) cells cultured in the presence of DMSO for 5 days. A decrease in cell proliferation is concomitant with a decrease in RNA content (red luminescence)

Changes in cellular DNA and RNA content during mitogenic stimulation of lymphocytes are shown in Fig. 2. Based on differences in RNA content, it is possible to distinguish nonstimulated, quiescent cells from cells entering the cell cycle. Their progression through the cell cycle is paralleled by further increase in RNA content. Correlated measurements of RNA and DNA offer a sensitive assay of lymphocytes' stimulation,

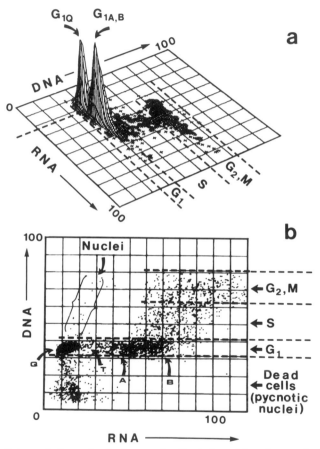

FIG. 2. Subpopulations of cells that can be discriminated in cultures of stimulated lymphocytes after differential staining of cellular DNA and RNA. (a) Lymphocytes stimulated in the allogeneic mixed cultures. Lymphocytes collected from two unrelated donors were mixed and cultured for 6 days prior to staining. (b) Phytohemagglutinin-stimulated lymphocytes on third day of stimulation. Several compartments of the cell cycle can be discriminated, including recognition of nonstimulated (noncycling; G_{1Q}) cells and cells progressing through various phases of the cell cycle such as cells in transition (T) and in early (A) or late (B) G_1 phase (Darzynkiewicz, 1984; Darzynkiewicz et al., 1980).

providing information regarding both the initial steps of stimulation (exit from G_0, RNA content increase) and cell cycle progression. Since stimulation of lymphocytes is a multistep process that does not always result in cell proliferation (e.g., in the absence of interleukin 2), the traditional assays based on radioactive thymidine uptake, in contrast to the present technique, cannot detect the early steps of the stimulation process and thus are useless in such situations.

Figure 3 shows RNA and DNA content of isolated cell nuclei from lung carcinoma. RNA content of hyperdiploid tumor cells is markedly increased compared to the diploid nuclei of normal host cells present in the sample.

Examples and application of the RNA–DNA staining technique are presented in several reviews (Darzynkiewicz, 1988; Darzynkiewicz and Kapuscinski, 1990; Darzynkiewicz and Traganos, 1990; Traganos, 1990).

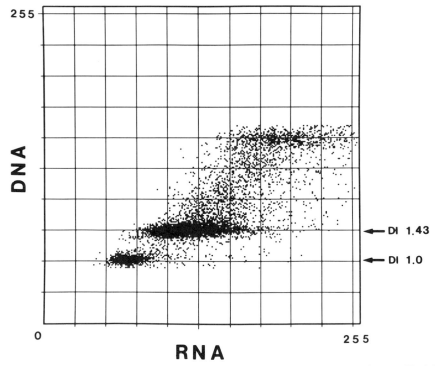

FIG. 3.   RNA and DNA content of nuclei isolated from human lung carcinoma. Nuclei of normal, host cells (DNA index, DI = 1.0) have low RNA content. Tumor cell nuclei (DI = 1.43) are more heterogeneous with respect to RNA content.

Acknowledgment

This work was supported by NCI grant CA 28704. I thank Mrs. Sally P. Carter for her assistance in preparation of the manuscript.

References

Bauer, K. D., and Dethlefsen, L. A. (1980). *J. Histochem. Cytochem.* **28,** 493–498.
Coulson, P. B., Bishop, A. O., and Lenarduzzi, R. (1977). *J. Histochem Cytochem.* **25,** 1147–1153.
Darzynkiewicz, Z. (1984). *In* "Growth, Cancer and Cell Cycle" (P. Skehan and S. J. Friedman, eds.), pp. 249–278. Humana Press, Clifton, N.J.
Darzynkiewicz, Z. (1988). *Leukemia* **2,** 777–787.
Darzynkiewicz, Z., and Kapuscinski, J. (1990). *In* "Flow Cytometry and Sorting" (M. R. Melamed, T. Lindmo, and M. L. Mendelsohn, eds.), pp. 291–314. Wiley-Liss, New York.
Darzynkiewicz, Z., and Traganos, F. (1990). *In* "Flow Cytometry and Cell Sorting" (M. R. Melamed, T. Lindmo, and M. L. Mendelsohn, eds.), pp. 469–501. Wiley-Liss, New York.
Darzynkiewicz, Z., Traganos, F., Sharpless, T., and Melamed, M. R. (1975). *Exp. Cell Res.* **95,** 143–153.
Darzynkiewicz, Z., Traganos, F., Sharpless, T., and Melamed, M. R. (1976). *Proc. Natl. Acad. Sci. U.S.A.* **73,** 2881–2884.
Darzynkiewicz, Z., Traganos, F., and Melamed, M. R. (1980). *Cytometry* **1,** 98–108.
Darzynkiewicz, Z., Traganos, F., Kapuscinski, J., Staiano-Coico, L., and Melamed, M. R. (1984). *Cytometry* **5,** 355–363.
Darzynkiewicz, Z., Kapuscinski, J., Traganos, F., and Crissman, H. A. (1987). *Cytometry* **8,** 138–145.
Frankfurt, O. S. (1980). *J. Histochem. Cytochem.* **28,** 663–669.
Kapuscinski, J., Darzynkiewicz, Z., and Melamed, M. R. (1982). *Cytometry* **2,** 201–211.
Shapiro, H. M. (1981). *Cytometry* **2,** 143–159.
Traganos, F. (1990). *In* "Flow Cytometry and Cell Sorting" (M. R. Melamed, T. Lindmo, and M. L. Mendelsohn, eds.), pp. 773–801. Wiley-Liss, New York.
Traganos, F., Darzynkiewicz, Z., Sharpless, T., and Melamed, M. R. (1977). *J. Histochem. Cytochem.* **25,** 46–56.
Traganos, F., Crissman, H., and Darzynkiewicz, Z. (1988). *Exp. Cell Res.* **179,** 535.

Chapter 28

Flow Cytometric Analysis of Double-Stranded RNA Content Distributions

OSKAR S. FRANKFURT[1]

Grace Cancer Drug Center
Roswell Park Memorial Institute
Buffalo, NY 14263

I. Introduction

The goal of this investigation was to develop a simple, reproducible, and specific procedure for staining of cellular RNA. The method is based on the well-known property of propidium iodide (PI) to form highly fluorescent intercalation complexes with double-stranded nucleic acids but not with other polyanions. This property is widely used for flow cytometric (FCM) analysis of DNA after RNA elimination. We suggest here the use of PI for RNA staining following DNA digestion with DNase. The highly specific character of PI binding to nucleic acids in different cell types, which is the basis for its wide application to DNA analysis, also provides a basis for quantitating double-stranded RNA (dsRNA) in DNase-treated cell populations (Frankfurt, 1980).

II. Materials and Methods

A. Reagents

Solution SMT (Sucrose–Magnesium–Tris)

1. Dissolve 2.43 g tris(hydroxymethyl) aminomethane in 800 ml distilled water (dH$_2$O).

[1] Present address: Oncology Laboratory, Cedars Medical Center, Miami, Florida 33136.

2. Adjust to pH 6.5 with concentrated HCl.
3. Add 85.6 g sucrose and 1.016 g MgCl$_2$.
4. Add water to 1 liter.

DNase Solution

Dissolve 100 mg DNase (Deoxyribonuclease I, Worthington) in 100 ml SMT solution. Store DNase solution frozen at −20°C in small portions.

Propidium Iodide Staining Solution

1. Dissolve 1.21 g Tris and 1.016 g MgCl$_2$ in 1 liter dH$_2$O.
2. Adjust to pH 7.4 with 6–7 ml 1 N HCl.
3. Dissolve 6 mg PI in 100 ml of above solution. The solution can be stored at 4°C in the dark for several months.

B.   Procedure

Fixation

1. Rinse cells in saline and resuspend in cold saline at the concentration of 1–2 × 10^6 cells/ml.
2. Add 2 ml cold absolute ethanol per 1 ml cell suspension dropwise with continuous mixing.
3. Store at 4°C at least overnight before staining.

Staining

1. Rinse fixed cells in 0.9% NaCl.
2. Resuspend cell pellet containing 1–2 × 10^6 cells in 0.5 ml DNase solution.
3. Incubate in the water bath at 37°C for 40 minutes.
4. Keep on ice for 10 minutes.
5. Add 2 ml of cold PI staining solution and keep on ice for 30 minutes before measurement. Prolonged incubation (up to 6 hours) of stained cells at 4°C before FCM analysis does not alter the fluorescence intensity.

To estimate the extent of nonspecific fluorescence, cell samples should be treated with solution containing both DNase and RNase (Ribonuclease A, Sigma, 1 mg/ml). Very low fluorescence intensity after treatment with both enzymes will indicate the specificity of the dsRNA staining.

C. Flow Cytometry

Cell fluorescence was measured on a FACS II (Becton-Dickinson) using an argon ion laser at 488 nm and a flow rate of 100–200 cells/second. This flow rate routinely gave a coefficient of variation (CV) of the G_1 peak in DNA histograms between 4 and 5%. The same flow rate was maintained for the dsRNA distribution analysis, which reduced the possibility of instrumental errors connected with nonoptimal flow rate.

For DNA analysis in RNase-treated cells, the instruments were adjusted so that position of the G_1 peak was in channel 40. For the dsRNA analysis in DNase-treated cells, the amplifier gain was increased in relation to the DNA analysis of the same cell type to place the peak of the broad dsRNA distributions in channel 60–70. Cell samples treated with both enzymes were analyzed at the same gain as DNase-treated cells to assess the intensity of nonspecific fluorescence.

Because of the low fluorescence intensity of the dsRNA-bound fluorochrome in normal bone marrow, the voltage supply of the photomultiplier tube was significantly increased for the measurements of these cells.

A total of 10,000 cells were analyzed for each histogram. The first 10 channels were thresholded to eliminate nonspecific signals.

III. Results and Discussion

The relationship between dsRNA content and cell proliferation was studied for different cell types. The dsRNA content distributions, of DNase-treated exponentially growing HeLa S3 cells were broad, unimodal, and slightly skewed to the right (Fig. 1). The fluorescence intensity of cells treated with both RNase and DNase (nonspecific fluorescence) was very low, indicating that fluorescence of DNase-treated cells shows the relative content of dsRNA. Comparison of the fluorescence intensity between DNase-treated and DNase plus RNase-treated cells showed that 93% of the fluorescence in DNase-treated cells was eliminated with RNase.

To establish the correlation between dsRNA content and proliferative activity, DNA histograms and dsRNA distributions were obtained from HeLa S3 cells 24–120 hours after plating (Table I). Maximal dsRNA content was observed in 48-hour cultures, when the percentage of cells in S phase was also maximal. The dsRNA content was minimal in late plateau phase (120-hour culture), when the number of cells decreased as a result of detachment.

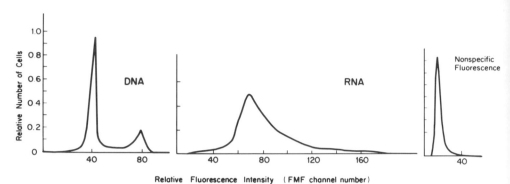

Relative Fluorescence Intensity (FMF channel number)

FIG. 1. Distributions of fluorescence intensity of PI-stained HeLa S3 cells treated with RNase (DNA histogram), DNase (RNA histogram), or both enzymes simultaneously (nonspecific fluorescence).

The dsRNA content in tumor cells from L1210 ascitic leukemia was maximal on the third day after transplantation (peak channel 73 ± 1.4) and significantly decreased on the seventh day ($43 \pm 1.5, p < .01$). The percentage of cells in the S phase (49.3%) was also maximal on the third day of growth and decreased to 12.3% on the seventh day after transplantation. Regeneration of mouse bone marrow after injection of 200 mg/kg cyclophosphamide (CPA) was accompanied by changes in the dsRNA distribution. A number of cells with high dsRNA content was increased 3–5 days after CPA injection during a period of active bone marrow regeneration. Such an increase was not apparent 7 days after CPA injection. Percentage of cells in S phase as determined from DNA histograms was 35% on the third day, 27% on the fifth day, and 14.1% on the seventh day after

TABLE I

PERCENTAGE OF CELLS IN S PHASE AND DSRNA CONTENT IN HELA S3 CELL CULTURE

Time of growth (hours)	Cell density (cells/cm² × 10⁴)	S Index (%)	dsRNA content	
			Peak channel	Mean content
0	4.6	—		
24	7.0	24.1	48 ± 1^a	52 ± 1
48	13.3	30.9	58 ± 1.1	64 ± 2.6
72	24.2	16.6	40 ± 0.5^a	49 ± 2.7
96	41.7	13.4	41 ± 1.7^a	49 ± 1.5
120	31.0	15.8	31 ± 1.4^a	40 ± 1

[a] Significantly different from values of 48 hour cultures ($p < .05$).

CPA injection in comparison with 17.5% in control. Thus, during bone marrow regeneration, an increased percentage of S-phase cells and high dsRNA content were observed at the same time.

Comparison of the dsRNA and DNA distributions showed that in three cell systems, the dsRNA content correlated with proliferative activity as characterized by the proportion of cells in S phase. In HeLa cell cultures, L1210 ascitic leukemia cells, and regenerating bone marrow cells, the maximal dsRNA content was observed at the same time as the maximal S index. In HeLa and L1210 cells, the dsRNA content was found decreased at the time of minimal proliferative activity. These data show that dsRNA content measured by FCM after PI–DNase staining may be a useful marker of proliferative activity. Correlation between RNA content measured by acridine orange (AO) technique and proliferation rate was also observed by Darzynkiewicz *et al.* (1979).

Analysis of dsRNA distribution of cells from normal thymus and thymic lymphoma showed significantly more intensive dsRNA-bound fluorescence in neoplastic cells (Fig. 2). Cells with fluorescence intensity greater than channel 67 comprised 2.8% of the total number in thymus and 50% in lymphoma. Barlogie *et al.* (1983) also observed markedly higher levels of dsRNA in tumor cells compared to normal cells in human leukemias and solid tumors.

Since dsRNA is localized predominantly in the nucleoli and a prominent nucleoli is a characteristic feature of malignant cells, it is possible that high levels of dsRNA may be used as a marker for malignant cells.

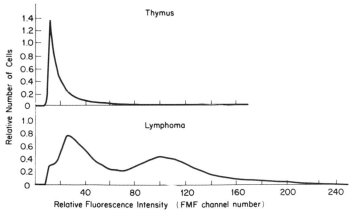

FIG. 2. The dsRNA content distributions of cells from thymus of C57B1 mouse and of cells from thymic lymphoma of AKR mouse. Cell suspensions were fixed in ethanol, treated with DNase, and stained with PI.

It should be mentioned that pyronin Y, at low concentration, is also a specific fluorochrome of dsRNA (Darzynkiewicz *et al.*, 1987) and can be used in conjunction with Hoechst 33342, to stain dsRNA and DNA differentially (Shapiro, 1981).

ACKNOWLEDGMENTS

This work was supported by NIH grants CA21071, CA13038, and CA24538.

REFERENCES

Barlogie, B., Raber, N. M., Schumann, J., Johnson, T. S., Drewinko, B., Swartzendruber, D. E., Gohde, W., Andreeff, M., and Freireich, E. (1983). *Cancer Res.* **43,** 3982–3997.
Darzynkiewicz, Z., Evenson, D. P., Staino-Coico, L., Sharpless, T. K., and Mylamed, M. (1979). *J. Cell. Physiol.* **100,** 425–438.
Darzynkiewicz, Z., Kapuscinski, J., Traganos, F., and Crissman, H. A. (1987). *Cytometry* **8,** 138–145.
Frankfurt, O. S. (1980). *J. Histochem. Cytochem.* **28,** 663–669.
Shapiro, H. M. (1981). *Cytometry* **2,** 143–150.

Chapter 29

Simultaneous Fluorescent Labeling of DNA, RNA, and Protein

HARRY A. CRISSMAN,* ZBIGNIEW DARZYNKIEWICZ,‡ JOHN A. STEINKAMP,* AND ROBERT A. TOBEY†

*Cell Biology Group
†Biochemistry and Biophysics Group
Los Alamos National Laboratory
Los Alamos, New Mexico 87545
‡Cancer Research Institute of the
New York Medical College at Valhalla
Valhalla, New York 10595

I. Introduction

DNA, RNA, and protein comprise the bulk of macromolecules in cells. Interrelationships in synthesis and accumulation of these moieties appear to play an important role in regulating cell cycle traverse capacity, cell division, growth and size. Consistency in the size range of the population volume distribution and the cycle generation time is controlled by transcriptional and translational processes that rigidly couple temporal metabolism of DNA, RNA, and protein.

Flow cytometry (FCM) provides a rapid and precise method of performing multiple biochemical measurements on single cells, thereby allowing for subsequent correlation of the various metabolic parameters. Simultaneous FCM analysis of cellular DNA and RNA (Darzynkiewicz et al., 1976) and DNA and protein (Crissman and Steinkamp, 1973, 1982) contents, for example, permits distinction between quiescent and cycling cells (Darzynkiewicz et al., 1908a,b), and cells in various stages of differentiation (Darzynkiewicz et al., 1976) as well as determination of the relationship between cell growth and the cell division cycle (Darzynkiewicz et al., 1982). Correlated measurement of DNA, RNA, and protein permits

METHODS IN CELL BIOLOGY, VOL. 33

assessment of the ratio of RNA to DNA and RNA to protein per cell, and such information can serve as a sensitive gauge on the metabolism of cells located at particular stages of the cell cycle under a variety of experimental conditions.

In this chapter we describe an FCM method for direct determination and correlation of DNA, RNA, and protein in individual cells (Crissman *et al.*, 1985a). The fluorescent labeling protocol incorporates modification of procedures for fluorochroming DNA with Hoechst 33342 (HO) (Arndt-Jovin and Jovin, 1977), RNA with pyronin Y (PY) (Tanke *et al.*, 1981; Shapiro, 1981; Pollack *et al.*, 1982), and protein with fluorescein isothiocyanate (FITC) (Crissman and Steinkamp, 1973; Gohde *et al.*, 1976). Analysis can be performed in a three-laser system such as that previously described (Steinkamp *et al.*, 1982).

II. Applications

In most instances the rates of RNA and protein synthesis in cells are constant, and increases in cellular content of both macromolecules are linear and proportional across the cell cycle of exponentially growing mammalian populations. However, a variety of cycle-perturbing agents, including drugs, induce a differential uncoupling of normal synthetic patterns, causing a disproportionate accumulation of these cellular constituents. These conditions may lead to states of unbalanced growth, loss of long-term viability, and eventual cell death.

In one study (Crissman *et al.*, 1985b), correlated analysis of DNA, RNA, and protein by FCM was used to determine the effects of adriamycin (AdR) on exponentially growing CHO cells. Treatment with AdR induced a differential response in metabolism of DNA, RNA, and protein in the cell population. At 15 hours after drug treatment 85% of the cells were arrested in the G_2 phase, and this subpopulation had mean ratio values for RNA/DNA and RNA/protein that were 44% and 31% elevated, respectively, above ratio values of control cells. These data indicated that the G_2-arrested, AdR-treated cells were in a gross state of unbalanced growth. Few cells are observed in S phase, but the cells remaining in G_1 phase (i.e., 15%) had RNA/DNA and RNA/protein ratio values almost identical to control values. Survival studies showed that, compared with the exponential control CHO cell population, the AdR-treated exponential population had only a 12% surviving fraction. Subsequent viable cell-sorting studies

(Crissman *et al.*, 1988), based on bivariate DNA and cell volume analysis, and plating efficiency assays confirmed the accuracy of the DNA, RNA, and protein measurements for assessing the state of unbalanced growth in the G_2-arrested, drug-treated subpopulation. In contrast, cells in the G_1 phase, which maintained normal levels of both RNA and protein, had survival values 40 times higher than the G_2 subpopulation. These results indicate that correlated analyses of DNA, RNA, and protein represent a very useful approach for studying mechanisms and control of the cell cycle.

In addition to the study just outlined, we have used this technique to (a) characterize granulosa cell subpopulations from preovulatory avian follicles (Marrone and Crissman, 1988), (b) correlate DNA, RNA, and protein contents in normal and rickettsia-infected L929 cells (Baca and Crissman, 1987), (c) study the effects of conditioned medium on proliferation and antimetabolite sensitivity of promyelocytic leukemia (HL-60) cells *in vitro* (Elias *et al.*, 1985), (d) determine the metabolic effects of three different DNA synthesis-inhibiting drugs on the entry of G_1-synchronized CHO cells into S phase (D'Anna *et al.*, 1985), as well as (e) monitor the state of balanced growth in CHO cells and human fibroblasts during cell synchrony experiments (Tobey *et al.*, 1988).

III. Materials

Hoechst 33342 (HO) (Calbiochem, San Diego, CA) is a benzimidazole derivative that has a high specificity for double-helical DNA and binds preferentially to A-T base regions, but does not intercalate. This dye emits blue fluorescence when excited by ultraviolet (UV) light at ~ 350 nm. Details on the DNA-specific Hoechst fluorochromes are presented by Crissman *et al.* (Chapter 9, this volume). Stock solutions of HO are prepared in distilled water (dH$_2$O) at 1.0 mg/ml and stored in foil-wrapped containers in a refrigerator for at least 1 month.

Pyronin Y (Polysciences, Warrington, PA), also known as pyronin G or gelb, is a red fluorescent dye that binds preferentially to ribosomal RNA. Only the highly purified grades of the dye should be used for cellular RNA staining. Brachet (1940) first introduced this compound for use in cytochemistry. Tanke *et al.* (1981), Shapiro (1981) and Pollack *et al.* (1982) have used PY for staining RNA in FCM studies. Studies (Traganos *et al.*, 1988) indicate that PY binds to double-stranded RNA (dsRNA) and more specifically to polyribosomal rather than to ribosomal RNA. Stock solutions of PY are prepared in dimethyl sulfoxide (DMSO) at 1.0 mg/ml and stored in a refrigerator for at least 1 year. Although DMSO may freeze at this temperature, rapid thawing of the solution at 37°C prior to each use does not appear to produce any noticeable effect on PY–RNA stainability.

Fluorescein isothiocyanate (FITC, isomer 1) (BBL, Microbiology Systems, Cockeyville, MD) is a green-yellow fluorescent dye that is often used as a fluorescent conjugate bound to antibodies. However, in this technique it is used in the free-dye form as a general cellular protein stain. Only FITC of good quality and purity should be used for cell staining. Previously, Gohde *et al.* (1976) and we (Crissman and Steinkamp, 1973, 1982) have used FITC in combination with ethidium iodide or propidium iodide, respectively, for simultaneous FCM analysis of DNA and protein using a single excitation source. Initially a small amount of FITC is prepared in 95% ethanol at 1.0 mg/ml by vortexing, and then this solution is diluted to ~ 10 μg/ml with dH$_2$O. A fresh solution should be made just prior to use, since FITC solutions do not appear to be stable for a long period of time.

The HO, PY, and FITC dyes are best preserved by storage in airtight containers in a refrigerator-freezer.

Saline GM and 95% ethanol are used to fix cells according to the procedure outlined by Crissman and Tobey (Chapter 10, this volume). Cell are routinely fixed for at least 1 hour in ethanol (final concentration, 70%). No detailed studies have been performed in our laboratory to determine the extent of any potential loss of either RNA or protein during prolonged fixation. Ideally, cells should be stained and analyzed as soon as possible; however, we have not noticed any gross deterioration in stainability of these components after 1 week in fixative.

IV.   Staining Procedure

Fixed cell samples are centrifuged and the ethanol fixative is removed by aspiration. A quantity of stain solution containing HO, PY, and FITC at concentrations of 0.5, 1.0, and 0.1 μg/ml in phosphate-buffered saline (PBS) is added to the cell pellet to provide a cell density of ~ 7.5 × 10^5 cells/ml. The sample may be vortexed slightly to resuspend the cells in the stain. Stained samples are allowed to stand at room temperature (RT) for 30 minutes prior to FCM analysis, which should be completed by 2 hours following staining. This procedure is a simplified version of our earlier protocol (Crissman *et al.*, 1985a,b).

V.   Critical Aspects

As in any multicolor, fluorochrome-labeling technique, the dye concentrations should be maintained as low as possible to prevent dye–dye fluorescence interference, and still the concentration of each dye should be sufficient to reflect quantitatively the cellular content of the component they are designed to label. Usually these criteria must be determined empirically for each cell type. In this protocol the amount of the green-FITC and the red-PY fluorescence will be proportional to the cell volume, and the dye concentrations must be adjusted accordingly to prevent red–green fluorescence interference, since the spectra for FITC and PY do overlap to some extent. For example, human squamous cells that have a large cytoplasm will require lower FITC and PY dye concentrations for staining than the comparatively smaller CHO cells. In order to minimize potential blue (HO), green (FITC), and red (PY) fluorescence interference, a three-laser sequential excitation scheme was devised, as will be described. Also, the 457.9-nm laser line, instead of the 488.0-nm line that is usually employed, was used for excitation of FITC in order to minimize excitation of PY. The PY was excited at 530.9 nm to minimize excitation of FITC. Details are described in Section VI (see also Crissman et al., 1985a,b).

VI.   Controls and Standards

In order to test the reliability of the protocol for quantitative determinations of cellular DNA, RNA, and protein, one must analyze cells stained with only one of each of the dyes and compare results with those obtained for the same cells stained with all three dyes in combination. In all cases cells should be excited by all three laser beams and fluorescence analysis be performed in the blue, green, and red channels to determine the degree of nonspecific fluorescence in each channel (see Section VII). Results of such a FCM analytical test for CHO cells are shown in Fig. 1.

Analysis of populations of CHO cells stained with all three dyes provided distribution profiles for DNA, protein, and RNA that were nearly identical to distributions obtained for cells stained with only one of each respective dye (Fig. 1A–C). Cells stained with all three dyes (Fig. 1D) showed a slight decrease in (FITC) green fluorescence and a slight increase in (PY) red fluorescence as compared with respective single dye-stained

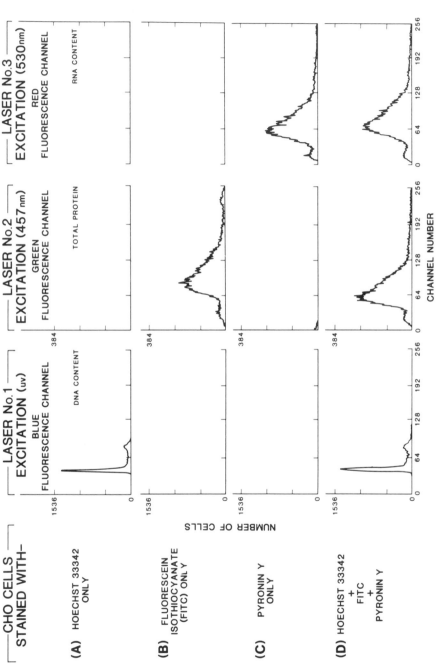

FIG. 1. Single-parameter frequency distributions for populations of ethanol-fixed CHO cells stained only with HO (0.5 μg/ml) for DNA; (B) with FITC (0.1 μg/ml) for protein; (C) with PY (1.0 μg/ml) for RNA; or (D) with a combination of all three dyes, as described in the text. All populations (A–D) were analyzed in the three-laser system at all three wavelengths indicated, and fluorescence was monitored in each channel for each stained cell population.

cells (Fig. 1, B and C, respectively). This, in all probability, is due in part to energy transfer from FITC to PY. The DNA profiles in Fig. 1A and Fig. 1D are similar, indicating no change in HO–DNA fluorescence intensity in the three-color stained cells. Data in Fig. 1 were all obtained using the same electronic gain and laser power settings for all three lasers. Cells pretreated with RNase prior to PY staining and analysis showed an 85–90% decrease in red fluorescence compared with distributions for non-RNase-treated cultures (Fig. 1C). Collectively data such as those provided are useful and necessary for demonstrating the sensitivity and reliability of multicolor staining and fluorescence analysis for accurate assessment of cellular DNA, RNA, and protein with the optimal staining and analysis conditions devised for each cell type.

VII. Instrument

For FCM analysis, the three-laser excitation flow system previously described (Steinkamp et al., 1982) was used with the laser beams tuned to the UV (333.6–363.8 nm), 457.9 nm, and 530.9 nm. The beams were focused onto the cell stream at three different points, with spacing of ~250 μm. The three-color fluorescence detector consisted of colored glass and dichroic color-separating filters for measuring blue (400–500 nm), green (515–575 nm), and red(>580 nm) fluorescence emission from each stained cell. These emission measurement regions were preselected for each dye so that HO-bound DNA (blue), FITC–protein (green), and PY-bound RNA (red) fluorescence, respectively, were monitored at the UV, 457.9-nm, and 530.9-nm excitation points on the cell stream. Fluorescence measurements were correlated on a cell-to-cell basis. Fluorescence signals and fluorescence ratios signals (Crissman and Steinkamp, 1973) were input to signal-processing electronics for subsequent storage (i.e., list-mode fashion) in a DEC PDP-11/23 computer, and subsequently displayed as single-parameter frequency distribution histograms and two-parameter contour diagrams (Salzman et al., 1981).

Histograms for forward-angle light scatter detected from 0.5° to 2.0°, three colors of fluorescence, and two fluorescence ratios were calculated from the data, which were collected and stored in list mode for a given stained population of cells.

Gated analysis, as described previously (Crissman and Steinkamp, 1973; Salzman et al., 1981), can be used to derive numerical data (i.e., relative values) for comparison of the ratios of RNA/DNA and RNA/protein for

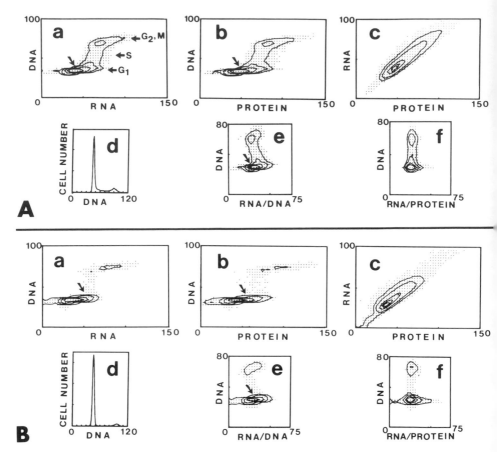

FIG. 2. (A and B). Contour isometric maps (a, b, c, e, and f) DNA frequency histograms (d) representing the distribution of cells with respect to DNA, RNA, and protein content of CHO cells from (A) exponentially growing cultures and (B) cultures deprived of isoleucine for 30 hours in which cells are synchronized in G_1 phase. The shaded areas and the sequential contours represent increasing isometric levels equivalent to 5, 10, 50, 250, 500, and 1500 cells, respectively; 4×10^4 cells were measured per sample. In the exponentially growing population 57, 24, and 19% of the cells were in G_1, S, and $G_2 + M$, respectively, compared to 95, 2, and 6% in the isoleucine-deprived G_1-arrested population. The arrows indicate the threshold RNA or protein content of G_1 cells; cells with RNA or protein below the threshold values did not immediately enter the S phase.

cells within specific regions of the cell cycle (Crissman *et al.*, 1985a,b). Using such analysis, a gate or window may be set between preselected regions (i.e., channel number x to channel number y) of the DNA histogram, encompassing, for example, cells in G_1 phase. Reprocessing of the data can then provide the numerical mean ratio value for cells within the

G_1 region of the cell cycle. Since data are collected and stored in list-mode fashion, data for any of the individual measurements can be retrieved in a similar manner for cells in any phase of the cell cycle. Potentially, gated analysis can be performed on any of the distributions (i.e., protein or RNA content) as well, and data can be correlated as described previously.

VIII. Results

This staining and analysis technique was used to characterize and compare cycling and noncycling (G_1-arrested) CHO cells, and to correlate cellular DNA, RNA, and protein, as well as the ratios of RNA/DNA and RNA/protein throughout the cell cycle for both populations (see Fig. 2). It is evident in Fig. 2A that G_1 cells with a low RNA or protein content do not enter the S phase. Detailed characterization of the critical RNA and protein thresholds for entry of G_1 cells into S phase are presented by Darzynkiewicz (Chapter 27, this volume). The G_1-arrested (noncycling) cells (Fig. 2B) show a much greater heterogeneity in RNA and protein compared with the cycling population; however, in both the cycling and noncycling populations (Fig. 2A,c and 2B,c), the RNA-to-protein contents are well correlated.

Analysis of the cellular RNA/DNA ratio in relation to DNA content (Fig. 2A,e) reveals a characteristic pattern reflecting changing rates of DNA replication and transcription during the cell cycle. Thus, during G_1 when DNA content is stable, cells accumulate increased quantities of RNA, but at different rates so that the G_1 phase is quite heterogeneous with respect to RNA. However, during progression through S phase, the rate of DNA replication exceeds RNA accumulation, giving rise to a nonvertical, negative slope of the S-phase cell cluster. Cells in G_2 + M have RNA/DNA ratios in the same range as the majority of the G_1 cells. (For details see Crissman et al., 1985a.)

ACKNOWLEDGMENTS

This work was supported by the Los Alamos National Flow Cytometry and Sorting Research Resource funded by the Division of Research Resources, National Institutes of Health (grant P41-RR01315) and the Department of Energy.

REFERENCES

Arndt-Jovin, D., and Jovin, T. M. (1977). J. Histochem. Cytochem. 25, 585–589.
Baca, O. G., and Crissman, H. A. (1987). Infect. Immun. 55, 1731–1733.
Brachet, J. (1940). C. R. Seances Soc. Biol. 133, 88–90.

Crissman, H. A., and Steinkamp, J. A. (1973). *J. Cell Biol.* **59**, 766–771.

Crissman, H. A., and Steinkamp, J. A. (1982). *Cytometry* **3**, 84–90.

Crissman, H. A., Darzynkiewicz, Z., Tobey, R. A., and Steinkamp, J. A. (1985a). *Science* **228**, 1321–1324.

Crissman, H. A., Darzynkiewicz, Z., Tobey, R. A., and Steinkamp, J. A. (1985b). *J. Cell Biol.* **101**, 141–147.

Crissman, H. A., Wilder, M. E., and Tobey, R. A. (1988). *Cancer Res.* **48**, 5742–5746.

D'Anna, J., Crissman, H. A., Jackson, P. J., and Tobey, R. A. (1985). *Biochemistry* **24**, 5020–5026.

Darzynkiewicz, Z., Traganos, F., Sharpless, T., and Melamed, M. R. (1975). *Exp. Cell Res.* **95**, 143–153.

Darzynkiewicz, Z., Traganos, F., Sharpless, T., and Melamed, M. R. (1976). *Proc. Natl. Acad. Sci. U. S. A.* **76**, 358–362.

Darzynkiewicz, Z., Sharpless, T., Staiano-Coico, L., and Melamed, M. R. (1980a). *Proc. Natl. Acad. Sci. U. S. A.* **77**, 6696–6699.

Darzynkiewicz, Z., Traganos, F., and Melamed, M. R. (1980b). *Cytometry* **1**, 98–108.

Darzynkiewicz, Z., Crissman, H., Traganos, F., and Steinkamp, J. (1982). *J. Cell. Physiol.* **113**, 465–474.

Elias, L., Wood, A., Crissman, H. A., and Ratliff, R. (1985). *Cancer Res.* **45**, 6301–6307.

Gohde, W., Spies, I., Schumann, J., Buchner, T., and Klein-Dopke, G. (1976). *In* "Pulse-Cytophotometry" (T. Buchner, W. Gohde, and J. Schumann, eds.), pp. 27–32. European Press, Ghent, Belgium.

Marrone, B. L., and Crissman, H. A. (1988). *Endocrinology* **122**, 651–658.

Pollack, A., Prudhomme, D. L., Greenstein, D. B., Irvin, G. L., III, Clafin, A. J., and Block, N. L. (1982). *Cytometry* **3**, 28–35.

Salzman, G. C., Wilkins, S. F., and Whitfill, J. A. (1981). *Cytometry* **1**, 325–336.

Shapiro, H. M. (1981). *Cytometry* **2**, 143–150.

Steinkamp, J. A., Stewart, C. C., and Crissman, H. A. (1982). *Cytometry* **2**, 226–231.

Tanke, H. J., Niewenhuis, A. B., Koper, G. J. M., Slats, J. C. M., and Ploem, J. S. (1981). *Cytometry* **1**, 313–320.

Tobey, R. A., Valdez, J. G., and Crissman, H. A. (1988). *Exp. Cell Res.* **179**, 400–416.

Traganos, F., Crissman, H. A., and Darzynkiewicz, Z. (1988). *Exp. Cell Res.* **179**, 535–544.

Chapter 30

Flow Cytometric Cell-Kinetic Analysis by Simultaneously Staining Nuclei with Propidium Iodide and Fluorescein Isothiocyanate

ALAN POLLACK

Department of Radiation Therapy
U. T./M. D. Anderson Cancer Center
Houston, Texas 77030

I. Introduction

The simultaneous staining of DNA with propidium iodide (PI) and nuclear protein with fluorescein isothiocyanate (FITC) provides information analogous to acridine orange (AO) staining of partially denatured DNA in whole cells (Darzynkiewicz *et al.*, 1982). In the procedure described, nonionic detergent is used to strip the nuclei of cytoplasm and the nuclei are then stained without washing. The technique is rapid and allows for the discrimination of multiple cell cycle compartments. Moreover, the analysis can readily be applied to solid tumors.

Propidium iodide fluorescence of stained nuclei is proportional to the DNA content under most conditions (Darzynkiewicz *et al.*, 1984). The addition of FITC staining provides a degree of separation of overlapping cell populations and cell cycle compartments. Fluorescein isothiocyanate covalently binds to the ε-amino group of lysine and the initial binding can be affected by various conditions (Ward and Fothergill, 1976). Although there are exceptions (Dyson *et al.*, 1987), under most conditions the binding appears to be proportional to protein content (Blair *et al.*, 1979; Roti Roti *et al.*, 1982; Dyson *et al.*, 1987; Pollack *et al.*, 1989). For simplicity, nuclear FITC staining and nuclear protein content will be used interchangeably.

315

The preparation of nuclei that are free of cytoplasmic tags is important for the accurate measurement of both DNA and nuclear protein. Since PI stains double-stranded nucleic acid, including RNA, the presence of cytoplasmic tags will cause an increase in the coefficient of variation (CV). Adequate preparation of nuclei results in CV values comparable to most single-parameter DNA-staining techniques, and treatment with ribonuclease is not necessary.

II. Materials and Methods

A. Reagents and Materials

1. Nuclear isolation buffer (NIB): 0.5% Nonidet P-40 (NP-40; Sigma, St. Louis, MO), 0.05 M NaCl, 1.0 mM EDTA, 0.05 M Trizma base: HCl, pH 7.4
2. Bicarbonate buffer: 0.03 M NaHCO$_3$, pH 8.0–8.3
3. PI Stock: 70 μg/ml PI (Calbiochem-Behring, La Jolla, CA) in bicarbonate buffer
4. FITC Stock: 3 μg/ml FITC (Miles Labs Inc., Elkhart, IN) in bicarbonate buffer
5. Teasing buffer: 0.1 M Trizma base, 5 ml/liter of 38% HCl, 0.08 M NaCl, pH 7.4
6. Nylon mesh: pore size 105 and 60 μm (Small Parts, Inc., Miami, Fl)

Note that the solutions containing the bicarbonate buffer are made up fresh every 2 weeks.

B. Staining Procedure for Single-Cell Suspensions

During the entire procedure the cells/nuclei are kept at 4°C. The cells are washed twice in Hanks' balanced salt solution, calcium and magnesium free (HBSS), to wash out any protein in the medium. For mitogen-stimulated lymphocytes, the solution is supplemented with 1.0 mM EDTA to minimize clumping.

1. Pellet 3–7.5 × 10^5 cells in a tube and decant medium. Vortex-mix the pellet and add 1 ml NIB.
2. Vortex-mix for 4 seconds or syringe several times through a 26-gauge needle to facilitate removal of cytoplasm.
3. In rapid succession, add 0.1 ml FITC stock, 0.9 ml bicarbonate buffer, and 1 ml PI stock.

4. Incubate on ice for at least 30 minutes. Samples should be stable overnight.
5. Vortex-mix for 4 seconds and examine under phase-contrast fluorescence microscopy for cytoplasmic tags.
6. Filter through 30–60 μm nylon mesh before FCM analysis.

C.   Staining Procedure for Solid Tumors

1. Tissue may be stored dry at −70°C or below for several months without adverse effect. Storage at −20°C or above results in extensive debris and broad CV values.
2. Thaw and add 0.5 ml teasing buffer to a small piece of tissue, and very lightly tease the tissue with forceps until the solution becomes slightly cloudy. Excessive teasing of the tissue will result in an increase in debris.
3. Filter the suspension twice, first through 105-μm and then through 60-μm nylon mesh.
4. Add 1.5 ml of NIB to the suspension, and vortex-mix for 4 seconds.
5. Remove 0.1 ml of nuclear suspension, dilute with 0.9 ml PI solution, and count the number of fluorescent nuclei using a hemocytometer.
6. Remove $3–5 \times 10^5$ nuclei and dilute to 1 ml with NIB.
7. Repeat steps 3–6 in the staining procedure for single-cell suspensions (Section II, B).

D.   Preparation of Chicken Red Blood Cells (CRBC) as an Internal Standard

1. Heparinized chicken blood is washed twice in a large volume of HBSS, and 1 drop of concentrated CRBC is aliquoted per tube.
2. Pellet CRBC, decant medium, and store frozen at −70°C.
3. Prior to staining, thaw and dilute pellet in NIB for counting (see Section II,C, step 5). The approximate amount of NIB to add will depend on the average number of cells aliquoted.
4. Adjust the CRBC concentration to 1×10^6/ml in NIB and add 20 μl to each milliliter of sample *prior to* staining.

E.   Instrumentation

A Coulter Electronic (Hialeah, FL) EPICS V flow cytometer with a 5-W argon ion laser at 488 nm (350–400 mW) was used. The fluorescence emission was filtered through a 515-nm long-pass filter, 560-nm dichroic filter, 630-nm long-pass filter (red photomultiplier tube for DNA), and a

530-nm short-pass filter (green photomultiplier tube for nuclear protein). Samples were kept at 4°C using the Coulter viable cell handling system. The analog signals were gated, using a gated amplifier triggered on red fluorescence prior to analog-to-digital conversion. This gating is necessary when nuclei are isolated and stained without removal of the cytoplasmic material. Under these conditions FITC stains the cytoplasmic debris, resulting in a large number of fluorescent signals that can overload the system. Gating on the red signal prior to analog-to-digital conversion alleviates this problem.

III.   Results

The simultaneous staining of DNA and nuclear protein permits the identification of multiple cell cycle compartments including G_{1Q}, G_{1A}, G_{1B}, S, G_{2A}, G_{2B}, and M. This discrimination is most clearly seen in cell lines cultured *in vitro* (Roti Roti *et al.*, 1982; Pollack *et al.*, 1984, 1985, 1986; Dyson *et al.*, 1987), but is also observed *in vivo* (Pollack *et al.*, 1984, 1986). A distinct G_{1Q} population has thus far only been observed in mitogen-stimulated lymphocytes.

A.   Compartmentalization of G_1

Several cell lines grown *in vitro* have been stained for DNA/nuclear protein, and each has shown a relatively broad range of nuclear protein in G_1 (Fig. 1). The subdivision of the histograms into low-protein, G_{1A}, and high-protein, G_{1B} compartments is based on studies by Darzynkiewicz *et al.*, (1982), Roti Roti *et al.*, (1982), and stathmokinetic experiments using the technique described herein (Pollack *et al.*, 1984). G_{1A} cells are defined as cells having nuclear protein levels below that of S-phase nuclei. G_{1B} cells are defined as having nuclear protein levels equal to or above S-phase nuclei. The mitotic block from colcemid treatment causes a reduction in G_{1A} prior to G_{1B}, and following colcemid removal G_{1A} is replenished first.

As mentioned before, low-protein G_{1Q} nuclei are seen in DNA/nuclear protein histograms of mitogen-stimulated lymphocytes. G_{1Q} nuclei have lower nuclear protein than G_{1A} or G_{1B} nuclei (Pollack *et al.*, 1984, 1985, 1986). During the early stages of the initiation of lymphocyte proliferation by mitogen, there are significant increases in nuclear protein prior to the onset of DNA synthesis. In these histograms G_{1Q} is separated from G_{1A} and G_{1B}; however, population overlapping makes it difficult to discern simultaneously all three G_1 compartments (G_{1Q}, G_{1A}, and G_{1B}).

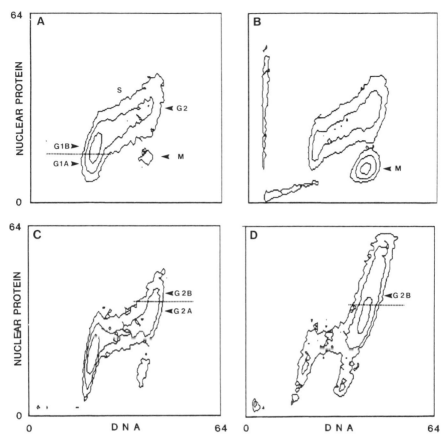

FIG. 1.   PI/FITC staining of (A) untreated log phase EL4 cell nuclei, (B) EL4 nuclei after a 6-hour exposure to colcemid (2 μg/ml), (C) untreated log phase HEP2 cell nuclei, and (D) HEP2 cell nuclei 12 hours after X irradiation of cells to 500 cGy.

B.   Compartmentalization of G_2

In most of the unperturbed cell lines tested in log phase growth, and in mitogen-stimulated lymphocytes, ~80% of the G_2 nuclei have nuclear protein levels equivalent to S-phase nuclei. The remaining G_2 nuclei have slightly higher nuclear protein levels (Fig. 1). Roti Roti *et al.* (1982) have shown through radiolabeling and sorting experiments that cells in early G_2 have low nuclear protein levels as compared to higher nuclear protein-containing late G_2 cells. Moreover, cells arrested in G_2 from ionizing radiation contain a subpopulation of G_2 cells with elevated nuclear protein levels. To characterize changes in G_2 nuclear protein levels, G_2 was empirically divided into low-nuclear protein G_{2A} (nuclear protein levels equal to S-phase nuclei) and high-nuclear protein G_{2B} compartments (Fig. 1). When proliferating lymphocytes are continuously labeled with 5 μCi/ml [^{3}H]thymidine (2 Ci/mmol), the cells accumulate first in G_{2A} and then G_{2B} (Pollack *et al.*, 1985). The number of cells in G_{2B} and the average nuclear protein content of G_{2B} cells increases with absorbed dose. These results, in combination with sorting experiments by Roti Roti *et al.* (1982), indicate that G_{2A} and G_{2B} represent early and late compartments of G_2, respectively.

Roti Roti *et al.* (1986) have shown that similar results are obtained when proliferating cells are exposed to external ionizing radiation. We subsequently observed (Pollack *et al.*, 1989) that X-irradiated log phase HEP2 cells exhibit transient G_2 arrest with concomitant transient G_2 nuclear protein elevation. The increase in nuclear protein was dose-dependent, transient, and specific for G_2 cell nuclei; G_1 cell nuclei following escape from G_2 arrest had normal protein levels.

C. Separation of G_2 from M

The bivariate DNA/nuclear protein histograms in Fig. 1 show that colcemid treatment of log phase EL4 cells results in a significant increase in the proportion of nuclei falling in the low-protein region below G_{2A} and late S phase. Manual microscopic counts of the number of mitotic cells (mitotic index, MI) demonstrate that both decreases and increases in MI correlate with the changes in the proportion of nuclei in this low-protein area of the histograms (M area). The proportion of M-area nuclei (referred to as nuclei despite the apparent lack of a nuclear membrane) averages 60–80% of the MI (Pollack *et al.*, 1985; Roti Roti *et al.*, 1986). It appears that some of the mitotic nuclei disaggregate. Sorting experiments indicate that there is less than a 10% overlap between G_2 and M (Pollack *et al.*, 1985). M-area nuclei have the cytologic appearance of cytoplasm-free chromosomes tightly packed together in a matrix. The M-area nuclei are distinctly different from sorted G_2 nuclei in terms of size, staining with Wright's Giemsa, and the shape of the nuclear border. Sorted M nuclei were smaller, stained darker, and had more irregularly shaped nuclear borders than G_2 nuclei.

D. Results in Solid Tumors

The DNA/nuclear protein histograms of solid tumors grown *in vitro* are similar to those of cell lines grown *in vivo* (Pollack *et al.* 1984, 1986).

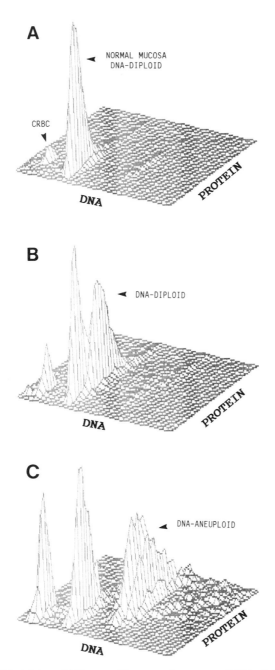

Fig. 2. PI/FITC staining of (A) normal human colon mucosa (DNA-diploid) with CRBC included as an internal standard, (B) a DNA-diploid moderately differentiated colon carcinoma, and (C) a DNA-aneuploid moderately differentiated colon carcinoma.

Multiple cell cycle compartments are identified, although G_{1Q} has not been observed thus far in solid tumors. The preparation of solid-tumor tissue may be complicated by substantial debris, depending on the cell type. Satisfactory histograms are obtained by staining the nuclei at low concentration ($<2.5 \times 10^5$ nuclei/ml).

Figure 2 shows some examples of human colon tissue prepared for DNA/nuclear protein analysis. In general, DNA-diploid colon carcinomas have higher nuclear protein than normal colon mucosa. By separating these overlapping cell populations, the number of contaminating normal diploid cells may be approximated.

IV.   Standards

Both internal and DNA-diploid standards are recommended, particularly if the technique is used for clinical samples. Internal standards such as CRBC are useful for standardization of DNA and are easily incorporated into the staining procedure. Nuclear protein content of CRBC is much lower than that of normal DNA-diploid cells and therefore is not an adequate standard for nuclear protein. The CRBC are added to the sample prior to staining.

Human spleen cells may be used as a DNA-diploid and protein standard. Small pieces of human spleen are stored frozen at $-70°C$ and are prepared in the same manner as solid tumor tissue. After thawing, the nuclei are isolated, counted, and added to one of the tumor samples prior to staining.

V.   Summary of Critical Aspects of the Procedure

The technique described is dependent on stripping nuclei free of cytoplasm. It is critical that this be monitored microscopically, since some cell types (e.g., epidermal cells) will not lyse in NIB. The concentration of NP-40 may be increased for some cell types that still contain cytoplasmic tags in 0.5% NP-40. In addition, similar results are obtained using other nuclear isolation protocols such as that described by Thornthwaite et al. (1980). The solution may be slightly hypotonic; the NIB used was 195 mOsm. Less hypotonic solutions will tend to promote M-nuclei disaggregation and nuclear protein loss that will minimize the differences observed in G_1 and G_2.

In the procedure described, nuclei are stained without washing to avoid clumping and selective loss of nuclei by separation techniques. As a consequence, there may be considerable low-level background fluorescence from the staining of the debris by FITC. This becomes a more significant problem in the preparation of solid tumors. The fluorescence signals from the debris are considerably less than the signals from intact stained nuclei and do not interfere directly with the acquisition of satisfactory histograms. It may be necessary to pregate signals, using a gated amplifier, prior to storing the digitized data in computer memory such that unnecessary low-level signals from the debris are not acquired.

Solid tumors were prepared by teasing the tissue in a teasing buffer and filtration through nylon mesh. This may result in some cell selection; however, I have observed that the use of wire screen mesh to disaggregate solid tumors results in significantly more debris and clumping. Since tumor tissue is in general more easily disaggregated than stroma, teasing may promote selection of the tumor cell component.

Under most conditions, FITC staining is proportional to nuclear protein content (Blair *et al.*, 1979; Roti Roti *et al.*, 1982, 1986; Dyson *et al.*, 1987; Pollack *et al.*, 1989). There is evidence, however, that the exposure of cells to certain treatments can cause an increase in the number of available FITC-binding sites, rather than nuclear protein levels. Dyson *et al.* (1987) showed that under certain conditions heat shock causes an increase in nuclear FITC staining without an increase in nuclear protein as determined by a biochemical assay.

REFERENCES

Blair, O. C., Winward, R. T., and Roti Roti, J. L. (1979). *Radiat. Res.* **78**, 474–484.

Darzynkiewicz, Z., Crissman, H., Traganos, F., and Steinkamp, J. (1982). *J. Cell. Physiol.* **113**, 465–474.

Darzynkiewicz, Z., Traganos, F., Kapuscinski, J., Staiano-Coico, L., and Melamed, M. R. (1984). *Cytometry* **5**, 355–363.

Dyson, J. E. D., McLaughlin, J. B., Surrey, C. R., Simmons, D. M., and Daniel, J. (1987). *Cytometry* **8**, 26–34.

Pollack, A., Moulis, H., Block, N. L., and Irvin, III, G. L. (1984). *Cytometry* **5**, 473–481.

Pollack, A., Moulis, H., Greenstein, D. B., Block, N. L., and Irvin, III, G. L. (1985). *Cytometry* **6**, 428–436.

Pollack, A., Moulis, H., Prudhomme, D. L., Block, N. L., and Irvin, III, G. L. (1986). *Ann. N. Y. Acad. Sci.* **468**, 55–66.

Pollack, A., and Ciancio, G. (1989). *In* "Flow Cytometry," II. (A. Yen, ed.), pp. 29–48. CRC Press Inc., Boca Raton, FL.

Roti Roti, J. L., Higashikubo, R., Blair, O. C., and Uygur, N. (1982). *Cytometry* **3**, 91–96.

Roti Roti, J. L., Kristy, M. S., and Higashikubo, R. (1986). *Radiat. Res.* **108**, 52–61.

Thornthwaite, J. T., Sugerbaker, E. V., and Temple, W. J. (1980). *Cytometry* **1**, 229–237.

Ward, H. A., and Fothergill, J. E. (1976). *In* "Fluorescent Protein Tracing" (R. C. Nairn, ed.), pp. 5–38. Churchill Livingston, London.

Chapter 31

Flow Cytometric Methods for Studying Isolated Nuclei: DNA Accessibility to DNase I and Protein–DNA Content

RYUJI HIGASHIKUBO, WILLIAM D. WRIGHT,
AND JOSEPH L. ROTI ROTI

Section of Cancer Biology
Radiation Oncology Center
Washington University School of Medicine
St. Louis, Missouri 63108

I. Introduction

Flow cytometric (FCM) studies of isolated nuclei can provide a wealth of information regarding the structure and function of this hereditary organelle. This information derives from two properties of isolated nuclei. First, large molecular probes that normally are unable to cross the cellular membrane and cytoplasmic barrier become readily accessible to nuclei. Second, cytoplasmic material does not obscure the macromolecular constitution of nuclei, so that its changes can be easily detected. With improvements in nuclear isolation techniques, it has become possible to use isolated nuclei in suspension for FCM analysis (Pollack *et al.*, 1984; Roti Roti and Winward, 1978; Roti Roti *et al.*, 1982). In fact, many techniques for the preparation of tissues for FCM analysis produce isolated nuclei rather than single cells. Therefore, it is important to consider the additional information that may be obtained by studying nuclei. Two experimental procedures involving isolated nuclei described here are (1) DNase I sensitivity of chromatin and (2) nuclear protein–DNA dual-parameter analysis.

325

The DNase I digestion kinetics reflect differences in certain types of chromatin structure based on changes in enzyme binding and digestion, and dye [ethidium bromide (EB) or propidium iodide (PI)] binding properties (Roti Roti *et al.*, 1985). These molecules cannot be transferred through the cell membrane and cytoplasm fast enough for this technique to be used in whole cells. Although fixation of cells with ethanol or treatment with detergents such as Triton X-100 makes the membrane permeable, the transportation of macromolecules through the cytoplasm under these conditions is still too slow. In addition, both ethanol fixation and detergent permeabilization cause changes in nuclear structure (Roti Roti and Wilson, 1984), which may obliterate the structural differences one might observe by this method. Thus, use of isolated nuclei is probably the "method of choice" to study the accessibility of DNA in chromatin to nuclease attack.

Bivariate nuclear protein–DNA analysis gives an added advantage over univariate DNA analysis in that it can provide a finer subdivision of cell cycle stages. In univariate DNA histogram analysis three phases (G_1, S, and G_2M) can be distinguished, whereas the bivariate method makes it possible to distinguish seven compartments by subdividing G_1, S, and G_2 into early and late compartments, and a subpopulation that correlates with M. In addition, nuclear protein content is a very sensitive indicator of cell growth and its perturbation. Not only does the nuclear protein content change according to cell cycle position (Roti Roti *et al.*, 1982), but the nuclei are also found to accumulate an excess amount of proteins when subjected to certain types of stress such as heat shock or osmotic shock (Roti Roti and Winward, 1978). The cells accumulated in G_2 following X irradiation have been shown to exhibit unbalanced growth manifested as an extraordinary accumulation of nuclear protein (Roti Roti *et al.*, 1986b). Thus this technique is useful in observing nuclear protein content changes, redistribution of a cell population through the cell cycle, and unbalanced growth.

II. Application

Changes in DNase I digestion kinetics have been used to observe changes in chromatin structure under various conditions. Exposing cells to hyperthermia or hypertonic shock causes an accumulation of excess nuclear protein, which results in an inhibition of digestion kinetics. When cells were treated with *n*-butyrate (Darzynkiewicz *et al.*, 1981; Roti Roti *et al.*, 1985) or a high dose of X irradiation, different phases of digestion kinetics were facilitated. Differences in the digestion kinetics were also observed between proliferating and quiescent populations of mouse mammary tumor cells.

Nuclear protein–DNA analysis has many applications. We have used the technique in two areas: (1) the redistribution of a population following cell cycle perturbation with drugs or ionizing radiation, and (2) changes in nuclear protein content as a result of hyperthermia. This latter feature appears to be useful in predicting the response of cell populations to hyperthermia under certain conditions.

III.   Materials and Methods

A.   Nuclear Isolation

Two nuclear isolation procedures have been used in our laboratory. One involves washing off the cytoplasm with detergent. The other involves swelling of the cytoplasm with a hypotonic solution followed by homogenization of the cytoplasm in a glass homogenizer and a subsequent detergent wash. The first method has been used successfully in isolating nuclei from HeLa cells grown in suspension, whereas the second method has been used extensively for the cells grown in monolayer (e.g., CHO, RIF-1, mouse mammary tumor 66,67) or cells grown *in vivo*. Some modifications in the procedures may be required to obtain optimal results for other cell types.

1.   Isolation of Nuclei from Suspension Cultures

Reagents

1. Spinner salt solution, pH 7.0; 270–300 mOs/kg*; contains per 1 liter:
   KCl, 0.4 g
   $NaHCO_3$, 2.2 g
   $MgSO_4$, 0.1 g
   NaH_2PO_4, 1.4 g
   NaCl, 6.8 g
   D-Glucose, 1.0 g

2. Triton X-100 solution (TX-100), pH 7.6*; contains, per 1 liter:
   NaCl, 4.68 g
   EDTA, 7.45 g
   TX-100, 10 ml
3. TMNP (Tris-buffered saline), pH 7.4**; contains, per 1 liter:
   Tris base, 1.58 g
   $MgCl_2 \cdot 6 H_2O$, 2.02 g
   NaCl, 0.58 g
   PMSF (phenylmethylsulfonyl fluoride), 1.0 ml of 0.1 M stock in
   95% ETOH

Items with one asterisk are not normally adjusted—just checked; those
with two asterisks must be adjusted.

Isolation Procedure

Note that all procedures are to be carried out at 4°C. Centrifugation
speeds are quoted for a Beckman J-6B refrigerated centrifuge with TY-JS-
4.0 swinging bucket rotor or equivalent.

The nuclear yield of this procedure ranges between 65 and 95% but
never below 50%. If it falls below 50%, some modification in the procedure should be in order. To arrive at a sufficient number of isolated nuclei,
it is advisable to start the procedure with twice the number of cells.

1. Place cells in medium in 15- or 50-ml conical screw-cap tubes depending on volume. The tubes should be full.
2. Centrifuge at 800 rpm for 5 minutes; decant growth medium.
3. Resuspend the pellet in one-fifth tube volume of spinner salt solution by repeated pipeting.
4. When a uniform suspension is obtained, refill the tubes to full
   capacity and centrifuge at 1500 rpm for 5 minutes.
5. Repeat steps 3 and 4 two to three times to remove growth medium
   and associated serum proteins completely.
6. Resuspend the pellets from the last spinner salt wash in TX-100
   solution (one-fifth tube volume) by vigorous pipeting.
7. At this stage, if you started with a 50-ml tube, transfer the suspension to a 15-ml tube.
8. Fill up the tube and centrifuge at 2000 rpm for 5 minutes; decant.
9. Repeat steps 6 and 8 once (the number of TX-100 washes depends
   on cell lines and their condition).
10. Resuspend in 10 ml TMNP by vigorous pipeting.
11. Centrifuge at 2000 rpm for 5 minutes. Decant.
12. Resuspend to 2×10^6 cells/ml in TMNP.

2. HOMOGENIZER METHOD

Reagents

1. Swelling buffer (TEM), pH 7.8; contains, per 1 liter:
   Tris base, 1.58 g
   EDTA, 0.38 g
   $MgCl_2 \cdot 6 H_2O$, 0.813 g
2. Lysis buffer, pH 7.8; contains, per 1 liter:
   Tris base, 1.58 g
   EDTA, 0.38 g
   $MgCl_2 \cdot 6 H_2O$, 0.813 g
   Sucrose, 85.58 g
   TX-100, 0.5 ml
3. TMNP

See the Suspension cultures method (Section III,A,1).

Isolation Procedure

Note that all procedures are to be carried out at 4°C. Centrifugation speeds are quoted for a Beckman J-6B refrigerated centrifuge with TY-JS-4.0 swing bucket rotor or equivalent.

The nuclear yield of this method usually exceeds that of the previous method. The total starting number of cells should be adjusted accordingly.

1. Wash cells three times as in the previous method to remove the growth medium completely.
2. Suspend in swelling buffer at 1×10^7 cells/ml and keep on ice for 5 minutes.
3. Transfer to a 0.15-mm-clearance tissue grinder (Corning 7725, Corning, NY), and disrupt with 15 strokes.
4. Add an equal volume of lysis buffer to the homogenate.
5. Let it stand for 15 minutes on ice.
6. Wash the nuclei twice in TMNP and resuspend to 2×10 nuclei/ml.

B. RNase A Digestion

Since both the DNase digestion and the nuclear protein–DNA assays use intercalating fluorochromes as DNA-specific dyes, digestion of nuclear RNA (especially double-stranded) is necessary to obtain suitable DNA histograms.

Reagent

RNase A stock solution: 1 mg/ml RNase A (Sigma, St. Louis, MO) dissolved in double-distilled water. Heat in boiling water for 5 minutes before each use in order to eliminate proteases and other undesirable enzymatic activities.

Procedure

1. Add RNase A stock solution to the nuclear suspension to a final concentration of RNase A, 25 μg/ml.
2. Digest for 30 minutes at 37°C.

C.   DNase I Digestion

Reagents

1. Enzyme, DNase I (Sigma, St. Louis, MO): 5 mg/ml in TMNP, prepared fresh before use. Keep on ice.
2. Staining solution: EB (Sigma, St. Louis, MO); 600 μg/ml TMNP with 0.2 M EDTA

Procedure

The overall digestion kinetics are as follows. Digestion will be carried out at 20°C. The nuclear suspension should be 2×10^6 nuclei/ml. Allow 1 ml of the suspension per time point.

1. Before adding the enzyme to the nuclear suspension, an undigested control sample should be analyzed. Take 0.9 ml nuclear suspension, add 0.1 ml of the staining solution, filter through a 20-μm monofilament filter (Small Parts, Inc., Miami, FL), and subject to FCM analysis. Depending on the instrument, it may take a few minutes for fluorescence intensity to reach equilibrium. Once the control data have been acquired, do not change the instrument conditions (e.g., photomultiplier tube voltage, amplifier gain).
2. Add the enzyme solution to the nuclear suspension to a final concentration of 90 μg/ml. Mix quickly. Take 0.9 ml of the sample and add to 0.1 ml staining solution. Analyze immediately.
3. At each time point thereafter, take 0.9 ml of sample and add to 0.1 ml staining solution.

The procedure for initial nicking kinetics is basically the same as the overall digestion procedure just described, with the following exceptions.

1. Enzyme concentration is one-tenth that of the overall digestion.
2. Digestion is carried out on ice.

Special Considerations

1. The undigested control histogram that will be used as a reference point should be placed such that a 20% increase in fluorescence intensity at the 0 minute point will not result in the DNA histogram going over the channel range. Of course, if the amplifier has reliable linearity between different gain settings there would be little error in making a correction for the change, and this consideration should not matter.

2. The 0 minute point, which will be taken immediately after the addition of the enzyme, may have ±10 seconds variation due to sample handling. This variation may result in the 0 minute point having a larger error than other points. On-line FCM, however (see Nooter *et al.*, Chapter 52, this volume), can diminish this error.

3. Although the staining solution contains a high concentration of EDTA, which should result in rapid cessation of the DNase I enzymatic activity, one should analyze the sample as soon as possible in order to eliminate further digestion by residual enzymatic activity.

Instruments

Ethidium bromide can be excited by either 488- or 514-nm lines of an argon laser, and fluorescence is detected through a 620- or 640-nm long-pass filter along with forward light scatter. The sample delivery rate should be adjusted to an optimum for the flow system in order to minimize data acquisition time.

D. Nuclear Protein–DNA Staining

Staining Solution

1. Fluorescein isothiocyanate (FITC) (Calbiochem-Behring, Los Angeles, CA), 30 μg/ml in TMNP. The FITC dissolves very slowly, requiring several hours at room temperature (RT) with constant mixing. Also, the solution is stable for only a week or two at 4°C in the dark.

2. Propidium Iodide (PI) (Sigma, St. Louis, MO), 70 μg/ml in TMNP. This solution should last up to 2 months when kept covered at 4°C.

Staining Procedure

1. Resuspend 2×10^6 RNase-treated nuclei in 0.8 ml TMNP.
2. Add 0.2 ml FITC staining solution. Stain for 30 minutes at 4°C.
3. Add 1.0 ml PI solution and stain for at least 30 minutes (preferably ≥ 1 hour). With a clean nuclear preparation the stained samples can be kept for several days before FCM analysis.

Special Considerations

1. Stained nuclei seem more stable than unstained ones. If FCM analysis cannot be done right away, stain the nuclei and keep them at 4°C.

2. In this technique it is vital that clean nuclei are obtained, especially when quantitative analysis of nuclear protein content is to be done. Thus, it is advisable to observe these stained nuclei under a fluorescence microscope. If numerous green cytoplasmic tabs or more than the usual debris is seen, it is considered an indication of inadequate sample preparation.

Instruments

1. The fluorescent dyes are excited with the 488-nm line of an argon laser. FITC fluorescence is detected through an 535-nm bandpass filter and PI is detected through a 640-nm long-pass filter. Data are collected as PI versus FITC dual parameters or, if possible, as list mode of LS, PI, and FITC.

2. There may be some FITC fluorescence getting into PI channels, resulting in a slight tilt in histograms. This may be corrected by use of a dual fluorescence compensation network.

IV.   Results

A.   DNase I Digestion

Typical overall and initial DNase I digestion kinetics of HeLa nuclei are shown in Fig. 1. The digestion kinetics are characterized by three phases. Immediately following the addition of the enzyme, an instantaneous increase in fluorescence of $\sim 20\%$ is observed, which is exemplified in detail in the initial kinetics. The fluorescence then decreases rapidly at first, reaching final phase where a very little or no decrease in fluorescence is observed. Differences in DNase I sensitivity of chromatin in this assay can be manifested by changes in (1) the level of the initial fluorescence increase, (2) the rate of flourescence decrease, (3) the final fluorescence level—or any combination of these. See Fig. 2 for some examples of these differences.

B.   Nuclear Protein–DNA Staining

A dot plot of a typical dual-parameter nuclear protein–DNA histogram and its subdivision into seven compartments is shown in Fig. 3. The subdivision is done by a simple mathematical computation described previously

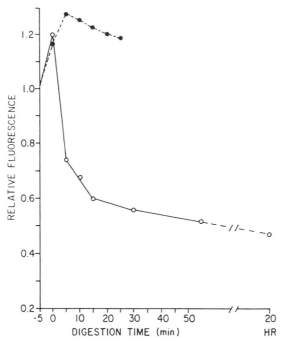

FIG. 1. Overall (○—○) and initial kinetics (●—●) of DNase I digestion of isolated HeLa nuclei. The overall kinetics were carried out at 20°C with an enzyme concentration of 7.5 μg/μg DNA, whereas the initial kinetics used one-tenth concentration of the enzyme at 0°C. Mean PI fluorescence from at least 5×10^4 nuclei was calculated and normalized to that of undigested nuclei. Reprinted from Roti Roti *et al.* (1985).

(Roti Roti *et al.*, 1982). Briefly, the SP_1 (see Fig. 3) was excluded from the analysis population by blocking it out. From three to five channels of mid-S (along the PI axis), mean FITC channel, and standard deviations were computed. The mean ± standard deviation gave the lines separating EG_1 and LG_1, and EG_2 and LG_2, respectively. Propidium iodide distributions of EG_1 and LG_2 were used to calculate G_1/S and S/G_2 boundaries at $EG_1 + 1$ SD and $LG_2 - 1$ SD, respectively. The halfway point between G_1/S and S/G_2 is used to separate S phase into two fractions. Figure 4 shows an example of this technique applied to cellular recovery following heat shock (45°C for 30 minutes) in HeLa cells. Immediately after heat exposure (0 hour), protein content of nuclei, seen as FITC fluorescence intensity, was 2.2 times that of control (Fig. 4A). The excess proteins were removed rapidly during the first 8 hours. Between 8 and 30 hours, the protein content remained almost constant, followed by a rapid accumulation of proteins, especially in G_2, indicating unbalanced growth. Figure 4B

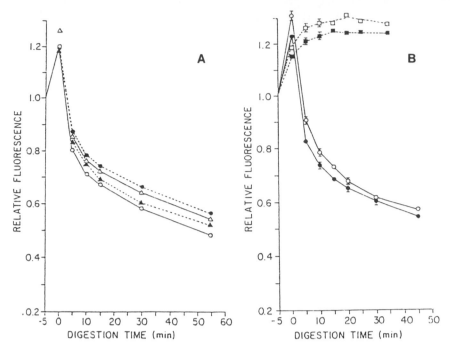

FIG. 2. Examples of DNase I digestion kinetics. (A) Changes observed in HeLa cells treated for 30 minutes at various hyperthermic temperatures 37°C, ○; 45°C, ▲; 46°C, △; 47°C, ●. (B) Changes observed in proliferating (P) and quiescent (Q) populations of mouse mammary adenocarcinoma 67 cells: Q cells at 0°C (□) and 20°C (○); P cells at 0°C (■) and 20°C (●). Higher heat dose makes chromatin less accessible to the enzyme and the dye, inhibiting the overall kinetics. Chromatin in the quiescent cells is more resistant to the enzyme digestion and there is more change in the initial kinetics. Reprinted from Roti Roti *et al.* (1985).

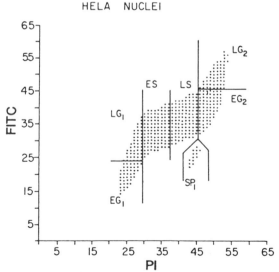

FIG. 3. A scattergram of dual-parameter nuclear protein–DNA distribution with lines separating them into seven compartments. Data from 100,000 nuclei were collected and the channels containing > 35 nuclei were shown. Reprinted from Roti Roti *et al.* (1982).

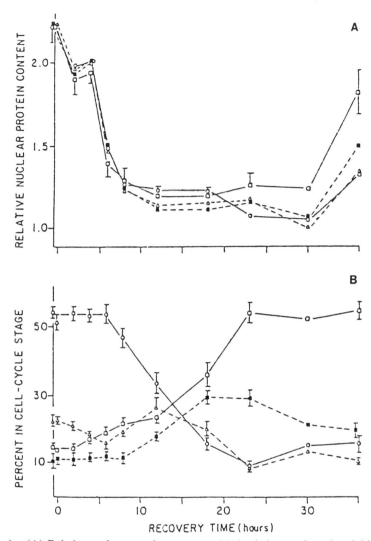

FIG. 4.   (A) Relative nuclear protein content and (B) relative number of nuclei in com partments G_1(○), ES (△), LS (■), and G_2(□) following exposure to 45°C for 30 minutes. HeLa cells were given the heat dose and returned to 37°C. At the designated time in 37°C the nuclear protein–DNA assay was carried out and data analyzed accordingly. A mean of eight experiments ± one standard error is plotted. (Figure from Roti Roti *et al.*, 1986a).

shows the distribution of cells in the cell cycle. The first 6 hours showed no significant changes in population distribution. Between 6 and 24 hours, a significant decrease in G_1 and a corresponding increase in G_2 were seen, clearly indicating redistribution of the population. Thus this technique is useful to observe cellular perturbations brought about by physical and chemical agents.

V. Summary

Two FCM methods utilizing isolated nuclei were described. A DNase I sensitivity assay, employing changes in binding and digestion kinetics of the enzyme as well as the binding of intercalating fluorochrome was used to observe structural changes of chromatin rendered by physical and chemical agents. A nuclear DNA–protein staining method was used to study changes in nuclear protein content, redistribution of populations in the cell cycle, and unbalanced growth, manifested as an extraordinary accumulation of nuclear protein brought about by physical and chemical perturbation.

ACKNOWLEDGMENTS

The work reported here was supported by NCI grants CA41102 and CA43198.

REFERENCES

Darzynkiewicz, Z., Traganos, F., Xue, S-B., and Melamed, M. (1981). *Exp. Cell Res.* **136,** 279–293.
Pollack, A., Moulis, H., Block, N.L., and Irvin, G. L. (1984). *Cytometry* **5,** 473–481.
Roti Roti, J. L., and Wilson, C. F. (1984). *Int. J. Radiat. Biol.* **46,** 25–33.
Roti Roti, J. L., and Winward, R. T. (1978). Radiat. Res. 74:159–169.
Roti Roti, J. L., Higashikubo, R., Blair, O. C., and Uygur, N. (1982). *Cytometry* **3,** 91–96.
Roti Roti, J. L., Wright, W. D., Higashikubo, R., and Dethlefsen, L. A. (1985). *Cytometry* **6,** 101–108.
Roti Roti, J. L., Uygur, N., and Higashikubo, R. (1986a). *Radiat. Res.* **107,** 250–261.
Roti Roti, J. L., Kristy, M. S., and Higashikubo, R. (1986b). *Radiat. Res.* **108,** 52–61.

Chapter 32

Acid-Induced Denaturation of DNA in Situ as a Probe of Chromatin Structure

ZBIGNIEW DARZYNKIEWICZ

*Cancer Research Institute of the
New York Medical College at Valhalla
Valhalla, New York 10595*

I. Introduction

Free DNA in aqueous solution at physiological pH and ionic strength has double-stranded conformation. Its treatment with heat, acid, or alkali causes the two strands to separate. This is known as denaturation, melting, or helix-coil transition, and is the result of destruction of hydrogen bonding between the paired bases of the opposite strands. Sensitivity of free DNA to denaturation depends exclusively on its guanine–cytosine to adenine–thymine (GC/AT) ratio, because the GC pair confers higher stability due to an additional hydrogen bond, in comparison with AT.

DNA in chromatin is stabilized by interactions with histones and other nuclear proteins (reviews, Darzynkiewicz, 1986, 1990). Studies on the stability of DNA *in situ* therefore provide an insight into chromatin structure, making it possible to discern the double helix-stabilizing interactions. The classical biochemical methods to study DNA denaturation in chromatin are based on measurements of ultraviolet (UV) light absorption changes (hypochromicity) during heating of the sample (Subirana, 1973). These optical methods require chromatin isolation, shearing, and solubilization. This destroys higher orders of chromatin structure and limits application of such methods in investigations of DNA *in situ*. The newer calorimetric approach (Touchette *et al.*, 1986) yields data that are difficult to interpret, because the method does not allow discrimination between heat-induced destruction of nucleic acid–protein bonding (e.g., dissociation

METHODS IN CELL BIOLOGY, VOL. 33

of the nucleosomal core histones from DNA, "melting" of nucleosomes; Darzynkiewicz and Carter, 1989) from DNA interstrand bond breakage (DNA melting). Furthermore, the calorimetric method, which can only be applied to whole-cell populations in bulk, cannot be used to assess intercellular heterogeneity.

A flow cytometric (FCM) method for measurement of DNA *in situ* sensitivity to denaturation, developed in our laboratory (Darzynkiewicz *et al.*, 1975, 1977, 1979), is based on the metachromatic property of the dye acridine orange (AO). This dye, under certain conditions, can differentially stain double-stranded (ds) versus single-stranded (ss) nucleic acids (review, Darzynkiewicz and Kapuscinski, 1990). Namely, AO intercalates into dsDNA and in this mode of binding, upon excitation in blue light, emits green fluorescence. In contrast, AO complexes with ss nucleic acids are characterized by red luminescence (phosphorescence).

In the original method, stability of DNA *in situ* was assayed by subjecting permeabilized (ethanol-fixed) and subsequently RNase-treated cells to heat or acid, followed by staining with AO (Darzynkiewicz *et al.*, 1975, 1977). After partial denaturation of DNA by heat or acid, AO, as mentioned, can stain nondenatured DNA sections in green (maximum fluorescence at 530 nm), whereas dye interactions with denatured sections (ssDNA) result in red luminescence (maximum emission at 640 nm). Thus, the relative proportions of red and green luminescence of cells stained this way represent portions of the denatured and native DNA, respectively.

This method has been adapted to unfixed cell nuclei freshly prepared from the cell nuclei of solid tumors (Kunicka *et al.*, 1987, 1989). Because isolation of nuclei from solid tumors is more convenient than isolation of whole intact cells, the adaptation extends the applicability of the method to the clinic.

Extensive literature exists describing studies in which AO was used to measure DNA denaturation in solution (e.g., Ichimura *et al.*, 1971) and *in situ*, in metaphase chromosomes (e.g., Bobrow and Madan, 1973) or nuclei of the permeabilized cells (review, Darzynkiewicz, 1990). The *in situ* studies show great differences in DNA stability between various cell types, cells in different phases of the cell cycle, differentiated versus nondifferentiated cells, or even within individual chromosomes—in the latter case, reflected as chromosome banding. In general, DNA sensitivity to denaturation in cells is closely correlated with the degree of chromatin condensation: the more condensed the chromatin, the more unstable the DNA. So far, the sole exception to this rule was observed for DNA in cells undergoing spermatogenesis, where the late stages of chromatin condensation in spermatozoa were seen to be paralleled by an increase rather than a decrease in DNA stability (Evenson *et al.*, 1980).

II. Application

This method can be applied in diverse studies designed to provide information about changes in nuclear chromatin. Altered sensitivity of DNA to denaturation, as mentioned, accompanies cell differentiation (e.g., Traganos *et al.*, 1979; Evenson *et al.*, 1980), parallels cytotoxic effects of chemotherapeutic drugs (intercalators) on target tumor cells (e.g., Darzynkiewicz *et al.*, 1981), or precedes cell death induced by such different agents as tumor necrosis factor (TNF; Darzynkiewicz *et al.*, 1984) or caffeine (Kunicka *et al.*, 1990). In the latter cases, the mechanism of cell death is most likely by apoptosis, and the nuclear changes typical of apoptosis (chromatin condensation) are manifested by dramatically reduced stability of the DNA helix *in situ*. The method therefore can be used to make an early estimate of the proportion of cells undergoing programmed cell death or those killed by other means, when the mechanism of cell death is apoptosis.

The variability in DNA stability to denaturation is also apparent when tumor cells are compared with normal stromal, or infiltrating, host cells, as well as between tumors of the same type but of different stage (Kunicka *et al.*, 1987, 1989). Further studies, however, are needed to evaluate the prognostic potential of this marker in the clinic.

The most common application of the DNA denaturability assay, however, stems from the differences in DNA stability related to cell position in the cell cycle. Very sensitive to denaturation is DNA in mitotic cells, as well as in quiescent cells characterized by condensed chromatin (Darzynkiewicz *et al.*, 1977, 1979). Conversely, the most resistant is DNA in late-G_1 (G_{1B}) and early-S- phase cells. The method thus can discriminate cells in traditional phases of the cell cycle, including distinction of mitotic cells, and be able to identify quiescent cells arrested in G_1 (G_{1Q}), S (S_Q), or G_2 (G_{2Q}) (Darzynkiewicz *et al.*, 1980). The possibility for rapid quantification of mitotic cells contributed to wide application of this technique to score mitotic indices, especially in stathmokinetic experiments in which cells are arrested in mitosis (Darzynkiewicz *et al.*, 1981). This was the base for development of a new method for analysis of cell cycle perturbations, such as those caused by various physical or chemical agents (Darzynkiewicz *et al.*, 1986; see also Traganos and Kimmel, Chapter 25, this volume).

It should be emphasized that discrimination of kinetically different compartments of the cell cycle offered by this method is based on a rather nonspecific phenomenon, namely DNA *in situ* stability to denaturation, reflecting chromatin condensation. Therefore, if in certain cell types or situations chromatin changes do not accompany cell transitions between these compartments, the technique may fail to recognize certain cell groups (e.g., noncycling or differentiated).

As in the case of differential staining of RNA versus DNA with AO (see Darzynkiewicz, Chapter 27, this volume), restrictions related to specificity of DNA staining with this dye also pervade the present method. Namely, the specificity of AO to stain nucleic acids is not absolute, and the dye stains other polyanions as well. Therefore, following staining with AO, cells that contain large amounts of glycosaminoglycans or proteoglycans, such as normal fibroblasts, mast cells, chondrocytes, and differentiated keratinocytes, all have unacceptably high fluorescence unrelated to DNA. On the other hand, most cell lines proliferating in cultures, especially of tumor origin, as well as lymphocytes, monocytes, leukemias, and lymphoma, or cells isolated from most solid tumors, exhibit good or at least adequate specificity of DNA stainability with AO. The degree of non-specific luminescence can be estimated by control incubations of the cells with DNase, prior to staining with AO. The assay based on isolated nuclei rather than whole cells can circumvent the problems of nonspecific stainability resulting from the presence of cytoplasmic constituents.

III. Materials

The following materials are needed for the assay of DNA denaturability in whole, fixed cells.

1. Hanks' balanced salt solution (HBSS): The solution should contain Mg^{2+} but no phenol red. Phosphate-buffered saline (PBS) containing 1 mM $MgCl_2$ may be used instead of HBSS.
2. Fixative: The cells may be fixed either in 70–80% ethanol or in a mixture of 80% ethanol and absolute acetone (1 : 1, v/v). The latter fixative causes a somewhat higher degree of cell aggregation.
3. Stock solution of AO: Dissolve AO in distilled water (dH_2O) to a final dye concentration of 1 mg/ml. Use only high-purity, chromatographically tested AO. Several batches of AO obtained from Polysciences, Inc. (Warrington, PA) have been found to be satisfactory for FCM application. The stock solution may be kept in the dark at 4°C for several months without deterioration.
4. 0.1 N solution of HCl (solution A).
5. The AO staining solution (solution B): To prepare the buffer, mix 90 ml of 0.1 M citric acid with 10 ml of 0.2 M Na_2HPO_4; the final pH of this buffer is 2.6. Add 0.6 ml of the AO stock solution (1 mg/ml) to

100 ml of this buffer; the final AO concentration is 6 μg/ml. Solution B is stable and can be stored for several months in the dark, at 4°C. Solutions A and B may be stored in automatic-dispensing pipetter bottles set at 0.5 ml for solution A and 2.0 ml for solution B—the latter in a dark bottle.

6. RNase A: The DNase-free RNase preparation should be used. Pure RNase can be purchased from Worthington Biochemical Corp. (Freehold, NJ, code name, RASE). Less pure preparations may be used following a short (1 minute) heating at 100°C to inactivate DNase.

For DNA denaturability assay in isolated, unfixed cell nuclei, in addition to the materials described already, the following solution is needed.

7. Nuclear isolation solution: Prepare a solution containing 10 mM Tris buffer (pH 7.6), 1 mM sodium citrate, 2 mM MgCl$_2$, and 0.1% (v/v) nonionic detergent Nonidet P-40 (NP-40).

IV.   Cell Preparation and Staining

A.   Cell Fixation

1. Cells growing in suspension, hematologic samples: Rinse once with HBSS, suspend in HBSS (10^6–10^7 cells per 1 ml).
2. Cells growing attached to tissue culture dishes: Collect cells by trypsinization and pool the trypsinized cells with cells floating in the medium (the latter consist mostly of detached mitotic and dead cells). Rinse once with medium containing serum (serum is present to inactivate the trypsin; other means of trypsin inactivation such as addition of trypsin inhibitors may also be used), suspend cells (10^6–10^7 cells per 1 ml) in HBSS.
3. Cells dissociated from solid tumors: Rinse free of any enzyme used for cell dissociation; suspend in HBSS.

For the final suspension in HBSS the cells should be well dispersed and not exceed a density of 10^7 cells per 1 ml. The cells are then rapidly fixed by admixture of 1 ml of this suspension into 15-ml glass tubes containing 10 ml of fixative [80% ethanol or 1 : 1 (v/v) mixture of 80% ethanol and absolute acetone], at 0°–4°C. Cell clumping is minimized if the cell suspension is rapidly injected (e.g., with Pasteur pipet or syringe) into the fixative rather than layered onto the surface and then mixed. The reverse order (i.e., addition of fixation into tubes containing cell suspensions in HBSS) results in extensive cell loss due to their adherence to the glass surface and aggregation.

The fixation time may vary from 12 hours to several months, at 4°C.

B.   Cell Staining

1. Centrifuge the fixed cells. Suspend the cell pellet (10^6–10^7 cells) in 1 ml of HBSS. Add 10^3 units of RNase A. Incubate at 37°C for 1 hour. Centrifuge and resuspend cells in 1 ml of HBSS.

2. Withdraw 0.2-ml aliquot of cell suspension in HBSS and transfer it to a small (e.g., 5 ml volume) tube. Add 0.5 ml of 0.1 M HCl (solution A). After 30 seconds add 2.0 ml of AO solution (B). Transfer this suspension to the flow cytometer and measure cell fluorescence. The fluorescence pattern remains stable for several hours.

   Both treatment with HCl and staining with AO should be done at room temperature (RT, 20–24°C). Therefore, prior to use, solutions A and B should be adjusted to RT. DNA denaturation is incomplete when the solutions are cold.

C.   Isolation and Staining of Unfixed Nuclei from Solid Tumors

This procedure has been modified after Kunicka *et al.* (1989).

1. Place the tissue in the nuclear isolation solution (Section III, step 7), and trim to remove the necrotic, fatty, and other undesirable areas. The tissues may be kept frozen (at −40° to 60°C) prior to nuclear isolation.

2. Transfer the trimmed tissue to a small volume of the fresh isolation solution and mince finely with scalpel or scissors. Mix small tissue fragments by vortexing and/or pipeting using a Pasteur pipet or syringe with a large-gauge needle. Observe the release of nuclei by sampling the suspension and viewing it under the phase-contrast microscope. If release of the nuclei is inadequate, transfer the minced tissue, suspended in the isolation solution, into a homogenizer with a glass or Teflon pestle. Homogenize by pressing the pestle several times; check the efficiency of nuclear isolation under a phase-contrast microscope. The nuclei should be clean, lacking cytoplasmic tags, yet unbroken.

Collect the nuclear suspension from above the remaining tissue fragments (allowed to sediment for ~1 minute) with a Pasteur pipet, and filter through 40–60 μm pore nylon mesh. There should be no more than 10^7 nuclei per 1 ml of the isolation solution. All isolation steps should be done on ice.

3. Add 10^3 units of RNase A per up to 10^7 nuclei in 1 ml of the solution. Incubate for 30 minutes at 24°C.

4. Withdraw a 0.2 ml aliquot of nuclei suspension from the RNase A incubation medium and treat with 0.5 ml of 0.1 M HCl for 30 seconds at 24°C. Add 2.0 ml of AO solution (B) at 24°C. Transfer this suspension to the flow cytometer and measure the fluorescence. Since the material is unfixed, the fluorescence pattern is unstable and the nuclei should be measured shortly (up to 10 minutes) after addition of solution B.

V. Critical Aspects of the Procedure

Differential staining of dsDNA versus ssDNA, at equilibrium with AO requires proper concentration of free (unbound) dye in the final staining solution and during the actual measurement, that is, at the moment of cell intersection, in flow, with the laser beam. The problems of variability of AO concentration due to excess of cell number or different geometry of flow channels ("dye diffusion problems") pertinent to the present method are common to other methods utilizing AO, and are discussed in detail in the description of the DNA and RNA staining method, in this volume (Darzynkiewicz, Chapter 27).

To initiate this methodology, one is advised to start with a cell population enriched in mitotic cells, for example, by staining exponentially growing cells treated for 2–3 hours with a mitotic inhibitor such as colcemid or vinblastine. It is easier then to identify the mitotic cell population, which should differ from interphase cells by increased sensitivity of DNA to denaturation. Under proper conditions of cell staining, luminescence excitation, and separation of the emission spectra, the results should be similar to those shown in Fig. 1. It is important, in particular, to obtain the mitotic cell cluster positioned on a diagonal axis with G_2 cells (on green vs. red luminescence display, Fig. 1A) and thus having the same total luminescence (but different α_t; see legend to Fig. 1) compared to G_2 cells, as evident on the total luminescence versus α_t displays (Fig. 1B). If DNA denaturation is inadequate (e.g., in situations where AO concentration is too low, as may be the case when the dye diffuses extensively from the cells to sheath flow in some sorting channels), cells in mitosis are less separated from G_2 cells than is the case shown in Fig. 1. Increased AO concentration in the staining solution can often compensate for the diffusion effects. On the other hand, if DNA denaturation is excessive, mitotic cells have a disproportionally high red, and an extremely low green, luminescence.

In essence thus, the most critical points of the procedure relate to a proper choice of AO concentration and adequate separation of the green and red emission components.

VI. Standards

Standardization of the staining and measurement procedure requires that as a result of DNA denaturation, the increase in red luminescence should be proportional to the decrease in green fluorescence. Thus, for the cells with the same DNA content but differing in DNA sensitivity to denaturation (e.g., G_2 vs M), the sum of red and green luminescence intensities (total luminescence; see Fig. 1) should be the same. Obtaining this optimal stainability pattern requires proper selection of (i) emission filters, (ii) AO concentration, and (iii) respective red- and green-photo-multiplier sensitivity (voltage) settings.

Nonstimulated human peripheral blood lymphocytes can be used as a convenient standard that allows one to compare the measured cells, for instance in tumor cell analysis. A batch of fixed lymphocytes can be stored at 4°C for several weeks without change in their DNA sensitivity to denaturation. These cells may then be used as an external and/or internal reference standard to be compared with the α_t value of the measured cell population. Under proper conditions of cell staining and filter selection, the photomultiplier sensitivity settings of the red- and green-channel photomultipliers should then be routinely set to obtain mean numerical values of red and green luminescence of lymphocyte population to be equal to each other; hence the mean α_t value of lymphocytes is designated to be 0.5.

VII. Instruments

Specifics of the instrumentation that may be of relevance in measurements of AO luminescence are provided in Chapter 27 of this volume, describing the method for differential staining of DNA and RNA with this dye. Briefly, the optimal excitation of AO is in blue light, at 470–500 nm.

Because following DNA denaturation the red luminescence of the cells is generally rather strong, there is no need to try using shorter wavelengths for excitation, as was stressed in the case of weak red-luminescence measurements, in cells or cell nuclei having low RNA content. Thus, the 488-nm line of the argon ion laser can be used, and blue excitation filters such as BG 12 can be applied in instruments illuminated by a mercury lamp, for all samples. The combination of emission filters and dichroic mirrors should separate green fluorescence at 530 + 15 nm and red luminescence at >630 or 640 nm.

The geometry of the flow channel that affects the staining pattern due to diffusion of AO from cells to the sheath flow, and methods that can compensate for the diffusion are described by Darzynkiewicz (Chapter 27, this volume).

The data can be recorded in a list-mode fashion either as red- and green-luminescence intensities of individual cells, or as total cell luminescence (red plus green) and ratio of red to total luminescence, so called α_t (Darzynkiewicz *et al.*, 1975). In the latter case the analog signals from green and red photomultipliers can be added and then respectively divided by a specially designed electronic circuit board (e.g., such as used in some flow cytometers for compensation of the signals when the measured emission spectra overlap), so that the total luminescence and the α_t ratio are then digitized and recorded in the list mode. It is also possible to develop software that transforms the originally recorded data expressed as red- and green- luminescence intensities of individual cells to the total luminescence and α_t values of the same cells; so the transformed file is subsequently analyzed.

VIII. Results

A. Exponentially Growing Cells.

Figure 1 illustrates the typical distribution of exponentially growing cells with respect to their green and red luminescence after partial denaturation of DNA and staining with AO. The data are recorded either as bivariate distribution of green and red luminescence values (Fig. 1A) or total cell luminescence (green plus red) and α_t value (red luminescence/total luminescence) (Fig. 1B and C). Total cell luminescence represents total cellular DNA while α_t, which can vary from 0 to 1.0, is a measure of a portion of denatured DNA (Darzynkiewicz *et al.*, 1975). The relative intensities of red and green luminescence for each cell correlate with the extent of DNA denaturation, which in turn reflects the degree of chromating condensation. Mitotic cells (M) exhibit maximal denaturation of DNA and

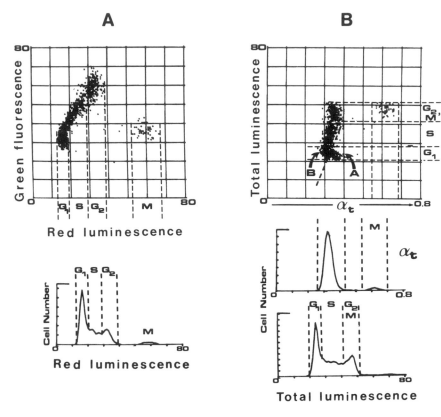

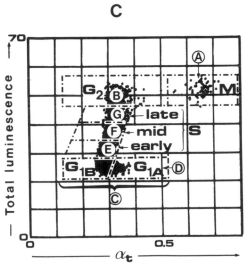

can be easily distinguished from interphase cells on the basis of their high red, and decreased (with respect to G_2 cells) green luminescence (Fig. 1A). The optimal adjustment of the staining conditions (primarily AO concentration and optical filters) and photomultiplier sensitivity settings should ensure that the imaginary line connecting M and G_2 cell clusters is diagonal (i.e., at a 45° angle with respect to red- and green-luminescence coordinates), and M cells have green fluorescence similar to that of G_1 cells.

The following populations can be discriminated by gating analysis, based on differences in total luminescence and/or α_t (Fig. 1C):

1. Cells in mitosis (M) have the highest α_t.
2. Cells in $G_2 + M$ form a typical $G_2 + M$ peak in total-luminescence (DNA content) histograms: After subtracting M cells, the number of G_2 cells can be estimated.
3. G_{1A} cells are classified as having chromatin (α_t) significantly different from cells in early-S phase. To this end, the gating window is at first located at the lowest quartile of the S population and the mean α_t and standard deviation (SD) from the mean values of these cells established. The threshold dividing G_{1A} from G_{1B} is then located on the α_t coordinate at the α_t value 2 SD above the mean α_t of these early-S cells.
4. Gating windows can be located along the S-phase cluster (total luminescence; DNA content) to identify cells in early-, mid-, and late-S phase.

Identification of all these populations is needed for detailed analysis of stathmokinetic experiments, which yield kinetic data on rates of cell and progression through different phases of the cell cycle (see Traganos and Kimmel, Chapter 25, this volume).

Based on differences in total cell luminescence, it is possible to distinguish G_1, S, and $G_2 + M$ cells. Based on differences in chromatin structure (α_t, DNA denaturability), the G_1 phase can be subdivided into G_{1A} and

---

FIG. 1. Luminescence of exponentially growing L1210 cells after removal of RNA, partial denaturation of DNA by acid, and subsequent staining with the metachromatic fluorochrome AO. The dsDNA stains green, while denatured ssDNA stains red. (A) Distribution of cells with respect to green versus red luminescence. (B) Total luminescence (red plus green) and α_t (red/total luminescence) of individual cells from the same culture as shown in (a). After transformation of the data the G_2 and M cell clusters are aligned along the horizontal line. Single-parameter frequency histograms of red luminescence, α_t, and total luminescence are shown below the scattergrams in (A) and (B). (C) Discrimination of various cell populations by gating analysis, based on differences in total luminescence and/or α_t. See discussion in text.

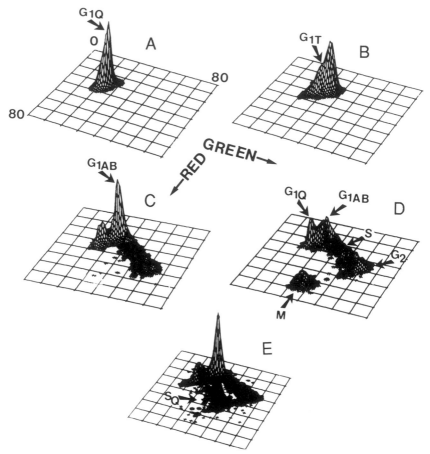

FIG. 2. Sensitivity of DNA *in situ* to acid-induced denaturation during stimulation of human lymphocytes (A–D) and in mononuclear cells from bone marrow of a patient with chronic myeloid leukemia during blastic crisis (E). Nonstimulated (A) and phytohemagglutinin (PHA)-stimulated lymphocytes 18 hours (B) and 3 days (C, D) after addition of PHA. The last culture (D) was additionally treated with colcemid for 8 hours before harvesting to arrest cells in mitosis. After staining with AO, the proportions of nondenatured and denatured DNA are represented by green and red luminescence, respectively. Subpopulations of cells in quiescence (G_{1Q}, S_Q), and during transition from quiescence to the cell cycle (G_{1T}, S, G_2, and M), can be distinguished as shown, respectively.

G_{1B} compartments, shown to have different kinetic properties (Darzynkiewicz *et al.*, 1980).

Staining of exponentially growing cells is most commonly utilized for the purpose of scoring cells in mitosis (e.g., to estimate mitotic indices in stathmokinetic experiments), to distinguish G_{1A} from G_{1B} cells, and in the case when cells grow at two ploidy levels, to distinguish G_1 cells of higher ploidy (4C DNA content) from G_2 cells of lower ploidy. Though having the same DNA content, these cells (G_1 vs G_2) differ in chromatin structure and under optimal staining conditions, can be distinguished by differences in their α_t.

B. Quiescent versus Cycling Cells

Nonstimulated and mitogen-stimulated lymphocytes are examples of quiescent and cycling cells, respectively. Distribution of these cells with respect to their red and green luminescence after partial DNA denaturation and staining with AO is shown in Fig. 2. The changes in stainability are a reflection of chromatin decondensation and cell progression through the cell cycle (DNA content increase) following mitogenic stimulation.

C. Nondifferentiated versus Differentiated Cells

Figure 3 illustrates stainability of Friend erythroleukemia cells growing exponentially (nondifferentiated) and differentiated. Their erythroid differentiation was induced by growth in the presence of dimethyl sulfoxide

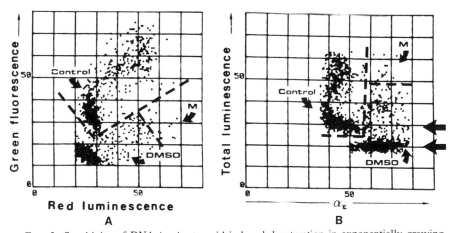

FIG. 3. Sensitivity of DNA *in situ* to acid-induced denaturation in exponentially growing (control) and differentiated (DMSO) Friend erythroleukemia cells. The cells were grown in the presence of 1.5% DMSO for 6 days, which resulted in their terminal erythroid differentiation. Cells from the control and DMSO-treated cultures were mixed in 1 : 1 proportion, fixed, and stained as described in the text. Positions of cell clusters, as marked by arrows, were identified from the unmixed populations. M, Cells in mitosis from the exponentially growing culture. (A) Red versus green luminescence; (B) total luminescence versus α_t. In (B) the positions of G_1 clusters are identified by thick arrows. Control cells have lower α_t and higher total luminescence; differentiated cells have decreased total luminescence and elevated values of α_t.

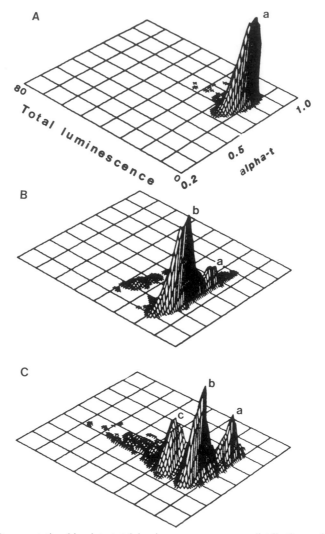

FIG. 4. Representative bivariate total luminescence versus α_t distributions of cell nuclei freshly isolated (unfixed), RNase- and acid-treated, and stained with AO. (A) Normal breast tissue; (B) fibroadenoma (benign, diploid tumor); (C) DNA aneuploid breast carcinoma. Major cell populations are labeled a–c: a, cells with diploid DNA content and high sensitivity of DNA to denaturation; b, diploid cells having DNA resistant to denaturation; c, aneuploid cells with DNA resistant to denaturation. Reprinted from Kunicka *et al.* (1989), with permission of the publisher.

(Traganos *et al.*, 1979). Differentiated cells can be distinguished as having lower total stainability of DNA (total luminescence) and increased sensitivity to denaturation, compared with their exponentially growing counterparts. During differentiation in other cell systems (other than erythroid differentiation), change in α_t is often observed but there is no change in total luminescence.

D. DNA Denaturation in Nuclei Isolated from Solid Tumors

The main cell populations that can be identified in samples of cell nuclei isolated from human tumors are shown in Fig. 4. In diploid tumors, DNA in tumor cells usually has lower sensitivity to denaturation compared to host stromal and infiltrating cells, and thus tumor cells can be distinguished as having lower α_t values. In aneuploid tumors both parameters, the total cell luminescence as well as the α_t, can discriminate tumor cells from normal cells (Kunicka *et al.*, 1987, 1989).

ACKNOWLEDGMENTS

This work was supported by NCI grant CA28704. I thank Mrs. Sally P. Carter for her assistance in preparation of the manuscript.

REFERENCES

Bobrow, M., and Madan, K. (1973). *Cytogenet. Cell Genet.* **12,** 145–153.
Darzynkiewicz, Z. (1986). *Int. Encycl. Pharmacol. Ther.* **121,** 1-98.
Darzynkiewicz, Z. (1990). *In* "Flow Cytometry and Sorting" (M. R. Melamed, T. Lindmo, and M. L. Mendelsohn, eds.) pp. 315–340. Wiley-Liss, New York.
Darzynkiewicz, Z., and Carter, S. P. (1989). *Exp. Cell Res.* **180,** 551–556.
Darzynkiewicz, Z., and Kapuscinski, J. (1990). *In* "Flow Cytometry and Sorting" (M. R. Melamed, T. Lindmo, and M. L. Mendelsohn, eds.) pp. 291–314. Wiley-Liss, New York.
Darzynkiewicz, Z., Traganos, F., Sharpless, T., and Melamed, M. R. (1975). *Exp. Cell Res.* **90,** 411–428.
Darzynkiewicz, Z., Traganos, F., Sharpless, T., and Melamed, M. R. (1977). *Cancer Res.* **37,** 4635–4640.
Darzynkiewicz, Z., Traganos, F., Andreeff, M., Sharpless, T., and Melamed, M. R. (1979). *J. Histochem. Cytochem.* **27,** 478–485.
Darzynkiewicz, Z., Traganos, F., and Melamed, M. R. (1980). *Cytometry* **1,** 98–108.
Darzynkiewicz, Z., Traganos, F., Xue, S-B., Staiano-Coico, L., and Melamed, M. R. (1981). *Cytometry* **1,** 279–286.
Darzynkiewicz, Z., Williamson, B., Carswell, E. A., and Old, L. J. (1984). *Cancer Res.* **44,** 83–90.
Darzynkiewicz, Z., Traganos, F., and Kimmel, M. (1986). *In* "Techniques in Cell Cycle Analysis" (J. W. Gray and Z. Darzynkiewicz, eds.). Humana Press, Clifton, New Jersey.

Evenson, D. P., Darzynkiewicz, Z., and Melamed, M. R. (1980). *Science* **210**, 1131–1134.
Ichimura, S., Zama, M., and Fujita, H. (1971). *Biochim. Biophys. Acta* **240**, 485–489.
Kunicka, J. E., Darzynkiewicz, Z., and Melamed M. R. (1987). *Cancer Res.* **47**, 3942–3947.
Kunicka, J. E., Olszewski, W., Rosen, P. P., Kimmel, M., Melamed, M. R., and Darzynkiewicz, Z. (1989). *Cancer Res.* **49**, 6347–6351.
Kunicka, J., Myc, A., Melamed, M. R., and Darzynkiewicz, Z. (1990). *Cell Tissue Kinet.* **23**, 31–39.
Subirana, J. A. (1973). *J. Mol. Biol.* **74**, 363–385.
Touchette, N. A., Anton, E., and Cole, R. D. (1986). *J. Biol. Chem.* **261**, 2185–2191.
Traganos, F., Darzynkiewicz, Z., Sharpless, T., and Melamed, M. R. (1979). *J. Histochem. Cytochem.* **27**, 382–389.

Chapter 33

Fluorescent Methods for Studying Subnuclear Particles

WILLIAM D. WRIGHT, RYUJI HIGASHIKUBO,
AND JOSEPH L. ROTI ROTI

Section of Cancer Biology
Radiation Oncology Center
Washington University School of Medicine
St. Louis, Missouri 63108

I. Introduction

Studies of subnuclear structure have begun to unravel its role in the organization and function of the eukaryotic genome. Titration studies with intercalating dyes have been shown to be sensitive probes of nuclear DNA supercoiling density (Roti Roti *et al.*, 1982). DNA–nuclear matrix attachment points appear to define domains for DNA supercoiling changes. Furthermore, studies of the nuclear matrix have implicated this structure in DNA synthesis, RNA transcription, processing, and transport (McCready *et al.*, 1980; Nelkin *et al.*, 1982; Cijek *et al.*, 1984; Alemaro *et al.*, 1983). The nature of these interactions involving DNA, chromatin, the nuclear matrix, and other nuclear structures, however, are poorly understood. We have developed two fluorescence assays of subnuclear protein–nucleic acid associations.

To investigate the effects of various environmental perturbations on the superhelical density of nuclear DNA, we have exposed salt-extracted (1 *M* NaCl) nuclei to varying concentrations of the intercalating, fluorescent dye propidium iodide (PI) and measured the resulting changes in the diameter of the fluorescent DNA "halo." This method is well suited to the detection of DNA damage by ionizing radiation and its repair as evidenced by the degree of superhelical rewinding occurring at high PI concentrations.

METHODS IN CELL BIOLOGY, VOL. 33

These particles are operationally defined as histone-depleted nuclei even though all proteins soluble in 1 M NaCl as well as histones will be depleted from these nuclei.

We have been able to analyze the nuclear matrix flow cytometrically by staining the isolated matrices with fluorescein isothiocyanate (FITC) for protein and PI for double-stranded nuclei acids. Results indicate the presence of an RNase-sensitive component vital to the structural integrity of the nuclear matrix that becomes protected under certain conditions. Resistance or sensitivity is defined by the residual PI fluorescence and the ability of intact fluorescent particles to be detected by the flow cytometer.

II. Applications

Studies of changes in the availability of DNA for supercoiling are indicative of the number and type of topological constraints on various supercoiled domains. Thus changes in the organization of DNA in chromatin at this level can demonstrate the increased association of DNA with immobile structures (as in the case of heat), the liberation from such constraints (as in the case of radiation), and the reestablishment of the control state of organization after perturbation. This technique is also applicable to sorted populations for determination of cell cycle-specific effects.

Flow cytometric (FCM) studies of subnuclear structure allow a rapid, sensitive means of determining the effects of perturbing agents on the structural integrity of a subnuclear particle. Present studies show differences between nuclear matrices from heat-shocked cells relative to control cells. These results may be relevant to the mechanism by which heat shock alters gene expression.

III. Materials: Histone-Depleted Nuclei

1. Lysis buffer: 2.0 M NaCl, 10 mM EDTA, 2 mM Tris pH 8.0, 0.5% Triton X-100. This solution may be stored indefinitely at room temperature (RT).
2. Propidium iodide stock solution: PI (Sigma, St. Louis, MO) is dissolved to a final concentration of 5 mg/ml in distilled water (dH$_2$O) and stirred overnight in a light-tight container. When stored at 4°C, occasional heating (60°C, 15 minutes) may be required to redissolve any precipitated dye.

3. Propidium iodide–lysis solutions: For a standard titration curve, the following dye–lysis solutions are made by adding an appropriate amount of PI stock to lysis buffer, the final volume being 10 ml : 1 μg/ml, 2 μg/ml, 4 μg/ml, 10 μg/ml, 15 μg/ml, 20 μg/ml, 70 μg/ml, and 100 μg/ml. These solutions are conveniently stored at 4°C in 15-ml screw-cap tubes wrapped in aluminum foil.
4. Lysis and viewing chambers: LabTek Tissue culture chamber/slides (four chambers; Miles Scientific, Naperville, IL) are useful vessels for cell lysis and staining.
5. Poly-L-lysine solution: Poly-L-lysine hydrobromide (Sigma, St. Louis, MO) is dissolved in dH$_2$O to a final concentration of 1 mg/ml and stored at −20°C.
6. Beckman J6B centrifuge with TY-JS-4 rotor: All centrifugation steps referred to here are carried out at 4°C. An equivalent alternative centrifuge may be used.

IV. Methods: Histone-Depleted Nuclei

Conditions for the optimal extent of cell lysis and chromosomal protein extraction that result in measurable nucleoids may vary among different cell types. The following method applies to HeLa S3 cells growing exponentially either as monolayers or in suspension. To decrease the mobility of negatively charged nucleoids near the slide surface, each chamber is rinsed with a solution of poly-L-lysine and air-dried ~ 10 minutes prior to cell lysis. The presence of poly-L-lysine does not affect the resultant halo diameter. Cells are removed from the growth medium by centrifugation at 800 rpm for 5 minutes and resuspended in an equal volume of Spinner salts solution (see Higashikubo et al., Chapter 31, this volume). After sedimenting the cells again at 1500 rpm for 5 minutes, they are resuspended to a concentration of 1–2 × 10^5/ml in Spinner salts solution. Then, 300 μl of this suspension is placed in one well of a LabTek slide for each dye concentration to be studied. An equal volume of the dye–lysis buffer concentration of interest is then carefully added to the cell suspension and the slide is rocked gently several times for mixing. The slides are left in the dark at RT for 10 minutes to ensure tthe completion of lysis. Some cell types may be more sensitive to the lysis procedure than others in terms of the length of time their nucleoids are stable in the lysis solution. In general, fluorescent nucleiods from HeLa cells are visualized and photographed within 30–40 minutes after lysis. The fluorescent nucleoids are visualized at 200 power under excitation with green (520–570 nm, 546-nm peak transmission) light on an inverted fluorescence microscope (Olympus). Fluorescence photomicrographs are taken of selected fields with an attached Olympus OM-2 camera body on automatic exposure mode at an ASA setting of 400 using Kodak Ectachrome ASA 400 daylight film (Eastman Kodak Co., Rochester, NY). The slides are then developed, labeled, and randomized prior to measurement. Size standardization is achieved for each film roll by using an image of a stage micrometer divided into 10-μm units.

V.   Results: Histone-Depleted Nuclei

The results from a typical set of experiments are presented in Figs. 1 and 2. The variation in halo diameter with increasing dye concentration can generally be superimposed over data obtained by sedimentation and can be resolved into two distinct phases (Fig. 1): (1) In low (<0.5 μg/ml) PI concentrations the HeLa nucleoid diameter is close to that of the intact nucleus ($\simeq$ 15 μm). As the dye concentration increases, more molecules are bound per unit DNA, thus unwinding endogenous right-handed super-helical domains, resulting in an increase in the fluorescent halo diameter. This increase in halo diameter continues until an equivalence point is

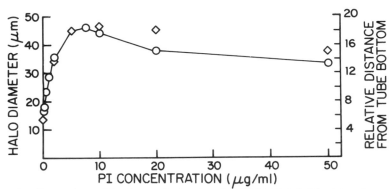

FIG. 1.   Comparison of the histone-depleted nuclear DNA "halo" (○) with sedimentation distance (◇) over a range of PI concentrations. The mean halo diameter ± 1 SE (left ordinate) from five pooled experiments is plotted against the corresponding PI concentrations. [14]CTdR-Labeled cells were lysed atop 15–30% linear sucrose gradients containing the indicated concentrations of PI. The histone-depleted nuclei were then sedimented at 10,000 rpm for 90 minutes and the peak fraction determined. When plotted as relative distances from the tube bottom (right ordinate), the variation in sedimentation distance with PI concentration correlates well with halo diameters obtained by methods described in the text.

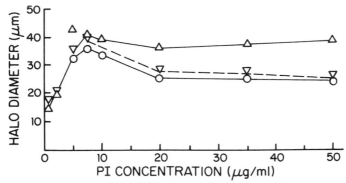

Fig. 2.   Induction and repair of DNA damage by ionizing radiation measured by the histone-depleted nuclear DNA halo technique. Cells were exposed to 10 Gy and either lysed immediately (△) or returned to 37°C for 30 minutes (▽). Radiation-induced DNA damage is demonstrated as the absence of the rewinding phase. In other words, at PI concentrations >7.5 the halo diameter remains constant rather than decreasing with increasing PI concentration. DNA repair is measured by the retuurn of the rewinding response and is complete by 30 minutes. Nonirradiated controls (○).

reached. (2) Increasing the dye concentration beyond this point results in the formation of left-handed superhelical domains that rewind the DNA helix causing a decrease in halo diameter. The "rewinding" phenomenon is dependent on a topologically constrained double helix. If the degree of constraint is reduced by the introduction of strand breaks by ionizing radiation (Fig. 2), the binding of additional dye molecules cannot force the DNA to rewind because the double helix can freely rotate around its axis between breakpoints. Thus, the degree of damage introduced by a given radiation dose can be estimated using this technique.

VI.   Materials and Instrumentation: Nuclear Matrix

1. TMNP (DNase I digestion buffer): 10 mM Tris pH 7.4, 5 mM MgCl$_2$, 10 mM NaCl, 0.1 mM phenylmethylsulfonyl fluoride (PMSF). PMSF is a serine protease inhibitor.
2. DNase I stock solution: Deoxyribonuclease I, type II (Sigma, St. Louis, MO) is dissolved in TMNP to a final concentration of 5 mg/ml and stored at −20°C.
3. High-salt buffer (HSB): 3.0 M NaCl, 10 mM EDTA, 10 mM Tris pH 9.0, 20 mM mercaptoethanol. The solution is stored at 4°C but warmed to room temperature (RT) before use.

4. Ribonuclease A stock solution: RNase A (Sigma, St. Louis, MO) is dissolved in TMNP (see Section VI, 1) to a final concentration of 1 mg/ml and stored at $-20°C$. Prior to use, the enzyme solution is placed in a 100°C water bath for 5 minutes to destroy any other nucleolytic activity.

5. Fluorescein isothiocyanate stock solution: FITC (Calbiochem-Behring, San Diego, CA) is dissolved to a final stock concentration of 30 μg/ml in TMNP and stored in a light-tight container at 4°C. This solution should be made fresh every 7–10 days.

6. Propidium iodide stock solution: PI (Sigma, St. Louis, MO) is dissolved into a final stock concentration of 70 μg/ml in TMNP and stored in a light-tight container at 4°C. This solution should remain stable for several months.

Particles stained with PI and FITC are excited with 300 mW at 488 nm of an argon laser. Fluorescence of FITC is detected through a 525-nm band-pass filter and PI fluorescence through a 640-nm long-pass filter. Histograms containing FITC, PI, and light scatter (LS) are obtained. Data are acquired at the lowest possible amplifier gain settings, especially for fluorescence, by adjusting the photomultiplier tube voltage. Data are stored in a computer and the means calculated. The mean values are corrected by the amplifier gain setting to obtain relative intensities.

VII.   Methods: Nuclear Matrix

A.   Initial Matrix Preparation

HeLa nuclei prepared as described previously (Higashikubo et al., Chapter 31, this volume) are suspended in TMNP at 2×10^6 nuclei/ml, DNase I is added to a final concentration of 25 μg/ml, and the digestion is allowed to proceed for 120 minutes at 37°C. DNA-depleted nuclei then are collected by centrifugation at 2000 rpm for 10 minutes and are washed once with the original volume of TMNP. The particles then are resuspended in 0.5 ml TMNP per 4 ml starting volume, and five volumes of HSB are added. This suspension is allowed to stand at RT for 10 minutes. The resultant DNA–histone-depleted particles are collected by centrifugation as before and washed two times with the original volume of TMNP.

B.   RNase Digestion

After the final TMNP wash, the matrix preparations are split into 15-ml conical centrifuge tubes equal to the number of RNase concentrations to be studied. The final yield of matrices is difficult to determine because the matrix particle is invisible by light microscopy and cannot be detected by an electronic particle counter. In general, to obtain enough matrix particles for FCM analysis (10^5 events/sample) of 10 RNase concentrations, a starting volume containing $2-3 \times 10^8$ cells is sufficient. Once split into the appropriate number of samples, RNase is added to the desired concentration from the boiled stock solution; the samples are mixed by vortexing and are placed on ice for 30 minutes. After digestion the particles are collected by centrifugation at 2000 rpm for 10 minutes and are washed once with 15 ml TMNP. The final pellet is resuspended in 0.8 ml TMNP.

C.   Staining

To each 0.8 ml sample, 200 μl of the FITC stock solution are added, the tubes vortexed, and allowed to stand on ice in the dark for 30 minutes. At this point, the samples are revortexed, 1.0 ml of PI stock solution added, and the tubes mixed again. After an additional 30 minutes on ice in the dark, the samples are vortexed, passed through a 20 μm filter (Small Parts, Inc., Miami, FL) and analyzed by FCM.

VIII.   Results: Nuclear Matrix

The LS, FITC, and PI profiles from a typical experiment are shown in Fig. 3. After DNase digestion and HSB extraction, the bimodal nuclear PI histogram becomes monodisperse and the intensity of FITC and LS signals is reduced. Similar results have been obtained by S. Papa, F. A. Manzoli, and co-workers (personal communication). These distributions are tested for resistance to RNase digestion over a range of 0.1–100 μg/ml of enzyme. RNase tolerance curves can be generated by plotting the mean of the PI distribution versus RNase concentration (Figs. 4 and 5) and thresholds for disruption collated under various conditions. The nuclear matrix is extremely sensitive to RNase, becoming virtually undetectable at concentrations as low as 0.1–0.3 μg/ml. By reversing the order of the isolation procedure by exposing DNase-digested particles to RNase prior to HSB extraction, it can be shown that the RNase-sensitive component in unheated cells is less accessible to the enzyme prior to HSB extraction. Due to the close proximity of residual protein and nucleic acid in the nuclear matrix, the presence of PI–FITC energy transfer could be a factor. However, quantification of PI and FITC fluorescence from heated and unheated samples shows an 8-fold increase in FITC fluorescence without

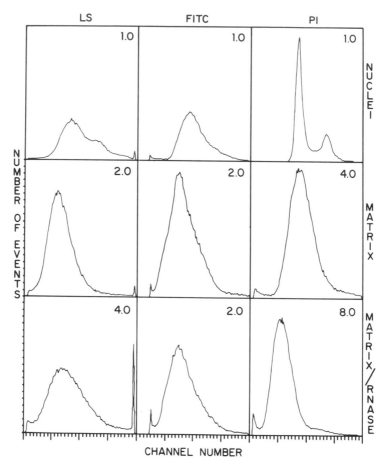

F I G. 3.   LS, FITC, and PI profiles of nuclei, nuclear matrices, and nuclear matrices after RNase digestion (5 μg/ml). These distributions, shown for illustration purposes, were derived from heat-shocked cells (45°C, 30 minutes). Amplifier gain settings are indicated in the upper righthand corner of each panel and are calculated relative to those of nuclei settings. After DNase I digestion and HSB extraction (matrix), the bimodal PI histogram becomes monodisperse and all three signals are reduced in intensity. RNase digestion further reduces fluorescence and LS intensities without discernible alteration in histogram shape. In identical experiments (not shown) using nuclei derived from unheated cells, exposure of matrices to RNase concentrations as low as 0.1 μg/ml result in the redistribution of all histograms into the debris peak.

any change in PI fluorescence, indicating that energy transfer is minimal. Further, the nuclear matrices from heated cells have increased resistance to RNase, presumably due to the sequestering of RNA (and/or) DNA sequences at the matrix.

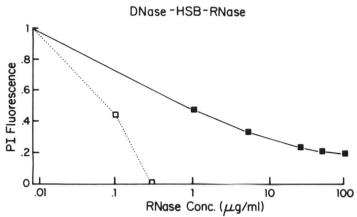

FIG. 4. RNase tolerance of nuclear matrices from heated and control cells. The mean of the PI fluorescence distribution relative to that of the undigested matrix (ordinate) is plotted against the RNase concentration. The results are shown for a typical experiment. The DNA-depleted, salt-extracted nuclear matrices prepared from cells exposed to heat (45°C, 30 minutes; ■) are more resistant to RNase disruption than those from control cells (□). The particles derived from unheated cells are virtually undetectable after exposure to 0.3 μg/ml RNase A, while those from heated cells show a distinct PI fluorescence histogram at RNase concentrations up to and exceeding 100 μg/ml.

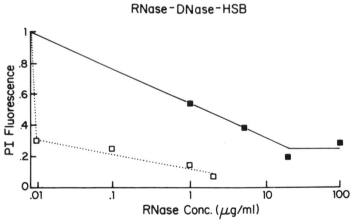

FIG. 5. RNase tolerance of nuclear matrices from RNase-digested nuclei. This experiment is similar to that for Fig. 4 except that the RNase digestion was done prior to DNase I digestion and HSB extraction. When nuclear matrices are made from nuclei previously exposed to varying concentrations of RNase, matrices from control cells (□) show an increased resistance to the enzyme while those from heated cells (■) seem unaffected. Taken together, these observations implicate a salt-extractable component in the protection from digestion of structurally important RNA species in the nuclear matrix. The sequence of steps in the isolation of nuclear matrices in each case is indicated above each panel for clarity.

IX.   Summary

Fluorescence assays can be used to reveal molecular interactions through rapidly demonstrable particle-associated events. The additional fact that in many cases fluorescent particles may be analyzed on a per-event basis lends credence to such techniques as probes for biologically significant perturbations and their resolution. Perhaps more importantly, the sorting capability of the flow cytometer enables detailed study of these events in cells in relation to their positions in the cell cycle. Further studies on the effect of drugs and other modalities on the organization of the genome and the nuclear matrix should prove of interest because the interactions of chromatin and this subnuclear particle could be predictive of the state of DNA metabolism under such conditions. With the additional ability of following such organizational changes through the cell cycle, the mechanisms of reversal of perturbing events might be elucidated.

ACKNOWLEDGMENTS

The work reported here was supported by NCI grants CA41102 and CA43198.

REFERENCES

Allmaro, H., Puvion, E., Kister, L., and Jacob, M. (1983). *EMBO J.* **2**, 93–96.
Cijek, E. M., Tsai, M-J., and O'Malley, B. W. (1984). *Nature (London)* **306**, 607–609.
McCready, S. J., Godwin, J., Mason, D. W., Brazell, I. A., and Cook, P. R. (1980). *J. Cell Sci.* **46**, 365–380.
Nelkin, B., Pardoll, D., Robinson, S., Small, D., and Vogelstein, B. (1982). *In* "Tumor Cell Heterogeneity: Origins and Applications" (A. H. Owens, Jr., D. S. Coffey, and S. R. Baylin, eds.). Academic Press, New York.
Roti Roti, J. L., and Painter, R. B. (1982). *Radiat. Res.* **89**, 166–175.

Chapter 34

Chromosome and Nuclei Isolation with the MgSO₄ Procedure

BARBARA TRASK AND GER VAN DEN ENGH

Biomedical Sciences Division
Lawrence Livermore National Laboratory
Livermore, California 94550

I. Introduction

The MgSO₄ isolation procedure (van den Engh *et al.*, 1984, 1985, 1986, 1988) employs Mg^{2+} to stabilize chromosomes in suspension. The procedures can be used successfully to produce high-resolution flow karyotypes from a variety of cell types (Fig. 1). Mitotic cells are swollen in a hypotonic buffer containing this stabilizing agent, HEPES as a buffer (pH 8.0), and dithiothreitol (DTT) as a reducing agent. This combination effectively reduces the number of chromosome clumps in the preparation. If propidium iodide (PI) is to be used as the DNA-specific stain, RNase is also added to the hypotonic buffer. It is essential that growth medium be completely removed from the mitotic cell pellet before addition of isolation buffer. This ensures sufficient hypotonicity for cell swelling. Chromosomes can then be released to form a monodispersed suspension with less rigorous mechanical shearing and thus, less chromosome damage. The swelling proceeds 10 minutes at room temperature (RT) before cell membranes are disrupted by the addition of Triton X-100. After 10 minutes on ice, the cells are mechanically disrupted by passage through a needle and syringe (in the case of fibroblast cells) or by vigorous vortexing (lymphoblast cells). Chromosome suspensions to which RNase has been added are then incubated 30 minutes at 37°C for RNA digestion. The chromosome suspension is then stained with DNA-specific fluorochromes. The stains should be

363

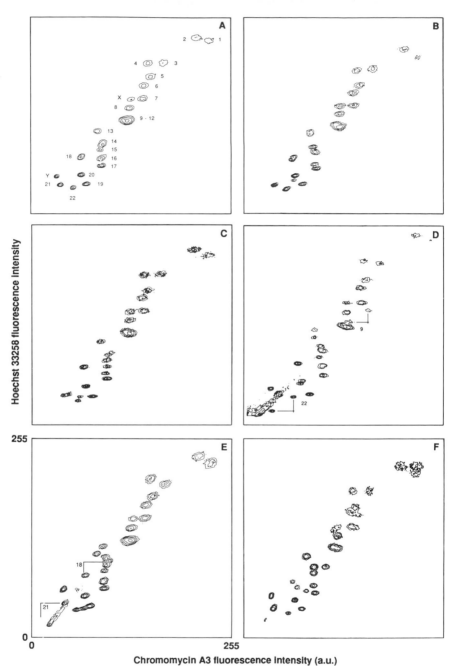

Hoechst 33258 fluorescence intensity

255

0

0   255

Chromomycin A3 fluorescence intensity (a.u.)

Fɪɢ. 1. Bivariate flow karyotypes of chromosomes isolated from a variety of cell types using the HEPES–MgSO₄ method and stained with Hoechst and chromomycin A3. (A) Fibroblast cell line HSF-7; (B) lymphoblast cell line GM131; (C) PHA-stimulated peripheral blood lymphocyte culture of a normal donor; (D) short-term culture of chronic myeloid leukemic blasts showing the products of the reciprocal 9;22 translocation; (E) amniocyte culture with a reciprocal 18;21 translocation; (F) erythroblast leukemic cell line HEL with multiple structural and numerical chromosome abnormalities. Numbers in (A) indicate the chromosome(s) responsible for the peaks in the flow karyotype. Reprinted with permission from Trask (1989).

allowed to equilibrate with the chromosomes before analysis; chromomycin binding can be considered complete after 1–2 hours. Chromosome resolution is greatly improved by the addition of 10 mM sodium citrate and 25 mM sodium sulfite (final concentration) at least 15 minutes before analysis. Chromosomes prepared with the MgSO$_4$ method may be stored at 4°C for 24–48 hours before analysis.

II. Isolation Procedure

1. Cell collection.
   a. For chromosome isolation from fibroblast or amniocyte cell lines, cultures are split 24–36 hours before colcemid addition. Cultures should be split at a dilution that results in 50–75% confluency at the time of colcemid addition. To block cells in mitosis, remove growth medium after shaking cultures gently. Add 10 ml growth medium containing 10% fetal calf serum (FCS) and 0.1 μg/ml colcemid (Sigma Chemical Co., St. Louis, MO) to each T-75 culture flask. Incubate cultures at 37°C for 4–6 hours (10–12 hours for human cells). Collect mitotic cells by shake-off.
   b. For chromosome isolation from human lymphoblast cell lines or phytohemagglutinin (PHA)-stimulated peripheral blood lymphocyte cultures, colcemid is added to exponentially growing cells in suspension culture to a final concentration of 0.1 μg/ml. Incubate cultures at 37°C for 10–12 hours. Collect all cells.
   c. For isolation of nuclei, fibroblast cultures are trypsinized to collect cells. Cells are washed twice with phosphate-buffered saline (PBS) before isolation. Lymphoblast cells are merely collected and centrifuged.
2. Centrifuge collected cells at 100–150 g for 10 minutes at RT in 15–50 ml tubes. Remove supernatant completely, drain tube on tissue, and wipe excess medium from inside of tube. Flick tube to loosen pellet.
3. For routine flow karyotyping, add 1 ml isolation buffer (see Section III) per 10^5 cells from a mitotic shake-off or 1 ml per 2–3 × 10^5 lymphoblast or lymphocyte cells. For large-scale chromosome isolation for high-speed sorting of chromosomes for the production of recombinant DNA libraries, we routinely increase the cell density in isolation buffer to 10^7/ml. In this case, KCl concentration in the isolation buffer is reduced to 40 mM for proper hypotonicity. For isolation of nuclei, isolation buffer is added to achieve 5 × 10^6

cells/ml. Flick the mixture gently to mix. Incubate at RT with lids on for 10 minutes. Pool aliquots, if necessary.

4. Add 0.1 ml Triton X-100 solution (2.5%) to each 1 ml of chromosome suspension. Flick to mix. Incubate on ice for 10 minutes.

5. Disrupt fibroblast cells by syringing 1-ml aliquots two to three times with 1-ml syringe or aliquots not larger than 2 ml with a 3-ml syringe and a 22.5-gauge needle. Disrupt lymphoblast cells or lymphocytes by vortexing at high speed ~ 10 seconds. For isolation of nuclei, vortex. Cell disruption may be monitored miscroscopically using Hoechst staining.

6. If PI is to be used as the DNA stain, boiled RNase (0.15 mg/ml final concentration, from 3 mg/ml stock, stored as frozen aliquots) should be added to the isolation buffer. The chromosomes are then incubated at 37°C for 30 minutes after syringing. Nuclei suspensions prepared for subsequent use in hybridization studies are treated with RNase.

7. Chromosomes can be stained at this point or stored at 4°C.

8. Chromosomes are stained with Hoechst 33258 (stock solution 200 μg/ml; final concentration 2 μg/ml) and chromomycin A3 (stock 1 mg/ml; final concentration 20 μg/ml or PI (400 μg/ml stock; final concentration 20 μg/ml). Concentrated chromosome suspensions are stained with twice these stain concentrations. Nuclei suspensions for hybridization studies are left unstained.

9. Before flow analysis of chromosomes, sodium citrate and sodium sulfite are added to final concentrations of 10 mM and 25 mM, respectively, from a 10x stock solution made just prior to use. This is best done at least 15 minutes before analysis.

III.   Isolation Buffer

Isolation buffer contains:
50 mM KCl
5 mM HEPES, pH 8.0
3 mM DTT or dithioerythritol
(optional: 0.15 mg/ml RNase)

Preparation

Isolation buffer is made with sterile distilled water the day it is to be used. Make 55 mM KCl + 5.5 mM HEPES stock solution and bring this to pH 8.0 with 0.1 M KOH. Nine volumes of this solution are added to one volume of 100 mM MgSO$_4$. Dithiothreitol is added from a 120 mM stock solution (stored frozen in small aliquots). RNase is added from a 3 mg/ml stock solution (boiled, stored frozen in small aliquots) for PI staining or isolation of nuclei. Buffer is then filtered through 0.22-μm disposable filter units before use.

REFERENCES

Trask, B. (1989) *In* "Flow Cytogenetics" (J. Gray, ed.), pp. 43–60. Academic Press, New York.

van den Engh, G., Trask, Cram, S., and Bartholdi, M. (1984). *Cytometry* **5**, 108–117.

van den Engh, G., Trask, B., Gray, J., Langlois, R., and Yu, L.-C. (1985). *Cytometry* **6**, 92–100.

van den Engh, G., Trask, B., and Gray, J. (1986). *Histochem.* **84**, 501–508.

van den Engh, G., Trask, B., Lansdorp, P., and Gray, J. (1988). *Cytometry* **9**, 266–270.

Chapter 35

Univariate Analysis of Metaphase Chromosomes Using the Hypotonic Potassium Chloride–Propidium Iodide Protocol

L. SCOTT CRAM, FRANK A. RAY, AND MARTY F. BARTHOLDI

Life Sciences Division
Los Alamos National Laboratory
University of California
Los Alamos, New Mexico 87545

I. Introduction

Analysis and sorting of individual metaphase chromosomes by flow cytometry (FCM) has become a valuable technique for cell and molecular biologists, cytogeneticists, and radiobiologists. Flow cytogenetic analysis can rapidly provide a quantitative karyotype of a mixed cell population. When used in conjunction with chromosome banding, enhanced information about chromosome structure and content can be obtained (Gray and Cram, 1989).

A goal for investigators using flow cytogenetic analysis is to resolve each chromosome type of the species of interest, and then, to sort and recover large numbers of a single chromosome type at high purity. Although sorting is not discussed in detail here, the ability to sort a single chromosome type at high purity is dependent on how well one can resolve one chromosome type from the other chromosomes. In addition to resolution, molecular weight of chromosomal DNA and recovery of sorted chromosomes is critical for most uses of flow sorted chromosomes (Van-Dilla *et al.*, 1986, Deaven *et al.*, 1986, Bartholdi *et al.*, 1987).

369

Several protocols for isolating and staining metaphase chromosomes for flow analysis have been reported in the literature. These procedures provide satisfactory results for resolving most chromosomes types from Chinese hamster and human cells. New procedures and modifications of existing protocols are continually being developed and provide new options and capabilities for specific applications. Improved results include resolution of additional chromosome types and homolog pairs (Trask et al., 1989), additional structural information (Bartholdi et al., 1989), improved recovery of sorted chromosomes, and higher molecular weight DNA.

Most analytical procedures rely on staining chromosomes with one or two DNA-specific fluorochromes. Univariate analysis (one-color staining) using propidium iodide (PI) is the focus of this chapter. The procedure was originally described by Aten et al. (1980). Bivariate flow karyotype analysis (two-fluorochrome staining) is most commonly used for resolving human chromosomes because of the somewhat unique adenine–thymine to guanine–cytosine (AT/GC) base pair ratio variation that exists among human chromosomes. Two different protocols using different swelling or isolation buffers ($MgSO_4$ and polyamines) are the methods of choice for bivariate chromosome staining. (See Trask and van den Engh, Chapter 34, and Cram et al., Chapter 36, this volume.)

Major steps consist of (1) cell culture, (2) increasing the mitotic fraction, (3) cell swelling, (4) chromosome isolation and stabilization, and (5) fluorescent labeling. Each of these steps is explained in detail. Each step is often modified to meet the demands of an experimental requirement. Therefore, the rationale behind each step is explained to facilitate modifications.

II. Applications

The hypotonic potassium chloride–propidium iodide (KCl–PI) procedure has been used to analyze chromosomes isolated from fibroblasts, peripheral lymphocytes, and lymphoblastoid cells derived from a variety of species, including human, Chinese hamster, mouse, and chicken. Cell growth rate and membrane permeability are two major factors that affect the direct applicability of the procedure provided here to a broad variety of cell types. These two factors influence the amount of time required to accumulate cells in mitosis and the hypotonicity of the swelling buffer.

III.   Materials

A.   Isolation Buffers

Solution A

75mM KCl
50 μg/ml propidium iodide

Solution B

75 mM KCl
50 μg/ml propidium iodide
1% Triton X-100
1 mg/ml RNase

B.   Preparation

The two isolation buffers are made up using distilled water and filter-sterilized. Stored at 4°C the solutions are stable for many weeks. RNase is made from a stock solution that has been boiled (to remove contaminating DNase). Both solutions are used at room temperature (RT), although some investigators prefer using 4°C solution A, and others prefer 37°C solution A.

IV.   Isolation and Staining Protocol

This protocol is specifically adapted for Chinese hamster fibroblast cultured in a 37°C, 5% CO_2 incubator using 20 ml of medium on a T-75 tissue culture flask.

1. Rapidly growing cells are subcultured at a 1 : 5 split ratio 2 days, or a 1 : 2 split ratio 1 day, prior to chromosome isolation. Cells are grown exponentially for two to four cell generation times (24–48 hours), and fibroblasts are not allowed to become more than 75% confluent at the time of colcemid addition. The cell cultures are then treated with 0.1 μg/ml colcemid and incubated for 3 hours to accumulate mitotic cells blocked at metaphase. The duration of colcemid incubation is dependent on cell cycle duration. Human fibroblasts are blocked for 10–12 hours.

2. Mitotic cells round up and become loosely attached and can therefore be differentially harvested by gentle shake-off to dislodge them from their attachment points; this leaves mostly interphase cells attached to the flask. Roughly $2–5 \times 10^5$ mitotic cells (contents of two T-75 cm^2 flasks) are optimum for the volumes given here. The cell culture media containing suspended mitotic cells from the two flasks are poured into a 50-ml centrifuge tube. A second block for an additional 3 hours and collecting cells from a second shake-off can improve results if the first shake-off is discarded. The fraction of mitotic cells collected can be determined at this point by removing a small aliquot of cells and staining with a fluorochrome that will diffuse through the cell membrane and stain DNA; 10 μM Hoechst 33342 works well for this assay.

3. The cell suspension is centrifuged at 100 g (800–1000 rpm) for 5 minutes at RT. The supernatant is aspirated, or poured off, leaving a small pellet at the bottom of the tube. Because of the small volume of hypotonic buffer that is used in step 4, it is critical to remove *all* the remaining tissue culture media from the cell pellet. Pouring leaves too much supernatant unless the inside surface is removed with a lint-free tissue.

4. Add 0.5 ml of solution A to the pellet and disperse the cells by gentle pipeting or flicking the tube. Cells are swollen in this hypotonic solution for 10 minutes. Propidium iodide is excluded from swelling cells as long as membrane integrity is maintained. Cell swelling is monitored in a phase/fluorescent microscope by monitoring the number that have swollen versus those that have taken up PI as a result of premature lysing. This is the most critical step of the procedure. Uniform cell swelling is required for the metaphase chromosomes to disentangle. Mitotic cells whose membranes permeabilize prior to chromosome disentanglement (these are PI-positive cells at the very earliest stages of swelling) will never completely disengage one from another or only a few chromosomes will be freed into suspension.

5. Add 0.5 ml of solution B and incubate for 3 minutes at RT. The detergent will dissolve cell membranes during this time. Propidium iodide serves two purposes: to stabilize the chromosome structure and stoichometrically stain DNA.

6. The metaphase chromosomes are dispersed by forcing the suspension through a $1\frac{1}{2}$-in., 22-gauge needle using a 5-ml syringe. Chromosome dispersal is monitored in a fluorescent microscope. Excessive clumps of chromosomes can sometimes be dispersed by additional syringing, but usually by five times one reaches a plateau and further syringing will damage chromosomes already in suspension.

7. Incubate the chromosome suspension at 37°C for up to 30 minutes for the RNase to act.

8. Samples should be filtered (60 μm nylon mesh) to remove large clumps and fibers that would disrupt the small sample stream (2–3 μm in diameter) necessary for high-resolution flow karyotype analysis.

Samples should be analyzed within about one week. Samples prepared in a sterile environment and stored at 4°C sometimes retain their characteristics for longer periods of time.

V.   Critical Aspects of the Procedure

A.   Cell Handling

Exponential cell growth, uniform cell swelling, and chromosome stabilization are the most critical steps of this chromosome isolation procedure. Cell swelling can also be monitored by adding 10 μM Hoechst 33342 to the cell culture 30 minutes prior to mitotic shake-off. This facilitates determining the mitotic fraction collected and monitoring swelling. The blue-stained chromosomes can be easily observed for uniform swelling and chromosome disentanglement during incubation in solution A (Section IV, step 4). At the first instance of membrane permeabilization, PI will enter the cell and the chromosomes will turn from blue to pink if ultraviolet (UV) excitation is used. Some cells require 50 or 60 mM KCl buffer for adequate swelling.

B.   Instrumental Analysis

Very careful optimization of instrumental variables is required to resolve each of the 11 Chinese hamster chromosomes (10 autosomes plus the X) or 18 of the 24 human chromosome types. The Chinese hamster Y chromosome is not resolved from chromosome 5. To achieve optimum resolution, factors of importance include laser power, light collection efficiency, filter selection, and sample stream diameter.

The results illustrated in Fig. 1 were achieved using 1–2 W of laser power at 488 nm, a fluorescence-collecting objective (40×) with a numerical aperture of 0.85, a good-quality 520-nm long-pass filter, and a sample stream diameter of 2–4 μm. Roughly equivalent data have been collected using several different commercial systems. Interpretation of the distributions has been described in detail by Gray and Cram (1989).

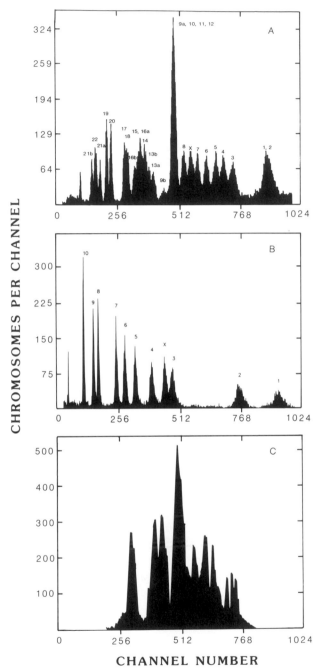

FIG. 1. Chromosomes isolated from (A) human fibroblasts, (B) cloned Chinese hamster cells, and (C) euploid mouse cell strain using the hypotonic KCl isolation buffer and stained with PI. The numbers indicate the chromosome type associated with each peak. Data from Gray and Cram (1990).

VI. Controls and Standard

Chromosomes isolated from normal, euploid cells are used as controls for DNA content and number of chromosomes of each type. Most cell lines contain some numerical and structural rearrangements and are therefore inadequate as controls for many experiments. Cell strains are also suitable and generally contain homogeneous cell types producing tighter flow karyotypes. Small 1.2-μm polystyrene microspheres are used to assess instrument performance. Run at high concentrations (5×10^7/ml), they are used to establish conditions necessary for a narrow sample stream and stable flow, and to align the sample stream and laser beam. Euploid Chinese hamster chromosomes isolated using the KCl hypotonic buffer are easily obtained, have a small number of peaks and a well-characterized and easily recognized flow karyotype (Cram *et al.*, 1983), and can be used to establish biological and instrument conditions prior to attempting more complex flow karyotypes.

VII. Results

Figure 1 illustrates three flow karyotypes of chromosomes prepared using the hypotonic KCl–PI procedure and isolated from a human fibroblasts (Fig. 1A), Chinese hamster cell (Fig. 1B), and a euploid mouse cell (Fig. 1C). Each of the peaks in the mouse flow karyotype contains multiple chromosome types. Flow karyotypes of normal euploid cells should have very few counts between well-defined peaks, and the coefficient of variation (CV) of peaks containing only single chromosome types should be in the 1–2% range as illustrated for Chinese hamster chromosomes in Fig. 1B. The two number 9 homologs of this particular Chinese hamster contain slightly different amounts of DNA, giving rise to two peaks in the flow karyotype.

The advantages of this procedure are its high resolution, ease of applicability, high molecular weight of sorted chromosome DNA, and the option of counterstaining sorted chromosomes for identification (VanDilla, Deaven, *et al.*, 1986).

ACKNOWLEDGMENTS

This work was performed under the auspices of the Department of Energy and the National Center for Research Resources of NIH (grant RR01315).

REFERENCES

Aten, J. A., Kipp, J. B. A., and Barendsen, G. W. (1980). *In* "Flow Cytometry", IV pp. 485–491.

Bartholdi, M. F., Meyne, J., Albright, K., Luedemann, M., Campbell, E., Chritton, D., and Deaven, L. L. (1987). *In* "Methods in Enzymology" (M. Goettsman, ed.), Vol. 151, pp. 252–267. Academic Press, Orlando, Florida.

Bartholdi, M. F., Meyne, J., Johnston, R. G., and Cram, L. S. (1989). *Cytometry* **10,** 124–133.

Cram, L. S., Bartholdi, M. F., Ray, F. A., Travis, G. L., and Kraemer, P. M. (1983). *Cancer Res.* **43,** 4828–4837.

Deaven, L. L., Van Dilla, M. A., Bartholdi, M. F., Carrano, A. V., Cram, L. S., Fuscoe, J. C., Gray, J. W., Hildebrand, C. E., Moyzis, R. K., and Perlman, J. (1986). *Cold Spring Harbor Symp. Quant. Biol.* **51,** 159–168.

Gray, J. W., and Cram, L. S. (1990). *In* "Flow Cytometry and Sorting" (M. Mendelsohn, T. Lindmo, and M. Melamed, eds.), pp. 503–530.

Trask, B., van den Engh, G., Mayall, B., and Gray, J. W. (1989). *Am. J. Hum. Genet.* **45,** 739–752.

Van Dilla, M. A., Deaven, L. L., Albright, K. L., Allen, N. A., Aubuchon, M. R., Bartholdi, M. F., Browne, N. C., Campbell, E. W., Carrano, A. V., Clark, L. M., Cram, L. S., Fuscoe, J. C., Gray, J. W., Hildebrand, C. E., Jackson, P. J., Jett, J. H., Longmire, J. L., Lozes, C. R., Luedemann, M. L., Martin, J. C., McNinch, J. S., Meincke, L. J., Mendelsohn, M. L., Meyne, J., Moyzis, R. K., Munk, A. C., Perlman, J., Peters, D. C., Silva, A. J., and Trask, B. J. (1986). *Biotechnology* **4,** 537–552.

Chapter 36

Polyamine Buffer for Bivariate Human Flow Cytogenetic Analysis and Sorting

L. SCOTT CRAM, MARY CAMPBELL, JOHN J. FAWCETT, AND
LARRY L. DEAVEN

Life Sciences Division
Los Alamos National Laboratory
University of California
Los Alamos, New Mexico 87545

I. Introduction

Of the three chromosomes isolation protocols included in this methods monograph, the polyamine buffer is best suited for obtaining high molecular weight chromosomal DNA (Silar and Young, 1981; VanDilla, Deaven *et al.*, 1986; Bartholdi *et al.*, 1987). The chromosome-stabilizing buffer uses the polyamines spermine and spermidine to stabilize chromosome structure (Blumenthal, 1979; Lalande *et al.* 1986, and heavy-metal chelators to reduce nuclease activity. This buffer has been the method of choice for the National Laboratory Gene Library Project—a joint project of Los Alamos National Laboratory and Lawrence Livermore National Laboratory to sort microgram quantities of all the human chromosomes and to construct chromosome-specific libraries (Deaven, VanDilla *et al.*, 1986; Deaven *et al.*, 1986). This program requires a massive chromosome-sorting effort that yields high molecular weight DNA for cloning.

The principles of chromosome isolation using the polyamine buffer are similar to those used with the hypotonic KCl procedure (Cram *et al.*, Chapter 35, this volume) and the $MgSO_4$ procedure (Trask and van den Engh, Chapter 34, this volume).

377

II.   Applications

This protocol has wide applicability; however, as with other protocols, the following variables must be optimized for each cell type: blocking time, cell concentration, type of hypotonic swelling buffer, swelling time, volume of hypotonic buffer, and vortexing time. The values provided work well for normal human fibroblasts and other cell types including somatic cell hybrids. Chromosomes prepared using this protocol are highly condensed and particularly difficult to band following sorting.

III.   Materials

A.   Stock Solutions

Stock Solution I

Prepare a 10× solution (stock solution I) containing Tris, EDTA, KCl, and NaCl.

150 mM Tris-HCl (Trizma hydrochloride, Sigma catalog no. T-3253)
 20 mM EDTA (tetrasodium salt, Sigma catalog no. ED455)
800 mM KCl
200 mM NaCl

Combine 2.36 g Tris-HCl, 0.760 g EDTA, 5.96 g KCl, and 1.16 g NaCl in 100 ml distilled water (dH$_2$O), filter-sterilize, store at room temperature (RT), and make fresh every month.

Stock Solution II

Prepare a 10× solution (stock solution II) of EGTA.

5 mM EGTA (Sigma catalog no. E-4378).

Add 0.190 g to 100 ml dH$_2$O, add as little concentrated NaOH as possible to get into solution, filter-sterilize, store at RT, and make fresh every month.

Polyamines

Make up 10 ml and freeze in 0.1-ml aliquots; do not refreeze after use.

0.4 M Spermine tetrahydrochloride (Sigma catalog no. 5677), use 139 mg/ml in dH$_2$O

1.0 M Spermidine trihydrochloride (Sigma catalog no. 56766), use 254 mg/ml in dH_2O)

B.   Preparation of Chromosome Isolation Buffer (Just Prior to Use)

1. Add 5.0 ml of each of the two 10× stock solutions to 40 ml dH_2O and adjust pH to 7.2 ± .05 with HCl. Add 50 μl of 2-mercaptoethanol.
2. To 25 ml of the above solution, add 30 mg digitonin and incubate at 37°C for 45 minutes.
3. Sterilize by filtering. Undissolved digitonin is removed in this step.
4. Add 12.5 μl of spermine and 12.5 μl of spermidine.
5. Place on ice. Use immediately.

IV.   Isolation and Staining Procedure: Human Fibroblasts

Portions of this procedure are similar to the hypotonic KCl (Cram *et al.*, Chapter 35) and $MgSO_4$ (Trask and van den Engh, Chapter 34) procedures described elsewhere in this volume. For additional discussion, these chapters should be consulted.

1. Block cells with 0.1 μg colcemid per 1 ml of tissue culture medium for 12–15 hours.
2. Prepare fresh chromosomes isolation buffer and place on ice.
3. Shake off mitotic cells and pool supernatant in 50-ml centrifuge tubes, at RT. Adjust supernatant volume to achieve a cell concentration of $8–13 \times 10^6$ mitotic cells per tube.
4. Centrifuge at 100 g (~ 800–1000 rpm) for 10 minutes at 4°C.
5. Aspirate supernatant, leaving as little as possible behind.
6. For swelling, add the appropriate amount of hypotonic buffer. The amount and hypotonicity vary depending on cell type and cell concentration. Amounts range from 2.5 to 5.5 ml per 10^7 cells. Choice of hypotonic buffers are 25–75 mM KCl or 55 mM Ohnuki's buffer (Ohnuki, 1965, 1968). Typically, 40–75 mM KCl is used for somatic cell hybrids and 25 mM KCl for human fibroblast cell strains. Swelling times can vary from 25 to 100 minutes, depending on the cell type and which swelling buffer is used. The longer swelling times are necessary with Ohnuki's buffer. See Chapter 35 for techniques to monitor and optimize cell swelling. Cell swelling is performed at RT.
7. Centrifuge at 100 g for 5 minutes at RT.
8. Aspirate supernatant and flick tube to loosen pellet.

9. Add 1.0 ml of cold chromosome isolation buffer to each tube. Mix gently.

10. Vortex each tube vigorously for 60 seconds. Place sample on ice for 10 minutes and observe chromosomes under microscope after staining with propidium iodide to determine degree of dispersion. If needed, repeat vortexing for 15 seconds and maintain sample on ice while rechecking dispersion. Repeat vortexing cycle as needed. It is better to vortex too little than too much.

11. Chromosomes can be stained at this point or stored at 4°C.

12. Staining is with Hoechst 33258 and chromomycin A3. Hoechst stock solution is made up in dH_2O (0.5 mg/ml) and 11.2 μl added to 1.0 ml of chromosomes (final concentration, 4.8 μM). Chromomycin A3 is made up in McIlvane's buffer pH 7, with 5 mM $MgCl_2$ (1 : 1); 300 μl of this solution are added to 1.0 ml of chromosomes at least 3 hours before analysis to establish equilibrium (final concentration, 125 μM).

V.   Critical Aspects of the Procedure

The protocol will yield about $2–5 \times 10^6$ chromosomes per 1 ml, when using fibroblasts or somatic cell hybrids. Even higher concentrations are advantageous for establishing a very narrow sample stream diameter and, in turn, to obtain higher resolution. The yield of isolated chromosomes based on the number of mitotic cells is between 2 and 5%. Chromosomes can be stored at 4°C for several months without loss of resolution. Once they are stained they should be analyzed within 24 hours; shelf life at 4°C is about 1 week.

For improved stability and enhanced sorting recovery, as well as for maintenance of high molecular weight DNA, we have found improved results using the polyamine buffer maintained at 4°C as the sheath fluid. When the buffer is used as sheath fluid, the mercaptoethanol and digitonin are omitted. Bartholdi *et al.* (1987) should be consulted for additional details about chromosome sorting.

VI.   Results

Figure 1 illustrates a bivariate flow karyotype of chromosomes isolated from GM 130 cells obtained from NIGMS Human Genetic Mutant Cell Repository (Camden, NJ). These cells serve as a useful "control" in that

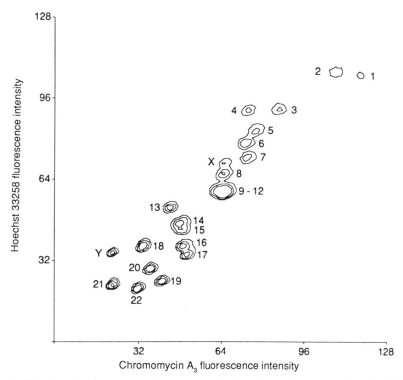

FIG. 1. Bivariate flow karyotype of chromosomes isolated from the human cell strain GM 130 using the polyamine buffer and stained with chromomycin A3 and Hoechst 33258.

they are readily available and yield reproducible results. These results were obtained using an EPICS V flow cytometer (Coulter Electronics, Inc.) equipped with two argon ion lasers tuned to 458 nm and the ultraviolet range, and a single photomultiplier to optimize light collection as described by Bartholdi *et al.* (1987). Additional information about data interpretation and analysis has been provided by Gray and Cram (1989).

ACKNOWLEDGMENTS

This work was performed under the auspices of the Department of Energy and the National Center for Research Resources of the NIH (grant RR01315).

REFERENCES

Bartholdi, M. F., Meyne, J., Albright, K., Luedemann, M., Campbell, E., Chritton, D., and Deaven, L. L. (1987). *In* "Methods in Enzymology" (M. Goettsman, ed.), Vol. 151, pp. 252–267. Academic Press, Orlando, Florida.

Blumenthal, A. D. (1979). *J. Cell Biol.* **81,** 255–259.

Deaven, L. L., Hildebrand, C. E., Fuscoe, J. C., and Van Dilla, M. A. (1986a). *In* "Genetic Engineering" (J. K. Setlow and A. Hollaender, eds.), Vol 8, pp. 317–332, Plenum.

Deaven, L. L., Van Dilla, M. A., Bartholdi, M. F., Carrano, A. V., Cram, L. S., Fuscoe, J. C., Gray, J. W., Hildebrand, C. E., Moyzis, R. K., and Perlman, J. (1986b). *Cold Spring Harbor Symp. Quant. Biol.* **51,** 159–168.

Gray, J. W., and Cram, L. S. (1990). *In* "Flow Cytometry and Sorting" (M. Mendelsohn, T. Lindmo, and M. Melamed, eds.), pp. 503–530.

Lalande, M., Donlon, T., Petersen, R. A., Liberfarb, R., Manter, S., and Latt, S. A. (1986). *Cancer Genet. Cytogenet.* **23,** 151–157.

Ohnuki, Y. (1965). *Nature (London)* **208,** 916–917.

Ohnuki, Y (1968). *Chromosoma* **25,** 402–428.

Silar, R., and Young, B. D. (1981). *J. Histochem. Cytochem.* **29,** 74–78.

Van Dilla, M. A., Deaven, L. L., Albright, K. L., Allen, N. A., Aubuchon, M. R., Bartholdi, M. F., Browne, N. C., Campbell, E. W., Carrano, A. V., Clark, L. M., Cram, L. S., Fuscoe, J. C., Gray, J. W., Hildebrand, C. E., Jackson, P. J., Jett, J. H., Longmire, J. L., Lozes, C. R., Luedemann, M. L., Martin, J. C., McNinch, J. S., Meincke, L. J., Mendelsohn, M. L., Meyne, J., Moyzis, R. K., Munk, A. C., Perlman, J., Peters, D. C., Silva, A. J., and Trask, B. J. (1986). *Biotechnology* **4,** 537–552.

Chapter 37

Fluorescence in Situ Hybridization with DNA Probes

BARBARA TRASK AND DAN PINKEL

Biomedical Sciences Division
Lawrence Livermore National Laboratory
Livermore, California 94550

I. Introduction

Fluorescence *in situ* DNA hybridization (FISH) can be used to label fluorescently specific nucleic acid sequences in cells or chromosomes (Bauman *et al.*, 1980; Langer-Safer *et al.*, 1982; Landegent *et al.*, 1984; Hopman *et al.*, 1986a,b). The FISH procedure can be used to reveal the location of these sequences and to quantify their copy number. The advantages of FISH over *in situ* hybridization with radioactively labeled probes include spatial resolution, convenience, and speed. In addition, FISH techniques approach the sensitivity of autoradiographic techniques. Several laboratories have reported the successful localization of single-copy sites of specific sequences using probes of 2–5 kilobase pairs (kbp) (Lawrence *et al.*, 1988; Pinkel *et al.*, 1988).

The FISH procedure can be broken down into several basic steps:

1. Denaturation of target nuclei or chromosomes to separate the double-stranded DNA into single strands.
2. Addition of denatured, chemically labeled DNA probe.
3. Incubation of target and probe molecules under proper conditions (temperature, buffer, time) to allow efficient and specific hybridization or reannealing of probe to complementary sequences in target.
4. Washing to remove mismatched and unhybridized probe molecules. In the procedure described here, the salt and formamide concentrations of the wash fluids are the same as for the hybridization.

383

The elevated temperature removes probe that is bound with poor homology.

5. Immunofluorescent detection of chemical labels on probe molecules.

In addition to these basic steps, other steps can be added to improve the accessibility of probe and detector molecules to the target DNA (proteinase K). The buffers and procedures are designed to minimize nonspecific binding of probe and detector molecules (RNase treatment, milk in buffer, for example).

During the 1980s, several schemes for probe modification and hybridization detection were described. These include (1) biotinylated probe detected with fluorescently labeled avidin or antibiotin antibody (Langer-Safer et al., 1982); (2) aminoacetylfluorene (AAF)-modified probe detected with an anti-AAF antibody (Landegent et al., 1984); (3) sulfonate-modified probe detected with an antisulfonate antibody (kits for this procedure are available from Organice Ltd., Yavne, Israel, and TMC Bioproducts, Rockland, ME); (4) mercurated probe detected by a sulfhydryl group linked to either a fluorescent ligand or a hapten, which is detected with an antihapten antibody (Hopman et al., 1986a,b); (5) digoxigenin-labeled probes detected with an antidigoxigenin antibody (kits available from Boehringer-Mannheim, Inc., Mannheim, FRG). In each case, the primary antibody is detected with a fluorescently labeled antiimmunoglobulin antibody.

Biotin and digoxigenin moieties can be introduced in the probe via nick translation or random-primed polymerase reactions. Aminoacetylfluorene, sulfonate, and mercury can be introduced in the probe via simple chemical reactions. The fluorochrome attached to avidin or antibody can be varied. We have had most success with fluorescein and Texas red in terms of signal

---

FIG. 1. Fluorescence *in situ* hybridization to metaphase spreads and interphase cells on slides. These hybridizations were done with biotin-labeled probes at 37°C. The bound probes were detected with avidin–fluorescein, and the signals in (A) and (B) were amplified using biotinylated antiavidin antibodies and a second layer of avidin–FITC. The DNA has been counterstained with propidium iodide in (A) and (B). These black-and-white photos were produced from color slides. (A) Cells from UV20HL21-27, a human–hamster somatic cell hybrid that has retained human chromosomes 8 and 12, hybridized with human genomic DNA (data from Pinkel et al., 1986); (B) human cells from a peripheral blood lymphocyte culture, hybridized with a chromosome-specific repetitive probe for chromosome 15 (pD15Z1, Higgins et al., 1985); (C) cells from ADE-C, a Chinese hamster cell line, hybridized with a set of cosmids containing in total 270 kb of the genome including and surrounding the Chinese hamster dihydrofolate reductase gene (Looney and Hamlin, 1987). This cell line has two copies of this region. Unlabeled Chinese hamster genomic DNA was added to the hybridization mixture in (C) to block binding of repetitive sequences in the cosmids to the target cells.

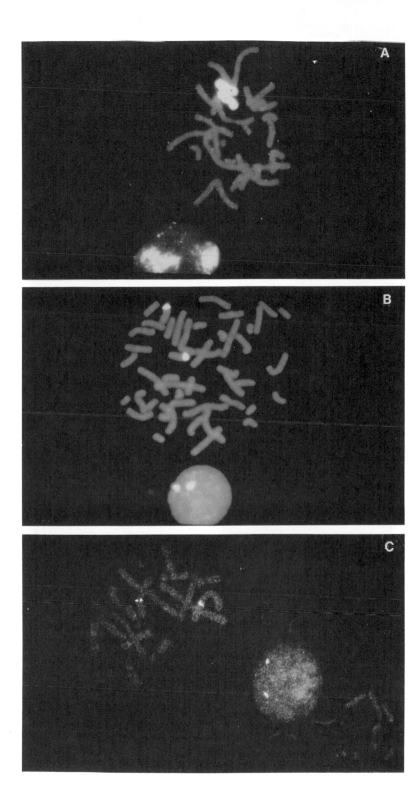

intensity and resistance to bleaching, and less success with phycoerythrin, presumably because of its large size. The AAF and biotin systems or mercury and biotin systems can be used simultaneously to label two DNA sequences with different fluorochromes (Hopman *et al.*, 1986b; Trask *et al.*, 1988).

A procedure for fluorescence hybridization to chromosomes or nuclei on slides is given in protocol 1 (Section II,A). Typical results are shown in Fig. 1. This procedure (Pinkel *et al.*, 1986) incorporates several modifications and simplifications into procedures for *in situ* hybridization developed

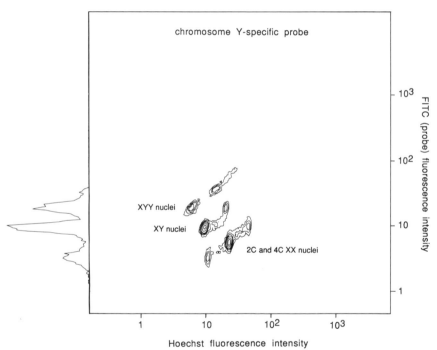

FIG. 2.   Y probe fluorescence from four different human cell types: Bivariate distribution of probe-related fluorescence versus Hoechst (DNA) fluorescence intensity of a mixture of nuclei from normal male (labeled XY), diploid female (2C XX) and tetraploid female (4C XX) human lymphoblasts, and 47, XYY human fibroblasts. Nuclei from these cell types were mixed before being hybridized in suspension to pY3.4A, a chromosome Y-specific repetitive sequence probe labeled with AAF. After hybridization, bound probe was labeled immunofluorescently with FITC, and the nuclei were stained for total DNA content with Hoechst 33258. The fluorescence intensities of ~ 10,000 nuclei were quantified using a dual-beam flow cytometer, accumulated in list mode, and displayed using a 256 × 256 channel matrix. Contour lines indicate the number of events in each channel. The distribution along the ordinate represents the FITC fluorescence distribution of the G_1 subpopulations of 46, XX, XY, and XYY nuclei. Data from Trask *et al.* (1988).

by Mary Harper (Harper *et al.*, 1981) and by David Ward and colleagues (Langer-Safer *et al.*, 1982). A technique is given for the simultaneous detection of two probes (one AAF-labeled and one biotin-labeled) using two different fluorochromes.

The AAF and biotin procedures have been adapted to label nuclei in suspension for quantification of bound probe in a flow cytometer (Trask *et al.*, 1985, 1988). The procedure for suspension labeling is given in protocol 2 (Section II,B), and results using this technique are shown in Fig. 2.

II. Protocols for Fluorescence *in Situ* Hybridization

A. Protocol 1: Hybridization on Slides

The procedure currently used in our laboratory for fluorescence hybridization is described in full here. With highly repetitive probes, good results are also obtained without RNase or proteinase K treatment (steps 2 and 3), with the hybridization incubation shortened to 2 hours (signals are visible after 10 minutes incubation), and with reduced washing steps.

1. SLIDE PREPARATION

Cells are fixed in several changes of methanol–acetic acid (3 : 1) and dropped on clean slides. The slides can be stored under N_2 gas at $-20°C$. The slides may be artificially aged by baking several hours at 65°C. The necessity for slide aging may vary with cell type.

Locate the best area on slide using low-power phase-contrast microscopy. The area where cells have been dropped can also be seen by breathing lightly over the slide. Mark the back of the slide with a scribe to indicate roughly where the boundaries of the coverslip should be. The hybridization and multiple cytochemical reactions are carried out under coverslips to minimize the amount of reagents required.

2. RNASE TREATMENT

Apply 50 μl of RNase (100 μg/ml in 2× SSC; see Section III for composition of the solutions) to each slide and cover with a coverslip (22 × 40–50 mm). Place slides in a moist chamber at 37°C for 1 hour. Rinse in three 2-minute changes of 2× SSC (pH 7) at room temperature (RT), dehydrate in an ethanol series (70, 90, 100%), and dry with an air jet.

3. Proteinase K and Paraformaldehyde Treatment

For single-copy) hybridization, the slides can be treated with proteinase K and paraformaldehyde. The proteinase K solution (0.3–0.6 μg/ml in 20 mM Tris-HCl (pH 7.5), 2 mM CaCl$_2$) is preincubated for 2 hours at 37°C before use. The amount of proteinase K needs to be adjusted for enzyme lot and cell type. The range given has been found suitable for methanol–acetic acid-fixed lymphocyte preparations. The proteinase K concentration is high enough that without paraformaldehyde fixation, the chromosomes would be destroyed in step 4. Incubate slides for 7.5 minutes at 37°C in proteinase K solution. Rinse slides in three 2-minute changes of 2× SSC at RT. Fix slides in 4% paraformaldehyde for 10 minutes at RT. Wash in 2× SSC.

4. Denature Target DNA

Immerse slides in denaturing solution (70% formamide, 2× SSC, pH 7) for 2 minutes at 70°C. Fresh denaturing solution is made weekly and is stored at 4°C between uses. A RT slide put into a 50-ml Coplin jar will cause the temperature to drop by ~1°C, so do only a few slides at a time. Transfer the slide quickly to chilled 70% ethanol for 1 minute. Agitate to rinse off the denaturing solution. Continue dehydration in 80, 90, 100% ethanol, and dry with an air jet.

5. Prepare Hybridization Mixture

We make several "Master Mixes" for hybridization at different stringencies. These solutions are stored at −20°C and can be used for several months. Typical recipes are as follow.

MM1:
    5 ml Formamide
    1 g Dextran sulfate
    1 ml 20 × SSC
MM2.1:
    5.5 ml Formamide
    1 g Dextran sulfate
    0.5 ml 20 × SSC

These solutions are heated to 70°C for several hours to dissolve the dextran sulfate, and then cooled. The pH is adjusted to 7.0, and the volume is brought to 7 ml. This volume is 70% of that used in the final hybridization mix. The remaining 30% of the volume is filled with probe, carrier DNA, and water if needed. Used in this way, MM1 gives a hybridization mix that is 50% formamide–10% dextran sulfate–2× SSC, while MM2.1 gives 55% formamide–10% dextran sulfate–1× SSC. The melting temperature of DNA is ~8°C lower in MM2.1 than in MM1. [MM1 can alternatively be made by mixing 5 ml formamide, 1 ml 20× SSC, and 2 ml 50% dextran sulfate (in H_2O).]

We tend to do all hybridization at 37°C and control stringency with the composition of the hybridization mix. MM2.1 has been found to yield good specificity with a broad sampling of chromosome-"specific" repetitive probes. These same probes, when hybridized in MM1, frequently hybridize to many chromosomes. A typical hybridization mix for a 2 × 2 cm area on a slide might be: 7 μl MM + 1 μl carrier DNA (stock = 500μg–10 mg/ml) + 2 μl probe mixture (diluted if necessary) = 10 μl.

Typical probe concentrations are 2 μg/ml for labeled genomic DNA and 0.4 μg/ml for chromosome-specific repetitive sequences. The biotinylated probe is prepared via a nick translation reaction using a BRL kit with biotin-11-dUTP in the reaction mixture in place of thymidine (see Appendix I, Section IV,A). Approximately 20% of the thymidines in the probe are replaced with biotinylated nucleotide. Probes labeled with AAF are produced according to the procedure given in Appendix II (Section IV,B).

The carrier DNA can be herring sperm DNA sonicated to 400-bp fragments for chromosome-specific repetitive probes. The hybridization mix is heated to 70°C for 5 minutes to denature the probe, and is cooled quickly on ice. If the biotinylated probe contains unique and repeated sequences and visualization of the unique sequences is desired, the carrier DNA is sonicated genomic DNA, which reanneals to and blocks hybridization of repetitive sequences to target cells. In this case, probe and carrier are denatured and prehybridized at 37°C for ≤1 hour before being applied to slides.

6. HYBRIDIZATION

Apply 2–3 μl hybridization mix per 1 cm^2 of coverslip. Thus, an 18-mm-square coverslip requires 10 μl of the mix. Apply the coverslip, and gently push out air bubbles. Seal the edges with rubber cement. Put the slides in a moist chamber and incubate overnight at 37°C (or minutes to hours for high-copy number probe) or at higher temperatures for higher stringency. Chromosome-"specific" repetitive probes give hybridization signals very rapidly. Thus, if the hybridization mix is placed on an RT slide, it may spend a substantial time before reaching the 37°C hybridization temperature. Binding to secondary target sites has a chance to occur, and this tends not to be reversible. We recommend doing step 6 on a slide warmer or in a 37°C environment (e.g., heated glovebox) to obtain optimal specificity from the probes.

7. WASH STEPS

Remove slides from incubator and peel off rubber cement with forceps. Do not allow the slide to dry at any point from now on. The resulting salt crystals cause nonspecific binding of the biotin detection reagents. Wash slides in three 3-minute changes of 50% formamide–2× SSC (pH 7) at 45°C. Agitate periodically. The coverslips will slide off easily in the first wash. Then wash in three 2-minute changes of 2× SSC at the same temperature. Place slides in PN buffer for storage (see Section III for PN composition).

8. FLUORESCENT STAINING OF BIOTINYLATED PROBES

a. Blocking. Remove a slide from PN buffer in which it has been stored, drain, and blot excess liquid from the edge. Apply PN buffer containing 5% nonfat dry milk (~5 μl/cm^2 of coverslip). Allow the slide to sit at RT for ~5 minutes. This is a blocking step designed to decrease background binding of avidin to the slide; the need for it has not been rigorously established.

b. Avidin–FITC Incubation. Remove coverslip and drain the excess liquid from the slide. Apply avidin–fluorescein isothiocyanate (FITC) (5 μg/ml in PN buffer with 5% milk; PNM) at ~5 μl/cm^2. The slide is then incubated for 20 minutes in a moist chamber at RT. The amount of avidin used here is in excess, so the dilution that occurs as it is applied to the wet slide is not critical. This concentration is sufficient even for the amplification steps (Section II,A,8,d), where there are many times more binding sites available. Alternatively, we have made staining tubs and slide carriers that allow processing of 25 slides at a time. The tub is filled with 90 ml of the avidin solution in PNM with 0.01% sodium azide and is stored at 4°C when not in use. The avidin and antibody in step 8,d can be stored for weeks this way. Each tub can process several hundred slides before the avidin is depleted.

c. Washing. Wash the slides in three changes of PN buffer at 45°C for 2 min each. Agitate occasionally. The slides can now be viewed (skip to step 11), or one can proceed to step 8,d to amplify the signal.

d. Amplification. Amplification of probe fluorescence is accomplished by applying a biotinylated goat–antiavidin antibody followed by another layer of avidin–FITC. This can be done even after observation of the slide. In this case, the slides are washed in several changes of PN buffer. We have "refreshed" old slides up to several weeks after initial observation. To apply the antibody, drain the excess liquid from the slide. Add PN buffer containing 5% milk (5 μl/cm^2 coverslip), cover with coverslip, and leave at RT for 5 minutes. Remove the coverslip, drain the excess liquid, and add the antibody solution (PN buffer, 5% milk; 5 μg/ml antibody). Incubate at RT for 20 minutes. Wash as in step 8,c in PN buffer. Add a second layer of avidin–FITC by repeating steps 8,a–c. This step may be repeated.

9. FLUORESCENT STAINING OF AAF-LABELED PROBES

a. Blocking. Remove a slide from PN buffer in which it has been stored, drain and blot excess liquid from the edge. Apply phosphate-buffered saline (PBS)–0.1% Tween containing 2% normal goat serum ($\sim$5 μl/cm^2 of coverslip). Allow the slide to sit at RT for 5 minutes. This is a blocking step designed to decrease background binding of immunofluorescent reagents to the slide; the need for it has not been rigorously established.

b. Anti-AAF Incubation. Remove coverslip and drain the excess liquid from the slide. Apply anti-AAF (provided by Dr. Robert Baan, Rijswijk, The Netherlands) at 1:750 in PBS–Tween–NGS at $\sim$5 μl/cm^2. The slide is then incubated for 30 minutes in a moist chamber at 37°C.

c. Washing. Wash the slides in three changes of PBS–0.1% Tween at RT for 5 minutes each. Agitate occasionally.

d. Goat Antimouse IgG–FITC Incubation. Remove the coverslip, drain the excess liquid, and apply the antibody solution: goat antimouse IgG–FITC (1:300–1:1000 depending on supplier) in PBS–0.1% Tween containing 2% NGS. Incubate at 37°C for 30 minutes. Wash as in step 9,c in PBS–Tween.

10. TWO-COLOR LABELING

For two-color labeling after simultaneous hybridization with two probes, one AAF-labeled and one biotin-labeled, incubate slides in 1:750 dilution of anti-AAF and 5 μg/ml streptavidin–Texas red (Molecular Probes) in PN buffer containing 5% nonfat dry milk. Incubate at 37°C for 30 minutes. Wash as in step 8,c. Apply FITC-conjugated second-layer antibody and wash as in step 9,d.

11. FLUORESCENCE MICROSCOPY

To view total DNA and probe fluorescence separately, we use DAPI (4′, 6-diamidino-2-phenylindole) as a counterstain, since it can be excited independently from fluorescein. To see total DNA and probe simultaneously, we use the red fluorescent DNA dye, propidium iodide (PI).

However, if probe fluorescence is dim, PI fluorescence may be so bright that it overwhelms the fluorescein. The DNA is counterstained by applying DAPI or PI to the slides in an antifade solution (0.25 μg/ml or <2.5 μg/ml, respectively; see Section III for antifade recipe). For clearest images with microscope, it is important to have only a very thin layer of material under the coverslip. Therefore, the slide is drained well, but not dried, and the antifade is applied (1.5 μg/cm^2). The slide can be viewed immediately. Each field can be observed for many minutes before substantial fading of the fluorescence occurs. Fluorescence due to DAPI (>425 nm) is viewed with ultraviolet (UV) excitation (360–370 nm); PI and fluorescein fluorescence (>515 nm) is viewed with excitation wavelengths 475–495 nm.

12.　Storage

Slides can be stored for several days with coverslips on, if drying of antifade solution is prevented, or they can be stored several weeks in a Coplin jar containing PN buffer at 4°C.

B.　Protocol 2: Hybridization to Nuclei in Suspension

1.　Nuclei Isolation

Cells are collected from suspension cultures or are removed from plates by trypsinization. Nuclei can be isolated using the MgSO$_4$ chromosome isolation procedure (van den Engh et al., 1986, see Trask and van den Engh, Chapter 34, this volume) at 5×10^6 cells/ml isolation buffer. Nuclei are treated with RNase as indicated in the procedure. Nuclear suspensions are vortexed rather than syringed to release nuclei, and are not stained. Alternatively, some cell types simply can be washed with PBS before ethanol fixation. Pool cells to achieve $4–5 \times 10^6$ cells/ml.

2.　Nuclei Fixation and Acid Treatment

In 50-ml tube, add cold 100% ethanol to isolated nuclei or PBS-washed cells while vortexing to achieve 70% final concentration. Let mixture stand on ice for 10 minutes. Centrifuge at 150 g for 10 minutes at 4°C. Add three times the original nuclei volume of 100% cold ethanol while vortexing pellet. Let mixture stand on ice for 10 minutes. Centrifuge again. Resuspend pellet in one-half the original nuclei volume of 0.1 N HCl–0.5% Triton X-100. Let stand at RT exactly 10 minutes. Fill the 50-ml tube with

IBM–0.25% Triton X-100. (IBM is 50 mM KCl, 10 mM MgSO$_4$, 5 mM HEPES, pH 8.0.) Centrifuge again. Repeat IBM–Triton X-100 wash. (Nuclei to be viewed by fluorescent microscopy and not analyzed in flow can be optionally fixed with paraformaldehyde: resuspend nuclei in original volume 2× SSC–0.1% Tween-20. Add an equal volume of 2% paraformaldehyde in 1× PBS–50 mM MgSO$_4$. Let stand 10 minutes at RT. Fill tube with IBM–Triton X-100. Centrifuge.) Resuspend nuclei in IBM–Triton X-100 at 10^8/ml. Count in hemocytometer after 50-fold dilution in IBM–Triton X-100 plus 2 μg/ml Hoechst 33258. Suspension should consist primarily of single, intact nuclei.

3.   HYBRIDIZATION IN SUSPENSION

1. Prepare hybridization mix:
   5 parts formamide
   1 part 20× SSC
   2 parts 50% dextran sulfate
   Bring pH to 7.0.
   (This can be stored at this point in freezer.)
   Add 1 part 10 mg/ml herring sperm DNA.
2. Mix 1 μl nuclei suspension (10^8/ml) with 18 μl hybridization mix. This can be done in bulk, and from this pooled set of nuclei 19 μl can be put in each 1.5-ml Eppendorf tube of the experiment. Each tube will then contain 10^5 nuclei.
3. Add probe to each tube: 100 ng/tube for AAF-labeled probes, 1–2 μl (or 20–40 ng/tube) for biotin-labeled probes. Probes are labeled with biotin using a commercial nick translation kit (Appendix I, Section IV, A). Probes labeled with AAF are produced using the procedure described by Landegent *et al.* (1984) (Appendix II, Section IV,B).
4. Denature nuclei and probe together by placing tubes at 70°C for 10 minutes.
5. Do not quench suspension on ice, but bring tubes quickly to 37°C or higher temperature depending on probe for optimal stringency. Incubate overnight.

4.   POSTHYBRIDIZATION WASH PROCEDURE

Add 1.25 ml 50% formamide–2× SSC (pH 7.0) at 42°C to each tube. Let stand at 42°C for 10–15 minutes. Mix by vortexing occasionally. Bring tubes to RT. Add 100 μl blood cells treated with dimethylsuberimidate (DMS; see below, stock solution, 10^7/ml). Mix, then centrifuge at 150 g at RT for 10 minutes. Flick tube to loosen pellet. Add 1.25 ml 2× SSC (pH 7.0) at 42°C, but let tubes stand at RT for 10–15 minutes. Centrifuge as before. Add 1.25 ml IBM–Triton X-100. Let stand at RT 5 minutes. Centrifuge.

DMS-Treated Erythrocytes

Suspend washed erythrocytes, from which serum and white blood cells have been removed by centrifugation, in physiological salt at a concentration of 10^8/ml. Treat with DMS three times. For each treatment, K_2CO_3 and DMS (Pierce Chemical Co.) are added to the erythrocytes from a 5× concentrated stock solution, mixed immediately before use. Final K_2CO_3 concentration is 20 mM. Final DMS concentrations for the three treatments are 3 mM, 10 mM, and 10 mM. (Additional adjustment of the pH to 9–10 with 100 mM K_2CO_3 may be required during the last two treatments.) After 15 minutes at 25°C, the pH is adjusted from 10 to 8 after each DMS treatment by the addition of 50 μl of 100 mM citric acid per 1 ml of suspension. The fixed erythrocytes are then centrifuged and resuspended in 2 × SSC at 10^8/ml. They can be stored at 4°C after the addition of 0.1% sodium azide for at least a year.

5.   FLUORESCENT DETECTION OF AAF PROBES

Add 200 μl PBS containing 0.05% Tween and 2% normal goat serum (NGS) to pellet. Vortex gently to mix. Let stand 10 minutes at RT. Add 20 μl of a 1 : 100 dilution of monoclonal antibody to AAF (4F clone, provided by Dr. Robert Baan, Rijswijk, The Netherlands). Incubate at 37°C for 45 minutes. Add 1.25 ml PBS–Tween. Let stand with occasional mixing at RT for 10 minutes. Centrifuge. Add 200 μl PBS containing 0.05% Tween–2% NGS. Vortex. Let stand 10 minutes at RT. Add 20 μl of a 1 : 100–1 : 300 dilution of goat antimouse–FITC, depending on supplier. Incubate at 37°C for 45 minutes. Add 1.25 ml PBS–Tween. Let stand with occasional mixing at RT for 10 minutes. Centrifuge.

6.   FLUORESCENT DETECTION OF BIOTIN PROBES

Add 200 μl 4 × SSC containing 0.1% Triton X-100 and 5% bovine serum albumin (BSA). After 10 minutes at RT, add 20 μl avidin–DCS–FITC (Section III) at 15 μg/ml. Incubate 30 minutes at 37°C. Wash once in

1.5 ml 4× SSC–0.1% Triton X-100 and once in 1.25 ml IBM–Triton X-100. Let stand at RT 10–15 minutes with occasional vortexing. Centrifuge.

7. FLUORESCENCE MICROSCOPY

Resuspend nuclei in 250 μl IBM–Triton X-100 containing 2 μg/ml Hoechst 33258. Mix by vortexing. For photography, add an equal volume of antifade solution (Section III) to the nuclei on slides. See protocol 1 for filter information (Section II,A).

8. FLOW CYTOMETRY

Resuspend nuclei in 750 μl IBM–Triton X-100 containing 2 μg/ml Hoechst 33258 for flow cytometry (FCM). Nuclei are measured on a dual-beam flow cytometer in which the two beam spots are spatially and chromatically separated. The nuclei and erythrocytes are first illuminated by 200 mW UV light (multiline 351–364 nm from Spectra Physics laser model 171, Mountain View, CA) for Hoechst 33258 excitation. Hoechst fluorescence is measured through a KV418 (Schott) filter. Electronic thresholds are set to select nuclei and exclude erythrocytes and small debris from further analysis. The second laser (Spectra Physics Model 171) produces 1 W light with a wavelength of 488 nm for FITC excitation. The FITC fluorescence of each particle identified as a nucleus by its Hoechst fluorescence is collected through a 530-nm bandpass filter (DF 530/30 nm, Omega Optical Inc., Brattleboro, VT).

III. Recipes and Ingredients

20× SSC

3.0 M NaCl
0.3 M Na citrate

2× SSC

Dilute above and bring to pH 7 with 1 N HCl.

Formamide

Bethesda Research Lab Ultra Pure or IBI is used directly.

PN Buffer

0.1 M phosphate buffer, pH 8.0
(Mix 0.1 M Na_2HPO_4 and 0.1 M NaH_2PO_4 to achieve proper pH)
0.5% Nonidet P-40 (NP-40)

RNase

Stock solution: 500 μg/ml (Sigma, Ribonuclease A) in 2× SSC. Put in boiling water for 10 minutes to inactivate DNase. Freeze in aliquots suitable for 1 : 5 dilution.

Dextran Sulfate

Dissolve 25 g dextran sulfate in water and make final volume 50 ml. This takes some time; heating to 70°C helps. The solution is very viscous. Aliquots can be frozen.

Carrier DNA

Dissolve DNA (herring sperm DNA) in autoclaved water at 10 mg/ml. Shear by sonication to average size of 500 bp as determined by agarose gel electrophoresis.

Paraformaldehyde

Dissolve 2 g paraformaldehyde powder in 25 ml of 100 mM $MgCl_2$. Add several drops of 1 N NaOH and heat to 70°C to speed this process. When dissolved, add 25 ml of 2× PBS (pH 7.5).

Detection Reagents

Fluorescein–avidin (DCS grade)
Biotinylated goat antiavidin
Normal goat serum (NGS)
(Available from Vector Laboratories, Inc., Burlingame, CA)
Rabbit polyclonal or mouse monoclonal anti-AAF (obtained from Dr. Robert Baan, Medical Biological Laboratory, Lange Kleiweg, Rijswijk, The Netherlands).

Antifade (Johnson and Nogueira, 1981)

Dissolve 100 mg p-phenylenediamine dihydrochloride (Sigma P 1519) in 10 ml PBS. Adjust to pH 8 with 0.5 M bicarbonate buffer (0.42 g $NaHCO_3$

in 10 ml water, pH to 9 with NaOH). Add to 90 ml glycerol. Filter through 0.22 μm filter to remove undissolved particulates. Store in aliquots with the desired counterstains in the dark at $-20°C$. This solution darkens with time, but still remains effective. It can be used after >6 months storage.

IV. Appendixes

A. Appendix I: Nick Translation to Incorporate Biotin-11-dUTP in DNA Sequence Probes

Nick translation reagent kit available from Bethesda Research Laboratories (catalog no. 8160SB). Biotin-11-dUTP available from Enzo Inc.

Proceed according to instructions with kit and biotin-dUTP, with several minor modifications, as follows.

1. Dry down 1 μl ^{3}H-ATP in 1.5-ml Eppendorf tube using a Speedvac centrifuge or by blowing N_2 gas over tube warmed to 68°C.
2. Prepare 16°C water bath (ice bucket of water with a handful of ice).
3. Add ingredients to tube, mixing gently with pipet tip after each addition. It is important to keep all ingredients on ice. Enzyme should be kept in freezer until it is ready to be used.

   DNA (1 μg), x μl
   H_2O, $35 - x$ μl
   dATP, dCTP, dGTP nucleotide mix, 5 μl
   Biotin-11-dUTP, 5 μl (note: this is double the amount called for by BRL)
   DNase-polymerase, 5 μl

   Note that enzyme is in glycerol, and reaction mixture must be stirred well with pipet tip after all ingredients are added.
4. Spin tubes briefly in Eppendorf centrifuge. Allow reaction to proceed at 16°C for 1 hour by floating tubes in water bath.
5. Stop reaction by the addition of 5 μl EDTA solution.
6. To determine the degree of biotin incorporation, remove 1 μl of the reaction mixture and place in a vial containing scintillation fluid.
7. Prepare spin columns to remove free nucleotides from DNA according to Maniatis *et al.* (1982). Columns contain Sephadex G-50 swollen in 50 mM Tris-HCl, 1 mM EDTA, and 0.1% SDS, pH 7.5. Centrifuge columns at $\sim$1600 rpm, 1 minute at RT in a tabletop

centrifuge. Add more G-50 as needed to fill. Spin columns 4 minutes to pack.

8. Apply sample to G-50 column. Spin 4 minutes. Collect volume into sterile Eppendorf tube. Determine volume collected. DNA concentration at this point is ~20 ng/μl.

9. Transfer 1 μl of DNA solution to second scintillation vial.

10. Calculate biotin incorporation:

$$\text{Fraction substituted} = \frac{\text{cpm recovered with DNA} \times 0.67 \times 100\%}{\text{cpm in reaction} \times [\text{AT}]}$$

where [AT] is the proportion of AT base pairs in the DNA being labeled (0.6 for human DNA). The counts are corrected for the difference in volumes before and after the column separation. The formula assumes labeling 1 μg DNA with 10^{-9} mol dATP (essentially all of which is cold) in the reaction mixture. These numbers are appropriate for the BRL kit. For successful hybridization, the modification should be > 10%.

11. Store biotinylated DNA preparations in freezer. Thaw and refreeze as needed.

B.   Appendix II: AAF-Labeling Procedure[1]

1. Example is for 20 μg DNA. Procedure can be scaled down to ~ 5μg if necessary. Bring DNA to a total volume of 80 μl with 1× TE, pH 7.6 (1× TE = 10 mM Tris-Cl–1 mM EDTA).

2. Add 20 μl cold 100% ethanol.

3. Warm tube to 37°C.

4. N-Acetoxy-2-aminoacetyfluorene (N-ACo-AAF) is a potent carcinogen. Stock solution of N-ACo-AAF in water-free dimethyl sulfoxide at 10 mg/ml should be made up in carcinogen protection facility (e.g. glovebox). Stock solution should be stored at −20°C in the dark. (N-ACo-AAF is commercially available from Chemsyn Sciences Laboratories, Lenexa, KS.) Note: Dispose of all waste in proper carcinogen receptacle and perform reaction in a fume hood. Keep N-ACo-AAF stock solution closed, as much as possible, dry, and in the dark. Any moisture that enters the tube will react with N-ACo-AAF and make the stock ineffective. N-ACo-AAF is light sensitive. Do all work in subdued light.

5. Warm N-ACo-AAF stock solution to 37°C.

---

[1] From Landegent *et al.* (1984).

6. Bring N-ACo-AAF and DNA to RT. Spin solutions briefly in Eppendorf centrifuge.
7. Add 5 μl N-ACo-AAF (10 mg/ml stock) to 20 μg DNA. Keep stock tube closed as much as possible. Mix briefly. Spin briefly.
8. Incubate DNA and N-ACo-AAF together at 37°C in the dark for 1 hour.
9. Add TE to 300 μl.
10. Extract three times with phenol–chloroform–isoamyl alcohol (25 : 24 : 1) that has been saturated with TE. DNA is in top clear phase. Transfer to new tube. Leave any white residue at interface with the phenol phase. The phenol removes free, unreacted, and still carcinogenic AAF. (See Maniatis *et al.*, 1982, for details on phenol extraction.)
11. Extract twice with TE-saturated ether. DNA is in bottom phase (see Maniatis *et al.*, 1982).
12. Let tube stand open at 37°C to allow ether to evaporate after last extraction. At this point, the unreacted carcinogen N-ACo-AAF has been removed. The labeled DNA can be handled on the bench and in normal lighting.
13. Dilute DNA to 1.5 ml with TE and transfer to 15-ml polycarbonate tubes. (Large tubes prevent sample loss due to splattering and aerosol formation during sonication with some microtips. If your sonicator does not cause severe sample loss, bring AAF-labeled DNA up to 400 μl and sonicate in 1.5-ml Eppendorf tube.)
14. Sonicate using microtip three times for 40 seconds each at maximum speed. (Clean sonicator tip with 100% MeOH). Between the 40-second sonications, put the tube on ice and carefully bubble N_2 gas through a pipet (plugged with sterile cotton) into the solution.
15. Spin to collect all liquid at bottom of 15-ml tubes.
16. Divide sample among several Eppendorf tubes for ethanol precipitation. Add one-tenth volume 2.5 M Na acetate, pH 5.2. Mix well. Add 2.5 volumes cold 100% ethanol. Incubate at −20°C for at least 30 minutes. Spin in Eppendorf centrifuge at 4°C for 30 minutes. Remove supernate. There should be a very small pellet. Wash pellet in 70% ethanol. Repeat spin and supernate removal. Use Speedvac centrifuge to remove remaining ethanol from pellet.
17. Resuspend pellet in TE to achieve a DNA concentration of ~ 250 ng/μl.
18. Measure optical density (OD) of DNA at 260 nm. A solution of 50 μg/ml gives an OD of 1.0 at 260 nm.
19. Run a sample of AAF-labeled DNA on a 1% agarose gel to determine if sonicated fragments are 200–400 bp. If fragments are still large, repeat steps 12–18.

REFERENCES

Bauman, J. G. J., Wiegant, J., Borst, P., and van Duijn, P. (1980). *Exp. Cell. Res.* **128**, 485–490.

Harper, M. E., Ullrich, A., and Saunders, G. (1981). *Proc. Natl. Acad. Sci. U.S.A.* **78**, 4458–4460.

Higgins, M. J., Wang, H., Shtromas, I., Haliotis, T., Roder, J. C., Holden, J. J. A., and White, B. N. (1985). *Chromosoma* **93**, 77–86.

Hopman, A. H. N., Wiegant J., Tesser, G. I., and van Duijn, P. (1986a). *Histochemistry* **84**, 169–178, 179–185.

Hopman, A. H. N., Wiegant, J., Raap, A. K., Landegent, J. E., van der Ploeg, M., and van Duijn, P. (1986b). *Histochemistry* **85**, 1–4.

Johnson, G., and Nogueira, G. (1981). *J. Immunol. Methods* **43**, 349–350.

Landegent, J., Jansen in de Wal, N., Baan, R., Hoeijmakers, J., and van der Ploeg, M. (1984). *Exp. Cell Res.* **153**, 61–72.

Landegent, J., Jansen in de Wal, N., van Ommen, G., Baas, F., de Vijlder, J., van Duijn, P., and van der Ploeg, M. (1985). *Nature (London)* **317**, 175–176.

Langer-Safer, P., Levine, M., and Ward, D. (1982). *Proc. Natl. Acad. Sci. U.S.A.* **79**, 6633–6637.

Lawrence, J. B., Villnave, C. A., and Singer, R. H. (1988). *Cell* **52**, 51–61.

Looney, J. E., and Hamlin, J. L. (1987). *Mol. Cell. Biol.* **7**, 569–577.

Maniatis, T., Tritsch, E. F., and Sambrook, J. (1982). "Molecular Cloning: A Laboratory Manual" Cold Spring Harbor Laboratory, Cold Spring Harbor, New York.

Pinkel, D., Straume, T., and Gray, J. W. (1986). *Proc. Natl. Acad. Sci. U.S.A.* **83**, 2934–2938.

Pinkel, D., Landegent, J., Collins, C., Fuscoe, J., Segraves, R., Lucas, J., and Gray, J. (1988). *Proc. Natl. Acad. Sci. U.S.A.* **85**, 9138–9142.

Trask, B., van den Engh, G., Landegent, J., Jansen in de Wal, N., and van der Ploeg, M. (1985). *Science* **230**, 1401–1403.

Trask, B., van den Engh, G., Pinkel, D., Mullikin, J., Waldman, F., van Dekken, H., and Gray, J. (1988). *Hum. Genet.* **78**, 251–259.

van den Engh, G., Trask, B., Gray, J. W. (1986). *Histochem.* **84**, 501–508.

Chapter 38

Flow Cytometric Analysis of Male Germ Cell Quality

DONALD P. EVENSON

Olson Biochemistry Laboratories
Department of Chemistry
South Dakota State University
Brookings, South Dakota 57007

I. Introduction

The testis of a fertile mammal is characterized by rapid proliferation of germ cells that undergo unique and complex patterns of differentiation. Testicular tissue is easily dissociated into a cellular suspension that is readily amenable to flow cytometry (FCM) studies.

We describe here dual-parameter (DNA, RNA) FCM measurements of acridine orange (AO)-stained cells (Darzynkiewicz *et al.*, 1976; see Darzynkiewicz, Chapter 27, this volume) used to assess the ratio of testicular cell types present, providing an indicator of testicular function. Because of differential DNA stainability and amounts of RNA present, seven to eight distinct populations of cells can be resolved by this technique (Evenson *et al.*, 1985). This measurement is very practical for animal studies but impractical for human and animal husbandry studies because of the invasive sampling procedures, although fine-needle biopsy samples are utilized by some laboratories (Thorud *et al.*, 1981). The same AO staining and FCM measurement technique used for testicular biopsy samples is also a rapid and practical method for measuring abnormal cells in semen (Evenson and Melamed, 1983). In addition to normal sperm, semen may include immature germ cells that have staining characterisitics similar or identical to testicular cells. Also, the presence of somatic cells (e.g., leukocytes) in semen can readily be detected by this same protocol (Evenson and

401

Melamed, 1983) and distinguished from morphologically similar round spermatids.

Several studies have shown an interesting correlation between sperm cell chromatin structure and exposure to toxic chemicals (Evenson *et al.*, 1985, 1986, 1989) and also a correlation with fertility (Ballachey *et al.*, 1987, 1988; Evenson, 1986, 1989; Evenson *et al.*, 1980). The FCM measurement of chromatin structure is based on the principle that abnormal sperm chromatin has a greater susceptibility to physical induction of partial DNA denaturation *in situ*. The extent of DNA denaturation following heat (Evenson, *et al.* 1980, 1985) or acid (Evenson *et al.* 1980, 1985) treatment is determined by measuring the metachromatic shift from green luminescence (AO intercalated into double-stranded (ds) nucleic acid) to red luminescence (AO associated with single-stranded (ss) DNA; Darzynkiewicz *et al.*, 1976). Apparently acid conditions that cause partial denaturation of protamine-complexed DNA in sperm with abnormal chromatin structure do not cause denaturation of histone-complexed somatic cell DNA (Evenson *et al.*, 1986). The FCM measurement of sperm chromatin structure as described here has been termed the sperm chromatin structure assay (SCSA) to distinguish it from other AO staining protocols. This protocol has also been subdivided into SCSA$_{acid}$ and SCSA$_{heat}$ to distinguish the physical means of inducing DNA denaturation.

II. Application

The primary applications of these techniques are in the fields of environmental toxicology, animal husbandry, and human fertility.

The described techniques provide for rapid, objective assessment for germ cell toxins that interfere with cell division and differentiation. Evenson and colleagues have shown that exposure of mice to toxic chemicals caused changes in the relative ratio of testicular cell types, (Evenson, *et al.*, 1985, 1986, 1989a) appearance of abnormal cell types in epididymi (Evenson *et al.*, 1989b), and increased sensitivity to acid (Evenson *et al.*, 1985, 1986, 1989) or heat-induced denaturation (Evenson, 1986; Evenson *et al.*, 1980, 1985) of sperm DNA. In studies exposing mice to 10 different toxic chemicals, the dose–response curves of FCM-derived α_t values were very similar in shape and sensitivity to the percentage abnormal sperm head morphology curves (Evenson *et al.*, 1985, 1986, 1989). Of added interest, several studies show that sperm cells arising from stem cells exposed to stem cell-specific mutagenic chemicals maintain for at least 45 weeks chromatin structural abnormalities detectable by these FCM methods (Evenson *et al.*, 1989).

The greatest impact of the SCSA technique may be for assessment of animal and human subfertility (Evenson, 1986; Ballachey *et al.*, 1987, 1988). Studies in this laboratory (Ballachey *et al.*, 1987) have shown a correlation of $-.58$ ($<.01$) between bull sperm chromatin structure and fertility ratings (FR) derived from 2×10^6 artificial insemination services and adjusted for handling, environment, cost of semen, quality of cows, and so on. It should be noted that this study measured one to eight *random* semen samples collected over several years of time. In contrast, a heterospermic study (Ballachey *et al.*, 1988) measured aliquots of the *same* semen samples used for fertility field trials; the correlation between heterospermic index and FCM data on sperm chromatin structure was 0.94 ($<.01$). Data for the relationship between human sperm quality and fertility is limited. In addition to our original report (Evenson *et al.*, 1980), studies have assessed chromatin quality following successful therapy for leukemia and testicular carcinoma (see Evenson, 1986, for review). Perhaps the most interesting cases have been studied in infertility clinics where semen quality was shown to be excellent by the standard criteria of sperm count, motility, and morphology and yet the sperm cell chromatin had a highly abnormal susceptibility to heat denaturation, leading to the speculation that sperm chromatin quality is related to clinical infertility (Evenson, 1986). All samples obtained from men of recent proven fertility have shown homogeneous AO staining profiles; evidence to date suggests that broadly heterogeneous profiles are consistent with subfertility.

III.   Materials

A.   Acridine Orange Staining Solutions

1. AO stock solution: Chromatographically purified AO (Polysciences, Inc. Warrington, PA) is dissolved in double-distilled water to a final concentration of 1 mg/ml and kept at 4°C for several months in darkness.
2. Acid–detergent treatment solution for first step of two-step acridine orange (TSAO) staining procedure: 0.15 M NaCl, 0.1% Triton X-100 (Sigma Chemical Co.), 0.08 N HCl in double-distilled water. The solution may be stored at 4°C up to several months.
3. AO-staining solution for second step of TSAO staining procedure: Mix 370 ml of 0.1 M citric acid buffer (kept as stock at 4°C) with 630 ml 0.2 M Na$_2$PO$_4$ buffer (kept as stock at 4°C); add 372 mg Na$_{(2)}$EDTA and 8.77 g NaCl. Adjust to pH 6.0. Now, 0.6 ml of AO stock solution is added to each 100 ml of stain solution. The stain

solution, kept in a glass amber bottle until use, is made fresh bi-weekly. Other details on the staining solutions are given in Darzynkiewicz (Chapter 27, this volume).

B.   Buffers and Other Materials

1. TNE buffer: 0.01 M Tris, 0.15 M NaCl, and 1 mM EDTA, pH 7.4
2. Hanks' balanced salt solution (HBSS): GIBCO Laboratories (Grand Island, NY)
3. RNase A: DNase-free RNase. Cooper Biomedical, Inc. (Malvern, PA).
4. Nylon filters: 53 and 153 μm mesh, 1 in. diameter (Tetko, Inc., Briarcliff Manor, NY)

IV.   Cell Preparation and Staining

A.   Testis Biopsy

A testicular biopsy sample ranging in size from 1 to 10 mm^2 is placed in a Petri dish containing HBSS and resting on a plate of steel on crushed ice. The tissue is minced with a razor blade or pair of curved scissors until a cellular suspension is obtained. This suspension is filtered through a 53-μm nylon mesh filter placed between a plastic tuberculin syringe and its protective end cap (tip cut off with wire cutter). The cellular suspension is transferred to the barrel of the syringe and the plunger is used to push the cells slowly through the filter.

A 0.20-ml aliquot of suspended cells $(1 \times 10^6/\text{ml})$ is stained by the TSAO technique of Darzynkiewicz et al. (1976) by first admixing with 0.40 ml of the Triton X-100–acid solution already described. After 30 seconds, 1.2 ml of the AO staining solution containing 6 μg AO/ml is admixed, placed in the flow cytometer sample chamber, and the sample flow initiated. Flow cytometric measurement is started 3 minutes after the sample is placed on the flow cytometer. The sample is kept at 4°C throughout this procedure.

B.   Sperm Samples

1.   Animal or Human Semen

Fresh or frozen, thawed semen is diluted with TNE buffer to ~1 × 10^6 cells/ml and stained by the TSAO procedure exactly as described for testicular cells. Frozen semen samples can be prepared by diluting semen to $\sim 2 \times 10^6$ cells/ml in TNE buffer plus 10% glycerol and placing in a $-70°$ to $-100°C$ freezer. For severe oligospermic samples, undiluted semen can be used directly; the acid–detergent solution used in the first step dramatically reduces any semen vicosity (Evenson and Melamed, 1983). Semen extended in milk for use in artificial insemination may also be used directly without producing apparent artifacts; nonclarified, egg yolk citrate extender causes some background noise that may occasionally present a problem. A single freezing and thawing has no effect on sperm chromatin structure (Evenson *et al.*, 1989a) but may disrupt early spermatid germ cells; these cells may be preserved by freezing the sample with techniques used for freezing tissue culture cells.

Whole sperm or nuclei isolated and purified through sucrose gradients (Evenson *et al.*, 1985) may also be fixed in 70% ethanol or 1:1 70% ethanol/acetone and then pelleted by centrifugation, suspended, and rehydrated in TNE buffer for 30 minutes at 4°C prior to acid treatment, staining with AO and measurement by FCM. The data from fixed samples are essentially similar to that obtained on fresh material (Evenson *et al.*, 1986); however, fresh or frozen samples are preferred.

2. EPIDIDYMAL OR VAS DEFERENS SPERM

For animal studies, a specific segment of the epididymis can be surgically removed from a killed animal and minced in TNE buffer as described before (Evenson *et al.*, 1986) for testis biopsies. The vas deferens may also be excised, placed in a 60-mm Petri dish containing TNE buffer, and the sperm removed by pressing a blunt-shaped probe along the length of the organ. The resulting sperm suspensions are filtered through 153-μm nylon mesh before analysis by the SCSA.

3. SONICATION AND RNASE DIGESTION OF SEMEN CELLS

For measurements of sperm chromatin structure that are dependent on association of AO with both dsDNA and ssDNA, some, or all, samples should be sonicated and/or treated with RNase to insure against any abnormally retained RNA producing a red luminescent signal unrelated to DNA. This is considered a precaution only because little or no significant differences have been observed between sonicated, RNase-treated or untreated cells.

Sperm suspended in TNE buffer in a test tube (Falcon 3033; Becton Dickinson Labware, Lincoln Park, NY) immersed in an ice-water slurry are sonicated for 30 seconds at a setting of 50 on low power (Bronwill Biosonik IV Sonicator, VWR Scientific, Inc., Minneapolis, MN), cooled for 30 seconds, and sonicated again for 30 seconds. The $\frac{1}{2}$-in. probe is placed just above the bottom of the tube. Optimal time and power required for sperm head–tail/cytoplasm separation varies between species.

The sonicate can be measured directly or the sperm heads can be purified if desired by centrifuging through a 60% sucrose solution. Addition of 1.5×10^3 RNase units/ml to the sonicate and incubation for 30 minutes at room temperature (RT) apparently has no effect on the AO staining distribution (Evenson *et al.*, 1985). Additional incubation may cause an increased red luminescence due to possible chromatin digestion by endogenous proteases. Whole unsonicated cells may also be incubated with RNase as before with the exception that 0.1% Triton X-100 is added to the suspension to permeate the cells.[1]

V.   Instrument

Blue laser light (488 nm) excitation of AO-stained cells at a power of ≥ 35 mW is optimal. Luminescence of individual cells is measured at wavelength bands of red (> 630 nm) and green (515–530 nm). Since mature, AO-stained mammalian sperm have very little red luminescence because they lack RNA and ssDNA, the red photomultiplier tube (PMT) gain may need to be set high enough that electronic noise may result with some instruments. Ortho Diagnostics engineers changed one resistor in the two preamplifier circuit boards to reduce background noise on our Cytofluorograf II for studies on abnormal sperm chromatin structure. The electronic hardware or the software of the interfaced computer must have the capability to generate the α_t data, which is the ratio of red to total (red + green) luminescence and is a measure of the extent of acid- or heat-induced DNA denaturation *in situ*. Other technical points relevant to cell staining with AO are discussed by Darzynkiewicz (Chapter 27, this volume).

VI.   Results

A.   Testis Biopsy

Figure 1 shows a typical distribution of AO-stained mouse testicular cells with respect to their green and red luminescence. Note that seven distinct cell populations can be discerned here in contrast to the four that can be distinguished by single-parameter DNA staining. This measurement is simple, rapid, and highly reproducible. The percentage total for each population, determined with computer assistance, may be related to specificity of the perturbing agent (Evenson *et al.*, 1986).

---

[1] Please see biohazard caution following references.

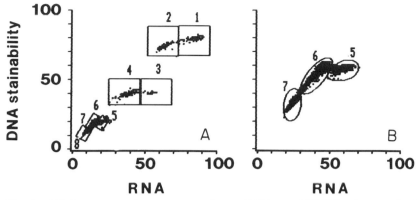

Fig. 1.   FCM cytograms of two-parameter (green/DNA vs red/RNA luminescent signals) distribution of two-step acridine orange-stained mouse testicular cells. (A) Boxes 1 and 2 correspond to 4n cells, boxes 3 and 4 to 2n cells, and boxes 5–7 to 1n cells. Region 8 corresponds to the staining position of mature sperm. (B) Computer software enlargement of the 1n cell data points that provides an increased resolution of the round, elongating, and elongated spermatids.

B.   Semen Samples Measured at PMT Gain Settings Used for Testicular Cells

Environmental toxins and stress may cause immature germ cells to be released from the testis prematurely. Measurement of germ cells from the epididymis, vas deferens, or ejaculated semen provides an easy determination of this potential abnormality. Figure 2 shows a two-dimensional frequency histogram distribution of AO-stained human semen samples. Figure 2A shows the typical narrow distribution for samples from fertile males. Figure 2B shows a majority of cells with staining characteristics of round spermatids and a minority of cells having characteristics of abnormal sperm, both populations being confirmed by light microscopy. Figure 2C shows a very heterogeneous population of cells ranging from near normal sperm to diploid cell staining characteristics. The primary feature of the protocol that allows complete resolution between mature sperm, elongated spermatids, and round spermatids, even with single-parameter DNA-staining measurements, is the use of acid extraction in the first step. Acid-extracted round spermatids and mature sperm have a 3.2- and 1.2-fold increase of DNA stainability, respectively, relative to non-acid-extracted, AO-stained cells. Similar differences are observed with other DNA dyes (see Evenson, 1989).

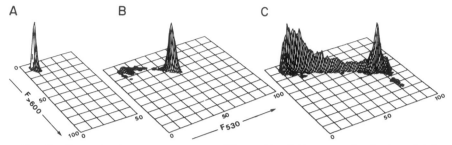

FIG. 2. Computer-drawn two-parameter (F_{530} vs $F_{>600}$) histogram distribution of AO-stained human semen cells. (A) Control sample obtained from a healthy fertile donor. (B) Sample obtained from an 18-year-old patient with stage III embryonal cell testicular carcinoma, 6 months after a unilateral orchiectomy and 3 months post-VAB 6 induction chemotherapy. (C) Sample obtained from a 28-year-old patient with stage I testicular cancer with teratoma, embryonal cell carcinoma, and seminoma, 11 months after unilateral orchiectomy. No normal mature sperm were seen and diploid cells were present. From Evenson and Melamed (1983).

C. Sperm Chromatin Structure Assay (SCSA)

The SCSA has proved useful for fertility assays and toxicology studies. Figure 3 shows the relation between fertility of human males and resistance of sperm nuclear DNA to acid-induced denaturation. The fluorescence distribution for DNA stainability is heterogeneous because of a known artifact (see discussion in Evenson, 1989), which has no effect on the α_t distribution of interest in this technique. Whereas early work (Evenson *et al.*, 1980, 1985) used heat to induce DNA denaturation in isolated nuclei, several later studies (Evenson *et al.*, 1986) showed that acid-induced denaturation produced the same results and is the method of choice in that it is more efficient, less time-consuming, and allows for the use of smaller samples.

Early studies used the mean of α_t ($\bar{x}\ \alpha_t$) values to quantitate the level of abnormality. While this may be of value, especially for human samples that typically are quite heterogeneous, most of our recent studies have utilized the variation of α_t as an indicator of abnormality. The standard deviation (SD) of α_t (SD α_t) obtained from computer analysis has been the most useful and most closely related to known fertility ratings (Ballachey *et al.*, 1987, 1988).

The SD α_t is very sensitive to small changes in chromatin structure, and studies using this parameter require very precise repeat settings of the PMT values for comparative measurements done on different days. The PMT values are set so that the $\bar{x}$ green luminescence of fertile sperm with high resistance to DNA denaturation is at about 50/100 channels and the $\bar{x}$ red luminescence is at about 15/100 channels. The most ideal situation is to measure all experimental samples at one time period; however, careful repeat settings of the red and green PMTs allow measurements of com-

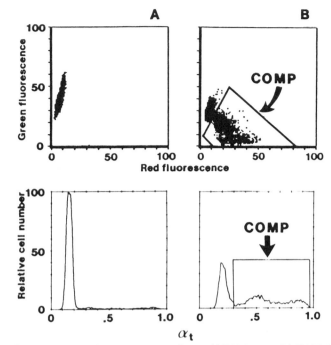

FIG. 3. FCM cytograms of two-parameter (green/dsDNA vs red/ssDNA luminescent signal) distribution of sperm from (A) a fertile human control and (B) a patient from an infertility clinic. The box marked COMP shows the cells outside the main population with an abnormal chromatin structure. The α_t distribution shows the extent of the abnormality.

pared samples over an extended period of time. The most precise repeat settings are obtained by using aliquots of a single semen sample that demonstrates heterogeneity of α_t. A semen sample is identified and then diluted with TNE buffer + 10% glycerol to a working concentration of 2×10^6 cells/ml. Several hundred aliquots (250 μl) of this dilution are placed into small snap-cap vials and immediately frozen at $-70°$ to $-100°$C. These samples are used to set the red and green PMTs to the same α_t index ($\bar{x}$ and SD α_t) as used previously. We typically measure a new calibration standard after every 5 or 10 samples to ensure that instrument settings have not drifted. For instrument calibration, a similar biological standard is preferred to fluorescent beads, since it also serves as a monitor for sample staining.

For comparison of fertile control samples with samples of questionable fertility, the following ratio provides an index of abnormality (AI) for the SCSA procedure:

$$AI = \frac{SD \ \alpha_t \ \text{subfertile}(?)}{SD \ \alpha_t \ \text{control}(s)}$$

At this time, it is difficult to define what values are incompatible with normal fertility (Evenson, 1986), since the interpretation of α_t parameters is still being explored. However, from our experience of measuring thousands of sperm samples derived from a variety of mammals, some of which had known fertility potential, the evidence suggests that a broadly heterogeneous pattern is indicative of either reduced or absent fertility.

ACKNOWLEDGMENTS

This work was supported by USDA grants 88-37242-4039 and 86-CRCR-1-2170. It is Publication No. 2137 from South Dakota State University Experiment Station.

REFERENCES

Ballachey, B. E., Hohenboken, W. D., and Evenson, D. P. (1987). *Biol. Reprod.* **36,** 915–925.

Ballachey, B. E., Saacke, R. G., and Evenson, D. P. (1988). *J. Androl.* **9,** 109–115.

Darzynkiewicz, Z., Traganos, F., Sharpless, T., and Melamed, M. R. (1976). *Proc. Natl. Acad. Sci. U.S.A.* **73,** 2881–2884.

Evenson, D. P. (1986). *In* "Clinical Cytometry" (M. Andreeff, ed.), pp. 350–367. New York Academy of Science, New York.

Evenson, D. P. (1989). *In* "Flow Cytometry: Advanced Research and Clinical Applications", (A. Yen, ed.), Vol. 1, pp. 217–246. CRC Press, Boca Raton, Florida.

Evenson D. P., and Melamed, M. R. (1983). *J. Histochem. Cytochem.* **31,** 248–253.

Evenson, D. P., Darzynkiewicz, Z., and Melamed, M. R. (1980). *Science* **240,** 1131–1133.

Evenson, D. P., Higgins, P. H., Grueneberg, D., and Ballachey, B. (1985). *Cytometry* **6,** 238–253.

Evenson D. P., Baer, R. K., Jost L. K., and Gesch, R. W. (1986). *Toxicol. Appl. Pharmacol.* **82,** 151–163.

Evenson, D., Baer, R. K., and Jost, L. K. (1989a). *J. Environ. Mol. Mutagen.* **14,** 79–89.

Evenson, D. P., Janca, F. C., Jost, L. K., Baer, R. K., and Karabinus, D. S. (1989b). *J. Toxicol. Environ. Health* **28,** 81–98.

Thorud, E., Clausen, O. P. F., and Abyholm, T. (1981). *In* "Flow Cytometry IV" (O. Lareum, T. Lindmo, and E. Thorud, eds.), pp. 175–177. Universitetsforlaget, Oslo.

NOTE ADDED IN PROOF. *Biohazard caution*: Since sonication as described above produces aerosols, we have modified our SCSA procedure for human semen which may contain infectious agents including HIV. Our current method utilizes a Branson Sonifier II, Model 450, coupled to a Branson Cup Horn (VWR Scientific, San Francisco, CA). The cup horn allows sonication of materials in a sealed test tube. Temperature of the sample is maintained by 4°C water flowing through the cup horn derived by using a peristaltic pump that drives water through a copper tube coil set in a flask containing an ice water slurry. For human samples, 0.5 ml of TNE buffer containing $\leq 1 \times 10^6$ sperm are placed into a 2-ml Corning cryogenic screw cap vial (VWR Scientific). The top end of this vial is placed into a large rubber stopper with a hole drilled through it that will hold the vial securely. This rubber stopper is then placed on top of the cup horn with the tube protruding into the cup horn and the bottom of the tube just off the bottom of the cup. Provided that the tubes are always placed the same distance into the rubber stopper, this arrangement likely provides for more repeatable sonication to the sample than holding the sample by hand. We are currently using 40 seconds of sonication pulsed for 70% of 1-second cycles at a setting of 3.0 of output power. This removes $\geq 95\%$ of heads from tails. Time and power settings may need to be varied for different sample volumes and for sperm of other species. Current work also indicates that sonication of human samples provides better results than the above RNAse procedure. All other human sampling procedures are done in a biological safety cabinet.

Chapter 39

Cell Preparation for the Identification of Leukocytes

CARLETON C. STEWART

Laboratory of Flow Cytometry
Roswell Park Memorial Institute
Buffalo, New York 14263

I. Introduction

Immunophenotyping is the term applied to the identification of cells using antibodies (Ab) to antigens expressed by these cells.[1] A monoclonal antibody (mAb) will react only with the specific epitope on an antigen, an epitope being the three-dimensional molecular structure to which the Ab binds (there may be several hundred different epitopes on a complex antigen). On the other hand, polyclonal Ab are actually the natural pool of several mAb produced within the organism, each binding to its epitope. Immunophenotyping using flow cytometry (FCM) has become the method of choice in identifying and sorting cells within complex populations. Applications of this technology have occurred in both the basic research and clinical laboratories. The current National Committee for Clinical Laboratory guidelines for FCM describe in detail the current recommendations for processing clinical samples (Landay, 1989). Single-color analysis

---

[1] Abbreviations used in this chapter: Ab, antibodies; APC, allophycocyanin; B-Ab, biotinylated antibodies; DC, DuoChrome; F(ab')$_2$, antigen-binding portion of antibody molecules; Fc, the fragment (F) of the antibody molecule that exhibits a nonvariable or constant (c) amino acid sequence within a given isotype and subclass; FcR, receptor on cells for the Fc portion of the antibody; F-GAM, fluoresceinated goat anti-mouse Ig; FL, fluorescein; FL-Ab, fluoresceinated antibodies; Ig, Immunoglobulin; mAb, monoclonal antibodies; PAB, phosphate-buffered saline containing 0.1% sodium azide and 0.5% bovine albumin; PBS, phosphate-buffered saline; PE, phycoerythrin; PE-Ab, phycoerythrinated antibodies; PE-GAM, Phycoerythrinated goat anti-mouse Ig; TR, Texas red.

411

using fluorescein-labeled antibodies (FL-Ab) has been used to immunophenotype cells.

Recently, advances in FCM instrumentation design, and the availability of new fluorochromes and staining strategies have led to methods for immunophenotyping cells with two or more Ab simultaneously (Bender *et al.*, 1989; Sneed *et al.*, 1989; Stewart, 1985; Stewart *et al.*, 1986a, b, 1988, 1989). With this progress has come the realization that the use of a single Ab can rarely suffice to identify a unique cell population, since cells within one population may have some epitopes in common with cells of another population, and so on. One can, however, take advantage of these characteristics to identify a population by the unique repertoire of epitopes that it expresses. Determining such a repertoire for any given cell population requires the simultaneous use of at least two Ab, each coupled to a marker that fluoresces at its individual wavelength.

Many problems and pitfalls can be encountered when labeling cells with Ab. These problems need to be recognized and addressed when they occur and solutions devised, if the data obtained are to be correctly interpreted. These problems, their probable solutions, and several strategies for staining cells with Ab labeled with different-color fluorochromes are the focus of this chapter.

II.   Antibodies

A brief review of the characteristics of Ab germane to their use in immunophenotyping is appropriate. Both mAb and polyclonal Ab are used to determine the immunophenotype of cells. Monoclonal Ab are derived *in vitro* from hybrids of mouse or rat origin, while polyclonal Ab are prepared *in vivo* by challenging the immune system of a suitable animal with the antigen and isolating the Ab produced from the blood.

The basic unit of all Ab comprises a heavy and light chain, the former defining the Ab isotype. As depicted in Fig. 1, the minimum functional Ab molecule consists of two heavy and two light chains linked together with disulfide bonds. The part of the molecule that contains only heavy chains is known as the Fc portion (see Table I), whereas the part that contains both heavy and light chains is the $F(ab')_2$ portion. The Fc receptor (FcR)-binding domain and the complement-binding and activation domain (Fig. 1) are in the Fc portion. Variations in the composition of the heavy chains lead to "isotypes", and there are "subclasses" within some isotypes. These variations are of practical importance because they define the reper-

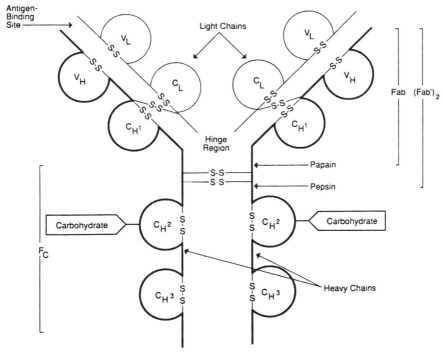

FIG. 1. Schematic representation of a monomeric Ab molecule. The Fab portion consists of the light chain and a fragment of the heavy chain. These two chains are held together by disulfide bonds. For any given Ab molecule there is a region of variable amino acid sequence (V_L) on the light chain and the heavy chain (V_H) that produces the epitope-binding site. Papain digestion produces two Fab fragments for each Ab monomer. There are also regions where the amino acid sequence is constant for Ab molecules of similar isotype and subclass. There is one constant region for light chains (C_L) and three for heavy chains (C_H1, C_H2, and C_H3). The heavy chains are held together by disulfide bonds. If pepsin is used to digest the Ig, a fragment [$F(ab')_2$] is produced containing two epitope-binding sites. The Fc fragment is that portion of the Ab posterior to the hinge region. From Pierce-Immunotechnology Catalog and Handbook, Section C (1989).

toire of cells to which the Ab will bind. Table I shows the immunoglobulins (Ig) and several of their properties that are due to variations in the Fc portion. In the $F(ab')_2$ portion, the light chains are of two varieties: κ and λ, and they are associated with the heavy chains by disulfide bonding; the designation "ab" refers to "antigen binding," and the numeral 2 tells us that there are two Ab-binding sites in the basic functional subunit. It is noteworthy that murine and rat Ab almost always have κ light chains and that most *in vitro* hybridoma-produced mAb are of the IgG or γ isotype.

TABLE I

Human Immunoglobulins[a]

Isotype and subclass	Serum concentration (mg/ml)	Complement fixing	Protein A reaction	Fc Receptor class on leukocytes			
				Mono-cyte	Neutro-phil	Eosino-phil	Baso-phil
IgG							
IgG$_1$	9	Yes	Yes	I, II, III	II, III	—	—
IgG$_2$	3	Yes	Yes	II	II	—	—
IgG$_3$	1	Yes	No	I, II, III	II, III	—	—
IgG$_4$	0.5	No	Yes	I, II	I	—	—
IgM	1.5	Yes	No	—	—	—	—
IgA							
IgA$_1$	3.0	No	No	—	II, III	—	—
IgA$_2$	0.5	No	No	—	II, III	—	—
IgD	0.3	No	No	—	—	—	—
IgE	0.00005	No	No	—	—	II	I

[a] Modified from Roitt *et al.* (1989).

For immunophenotyping using a single color, it is customary first to bind cells with a primary Ab, usually a mAb, directed to the epitope of interest. A secondary polyclonal Ab that is specific to epitopes on the primary Ab (the primary Ab therefore acting as an antigen in this reaction) and which is covalently coupled with fluorescein (FL) dye, is then bound to the primary Ab, effectively staining it and hence labeling the cell. Whenever possible the F(ab')$_2$ fragment of an affinity-purified second Ab should be used. It is of course implicit in this scheme that the second Ab does not react with the cellular epitope.

A typical second Ab might be labeled "FITC-conjugated goat F(ab')$_2$ anti-mouse IgG antibody (γ and light chain-specific, purified by affinity chromatography." This label contains much information in abbreviated form. "Goat anti-mouse IgG" means that the Ab was prepared by immunizing goats with mouse Ig. "Purified by affinity chromatography" means that the goat serum was passed over an affinity column (usually sepharose beads) to which mouse IgG was bound and, after elution from the column, the F(ab')$_2$ fragments were prepared and conjugated with fluorescein isothiocyanate ("FITC-conjugated").

Since this goat polyclonal second Ab was prepared using a mouse IgG affinity column, this preparation contains Ab that are specific for the heavy chain of mouse IgG. Because all isotypes found in serum (i.e., IgG, IgM, and IgA) have light chains, they will also bind this second Ab because they are "light chain-specific." Therefore, this polyclonal second reagent is not specific for mouse IgG. In order for a second Ab to be specific for IgG, it must have no light-chain activity. If the antibody were only heavy chain-specific, the label would read "γ-specific."

Any polyclonal second Ab contains all the isotypes found in the serum of the animal used to produce it (e.g., IgM, IgG, IgA). A simple way to determine if a second Ab to murine IgG has light-chain activity is to stain murine spleen cells with it; none should be positive. If there are positive cells, the reagent is defective and should not be used in applications where heavy-chain specificity is required.

III. Staining Cells

There are several different ways to stain cells with Ab, each having its advantages and disadvantages. The "primary Ab" is the Ab with specificity for the desired cellular epitope. The second Ab is the Ab with specificity for the epitopes on the primary Ab.

A. Primary and Second Antibody

This is the most often used method to immunophenotype cells. Cells are incubated with a primary Ab, and then they are stained using a fluorescein-ated second antibody. This method is most common because a single second Ab can be used for staining many different primary Ab. There are advantages to focusing a second Ab over directly conjugated primary Ab. The latter are more expensive and may not be available, while the second Ab provides an amplification of fluorescence that may be useful for resolv-ing cells that have only a few epitopes.

There are problems associated with this approach that may compromise the interpretation of results. Some problems are illustrated in Fig. 2. In a mixture of cells, some may have both the desired epitope and FcR (Fig. 2A), some cells may have only FcR (Fig. 2B), other cells may have only epitopes (Fig. 2C), and some cells may have neither (Fig. 2D). Only the cells with the desired epitope (Fig. 2A,C) should be identified by the mouse mAb and the second Ab described [i.e., fluoresceinated goat anti-mouse Ig (FGAM)]. When the mAb is added to the cells, it can bind to the desired epitope via the $F(ab')_2$ portion (Fig. 2A,C), or to the cell via the Fc portion (Fig. A,B). If the sample contains an equal portion of each cell type, three-fourths of the cells present will bind the mAb. Since the second Ab is a fluoresceinated $F(ab')_2$ goat anti-mouse Ig, it binds to the murine mAb (but not to any FcR, Fig. 2B.) Since one-third of the

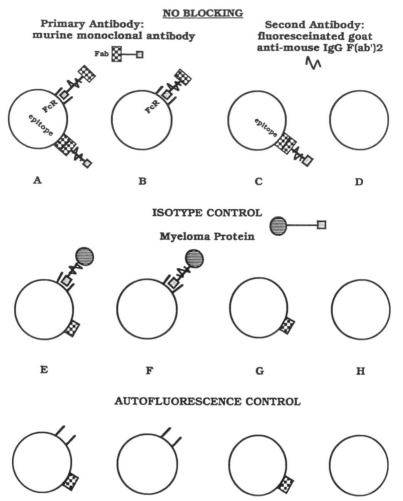

Fig. 2. Indirect immunofluorescence staining. The primary Ab is shown as a stick symbol that binds to cells with epitopes shown by the patterned rectangle. The Fc portion binds to its receptor, shown by the square. The second Ab binds to the primary Ab. The stick symbol for the isotype control has the same shape for the FcR but a circle for the F(ab')₂ portion for which there are no epitopes on the cells.

cells have only FcR, these cells are inappropriately labeled and counted as epitope-positive, thereby overestimating the percentage of epitope-positive cells by one-fourth.

    To account for this problem, the common practice is to use an "isotype" control. This control consists of the same cells, incubated with a myeloma protein having no epitope specificity, but having the same isotype and subclass as the primary Ab. As shown in Fig. 2, this myeloma protein is presumed to bind to cells having FcR (Fig. 2E,F). The second Ab is added, as before, to stain cells that have bound the isotype protein. The percentage of positive cells revealed by the isotype protein is then subtracted from the percentage obtained using the primary Ab. This procedure leads to an underestimate of the epitope-positive cells because one-quarter of the cells express both FcR and epitopes. These epitope-positive cells would be subtracted from the total positive. Thus, isotype controls may not be appropriate in some instances. The autofluorescence control should always be analyzed. It contains no Ab and is otherwise processed the same way as the other samples. This control provides a baseline to determine the minimum fluorescence above which positive cells are identified. Ideally, the isotype control and autofluorescence control would give identical results.

There is yet another problem when a myeloma protein is used as the isotype control. The myeloma protein, like the mAb, is derived from neoplastic cells. These cells produce structurally abnormal Ig molecules. Because they are abnormal, myeloma proteins of the same isotype and subclass may behave differently in their ability to bind to cells bearing FcR. This abnormal binding is illustrated in Fig. 3. Here, murine peritoneal

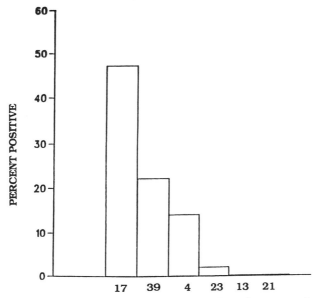

FIG. 3. Myeloma protein binding to macrophages. Macrophages were incubated with 10 μg of each myeloma protein for 15 minutes, washed and stained with F-GAM.

macrophages were stained for 10 minutes with six different IgG_1 myeloma proteins. IgG_1 was selected for this illustration because it has a weak affinity for the IgG_{2b} FcR. Two myeloma proteins did not bind to the macrophages; others bound to only a small percentage of them, whereas others stained a high percentage of the macrophages. Which one should be chosen for the isotype control protein? Which one will behave like the primary mAb antibody? The best way to answer these two questions is to find a better labeling system that does not exhibit these problems.

The Fc binding of Ab, as well as other nonspecific binding, can be blocked by first treating the cells with normal Ig. For the example shown in Fig. 4, a murine mAb and FGAM was used. Because the second Ab is derived from the goat, the blocking Ig is also derived from the goat. The cells are incubated for 10 minutes with an excess of goat Ig where the Ig binds the FcR (Fig. 4A,B). Without washing the cells, the primary murine mAb is added, whereupon it binds only to its epitopes (Fig. 4E,G). Because the labeled second Ab was made against murine IgG in the goat, it binds to the desired mAb but not the blocking goat Ig. To be sure that the block is effective, the "isotype control" Ab can be used in place of the primary Ab. Even if it did bind to unblocked cells, it will not bind to the blocked ones. Goat Ig was used in this example to block with because our second Ab was made in the goat. The general rule is: For indirect staining, block with the Ig fraction from the same species in which the second antibody was made.

It is best to use the Ig fraction rather than serum to block with because the serum may not contain sufficient Ig of the correct isotype and subclass to block FcR effectively. We have found the effective blocking concentration is 30 μg IgG per 10^6 cells in a volume of 100 μl. This concentration provides enough IgG of each subclass to block effectively all FcR binding as well as "nonspecific" binding. In Table I, the approximate amounts of Ig isotypes and IgG subclasses in the serum of most mammals is shown. If 10% serum was used and a concentration of 30 μg (3 mg)/ml/10^6 cells is the effective blocking concentration; 10% serum would not be an effective blocking reagent for IgG_2, IgG_3, IgG_4, or IgM. The problem would be even worse if lower serum concentrations were to be used.

B.   Directly Labeled Antibodies

An alternative approach to using a second Ab is to use a directly labeled primary Ab. Primary Ab are currently available labeled with fluorescein (FL-Ab), phycoerythrin (PE-Ab), or biotin (B-Ab). It is likely that Ab

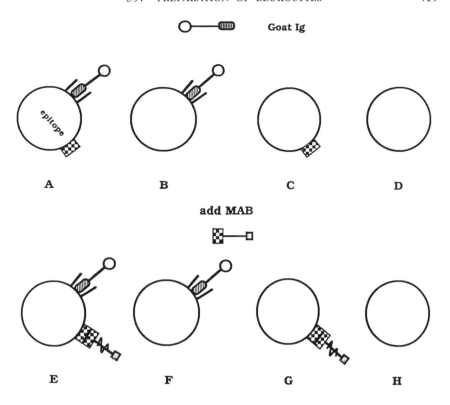

Goat Ig

add MAB

add fluoresceinated goat anti-mouse IgG F(ab')2

FIG. 4. Blocking Fc receptors. Cells are incubated for 10 minutes with goat Ig, shown as a stick symbol with its epitope-binding site a circle. The murine mAb is then added (shown as a patterned rectangle), and the cells are incubated for 15 minutes. After washing, the pelleted cells are labeled with F-GAM (shown as a jagged line), incubated 15 minutes, and again washed.

labeled with other fluorochromes will become available. Using a directly labeled pimary Ab, the cells can be preincubated with murine Ig as the block prior to staining with a labeled murine mAb. Rat Ig would be used if the labeled primary mAb were of rat origin. When using a B-Ab, labeled avidin is used as the second reagent. Avidins can be purchased with FL, PE, Texas red (TR), allophycocyanin (APC), and other newly developed fluorochromes attached.

IV. Cell Preparation

When working with cell suspensions that contain erythrocytes (e.g., blood or bone marrow) it has been the common practice first to pass the suspension over Ficoll–Hypaque to remove both erythrocytes and granulocytes (or to use neutrophil isolation medium if granulocytes are also desired). Whereas preparation of the cells in this way is often desirable, whole-blood (or bone marrow) lysis can be used to eliminate this time-consuming step, and the potential of selecting populations with ficoll separation is eliminated.

We do not recommend using the microtiter plate method for staining cells unless gradient-separated cells or prelysed cells are used and a microtiter sampling attachment to the flow cytometer is available. If whole blood (or bone marrow) is used, the volume of lysing buffer needed to lyse the erythrocytes is larger than that of the microtiter well and cells have to be transferred to a 10×75 mm tube to be analyzed by the flow cytometer. Whereas prelysed cells can be used, there is a tendency for these cells to aggregate during the washing steps. Therefore, postlabeling lysis is favored.

In the following procedures a "wash" means to add 4 ml phosphate-buffered saline containing 0.1% sodium azide and 0.5% bovine albumin (PAB) to the tube, centrifuge the cells at $1200\ g$ for 3 minutes at $4°C$ and decant the supernatant by inverting the tubes and blotting it by touching its rim to an absorbent towel. Thoroughly resuspend the cells in the residual buffer that has drained back to the bottom of the tube (not to exceed $100\ \mu l$). All labeling procedures are carried out on ice.

Each antibody should be titered prior to use so that the correct amount to stain the cells optimally is used. Our general practice is to have the optimal amount in $10\ \mu l$. Commercially available Ab that have already been titered may be diluted differently by the supplier so that more or less than $10\ \mu l$ is used. In these cases, use the supplier's recommended amount.

V. Labeling Procedures

For the following procedures, put $50\ \mu l$ of a cell suspension that is at 20×10^6 cells/ml of blood, bone marrow, nuclear-isolation medium-separated cells, or other cells into 10×75 mm tubes containing $30\ \mu g$ of blocking Ig. Do not block when measuring FcR.

A. Indirect Labeling with One Antibody

1. Add 10 μl goat Ig (3 mg/ml) for 10 minutes.
2. Add first Ab for 15 minutes.
3. Wash.
4. Add fluoresceinated goat F(ab')$_2$ second Ab for 15 minutes.
5. Wash.

For the control tube, add the isotype control Ab instead of the first Ab in step 2. Also prepare a tube containing only cells that will be used to measure cellular autofluorescence.

B. Direct Labeling with One Antibody

1. Add 10 μl XIg (3 mg/ml) for 10 minutes.[2]
2. Add FL-Ab for 15 minutes.
3. Wash.

For direct labeling, block with the Ig fraction from the same species as the Ab. Never block if the epitope being measured is the FcR. For example, CD16 is the FcR III receptor on granulocytes, monocytes, and natural killer (NK) cells. The Ab will not bind to this receptor if it is blocked.

C. Two Directly Labeled Antibodies

1. Add 10 μl XIg (3 mg/ml) for 10 minutes.
2. Add FL-Ab and PE-Ab for 15 minutes.
3. Wash.

D. One Directly Labeled Antibody with One Unlabeled Antibody

1. Add 10 μl goat Ig (3 mg/ml) for 10 minutes.
2. Add unlabeled Ab for 15 minutes.
3. Wash.
4. Add PE-goat anti-mouse F(ab')$_2$ second Ab for 15 minutes.
5. Wash.
6. Add 10 μl XIg (3 mg/ml) for 10 minutes.
7. Add FL-Ab for 15 minutes.
8. Wash.

---

[2] X is the species of Ig that is the same as the Ab. For murine Ab, block with normal mouse Ig; for rat Ab, block with normal rat Ig. Use both if an Ab from each is used.

It is absolutely essential in step 6 that the blocking Ig be from the same species as the unlabeled Ab. This is because the second Ab will have free binding sites that must be blocked so that it will not bind the directly labeled Ab added subsequently in step 7. For example, suppose that both the unlabeled and labeled Ab are murine mAb. After step 5, cells that have bound mAb1 and PE-GAM to them may bind the FL-mAb when it is added because free binding sites remain on the PE-GAM. To prevent this binding, these free sites must be blocked by mouse Ig, as shown in step 6.

If a rat mAb had been used, then rat Ig would be used for the block. If one mAb is rat and the other mouse, then the appropriate second Ab and block would be used: for unlabeled rat, use PE-GAR and block with rat Ig; for unlabeled mouse, use PE-GAM and block with mouse Ig. Fluorescein-labeled goat anti-mouse or anti-rat Ig could also be used in combination with PE-labeled Ab.

Note for two antibodies procedure:

1. Always treat with the unlabeled Ab followed by the labeled second Ab.
2. Block with normal Ig of the first Ab species before adding the directly labeled Ab.

One Antibody Labeled with Fluorescein, One with Biotin

1. Add 10 μl XIg (3 mg/ml) for 10 minutes.[3]
2. Add FL-Ab and B-Ab for 15 minutes.
3. Wash.
4. Add PE-avidin for 15 minutes.
5. Wash.

E. Three Antibodies, Two Directly Labeled and One Unlabeled (Requires Two Lasers)

1. Add 10 μl goat Ig (3 mg/ml) for 10 minutes.
2. Add unlabeled Ab for 15 minutes.
3. Wash.
4. Add TR goat F(ab')$_2$ second Ab for 15 minutes.
5. Wash.
6. Add 10 μl XIg (3 mg/ml) for 10 minutes.[4]
7. Add FL-Ab and PE-Ab for 15 minutes.
8. Wash.

[3] XIg is derived from the species in which the labeled Ab were produced.
[4] XIg is derived from the same species as the unlabeled Ab used to block any uncombined sites on the second Ab.

F. Three Antibodies: Two Labeled and One Unlabeled (Requires One Laser)

Currently there are no second Ab available that are labeled with a fluorochrome that can be excited at 488 nm and resolved from FL and PE. DuoChrome is a PE–TR complex conjugated to avidin. It can be used to obtain a third color using a single laser operating at 488 nm. The following procedure can be used to label cells with DuoChrome using an unlabeled antibody.

1. Add 10 μl goat Ig (3 mg/ml) for 10 minutes.
2. Add unlabeled Ab for 15 minutes.
3. Wash.
4. Add biotinylated goat F(ab')$_2$ second Ab for 15 minutes.
5. Wash.
6. Add XIg for 10 minutes.[5]
7. Add FL-Ab, PE-Ab, and DuoChrome for 15 minutes.
8. Wash.

G. Three Antibodies, Two Directly Labeled and One Biotinylated Antibody

1. Block with XIg for 10 minutes.[6]
2. Add FL-Ab, PE-Ab, and B-Ab.
3. Wash.
4. Add DuoChrome.
5. Wash.

H. Four Antibodies: Two Directly Labeled, One Unlabeled and One Biotinylated (Two Lasers Required)

1. Add 10 μl goat Ig (3 mg/ml) for 10 minutes.
2. Add unlabeled Ab for 15 minutes.
3. Wash.
4. Add TR–goat F(ab')$_2$ second Ab.
5. Wash.
6. Block with XIg for 10 minutes.[7]

[5] XIg is derived from the same species as the unlabeled Ab.
[6] XIg is derived from the same species as the three directly labeled Ab.
[7] XIg is derived from the same species as the unlabeled Ab.

7. Add FL-Ab, PE-Ab, and B-Ab for 15 minutes.
8. Wash.
9. Add APC-avidin for 15 minutes.
10. Wash.

VI.   Additional Procedures and Solutions

A.   Lysing Erythrocytes

1. Thoroughly resuspend cells in residual PAB.
2. Add 2 ml of lysing buffer and agitate cells for 5 minutes (rock, roll, or tumble).
3. Wash cells twice.

B.   Detecting Dead Cells

Unfixed Cell Suspensions

Resuspend cell pellet in 1 ml PAB containing 1 μg propidium iodide (PI) per 1 ml and analyze samples. An alternative is to add 10 μl of PI stock solution to 1 ml of cell suspension. Viable cells exclude PI, whereas dead cells are stained with this dye.

Fixed Cell Suspensions

1. Resuspend cell pellet (prior to fixation) in residual PAB and add enough ethidium monoazide (EMA) for a final concentration of 5μg/ml.
2. Put samples 18 cm from 40 W fluorescent light for 10 minutes.
3. Wash cells twice.

More extensive description of this procedure is provided in Chapter 40 of this volume, by Stewart.

C.   Fixation

1. Thoroughly resuspend cells in residual PAB.
2. Add 2 ml 1% paraformaldehyde.
3. Analyze samples within 5 days.

If desired, cells may be centrifuged and resuspended in less paraformaldehyde to have them more concentrated for data acquisition.

D. Solutions

PAB

Phosphate-buffered saline (PBS) containing no calcium or magnesium is prepared containing 0.1% sodium azide and 0.5% bovine serum albumin (BSA) fraction V (Sigma Chemicals, St. Louis, MO). The pH should be adjusted to 7.2. Be sure to check osmolality of the final solution (290 ± 5 mOsm). It is important to handle sodium azide with extreme caution.

1% Paraformaldehyde

1. Dissolve 10.7 g sodium cacodylate in 1 liter deionized water.
2. Adjust pH to 7.2 with 1 N HCl (~2 ml).
3. Add 6.0 g NaCl. Check osmolality.
4. Add more NaCl until osmolality = 290 mOsm (~6.0–6.8 g final).
5. Perform this task in a fume hood. Dissolve 10 g paraformaldehyde in the above buffer by heating until it just begins to boil.
6. Filter through 0.2-μm nalgene filter.

Lysing Reagent

1.6520 g Ammonium chloride
0.2000 g Potassium bicarbonate
0.0074 g EDTA (tetrasodium salt)

Add distilled water to 200 ml, use at room temperature. Store at 4°C, stable for 2 weeks.

Propidium Iodide Stock Solution

Handle PI with extreme caution. Prepare 7.4×10^{-5} M (20 mg/100 ml) PI (Sigma Chemicals, St. Louis, MO) in PBS. Store in a dark bottle or foil-wrapped container to protect from light (MW 668).

Ethidium Monoazide Stock Solution

Ethidium monoazide (Molecular Probes, Inc., Eugene, OR) is very light-sensitive and must be stored at −20°C in a dark vial or foil-wrapped container. The stock solution is prepared in PBS at 5 mg/ml. This can then be diluted further into small aliquots of 50–100 μg/ml and frozen at −20°C. These small aliquots can then be thawed one at a time as needed to do the assays. Discard remaining EMA after thawing.

REFERENCES

Bender, J. G., Stewart C. S., Van Epps D. E., and Walker, D. R. (1990) In Press.

Landay, A. (1989). *NCCLS.* **9,** 13.

Pierce-Immunotechnology Catalogue and Handbook, Section C (1989).

Roitt, I., Brostoff J., and Male D. (1989). "Immunology," Chap. 5, pp 5.1–5.11. Mosby, St. Louis.

Sneed, R. A., Stevenson A. P., and Stewart C. C. (1989). *J. Leuk. Biol.* **46,** 547–555.

Stewart, C. C. (1985). In "Handbook of Experimental Immunology" (D. M. Weir, ed.), Vol 2, pp. 44.1–44.17. Blackwell, Oxford.

Stewart, C. C., Lehnert B. E., and Steinkamp J. A. (1986a). *In* "Methods of Enzymology" (G. Di Sabato and J. Ererse, eds.), Vol. 132, pp. 183–192. Academic Press, Orlando, Florida.

Stewart, C. C., Stevenson A. P., and Steinkamp J. A. (1986b). *In* "Applications of Fluorescence in the Biological Sciences " (D. L. Taylor, A. S. Waggoner, F. Lanni, R. F. Murphy, and R. Birge, eds.), pp. 585–597. Liss, New York.

Stewart, C. C., Stevenson A. P., and Habbersett, R. C. (1988). *Int. J. Radiat. Biol.* **53,** 77–87.

Stewart, C. C., Stewart, S. J., and Habbersett, R. C. (1989). *Cytometry* **10,** 426–432.

Chapter 40

Multiparameter Analysis of Leukocytes by Flow Cytometry

CARLETON C. STEWART

Laboratory of Flow Cytometry
Roswell Park Memorial Institute
Buffalo, New York 14263

I. Introduction

Multiparameter flow cytometry (FCM) can be used to define the unique repertoire of epitopes on cells using forward scatter (FCS), side scatter (SSC), and several antibodies labeled with a different-colored fluorochrome (Shapiro, 1988; Stewart, 1989; Terstappen *et al.*, 1990; Willman and Stewart, 1989).[1] No other analysis system can obtain this information as rapidly on large numbers of cells while sorting the unique population. Using more than one antibody simultaneously permits two kinds of analysis to be performed: single-antibody histogram and multiparameter.

---

[1] Abbreviations used in this chapter: APC, allophycocyanin fluorescence; B, biotin; BP, Bandpass; CD3, antibody to T cells; CD4, antibody to helper T cells; CD8, antibody to suppressor T cells; CD14, antibody to monocytes; CD45, antibody to leukocytes; DC, DuoChrome fluorescence; dc, dead cells; DM, dichroic mirror; EMA, ethidium monoazide; F(ab')$_2$, part of antibody that has the epitope-binding site; FITC, fluorescein isothiocyanate; FL, fluorescein fluorescence; FL-BrdUrd, fluoresceinated antibody to bromodeoxyuridine; FSC, forward scatter; Ig, immunoglobulin; IgG, immunoglobulin isotype γ; lc, live cells; L, lymphoid cells; LLy, large lymphocytes; LP, long-pass; Ly, small lymphocytes; M, monocytes; mAb, monoclonal antibody; mIg, mouse immunoglobulin; MNC, mononuclear cells; NG, nongranulated granulocytes; NIM, neutrophil isolation medium; NK, natural killer cells; OE, other events; PAB, phosphate-buffered saline, supplemented with 0.1% sodium azide and 0.5% bovine albumin; PE, phycoerythrin fluorescence; PI, propidium iodide; SP, short-pass; SSC, side scatter; TR, Texas red fluorescence; TT, trigger threshold.

427

The simplest is single-antibody histogram analysis, in which each of the individual fluorescence histograms is treated as single-parameter data. Data analysis is rapid because only the percentage of positive cells for each marker is considered, and fewer samples are required because two or three antibodies are used in each sample. Multiparameter analysis is much more complex and time-consuming, because populations that are labeled with more than one antibody must be individually identified. Each file must be analyzed separately because current software is not capable of analyzing data automatically in a reliable fashion. This is because the relative positions of populations resolved by multiple antibodies can move their positions relative to one another in different samples. Regions established to resolve a population in one sample may have to be moved slightly to resolve the same population in another sample. Multiple antibodies, however, provide the opportunity to resolve cell populations that cannot otherwise be identified. These populations may be most important in understanding normal and disease processes.

Strategies for analyzing these types of data and the pitfalls encountered in multiparameter sample and data analysis will be discussed. Whereas the various aspects of analysis will be illustrated using human peripheral blood, the principles discussed can be applied to any type of cell population.

II.   List-Mode Data

For purposes of discussion, we use "sample" analysis to mean the acquisition of FCM data of labeled cells using a flow cytometer. We use "data" analysis to mean the analysis of those data. List-mode data should always be acquired when performing multiparameter sample analysis. By doing so, retrospective data analysis can always be performed. A simple list-mode file is illustrated in Table I. Suppose we wish to acquire forward scatter (FSC) that reflects cell size, and one color of fluorescence. When cells are analyzed, the measured values are recorded sequentially in the computer. For the first cell, the first number is FSC, the second number is fluorescence. The second cell produces the third and fourth numbers in the list, the third cell produces the fifth and sixth numbers, and so forth until all the cells have been measured. In this example, if 10,000 cells had been measured there would be 20,000 numbers in the list. The first and every third number thereafter would be FSC, the second and every fourth number thereafter would be fluorescence. Thus, measured parameters for each cell can be reanalyzed by the computer over and over again in the same order that the cells were measured the first time by the flow cytometer.

TABLE I

Simple List-Mode Data File

Data list	Explanation[a]
200	FSC cell no. 1
400	Fluorescence cell no. 1
100	FSC cell no. 2
700	Fluorescence cell no. 2
200	FSC cell no. 3
500	Fluorescence cell no. 3
Sn	FSC cell no. n
Fn	Fluorescence cell no. n

[a] FSC, Forward scatter.

By selecting regions utilizing one set of parameters that contain the values for a particular cell population, we can find and display the location of these cells in the other parameters. This action is often referred to as "reprocessing" the list-mode data file, and it will be illustrated in more detail later.

III. Single-Label Analysis

When a second labeled antibody is used, it is important to use the $F(ab')_2$ fragment. In Fig. 1, blood mononuclear cells (MNC) labeled either with the $F(ab')_2$ of fluoresceinated goat anti-mouse IgG (top panel) or intact fluoresceinated goat anti-mouse IgG (bottom panel) were analyzed. Both analyses were performed in the absence of any blocking Ig. A population of small cells and large cells are resolved in the histograms on the top of each view. The antibody staining for the cells is shown in the histogram along the side of each view. By projecting each histogram event into the bivariate display, the correlation of cell size with antibody staining can be visualized. Almost all the large cells and a small proportion of small cells are stained by the intact anti-IgG but not the $F(ab')_2$. This simple test should always be performed when using a new second antibody, and only those that behave like the $F(ab')_2$-marked cells illustrated in Fig. 1 should be used.

The effect of blocking is shown in Fig. 2. In this example, directly fluoresceinated anti-CD3 (FL-CD3) was used to label MNC. When cells are incubated with FL-CD3 many large cells are dimly stained (Fig. 2A).

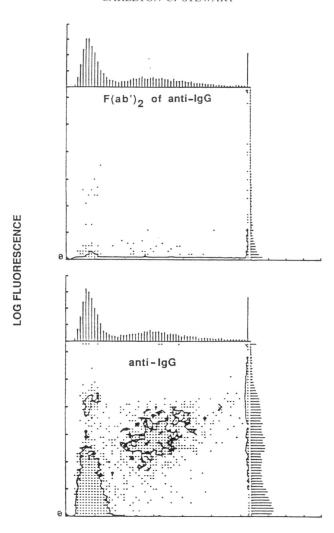

CELL VOLUME

FIG. 1. Second-reagent quality. Human mononuclear cells (MNC) were isolated from blood using neutrophil isolation medium (NIM; Los Alamos Diagnostics). Cells were adjusted to 20×10^6/ml in PAB and 50 μl were used for labeling. The MNC were incubated for 15 minutes at 4°C with either fluoresceinated goat $F(ab')_2$ anti-mouse IgG or fluoresceinated goat anti-mouse IgG. The cells were washed, resuspended, and analyzed. The cell volume histograms are shown on top and the antibody-staining histograms are shown alongside of each bivariate distribution.

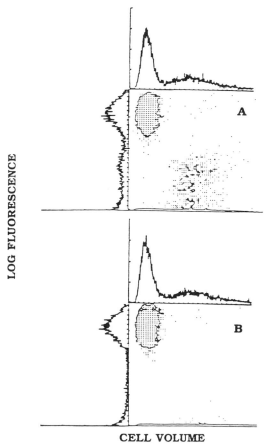

CELL VOLUME

FIG. 2. Effect of blocking on mAb binding to mononuclear cells (MNC). Human cells were isolated from blood using neutrophil isolation medium (NIM; Los Alamos Diagnostics). The cells were adjusted to 20×10^6/ml in PAB and 50 μl were used for labeling. (A) Cells were incubated with fluoresceinated CD3 for 15 minutes. (B) Cells were first incubated for 10 minutes with 30 μg mouse Ig and then fluoresceinated CD3 was added for 15 minutes. After washing, the samples were analyzed.

When the cells are incubated for 10 minutes with mouse IG (mIg) prior to the addition of the FL-CD3, no large cells are dimly stained (Fig. 2B). Thus, blocking the cells with mIg before staining them had a profound effect on the type of cells that were stained by the antibody.

For immunophenotyping cells, single-label analysis using the indirect method or directly labeled primary antibodies is most often used. Figure 3 shows human peripheral blood MNC separated using neutrophil isolation

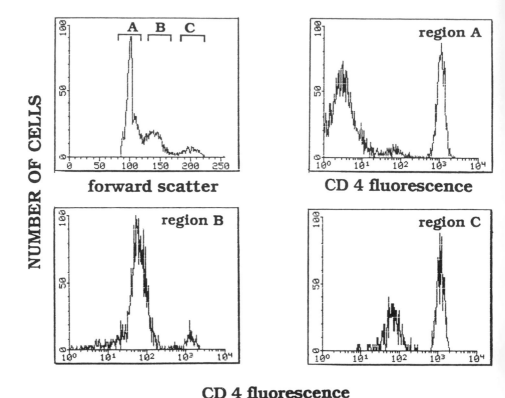

CD 4 fluorescence

FIG. 3. Single-color analysis. Human cells were isolated from blood using neutrophil isolation medium (NIM; Los Alamos Diagnostics). The cells were adjusted to 20×10^6/ml in PAB and 50 μl were used for labeling. The mononuclear cells were incubated 10 minutes with 30 μg mouse Ig, then for 15 minutes with fluoresceinated CD4, washed, and analyzed. Using forward light scatter, region A was selected to include small cells, region B intermediate cells, and region C large cells. The CD4 staining characteristics of the cells in each of these regions is shown in the appropriately labeled histograms.

medium (NIM) and stained with fluoresceinated anti-CD4 (FL-CD4). The blood was obtained from a patient who had been infused with interleukin 2 (IL-2). In the top left panel, three populations of cells with different FSC properties can be seen. The FSC measurement reflects cell size: events to the left represent the smallest and those to the right represent the largest cells. Those cells in region A are smallest, and when the fluorescence histogram of cells in this region is displayed by reprocessing the list-mode file for fluorescences associated with cells of this size, a brightly stained population and a negative population are resolved. This region has been termed the "lymphogate." Usually lymphoid immunophenotypes are obtained for cells that fall in this region only. The unstained cells that are in the left peak of region A are the unstained cells that have a low innate fluorescence called autofluorescence. A small population of cells between the two major peaks are dimly stained with CD4. Cells in region B are larger than those in region A and these cells are dimly stained with CD4, but a few brightly stained cells are also found in this region.

There is a population of very large cells in region C. Most of these cells are brightly stained with CD4 characteristic of the positive cells in region A. While this population of large lymphocytes (LLy) is not a prominent one in normal individuals, it is found in patients who receive intravenous IL-2. Had the common practice of analyzing only the small cells found in region A been employed, this population of large cells would not have been found. When the cells in each of these populations are sorted and stained with Wright's blood stain, the cells in region A are lymphoid (L), those in region B (with intermediate fluorescence intensity) are monocytes (M), and those in region C (with bright fluorescence intensity) are LLy.

The presence in each of the populations of cells within a size region with fluorescence characteristic of the dominant population in another size region is due to the overlap of populations. Using both size and fluorescence together can significantly reduce contamination of one population by the other. The use of bivariate analysis to accomplish this will be described later.

IV.   Optical Filtration

Before considering the use of more than one antibody, we should address the problem of overlapping fluorescence spectra of fluorochromes used to label monoclonal antibodies (mAb). As shown in Fig. 4, the emission from fluorescein (FL) overlaps that of phycoerythrin (PE), whereas PE emission does not significantly overlap FL. Emission spectra of PE, Texas red (TR), and allophycocyanin (APC) all overlap one another. The emission from each fluorochrome can be optimized to reduce unwanted fluorescence from a second fluorochrome using optical filtration. Table II lists the optical filters commonly used for fluorescence detection (Hoffman, 1988).

Optical filters are identified by the wavelengths they process. Long-pass (LP) filters will transmit wavelengths that are longer than the number given, whereas short-pass (SP) filters transmit shorter wavelengths. For a bandpass (BP) filter the first number refers to the wavelength that is

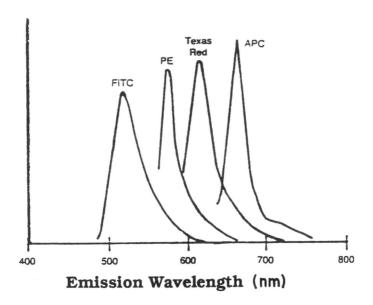

Emission Wavelength (nm)

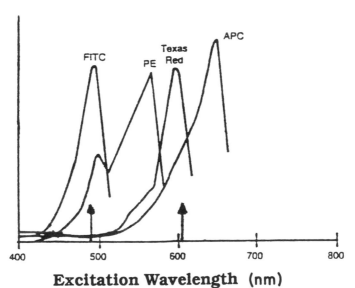

Excitation Wavelength (nm)

Fig. 4.   Emission spectra of common fluorochromes for labeling antibodies. Fluorescein (FITC) and phycoerythrin (PE) are usually excited at 488 nm. Texas red (TR) can be excited at 567 nm using the krypton line and optimally at 595 nm using a dye laser. Allophycocyanin (APC) is optimally excited using the 635-nm line from a He–Ne laser, but this wavelength cannot be used to excite TR. To excite TR and APC together, a dye laser tuned to 605 nm is best. Adapted from Hoffman (1988).

TABLE II

FILTERS COMMONLY USED FOR IMMUNOFLUORESCENCE MEASUREMENTS

Fluorochrome	Excitation (nm)	Dichroic mirror (nm)	Bandpass filter (nm)	Long-pass filter (nm)
Fluorescein	488	LP550	530–30	505
Phycoerythrin	488	LP600	575–30	560
DuoChrome	488	—	—	650
Texas red	605	LP650	630–20	620
Allophycocyanin	605	—	665–30	—

transmitted at or near 100% and the second number refers to the band-width (50% transmission). Thus, 530–30 means that the filter transmits light from 515 to 545 nm (30-nm bandwidth), and the maximum transmission is at 530 nm. Dichroic mirrors (DM) are used to change the direction of light along the optical path while at the same time transmitting or reflecting selective wavelengths. An LPDM transmits longer and reflects shorter designated wavelengths, while an SPDM does the opposite.

V. Compensation

Because the optical filters cannot remove all the overlap from the un-wanted fluorochrome, electronic subtraction, called spectral compensation, is used to eliminate the remaining fluorescence. Figure 5 shows a schematic diagram of how this is accomplished. The signal from the fluorescence 1 (FITC) photomultiplier tube is routed directly into the positive side of the fluorescence 1 differential amplifier and via an adjustable circuit into the negative side of the fluorescence 2 (PE) differential amplifier. Similarly, the fluorescence 2 PMT signal is routed into the positive side of its differential amplifier and, via an adjustable circuit, into the negative side of the fluorescence 1 differential amplifier. Adjusting the amount of signal fed into the negative side of the appropriate differential amplifier from particles labeled with a single fluorochrome allows removal of the unwanted fluorescences.

A bivariate distribution of a mixture of FL-labeled, PE-labeled, and unlabeled microspheres is shown in Fig. 6 before and after compensation. Adjusting the high voltage on the PMTs to increase or decrease amplification permits positioning of the unlabeled microspheres in the region of the

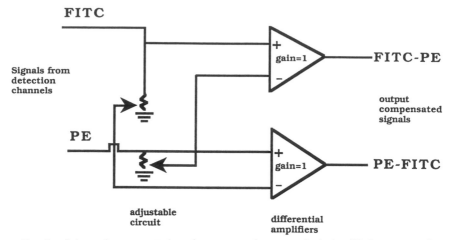

FIG. 5. Schematic representation of a compensation network. A simplified representation of the circuit for electronic compensation.

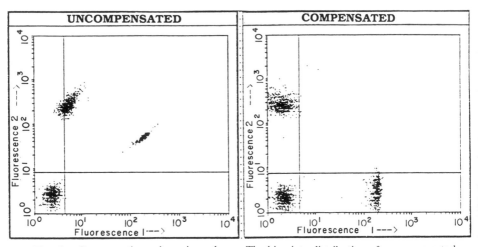

FIG. 6. Compensation using microspheres. The bivariate distribution of uncompensated (left panel) and properly compensated (right panel) microspheres is shown. The PMT high voltage was adjusted so that the unlabeled microspheres were in the lower left corner. The fluorescence 1 boundary limit is shown by the line parallel to the x axis and the fluorescence 2 boundary limit by the line parallel to the y axis.

bivariate plot bounded by 10×10 relative fluorescence units. When analyzed without compensation, the PE-labeled microspheres (fluorescence 2) straddle the fluorescence 1 boundary limit, while the FL-labeled microspheres (fluorescence 1) have significant emission detected by the fluorescence 2 channel well outside the fluorescence 1 boundary limit. Thus, PE has little fluorescence emission detected by the fluorescence 1 PMT, but fluorescein has considerable emission detected by the fluorescence 2 detector. After compensation has been adjusted, both fluorochromes are within their respective boundary limits. Follow the instrument instructions for making the PMT high-voltage and compensation adjustments.

A properly compensated instrument will provide good resolution of double-labeled cells independent of the fluorochrome's intensity. Improper compensation can lead to false positive or false negative cell populations. In Fig. 7, the effect of overcompensation is illustrated using lymphocytes labeled with either FL-CD4 (FL-1), or PE-CD4 (FL-2), separately or both together. In Fig. 7A, the negative cells are separated from the x and y axis, as are the labeled cells, and—illustrating good compensation—their leading edge is in line with the negative-control boundary limits (dashed line). (Some instruments do not have an offset, so that all events lie along the x or y axis and thus, cannot be seen. These instruments are difficult to compensate properly). As shown in Fig. 7B, cells stained brightly with one antibody (PE-CD4) and dimly with another are well resolved, and in Fig. 7C, the histogram shows the dim cells clearly resolved. Overcompensation can result in artifacts. As shown in Fig. 7D, the negative cells are correctly positioned, but too much compensation has been used on the stained cells so many of the cells are against the axis itself. As shown in Fig. 7E, when the same cells shown in Fig. 7D that were stained brightly with one mAb and dimly with a second mAb are reanalyzed they actually appear as negatively stained cells for the second mAb. The phenomenon results in a significant loss of the dimly stained cells, illustrated in the histogram shown in Fig. 7F.

Often propidium iodide (PI) has been used in combination with a fluoresceinated antibody in fixed preparations. One example of this combination is the measurement of cellular incorporation of bromodeoxyuridine (BrdUrd) into S-phase cells using a fluoresceinated antibody (FL-BrdUrd) and PI to measure DNA content (Stevenson *et al.*, 1985). To stain cells for DNA content, sufficient PI (usually > 5 $\mu g/ml$) must be used to obtain good coefficients of variation (CV) in the DNA histogram. This amount stains cells very brightly, and there is a significant amount of PI leakage into the fluorescein channel. This leakage can be as bright as the fluorescein from the FL-BrdUrd itself. It has been the practice to compensate for this PI leakage in the fluorescein channel.

PROPER COMPENSATION

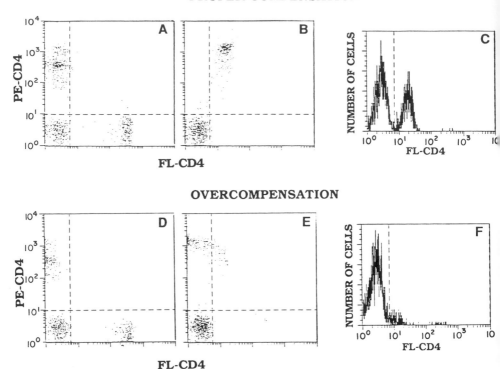

OVERCOMPENSATION

FL-CD4

FIG. 7.   Effect of overcompensation. Mononuclear cells (MNC) were labeled separately with FL-CD4 or PE-CD4 and mixed together. To obtain double-labeled cells with bright PE and dim FL fluorescence, a mixture of nine parts of PE-CD4 and four parts of FL-CD4 was prepared and the combination was used to label the cells. Cells were analyzed using a "lymphogate." The analysis of a mixture of cells that were separately labeled and mixed together is shown using (A) a proper compensation and (D) overcompensation. (B) Cells stained brightly with PE-CD4 and dimly stained with FL-CD4 analyzed with proper compensation; (C) histogram of fluorescence 1. (F) Histogram of fluorescence 1 of the overcompensated cells.

As mentioned before, the purpose of compensation is to remove the one fluorochrome's fluorescence in the other's channel, but when the second fluorochrome is present (i.e., cells have both fluorochromes present), one needs to detect them as double labeled. Since all cells are labeled with PI, we are now using compensation for a different purpose: to actually subtract one fluorochrome from the undesired channel when it is actually present on the cells. When this is done, the sensitivity for detecting dimly fluorescent cells in the target channel labeled with the appropriate fluorochrome for that channel—in this case, the cells labeled with FL-BrdUrd—can be significantly reduced.

In order to prepare fluorochromes with unique excitation and emission properties, tandem complexes of two fluorochromes have been produced. One commercially available tandem complex is DuoChrome (DC), a complex of TR and PE that is coupled to avidin (Stewart *et al.*, 1989). This molecule is designed to increase greatly the emission wavelength so that a single laser can be used to excite three fluorochromes: FL, PE, and DC. The system works through energy transfer. In the DC, the PE is excited at 488 nm and its emission photons are largely absorbed by the TR, thereby exciting it. The TR emits at a longer wavelength than PE, and its emission spectrum can be resolved from the PE emission by optical filtration and compensation.

In practice, however, there is considerable leakage of PE photons, resulting in a significant PE emission associated with the DC complex. Each batch may have slightly different leakage properties. Accordingly, all cells labeled with DC will have both a TR and PE emission component. Because of this leakage, compensation between PE and DC is operationally different from compensation between two completely different fluorochromes because PE emission must be subtracted from itself (similar to the PI problem described before). In addition, PE emission significantly overlaps the TR emission spectrum. These two problems result in a significant fluorescence heterogeneity.

The PE and DC fluorescence heterogeneity can make compensation difficult, and artifacts are easily introduced. The problem associated with undercompensation is shown in Fig. 8. This undercompensation can be especially troublesome when an antibody such as PE-CD4 that brightly stain cells is used in combination with one that stains less brightly and/or also stains a second population dimly like DC-CD8. In Fig. 8A, the bivariate distribution of lymphocytes separately labeled with PE-CD4 significantly overlaps the boundary set for DC-CD8$^+$ cells. The histogram showing the distribution of DC-CD8$^+$ cells suggests there is a high proportion of CD8-dim cells (filled region of histogram in Fig. 8B), but many are actually PE-CD4$^+$ cells because of the overlap of PE-CD4 fluorescence into the DC channel caused by undercompensation. This is shown in Fig. 8C and 8D, where only DC-CD8-stained cells were analyzed and the true proportion of CD8-dim cells are resolved.

The various problems just described must be recognized and corrected prior to sample analysis. Two simple rules should be followed to prevent these problems from occurring and to establish the correct spectral compensation:

1. Use particles that have lower autofluorescence compared to cells that are to be acquired; adjust the PMT voltages so that the events are in the 10 × 10 relative fluorescence unit region on the bivariate display.

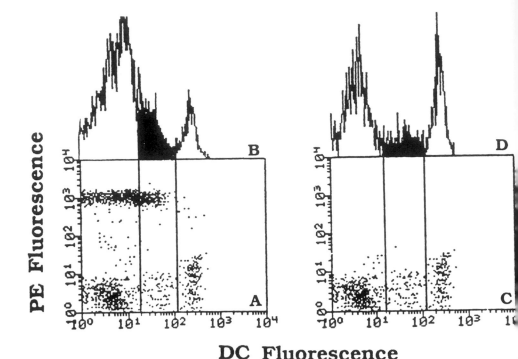

DC Fluorescence

FIG. 8.   Artifacts of undercompensation. Human cells were labeled with PE-CD4 and DC-CD8 and analyzed. (A) PE-CD4 versus DC-CD8 bivariate distribution; (B) projection of the DC-CD8 histogram shown above (A). The lines parallel to the ordinate show the positions of the CD8-dim and CD8-bright cells. From the histogram the percentage of CD8-dim is 27% and of CD8-bright is 10%. (C) The same cells were stained with only DC-CD8, with (D) the projection of the DC-CD8 histogram shown above it. From this histogram the percentage of CD8-dim is 4% and CD8-bright is 9%.

2. Compensation is adjusted using particles that are as bright or brighter than any stained cells that are to be analyzed.

The following procedure is recommended for adjusting instrument compensation:

1. Enter settings from a standard file (previously generated).
2. Using microspheres, evaluate the instrument performance and adjust as necessary.
3. Reacquire the previous day's cellular compensation mixture and compare.
4. Acquire the newly produced cellular compensation mixture and compare.

The cellular compensation mixture is prepared using normal blood (or other cells commonly used in the laboratory). The following antibodies consistently produce the brightest stained cells for the particular fluorochrome attached to them: FL-CD45, PE-CD4, B-CD8 with DC (using procedures described in Stewart, Chapter 39, this volume). A properly compensated instrument using a mixture of one part of FL-CD45 and one part of PE-CD4, and two parts of DC-CD8 and one part unlabeled cells to produce the cellular compensation mixture for that day is shown in Fig. 9.

The standard file is determined after experience. The criteria for selection of this file are that it contains the least autofluorescent cells and the brightest stained cells the laboratory has historically analyzed. One such file is shown in Fig. 9. A lymphocyte gate is used for acquisition to ensure the best homogeneity of each measurement. The FSC and side scatter (SSC) should be adjusted so that small cells that are predominantly lymphocytes appear in the predetermined gate each day. The SSC is a measurement that is particularly sensitive to cellular granularity. Excessive adjustment indicates a possible instrument problem. The standard file is found using the cellular compensation mixture and updating the standard file whenever autofluorescent cells in a new mixture are found that are dimmer than those obtained for the current standard file. It is essential that the instrument itself is in proper alignment when the standard file is created or updated.

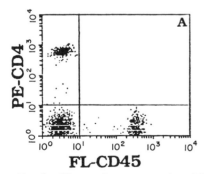 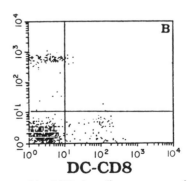

FIG. 9. Three-color compensation. Aliquots of human blood (50 μl total) were measured into four tubes containing FL-CD45 or PE-CD4 or B-CD8 or no reagent, and incubated 15 minutes at 4°C. The cells were washed and DuoChrome (DC) was added to the tube containing B-CD8 for 15 minutes and washed. Erythrocytes were lysed and the cells were then washed and fixed in 1% paraformaldehyde. One part of each cell suspension, except for two parts of CD8-stained cells, were mixed together and analyzed. The boundaries chosen for these single-labeled cells are shown for (A) FL versus PE and (B) DC versus PE.

There are several reasons for preparing the cellular compensation mixture daily.

1. If, after acquisition, the new mixture does not compare favorably with the standard file as well as the mixture from the previous day, then there is an instrument or staining problem.
2. If the mixture from the previous day looks like it did before, but the new mixture looks poor, the new compensation mixture is faulty.
3. By comparing the minor changes in settings over time, using both microspheres and the cellular compensation mixture, instrument performance can be monitored daily, trends established, and deterioration in instrument or sample preparation performance detected.
4. The use of a single large cellular compensation mixture that can be used over a period of time as a compensation standard is not recommended, because fixation results in a slowly increasing autofluorescence with time (described later). Preparing the mixture daily avoids this problem. The increase in cellular autofluorescence can be documented by comparing an old (14 days) cellular compensation mixture with a freshly prepared mixture or with the more stable microsphere standards used to evaluate instrument performance.

VI. Two-Color Fluorescence

To illustrate the amount of information that can be generated from a single sample using two antibodies simultaneously, a leukocyte differential can be obtained by staining whole blood with FL-CD45 and PE-CD14.

FIG. 10. Two-color fluorescence. Blood (50 μl) was incubated 10 minutes with 30 μg mouse Ig, and then FL-CD45 and PE-CD14 were added for 15 minutes. The cells were washed, fixed in 1% paraformaldehyde, and—without lysing the erythrocytes—were analyzed using the trigger threshold (TT) shown in (B). (A) Small cells (R1), large cells (R2), granular cells (R3), and other events (OE) outside R1 or R2 or R3, consisting of erythrocytes, aggregated platelets, dead cells, and debris. (B) The FL-CD45 bivariate distribution shows CD45-dull cells (R4). CD45-dim cells (R5), CD45-bright cells (R6), and CD14$^+$ cells (R7). (C) The forward-versus side-scatter distribution of CD45-dull events in R4, shown in (D). (E, F) Events coincident in both R3 and R5; these cells are neutrophils. (G, H) Events coincident in both R3 and R6; these cells are eosinophils. (J) The fluorescence of small events in R1, (I), including the location of small lymphoid cells (Ly), monocytes (M), nongranulated granulocytes (NG), and other events (OE). (L) The fluorescence of large events in R2 (K). Note CD45-bright large lymphoid cells (LLy).

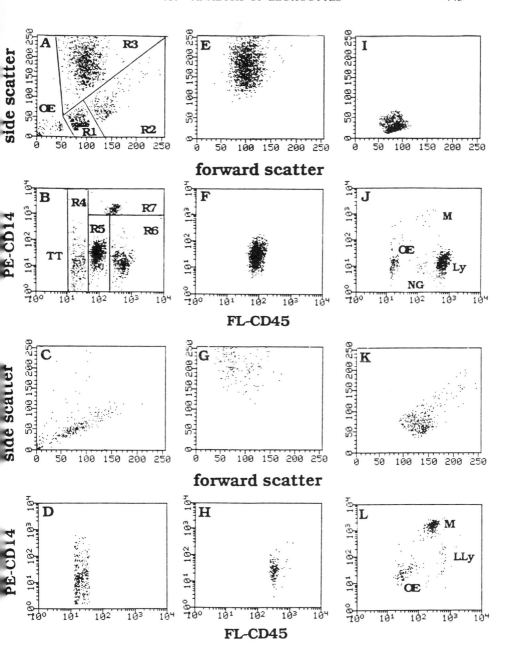

Although erythrocytes are usually lysed prior to analysis, this is not necessary in the presence of FL-CD45 because the sample can be analyzed by triggering the flow cytometer on CD45 fluorescence. A whole-blood sample in which FSC, SSC, FL-CD45, and PE-CD14 were all measured after triggering on CD45 fluorescence is shown in Fig. 10. In Fig. 10A, the FSC/SSC bivariate distribution for CD45$^+$ events above the trigger threshold (TT), shown in Fig. 10B, is displayed; region R1 contains predominantly small cells, R2 large cells, and R3, granulated cells. Note that there are not many other events (OE) in the lower left corner of this display. The bivariate distribution shown in Fig. 10B is also sectioned into regions: R4 containing CD45-dull cells, R5 containing CD45-dim cells, R6 containing CD45-bright cells and R7 containing CD14-bright cells. The FSC/SSC distribution of the CD45-dull cells after reprocessing the events in Fig. 10D are shown in Fig. 10C. These events are erythrocytes, aggregated platelets, dead cells, and debris. Some leukemias may express dull CD45 fluorescence and may appear in this region.

Analyzing the events that are coincident in R3 and R5 permits resolution of neutrophils as highly granulated cells (Fig. 10E) with the dim CD45 fluorescence (Fig. 10F). Similarly, eosinophils can be resolved by requiring coincidence with R3 and R6. Eosinophils have granularity similar to neutrophils (Fig. 10G), but they exhibit a higher CD45 fluorescence than do neutrophils (Fig. 10H).

It is common to use a lymphogate to exclude large and granulated cells from the analysis. Events found in R1, a typical lymphogate, are displayed in Fig. 10I and 10J. The events in this region are small lymphocytes (Ly) expressing bright CD45 fluorescence. There are also nongranulated granulocytes (NG), predominantly basophils (Terstappen et al., 1990; Loken et al., 1990), small M, and OE. Similarly, as shown in Fig. 10K and 10L, the large cells from R2 are not all M, as natural killer cells, large lymphoid cells (LLy), and OE are also present.

In practice, whole blood is generally lysed prior to data acquisition. Otherwise, CD45 must be present in every sample for triggering. It is more common to select a region for an acquisition gate based on FSC and SSC as illustrated in Fig. 11. In this example, all MNC have been acquired and small cells are found in R1 and large cells in R2. Granulated cells have been excluded. In Fig. 11B the bivariate distribution of FL-CD45 versus PE-CD14 is reprocessed for all events in the acquisition gate (R1 + R2). Besides L and M, there are NG and OE that contaminate this gate.

Thus, the gate (R1 or R2) that had been established for analyzing only the MNC (small or large) and whose number will form the denominator in calculating the percentage of positive cells for a particular mAb actually contains events that are not MNC; rather, they are, undesired events

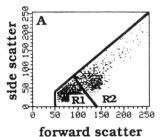

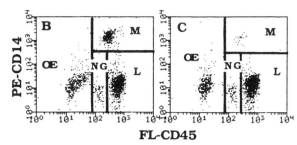

forward scatter   **FL-CD45**

FIG. 11.   Using CD45 to determine cellular events. Blood cells were stained as described in Fig. 10, except that the erythrocytes were lysed. The sample analysis was performed using the mononuclear cell gate (A). The gated data were divided into small events in R1 and large events in R2. (B) The FL-CD45 versus PE-CD14 fluorescence of all events. CD45$^+$, CD14$^-$ events are lymphoid cells (L); CD45$^+$, CD14$^+$ events are monocytes (M); CD45-dim, CD14$^-$ are nongranulated granulocytes (NG); and CD45-dull, CD14$^-$ are other events (OE), consisting of dead cells, aggregated platelets, and debris. This sample was chosen because it contains a high percentage (25%) of undesired events consisting of OE (21%) and a typical percentage of NG (4%). (C) The cells falling in region R1, the typical lymphogate.

represented by NG or OE. The extent to which these undesired events contaminate the small or large-cell region determines the accuracy of the calculated positive cells for a particular antibody. This error is consistently larger (5–25%), for whole-blood lysed cells than for gradient-separated (e.g., Ficoll-Hypaque) cells (4–10%). As mentioned before, it is common to use a lymphogate, R1, rather than an MNC gate. The events in R1, reprocessed in Fig. 11C, are similarly contaminated by the undesired events OE and NG.

This particular sample was selected for illustration because it contains an unusually high proportion of undesired events; nevertheless, all samples are contaminated to some degree with these undesired events. For several reasons, we do not recommend using a correction factor to increase artificially the calculated percentage of a particular cell population that is underestimated because of these undesired events.

1. In clinical immunophenotyping, malignant cells may not express CD45 to the same extent that their normal cell counterparts do, and they could be mistaken for undesired events. The correction factor would be difficult to define in these cases.
2. Undesired events may have clinical significance (e.g., frequency of NG), and if they are not reported, the clinical picture is biased.

Accordingly, we recommend that these events are reported as part of the sample report. Table III indicates an example.

TABLE III

ACCOUNTING FOR UNDESIRED EVENTS

Event description	Mononuclear cells (%)	Lymphogate (%)
CD45⁻ (OE)	21	18
CD45 dim	4	5
CD45 bright (L)	57	76
CD14⁺ (M)	18	1

VII.   Three-Color Fluorescence

By designing an appropriate combination of antibodies, specific cell subsets can often be explicitly resolved. The measurement of all the major T-cell subsets is one such example, and the results are shown in Fig. 12. The antibody combination used for this panel was FL-CD3, PE-CD4, and DC-CD8.

The bivariate plot of CD3/CD4 is shown in Fig. 12A. Region 1 is formed to include only $CD3^+$ cells. Not included in the region are unlabeled L and undesired events, M characterized by some fluorescence 1 autofluorescence and dim CD4 fluorescence and the contaminating NG that are also autofluorescent. The undesired events shown in Fig. 11 cannot be resolved without CD45 but they do not label with CD3. This can be confirmed by analyzing cells with CD45 and CD3. The process of examining fluorescence first in contrast to using FSC versus SSC first, has been called "backgating" (Stewart *et al.*, 1989; Loken *et al.*, 1990). Basically, the method allows one to see the measured FSC and SSC characteristics of a particular fluorescent antibody-stained cell population.

A bivariate plot of FSC versus SSC (Fig. 12B) is obtained of the $CD3^+$ cells found in region R1. Since all the other cells are excluded and only $CD3^+$ T cells are displayed, resolution of both small and large T cells is possible. The small T cells are bounded by R2 and the large T cells are bounded by R3. In order to obtain the phenotype of both the small and large T cells, a bivariate plot of CD4 versus CD8 is produced for cells coincident in both R1 and R2 (small $CD3^+$ cells, Fig. 12C) or R1 and R3 (large $CD3^+$ cells, Fig. 12D), respectively. Five additional regions are created in these two bivariate plots to include, moving clockwise from the upper left corner, $CD4^+$, $CD8^-$(R4), $CD4^+CD8^+$(R5), $CD4^-CD8^+$

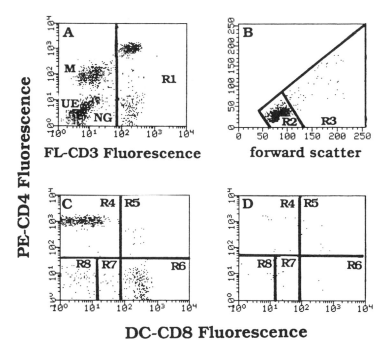

FL-CD3 Fluorescence   **forward scatter**

DC-CD8 Fluorescence

FIG. 12.   Three-color fluorescence. Human blood, 50 μl from the same individual shown in Fig. 11, was incubated 10 minutes with 30 μg mouse Ig, then for 15 minutes with FL-CD3, PE-CD4, and B-CD8. The cells were washed, incubated 15 minutes with DuoChrome, and washed; the erythrocytes were then lysed and analyzed. Using the concept of backgating, the region R1 was formed to include CD3$^+$ cells. Monocytes (M), nongranulated granulocytes (NG), and unstained events (UE) are also shown. (B) The forward versus side scatter of events in this region. Regions are created around small (R2) and large (R3) cells. (The small cells represent 67% of the total number of events analyzed.) The CD4 versus CD8 fluorescence (C) of small cells in R2 and (D) of large cells (R3). The following small and large CD3$^+$ cell populations are resolved: R4—CD4$^+$ CD8$^-$; R5—CD4$^+$CD8$^+$; R6—CD4$^-$ CD8$^+$; R7—CD4$^-$CD8 dim; R8—CD4$^-$CD8$^-$.

(R6), CD4$^-$CD8 dim (R7), and CD4$^-$CD8$^-$(R8). Using this panel, up to 10 T-cell subsets (5 small and 5 large cells) can be resolved.

It has been the common practice to gate only on small cells, using only the "lymphogate," and the immune phenotype of LLy has been largely ignored by the FCM community. While this population of cells is normally very small, as in the example shown in Fig. 12, it may be increased in individuals with immunologic disorders as shown in Fig. 3. There is no information, however, on the significance of the changes in this population because the cells within it have not been measured.

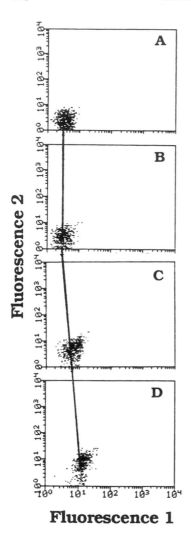

Fluorescence 2

Fluorescence 1

Fɪɢ. 13.   Storage of lysed whole blood in 1% paraformaldehyde. Human blood was lysed and fixed in 1% paraformaldehyde. The same sample was analyzed (A) 1 hour, (B) 7 days, (C) 14 days, and (D) 21 days after fixing, using the same instrument settings. There is a 5-fold increase in the mean autofluorescence from day 7 to 21.

VIII.   Fixing Cells

It is often important to fix human material prior to FCM analysis because of the health risk to the technical staff (Lipson *et al.*, 1985). A 1% solution of buffered paraformaldehyde is the fixative of choice. When cells are fixed, they should be analyzed as soon as possible but after no more than 1 week of storage. As shown in Fig. 13. prolonged storage of cells in paraformaldehyde causes an increase in autofluorescence. This problem can be especially noticeable if lysed whole blood is used because the fixed leukocytes bind fluorescent products released by the lysed erythrocytes. This problem is not as noticeable for cells that have been separated from erythrocytes.

IX. Detecting Dead Cells

Dead cells can bind antibodies nonspecifically, leading to an erroneous percentage of positive cells for some antibodies. Shipping cells, overnight storage, and cryopreservation are a few procedures that can result in increased cell death. It is important to account for these dead cells.

As shown in Figure 14A, dead cells stain with PI but live cells do not. The addition of PI to samples 5 minutes prior to analysis provides excellent dead-cell resolution. [Propidium iodide can be analyzed using the fluorescence 2 detector for instruments with two PMT detectors (in combination with a single antibody) or in combination with two antibodies in the fluorescence 3 position for instruments with three detectors.] Either gating against the dead cells during sample analysis or reprocessing the file on live cells will eliminate any PI overlap into the channels detecting antibody.

Because of the health risk in analyzing viable human cells, fixation is recommended. Fixed cells will all be labeled with PI, making it useless for

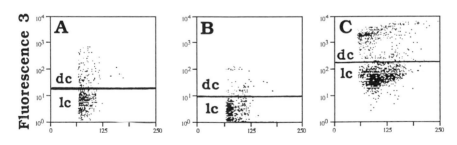

forward scatter

FIG. 14. Resolving dead cells. (A) Human blood cells were stained with 1 μg/ml propidium iodide (PI) and analyzed 5 minutes later. (B) Cells were stained with EMA and then fixed in 1% paraformaldehyde for 1 hour prior to analysis. Dead cells (dc) are found above the line and live cells (lc) below it. EMA is not as bright as PI because the emission maximum of EMA is at 600 nm and the measured fluorescence was above 650 nm. (C) Leukemic bone marrow was stained with four times the optimal concentration of EMA. Whereas dead cells (dc) are resolved better, there is a considerable amount of nonspecific staining of the live cells (lc) by EMA.

evaluating cell viability in fixed samples. Terstappen *et al.* (1988) reported that adding L-256 directly to the fixed sample results in labeling only the cells that were dead prior to fixation. Ethidium monoazide (EMA) can also be used in fixed cells (Riedy *et al.*, 1990). The EMA is first added to the viable cell suspension. The cell suspension is exposed to fluorescent light so that the photoactive EMA can irreversibly bind to DNA of dead cells. Unbound EMA is washed out of the cells and they are fixed. As shown in Fig. 14B and 14C for samples containing different numbers of dead cells, dead cells are resolved from live cells by EMA fluorescence.

It is important to recognize that dying cells dimly stain with PI or EMA. Therefore, the same fluorescence channel cannot be used to measure both an antibody and EMA (or PI) fluorescence by utilizing fluorescence intensity of PI or EMA staining to discriminate live and dead cells. Thus, cells exhibiting dim fluorescence for PI or EMA will appear in the same region as antibody-positive cells unless these dim cells are explicitly resolved in a separate dedicated channel.

The FCM described can all be performed using a single laser. For analyzing four or more colors of fluorescence, a second laser is required. An in-depth discussion of these procedures goes beyond the scope of this chapter.

Each laboratory has established its own experience and procedures. The intent of this discussion has been to illustrate the procedures that will lead to good FCM data and to illustrate problematic areas. The most important rule of all is to recognize when there is a problem. It is hoped the information provided herein will be of help in the recognition of problems.

References

Hoffman, R. A. (1988). *Cytometry Suppl.* **3**, 18–22.
Lipson, J. P., Sasaki, D. J., and Singleman, E. G. (1985). *J. Immunol. Meth.* **186**, 143–149.
Loken, M. R., Brosnan, J. M., Back, B. A., and Ault, K. A. (1990). *Cytometry* (in press).
Riedy, M. C., and Stewart, C. C. (1990). *Cytometry* (in press).
Shapiro, H. M. (1988). "Practical Flow Cytometry." Liss, New York.
Stevenson, A. P., Crissman, H. A., and Stewart, C. C. (1985). *Cytometry* **6**, 578-583.
Stewart, C. C. (1989). *SPIE* **1063**, 153–160.
Stewart, C. C., Stewart, S. J., and Habbersett, R. C. (1989). *Cytometry* **10**, 426–432.
Terstappen, L. W. M. M., Shaw, V. O., Conrad, M. P., Recktenwald, D., and Loken, M. R. (1988). *Cytometry* **9**, 447–484.
Terstappen, L. W. M. M., Hollander, Z., Meiners, H. and Loken, MR, (1990). *J. Leuk. Biol.* (in press).
Willman, C. L., and Stewart, C. C. (1989). *Semin. Diagn. Pathol.* **6**, 3–12.

Chapter 41

Identification and Purification of Murine Hematopoietic Stem Cells by Flow Cytometry

JAN W. M. VISSER AND PETER DE VRIES

Radiobiological Institute TNO
Rijswijk, The Netherlands

I. Introduction

All the different blood cell types are derived from a common ancestor, the pluripotent hematopoietic stem cell (PHSC). This cell not only has the capacity to differentiate into committed progenitor cells of all the various blood cell lineages, but can also renew itself, thus maintaining the hematopoietic organ at a steady-state level throughout life. The production of mature blood cells out of stem cells and committed progenitors requires >10 cell divisions. As a consequence, the stem cells are rare cells, even in hematopoietic organs. Their incidence in adult mouse bone marrow is estimated to be between 5 per 1000 and 2 per 100,000 (Visser and de Vries, 1988). No stem cell-specific cytochemical staining has been described yet. Stem cells are detected by examination of their offspring after *in vitro* culture or after transplantation into often lethally irradiated syngeneic or congenic recipients. There exists no consensus about the different culture and *in vivo* transplantation methods with regard to their specificity for pluripotent stem cells. Consequently, the quantitation of stem cells by those methods may lead to widely different results. Also consequently, identification of PHSC by flow cytometry (FCM) should be combined with cell culture and transplantation assays for verification. Therefore, FCM methods should employ supravital staining and labeling, which can be combined with sorting, cell culture, and transplantation procedures.

Flow cytometry and cell sorting have contributed significantly to the purification of PHSC and, concomitantly, this introduced uncertainty about

451

the validity of existing methods for stem cell enumeration. For many years the spleen-colony assay (CFU-S: colony-forming unit spleen) has been regarded as the clonogenic test for stem cells. Colonies of hematopoietic cells, which appear on the surface of spleens of mice 8–14 days after lethal irradiation and bone marrow transplantation, were thought to be each derived from a pluripotent stem cell in the graft that based itself in the spleen. Sorting experiments with Hoechst 33342 (Baines and Visser, 1983), and with antibodies against class I major histocompatibility complex (MHC) antigens (Harris et al., 1984; Visser et al., 1984) have for the first time indicated that early-appearing spleen colonies (day 8 CFU-S) are derived from other cells than late (day 12–14)-appearing ones. In addition, it could be shown, that "true" stem cells with extensive repopulating ability in all blood cell lineages belong to the sorted populations that give day 12 CFU-S. Subsequently, Bertoncello et al. (1985) indicated that these day 12 CFU-S can be sorted into two groups using rhodamine 123 (Rh123), a dye that stains the mitochondria. They observed that marrow-repopulating ability (MRA), a quality of true stem cells, was only present among the Rh123-dull cells, which contain about one-third of the total amount of day 12 CFU-S. Suspensions of Rh123-bright cells contained the other day 12 CFU-S, all day 8 CFU-S (Mulder and Visser, 1987), but no MRA. In addition, it was found that the stem cell type giving 30-day radioprotection was not present in the Rh123-bright fraction, but in the Rh123-dull fraction, together with the one responsible for thymus repopulation (Mulder and Visser, 1987; Visser and de Vries, 1988). Because of these observations the spleen-colony assay has become a measure of stem cells that should be interpreted with much caution. However, no other clonal assays are readily available.

Various protocols for identifying and sorting spleen colony-forming cells have been described in recent years. Most of these employed FCM in combination with other physical and immunological cell separation techniques, which served to remove the bulk of mature cells. Since the successful stem cell separations by Van Bekkum et al. (1971), equilibrium density centrifugation is the method of choice for preenrichment of the PHSC from murine bone marrow. In the flow cytometer, forward and perpendicular light scatter (FLS, PLS) measurements have proved to be useful to further eliminate mature cells from analysis and sorting procedures.

In addition, cells can be labeled in a viable state with fluorescent lectins and monoclonal antibodies (mAb) directed against cell surface molecules, some of which are differentiation antigens. The fluorescence of individual cells due to the binding of lectins and antibodies can be determined quantitatively by FCM. We and others successfully employed the lectin wheat germ agglutinin (WGA) to sort the stem cells (Visser and Bol, 1981;

Visser *et al.*, 1984; Lord and Spooncer, 1986). The use of WGA for this purpose was indicated by results obtained with free-flow electrophoresis and neuraminidase treatment to study membrane constituents of the stem cells (Bol *et al.*, 1981), and by subsequent analysis of WGA-labeled bone marrow cells using FCM (Visser *et al.*, 1981). A number of antibodies have been successfully employed for stem cell separation. Two strategies could be followed here; either it was attempted to find a specific stem cell label (positive selection), or all committed and fully differentiated cells were labeled (negative selection). Useful antibodies for labeling stem cells were found to be anti-H-2K^k (Van den Engh *et al.*, 1983; Visser *et al.*, 1984) and anti-Qa-m7 (Bertoncello *et al.*, 1986), both directed against class I MHC antigens. Also anti-Thy-1 (Muller-Sieburg *et al.*, 1986) and an antibody against a putative stem cell antigen (Sca-1; Spangrude *et al.*, 1988) have been reported to be of use for stem cell sorting. A diversity of cocktails of mAb against mature cells has been described and used to select stem cells negatively (Hoang *et al.*, 1983; Muller-Sieburg *et al.*, 1986; Spangrude *et al.*, 1988; Watt *et al.*, 1987). In addition, as mentioned earlier, supravital staining using Hoechst 33342 or Rh123 was also found to be useful for stem cell identification and sorting. It has become clear that a combination of a number of labels and dyes is needed to distinguish unambiguously the PHSC from the other cells in the heterogeneous bone marrow. In this chapter we describe a combination that has been of use to demonstrate heterogeneity of day 12 CFU-S, and that therefore may be of help to unravel further the true identity of the pluripotent stem cell.

II. Applications

At present the analysis and sorting of PHSC is primarily employed in experimental bone marrow transplantation and gene therapy, and to some extent in studies concerning the regulation of differentiation and proliferation.

The application for bone marrow transplantation arises from the concept that no malignant cells should be returned to the patient in the case of autologous transplantations and that preferably no lymphoid cells causing graft-versus-host reaction should be given in allogeneic grafts. As the malignant cells are different in most patients whereas the stem cells are probably similar, it is strategically of advantage to aim at developing a sorting procedure for the latter. Such a procedure would then also be of use for the separation of the lymphoid cells from the PHSC for allogeneic grafts. Transplantation of purified pluripotent stem cells, which have self-renewal capacity, should completely reconstitute the immunohematological system in patients after radiotherapy and chemotherapy, as well as in victims of radiation accidents. Application of FCM for the engineering of human bone marrow grafts may have to wait until high-speed sorting techniques are implemented. It is envisaged, however, that a combination of cell separation techniques, including FCM, could be successfully employed for that purpose. The present methodology for preparing those grafts already profits considerably from the analytical studies and preliminary FCM sorting of murine and monkey cells. In addition, the analysis of the bone marrow by using specific stem cell staining and labeling and FCM will be of use for monitoring the extent of effects of therapy on normal stem cells, which are among the most radiosensitive cells in the body, and which are essential for recovery after therapy.

A new application of purified stem cells concerns experimental gene therapy. Introduction or modification of genes in stem cells is expected to be of use for treating enzyme deficiencies and other malignancies by bone marrow transplantation. In order to evaluate the efficiency and extent of the DNA modifications in the stem cells, it is necessary to isolate them in sufficient quantities from the extremely heterogeneous suspension of bone marrow cells. Furthermore, to insert genes for therapy, it is also necessary that the stem cells remain pluripotential. Therefore, it is of importance to remove growth factor-producing cells, which naturally occur in the bone marrow, before the stem cells are cultured; otherwise, the PHSC will differentiate and mature.

In addition, growth factor-producing cells should also be removed before culturing stem cells, if the effects of the readily available, newly purified, and recombinant growth factors on the regulation of PHSC is studied.

The hematopoietic organ is an attractive general model system for studying the regulation of differentiation and proliferation. Because of their pluripotency, the most interesting cells for this purpose are the PHSC; because they are rare it is preferable to analyze them by sorting. The application of purified PHSC resulted in several findings. For instance, it was found that PHSC do not take hold in the thymus after transplantation, whereas some of the committed daughter cells of the PHSC, the prothymocytes, do so (Mulder and Visser, 1988); that osteoclasts arise from PHSC (Scheven et al., 1986); and that stem cells are sensitive to erythropoietin (Migliaccio et al., 1988).

It can be expected that the use of the analytical capabilities of FCM for stem cell studies will become most important if methods for the specific labeling of those rare cells are improved. Concomitantly, the definition of pluripotent stem cells should be sharpened as a result of the observation that the present stem cell tests yield heterogeneous results with sorted bone marrow cell fractions. The labeling method described here may be considered as one of the steps in this iterative process.

III.   Materials

A.   Buffers and Media

Bone marrow cell suspensions are prepared in Hanks' balanced salt solution (HBSS) (Laboratoires Eurobio, Paris) buffered at pH 6.9 with 10 mM HEPES buffer (Merck). For mouse bone marrow cells the osmolarity has to be 300–305 mOsm/kg. In general, HBSS without phenol red is preferred in order to avoid optical side effects of the medium in the flow cytometer. This relates in particular to the sheath fluid in the sorter. Cell suspensions are washed before and between labeling with mAb and conjugates with HBSS–HEPES, which we will further abbreviate as HH, containing 5% (v/v) of either fetal calf serum (FCS, Seralab) or newborn calf serum (NCS, Seralab), and 0.02% (v/v) sodium azide (Merck), which together will be further abbreviated as HSA (HH + serum + azide). After the final labeling, cells are resuspended in HH. Cells and media are kept at 0°–4°C unless indicated otherwise.

B.   Metrizamide

As the first step in stem cell purification procedures, bone marrow cells are separated using a discontinuous metrizamide (Nyegaard, Oslo, Norway) density gradient as described earlier (Visser *et al.*, 1984). The metrizamide (MA) is dissolved in HH + 1% (v/v) bovine serum albumin (BSA, fraction V, Sigma), adjusted first at pH 6.7, and subsequently at 300 mOsm/kg by appropriate dilution with water. Two solutions are prepared: a dense one of ~1.10 g/cm^3 and a light one of ~1.05 g/cm^3. Metrizamide solutions of desired intermediate densities (1.078 g/cm^3) are prepared by mixing these two solutions.

C.   Wheat Germ Agglutinin

Fluoresceinated wheat germ agglutinin (WGA/FITC: Polysciences, Inc., Warrington, PA) is stored as a stock solution of 1 mg/ml at −20°C. Thawed aliquots can be stored at 0°–4°C for up to several months. During prolonged storage (>12 months), fluorescein isothiocyanate (FITC) and WGA may become separated from each other. The solution is then processed on a G-25 Sephadex column to remove free FITC (Bauman *et al.*, 1985).

D.   *N*-Acetyl-D-Glucosamine

Wheat germ agglutinin can be removed from cells by incubation with the competing sugar *N*-acetyl-D-glucosamine (GlcNAc). For this purpose, GlcNAc (Polysciences, Inc.) is dissolved at a final concentration of 0.2 *M* in HH, which is diluted with water (two-thirds HH, one-third water) to compensate for the contribution of the sugar to the osmolarity. The solution is stored frozen; it is set out at 37°C shortly before use.

E.   Monoclonal Antibodies, Biotin and Avidin Conjugates

A biotinylated mAb against the H-2K^k antigen is obtained from Becton Dickinson (Becton Dickinson Immunocytometry Systems, Mountain View, CA). Aliquots are stored frozen and diluted 20-fold in HSA before use. Aliquots of FITC-labeled avidin (av/FITC, Sigma, Deisenhofen, FRG) and avidin conjugated to phycoerythrin (av/PE, Becton Dickinson) are stored frozen and used at a 100-fold or a 50-fold dilution. The dilutions are stored at 0°–4°C for several weeks without noticeable negative effects on the labelings.

Monoclonal antibody 15-1.1 and its subclone 15-1.1.4 were raised by intrasplenic immunization of a Brown Norway rat with suspensions that were highly enriched for CFU-S. However, the mAb do not bind to CFU-S but react strongly with bone marrow monocytes, granulocytes, and their immediate precursors (De Vries, 1988). The antibodies are available from the authors, upon request. The antibodies are conjugated directly to FITC in the following way. Equal volumes of saturated ammonium sulfate and hybridoma culture supernatant containing mAb 15-1.1 or 15-1.1.4 are mixed. After 1–2 hours at room temperature (RT; with occasional shaking), the suspension was centrifuged at 12,000 *g* for 30 minutes and the supernate removed. The pelleted precipitate was dissolved in 40% ammonium sulfate and centrifuged again. The pellet was then resuspended in distilled water at one-tenth to one-fifth of the original (supernatant) volume, and dialyzed against distilled water for 48 hours at 4°C. The content of the dialysis bag was centrifuged at 1200 *g* for 10 minutes and the supernate was collected and lyophilized. A solution was made containing 5 mg lyophilized protein per 1 ml phosphate-buffered (pH 7.2) saline (PBS, Bloedbank, Amsterdam). To each 1 ml of protein solution, 0.1 ml of a 0.5 M NaHCO$_3$ solution (pH 9.5) was added; the pH was checked and, if necessary, adjusted to pH 9.0 by adding NaHCO$_3$.

Under continuous stirring, 10 μl of a stock solution of FITC isomer-1 [1 mg/ml dimethyl sulfoxide (DMSO), Nordic Immunological Laboratories, Tilburg, The Netherlands] was slowly added per 1 mg protein. The mixture was incubated (continuous stirring) for 3 hours in the dark at RT. During this period, the pH was repeatedly checked and, if necessary, adjusted to pH 9.0. The mixture was then loaded on a PD-10 Sephadex column (Pharmacia 17-08500) and eluted with PBS.

F.   Rhodamine 123

The supravital dye Rh123 (Eastman Kodak, Rochester, NY) was dissolved in distilled water at a stock concentration of 1 mg/ml and stored at 0°–4°C in the dark. Shortly before use, 1 μl of this solution was diluted in 10 ml HH + 5% FCS. This dilution was prewarmed to 37°C. Alternatively, 1 μl Rh123 stock solution was added to 10 ml of the 0.2 M GlcNAc solution already described. Then, FCS (5% v/v) was also added to the solution.

IV.   Instruments

Flow cytometry for the analysis and sorting of PHSC can be performed efficiently using an instrument with only one laser. Since multiple labelings of the cells are generally necessary, some time can be saved if a two- or three-laser instrument is employed. We normally use a FACS II (Becton Dickinson Immunocytometry Systems, Mountain View, CA), which is slightly modified. The laser light-blocking bar for the PLS is broadened to 2 mm. Logarithmic amplification (T. Nozaki, Stanford, CA) is added. And, recently, a so-called Datalister (Becton Dickinson Immunocytometry Systems, Erembodegem, Belgium) with a Hewlett-Packard 310 computer system is added to provide list-mode data storage and analysis. This system replaces the old pulse height analyzer. The counters of the FACS are modified to allow single-cell sorting in Terasaki trays (Visser et al., 1984).

Two- and three-laser experiments are performed using the RELACS II and III, respectively (acronym for Rijswijk experimental light-activated cell sorter). These are standardly equipped with at least one argon ion laser (4–5 W visible), and with either a rhodamine 6G dye laser (type 375, Spectra-Physics, Mountain View, CA) or a 40–50 mW He–Ne laser (Shanghai Institute of Laser Technology, Shanghai, PRC). The RELACS instruments have parallel processing of up to eight parameters (Van den Engh and Stokdijk, 1989), which may be amplified linearly or logarithmically (four decades) at choice. The signals can be corrected for spectral overlap by electronic compensation. The RELACS II is equipped with Becton Dickinson hardware (originally FACS 440) for droplet formation and sorting. The RELACS III is used for analysis only. In both instruments data are recorded in list mode by Hewlett-Packard 9000-220 computers, and analyzed using the ELDAS software package (R. Jonker, TNO, Rijswijk, The Netherlands).

V.   Staining and Sorting Procedure

A.   Density Gradient Centrifugation

Discontinuous gradients are prepared by first pipeting 1 ml of a high-density MA solution (1.100 g/cm^3) containing 5–6 × 10^7 bone marrow cells in a round-bottomed tube (Falcon, type 2057). On top of this solution, 3 ml of a MA solution with intermediate density (1.078 g/cm^3) is layered. To label the cells during the density separation, and so as to save time, this MA solution contains WGA/FITC at a final concentration of 0.1 μg/ml. Finally, 1 ml of a low-density MA solution (1.055 g/cm^3) is put on top of the intermediate solution. The tube is centrifuged for 10 minutes, at 4°C, at 1000 g. The cells in the low-density fraction and from the interface between the top and intermediate layers (low-density cells) are collected, washed, centrifuged, and counted. This fraction typically contains 7–13% of the cells loaded onto the gradient.

B.   WGA Sort

The low-density cells are subsequently analyzed and sorted using FLS, PLS, and WGA/FITC fluorescence intensities as parameters. The FLS and PLS are measured using linear amplification; FITC fluorescence is measured using logarithmic amplification. The laser is set at 488 nm, 0.5 W. The laser light-blocking bar for FLS is set as narrow as possible (e.g., 1 mm in the FACS II); the one for PLS is taken broader than usual, (e.g., 2 mm in the FACS II), in order to block the optical disturbances generated by the droplet formation. The PLS is measured (preferably through a 488-nm laser line narrow-bandpass filter; for example, Melles Griot, Irvine, CA) by an S11 or S20-type photomultiplier. The FITC fluorescence is measured by an S20-type photomultiplier that is blocked by a filter combination consisting of a 520-nm long-wave pass (Ditric Research Optics, Hudson, MA) and a 520–550 nm bandpass filter (Pomfret, Stamford, CT).

The FLS signal is used to trigger the flow cytometer. The threshold is set such that the erythrocytes are not detected, whereas the lymphocytes are (Fig. 1). The sort windows are set as follows. First, the fluorescence window is set to include only positive cells. Those cells generally have a homogeneous FLS distribution. The FLS window is then set to exclude the larger cells and aggregates (Fig. 1). The PLS distribution of the cells within the FLS/fluorescence windows is then examined; it is generally broad with a level top. The PLS window is set to exclude the half of the cells with the brightest PLS using that distribution (Fig. 1). Between 5 and 10% of the low-density cells normally are selected in this combination of windows. We call the FLS/PLS window the "blast window." The cells are sorted with an analysis rate of 2500–3000 per second using a 50-μm nozzle (36,000 drops per second; 3 drops deflected per cell).

Deflected droplets are collected onto the wall of a 15-ml glass tube, which is coated with HH containing 5% BSA prior to the sorting. After ~ 2 hours, 2×10^7 cells are analyzed and ~ 10^6 WGA$^+$ blast cells are collected. The sample and collection tubes are cooled to 2–4°C during the sorting.

C. Removal of WGA/FITC and Labeling with 15-1.1 or Anti-H-2K

After the WGA sort, 10 ml of a 0.2 M solution of NAcGlc is added to the sorted cells (2–3 ml); subsequently, they are incubated for 15 minutes at 37°C in order to remove the WGA/FITC from the cells. Then they are

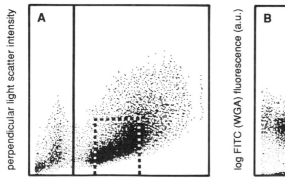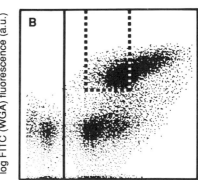

forward light scatter intensity

Fig. 1. Bivariate distribution of (A) the forward (FLS) and perpendicular (PLS) light scatter intensities, and (B) the FLS and WGA/FITC fluorescence intensities of mouse bone marrow cells. The straight vertical line indicates the FLS threshold for the triggering of the flow cytometer. The square window in (A) encloses the "blast" cells.

centrifuged (10 minutes, 400 g), and the pellet, resuspended in 1–2 ml of HSA, is transferred to a sample tube (Falcon 2058). The cells are centrifuged again and the pellet is now resuspended in 50–100 μl of the appropriate dilution of either 15-1.1/FITC or anti-H-2K^k/biotin. This suspension is incubated for 30–45 minutes at 0°–4°C, and again centrifuged. If H-2K^k labeling is performed, the pellet is then resuspended in 50 μl of an appropriate dilution of av/FITC, and again incubated for 30–45 minutes at 0°–4°C. After labeling with 15-1.1/FITC or av/FITC and centrifugation, the pellet is resuspended in HH (1–2 ml) and again analyzed by the flow cytometer.

D.   15-1.1 or Anti-H-2K Sort

Bound antibodies are detected using 0.5 W, 488-nm argon laser light for FITC conjugates, and 0.3 W, 590-nm rhodamine 6G dye laser light for Texas red (TxR) and 600 nm for allophycocyanin (APC). The photomultiplier that measured these latter conjugates is provided with 2-mm, 630-nm long-pass filters (Schott, Duryea, PA) for TxR or with a 1-mm, 660-nm one (Schott) for APC. The signals are logarithmically amplified. Cells are now analyzed with the FLS trigger threshold just above the noise and debris level, permitting detection of the erythrocytes that passed the WGA sort as passengers in deflected droplets containing desired cells. The FLS/PLS blast cell window is maintained for the sorting. The fluorescence window is set to sort the 15-1.1$^-$ cells (Fig. 2A) or the brightest 30–50% H-2K^{k+} cells (Fig. 2B). At this stage the window can be set using the log fluorescence histogram, which shows the wanted cells as a separate subpopulation.

E.   Rhodamine 123 Sort

After the WGA sort, subsequent incubation in GlcNAc, and the 15-1.1 or anti-H-2K^k labeling and sorting, the cells (generally between 10^5 and 3×10^5) are incubated in 2–3 ml of a solution of 1.0 μg/ml Rh123 in HH plus 5% FCS, for 15 minutes at 37°C. The cells are then washed once and resuspended in HH (1 ml) for a final sort run. The flow cytometer is equipped with the same filters as for the FITC measurements, only the photomultiplier power supply is reduced somewhat (800 V for Rh123 instead of 900 V). The FLS/PLS blast window now contains >50% of the cells. Two populations of cells can at this stage be recognized: one is dull fluorescent and the other bright fluorescent. They are clearly separated in the fluorescence histogram (Fig. 3). Both fractions are sorted and further studied.

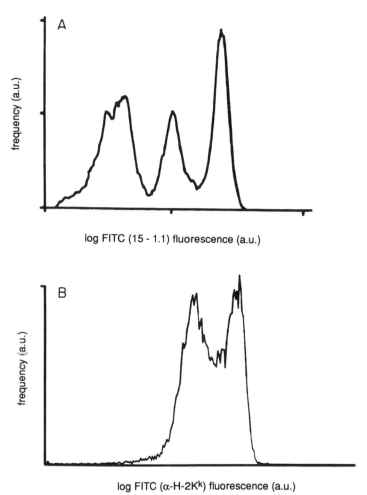

FIG. 2. Frequency distributions of the fluorescence intensities of sorted mouse bone marrow cells labeled with (A) 15-1.1/FITC and (B) anti-H-2K^k/biotin avidin/FITC. "Blast" cell population after a density cut and selection of the brightly WGA-fluorescent cells.

F.   Double and Triple Labeling

If a two- or three-laser flow cytometer is used, the lectin and antibody labelings can be combined. WGA/1-pyrene-butyryl (WGA/pyrene; Molecular Probes, Junction City, OR) instead of WGA/FITC is then used if ultraviolet (UV) lines (351–363 nm argon laser) are additionally available. And if a dye laser is added, WGA/TxR (Molecular Probes, Eugene, OR)

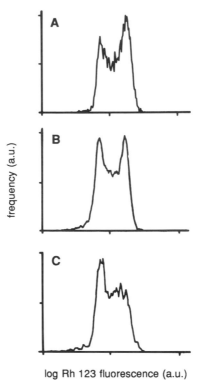

FIG. 3.   Frequency distributions of the fluorescence intensities of sorted mouse bone marrow cells labeled with rhodamine 123 (Rh123). "Blast" cell population after a density cut and (A) double WGA/FITC sort (Lord and Spooncer, 1986); (B) one WGA/FITC sort and subsequent 15-1.1/FITC (De Vries, 1988); (C) WGA and anti-H-2K sort (Visser *et al.*, 1984).

can be used instead. Accordingly, av/FITC may be replaced by av/PE or by avidin/allophycocyanin (av/APC; Becton Dickinson) to detect anti-H-2K^k/biotin.

VI.   Critical Aspects of the Procedure

A.   Osmolarities, pH Values, and Densities

Since it is necessary to analyze and sort living cells, the osmolarity of the media and reagents is of importance. It is therefore useful to have easy access to an osmometer, so that the reagents can be checked regularly.

This remark concerns all reagents. The right osmolarity is of utmost importance for the MA of the density gradient, since the cells may swell or shrink as osmolarity changes, and, thereby, their density may change yielding a different result after centrifugation. Metrizamide, however, tends to dimerize at higher concentrations, so that its contribution to the osmolarity is not linearly related to its concentration. Therefore, we prepare two densities of isotonic MA solutions and mix those to obtain intermediate densities. The osmolarity gradient is then negligible.

If the number of bone marrow cells recovered in the low-density fraction is < 7% or > 13% of the loaded cells, the MA solution of the intermediate density is adjusted by adding aliquots of the high-density or the low-density solution, respectively, such that the recovery will be between 7 and 13% again—ideally 10%—in later experiments with the same reagents.

Other materials may be used for the density gradient, such as BSA, Nycodenz (Nyegaard, Oslo, Norway), or Percoll (Pharmacia Fine Chemicals, Uppsala, Sweden), as long as they yield ~ 10% of the loaded cells in the low-density fraction. With BSA, care should be taken that the osmolarity is correct (use dialyzed BSA). Furthermore, with BSA the pH of the gradient will be low (near values of 5.2–5.5), and this affects the density of hematopoietic cells, such that lower densities of BSA solutions should be used to obtain 10% of the cells in the top fraction. With Percoll the pH is increased and in our hands not stable during storage: it tends to approach a value of 7.8–8.0 after storage for one night, whereas we prefer a pH of ~ 7.0. A pH of > 7.5 is toxic for the *in vitro* clonogenic hematopoietic progenitor cells. Percoll solutions should therefore be freshly prepared or adjusted shortly before use.

The amount of cells loaded on the density gradient may also affect its physical properties. If we load the bone marrow cells in the bottom fraction (with density 1.10 g/cm^3), only 10% of the cells will move upward, and up to 5×10^8 cells can be loaded without causing streaming. If the cells are loaded in the top fraction (of 1.05 g/cm^3), 90% of the cells will move downward, so that only 5×10^7 cells can be loaded per gradient tube. It is advantageous to divide the bone marrow cells at the start of the separation procedure over two or more density gradients, so that occasional small accidents with one of the tubes do not necessarily result in a complete failure.

Erroneous osmolarities, which have caused us trouble in the past (apart from those in density materials), were found in the sheath fluid (commercially available salt solutions, whether or not supplemented with azide, and adjusted for human cells), in antibody solutions both from ascites and from hybridoma cell supernatant, and in GlcNAc solutions.

B. Sterility of Materials and Instruments

Since the cells usually have to be cultured after the sorting, contamination of the suspension should be avoided. Normal precautions are taken during the preparation and handling of the cell suspensions. In addition, the sorter is sterilized: the sheath fluid is first replaced by water, which is run through the tubing to the nozzle and backward to the sample tube position, in order to remove salt or other crystals. Next, 70% ethanol is run through the tubing and preferably is kept in the tubings overnight. Subsequently, the ethanol is replaced by sterile water; finally this is replaced by the sheath fluid of choice (sterile, colorless HH).

C. Avidin and the Opsonization of Transplanted Stem Cells

If the stem cells are studied *in vivo* after the sorting, they should not contain labels that trigger the immune system to remove them. A finding by Bauman *et al.* (1985) is of much help for the study of sorted stem cells: *in vivo* opsonization of biotinylated-antibody-labeled stem cells can be prevented by additional labeling with avidin prior to transplantation. This is particularly useful, if anti-H-2K^k biotin is employed. For some mouse strains the appropriate antibody against the H-2K^k haplotype is hard to obtain, or sometimes the biotinylation is suboptimal, causing loss of stem cells upon transplantation. One should avoid these conditions, and, if necessary, compensate for them by additional incubation with avidin (Van den Engh *et al.*, 1983; Bauman *et al.*, 1985; De Vries, 1988). Negative selection of the stem cells using the mAb 15-1.1 avoids these problems (De Vries, 1988).

D. WGA/FITC, WGA, and FITC

After prolonged storage, FITC may be released from WGA/FITC. If bone marrow cells are incubated with free FITC, they will all become fluorescent. Therefore, free FITC has to be removed from old solutions of WGA/FITC. A G-25 Sephadex column is used for that purpose (Bauman *et al.*, 1985). This does not, however, separate WGA/FITC from WGA, so that after several runs through the column, a higher concentration of the solution has to be added to the cells in order to obtain the same fluorescence intensity. This, however, may lead to concentrations of WGA that are sufficient for agglutination. Such high concentrations should be avoided for FCM.

E.   Coated and Cooled Collection Tubes

The sorting of sufficient numbers of stem cells takes several hours. In addition, the frequency of stem cells is low. Therefore, it is useful to coat the wall of the collection tube with serum or BSA, so that the deflected droplets efficiently slide to the bottom of the tube, and the sort recovery is close to 100%. The added serum or BSA will also help to maintain the viability of the sorted cells. On the other hand, it prevents the adherence of monocytes and other adherent cells to the wall of the tube, and, therefore, these have to be removed by other methods (e.g., using the mAb 15-1.1). Furthermore, although the stem cell are physically relatively stable cells that are not killed by keeping them for up to 8 hours at RT, it is useful to cool the sample and collection tubes, in order to preserve cellular molecules with a short half-life, if those are to be studied after the sorting, or, to shorten the lag phase at the culturing of the sorted cells.

VII.   Results and Discussion

A.   Sorting of Stem Cells Using WGA, 15-1.1, H-2K, and Rh123

Table I gives the numbers of bone marrow cells and of CFU-S at the consecutive separation steps. The CFU-S have long served as a measure for pluripotent stem cells. It was assumed, with good reason, that each of the spleen colonies in irradiated and transplanted mice represented between 10 and 20 transplanted pluripotent stem cells. It was assumed that only 5–10% of the tranplanted stem cells took hold in the spleen to form a colony (Visser et al., 1984). This seeding efficiency factor (the so-called f-factor) was deduced from serial transplantation experiments. According to these values, some of the sorted fractions summarized in Table I should be >100% pure stem cells. However, the f-factor was not determined for these fractions. Besides, the fractions differ with respect to the time of appearance of the spleen colonies. The Rh123-dull cells give virtually no colonies at 8 days after irradiation and transplantation, but high numbers at 14–16 days. In addition, the Rh123-dull fraction contains the cells with MRA, whereas the Rh123-bright fraction does not. In these two aspects the Rh123-dull fraction resembles pluripotent stem cells in mice 1–2 days after 5-fluorouracil treatment. These have been shown to take hold initially in the bone marrow after transplantation and not in the spleen (Van Zant, 1984). Part of their offspring may then migrate to the spleen to form

TABLE I

ENRICHMENT FOR CFU-S

Suspension sorted[a]	Number of cells (% of unfractionated bone marrow cells)	Number of spleen colonies per 10^5 transplanted cells	
		Day 8 CFU-S	Day 12 CFU-S
Unfractionated bone marrow	100	28.3	30.6
LD cells	10	121	128
WGA$^+$ LD cells	1	1830	2040
WGA$^+$/WGA$^+$ LD cells	0.5	2340	3870
WGA$^+$/anti-H-2K^{k++} LD cells	0.1	2100	7400
WGA$^+$/15-1.1$^{-/+}$ LD cells	0.1	3900	5300
WGA$^+$/WGA$^+$/Rh123$^+$ LD cells	0.1	240	3910
WGA$^+$/WGA$^+$/Rh123^{++} LD cells	0.4	4810	3790
WGA$^+$/anti-H-2K^{k++}/Rh123$^+$ LD cells	0.06	360	3600
WGA$^+$/anti-H-2K^{++}/Rh123^{++} LD cells	0.04	3700	4500
WGA$^+$/15-1.1$^{-/+}$/Rh123$^+$ LD cells	0.05	1060	8650
WGA$^+$/15-1.1$^{-/+}$/Rh123^{++} LD cells	0.05	6330	7840

[a] LD, Low density.

late-appearing colonies. The spleen-colony assay and the seeding efficiency factor, therefore, do not seem to be the appropriate tools to quantitate these stem cells. Since no other colony assay for PHSC is established, the absolute purity of the Rh123-dull fraction cannot be determined at present.

The Rh123-bright fraction contains cells that form many day 8 and many day 12 spleen colonies. Enrichment factors for those types of CFU-S of 200–250 are regularly obtained in this fraction. Although this seems to be inferior to those reported by Spangrude and co-workers (1988), the final result is better: the mice used by Spangrude have a very low content of CFU-S, ~10 times less than the mice in most other laboratories, including ours. They report an enrichment factor of 750, only three times more than we do; consequently, the Rh123-bright fraction contains ultimately a higher frequency of day 12 CFU-S than the cell suspensions sorted by Spangrude et al. (1988). Unfortunately, the MRA is completely lost in this fraction, while the 30 day radioprotective activity of pluripotent stem cells is not found in the Rh123-bright fraction. The day 12 CFU-S in this Rh123-bright fraction, therefore, do not represent pluripotent stem cells. However, the MRA and the radioprotective activity are found in Rh123-dull fraction, which, therefore, should be further studied to elucidate the nature of the hematopoietic stem cell.

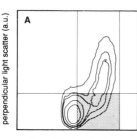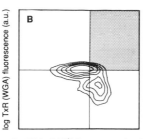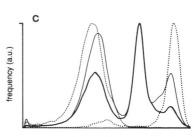

FIG. 4. Bivariate distributions of the light scatter (A) and fluorescence intensities (B) of mouse bone marrow cells labeled with WGA/TxR, anti-H-2K^k/biotin, av/PE, and the frequency distribution (C) of selected cells simultaneously labeled with 15-1.1/FITC: (—), all cells; (—), cells from blast cell window [marked square in (A)]; (·····), WGA/anti-H-2K$^+$ blast cells [squares of (A) and (B)]. Control sample (----) was incubated without mAb.

B. Multicolor Labeling of Hematopoietic Stem Cells

The subpopulation of interest for further studies can be identified in a two- or three-laser experiment with multiple labels. This is applied to screen candidate (newly raised) antibodies with anti-stem cell specificity (De Vries, 1988). An example is shown in Fig. 4. Low-density bone marrow cells labeled with WGA/TxR, 15-1.1/FITC, anti-H-2K^k/biotin, and av/PE show a cluster of rare cells that have FLS/PLS and fluorescence properties as expected from stem cells with these labels. The cluster contains ~0.4% of the low-density cells. New hybridoma cultures can be rapidly screened by this procedure for the presence of anti-stem cell antibodies. This avoids the laborious culturing of nonrelevant hybridomas. More importantly, this procedure ultimately will provide an alternative to detect and enumerate PHSC that will make the culture techniques for stem cell colony formation as well as the *in vivo* transplantation and repopulation methods superfluous. Such a stem cell-specific staining will be of use to diagnose the status of the PHSC compartment of the hematopoietic system. It is clear that this specific staining is likely to be a multilabel, multicolor one. The procedures described here may contribute to its eventual development.

REFERENCES

Baines, P., and Visser, J. W. M. (1983). *Exp. Hematol.* **11,** 701–708.
Bauman, J. G. J., Mulder, A. H., and van den Engh, G. J. (1985). *Exp. Hematol.* **13,** 760–767.
Bertoncello, I., Hodgson, G. S., Bradley, T. R., Hunter, S. D., and Barber, L. (1985). *Exp. Hematol.* **13,** 999–1006.

Bertoncello, I., Bartelmez, S. H., Bradley, T. R., Stanley, E. R., Harris, R. A., Sandrin, M. S., Kriegler, A. B., McNiece, I. K., Hunter, S. D., and Hodgson, G. S. (1986). *J. Immunol.* **136**, 3219–3224.

Bol, S. J. L., van Vliet, M., and van Slingerland, R. (1981). *Exp. Hematol.* **9**, 431–443.

De Vries, P. (1988). Ph. D. thesis, Erasmus University, Rotterdam.

Harris, R. A., Hogarth, P. M., Wadeson, R. J., Collins, P., McKenzie, I. F. C., and Pennington, D. G. (1984). *Nature (London)* **307**, 638–641.

Hoang, T., Gilmore, D., Metcalf, D., Cobbold, S., Watt, S., Clark, M., Furth, M., and Waldmann, H. (1983). *Blood* **61**, 580–588.

Lord, B. I., and Spooncer, E. (1986). *Lymphokine Res.* **5**, 59–72.

Migliaccio, A. R., and Visser, J. W. M. (1986). *Exp. Hematol.* **14**, 1043–1048.

Migliaccio, G., Migliaccio, A. R., and Visser, J. W. M. (1988). *Blood* **72**, 944–951.

Mulder, A. H., and Visser, J. W. M. (1987). *Exp. Hematol.* **15**, 99–104.

Mulder, A. H., and Visser, J. W. M. (1988). *Thymus* **11**, 15–27.

Muller-Sieburg, C. E., Whitlock, C. A., and Weissman, I. L. (1986). *Cell* **44**, 653–662.

Scheven, B. A. A., Visser, J. W. M., and Nijweide, P. J. (1986). *Nature (London)* **321**, 79–81.

Sprangrude, G. J., Heimfeld, S., and Weissman, I. L. (1988). *Science* **241**, 58–62.

Van Bekkum, D. W., van Noord, M. J., Maat, B., and Dicke, K. A. (1971). *Blood* **38**, 547–558.

Van den Engh, G., and Stokdijk, W. (1989). *Cytometry* **10**, 282–293.

Van den Engh, G., Bauman, J., Mulder, A., and Visser, J. W. M. (1983). In "Haemopoietic Stem Cells" (Sv. Aa. Killmann, E. P. Cronkite, and C. N. Muller-Berat, eds.), pp. 59–74. Munksgaard, Copenhagen.

Van Zant, G. (1984). *J. Exp. Med.* **159**, 679–690.

Visser, J. W. M., and Bol, S. J. L. (1981). *Stem Cells* **1**, 240–249.

Visser, J. W. M., and De Vries, P. (1988). *Blood Cells* **14**, 369–384.

Visser, J. W. M., Bol, S. J. L., and van den Engh, G. (1981). *Exp. Hematol.* **9**, 644–655.

Visser, J. W. M., Bauman, J. G. J., Mulder, A. H., Eliason, J. F., and De Leeuw, A. M. (1984). *J. Exp. Med.* **59**, 1576–1590.

Watt, S., Gilmore, D., Davis, J. M., Clark, M. R., and Waldmann, H. (1987). *Mol. Cell. Probes* **1**, 297–326.

Chapter 42

Fluorescent Cell Labeling for in Vivo and in Vitro Cell Tracking

PAUL KARL HORAN, MERYLE J. MELNICOFF, BRUCE D. JENSEN, AND SUE E. SLEZAK

Zynaxis Cell Science, Inc.
Malvern, Pennsylvania 19355

I. Introduction

A. Background

In cell biology a valuable goal is to arrive at a better understanding of how specific cell types perform *in vivo*. However, to study the properties of cells in the intact animal presupposes the ability to track cells within the animal's body. What makes any cell migrate to a disease or injury site, as part of the host defense mechanisms, is an important aspect of mammalian biology and may provide useful information in the management and treatment of disease.

A great deal of attention has been given to the molecular nature of intercellular recognition processes, an example of which is the lymphocyte T-cell receptor. While these receptors are important, they are functional only at close range. The underlying question is: "What causes the specific T cell to migrate to the disease site?" Is it a receptor-based function or phenomenon, or are there chemotaxis-like signals that cause immune cells to move into disease sites? The ability to track the *in vivo* migration of labeled cells could also provide insight into differentiation processes in tissues, such as thymus or bone marrow. Cell-tracking technology could be applied to tumor biology, to determine whether tumor-infiltrating lymphocytes migrate to metastatic sites. The ability to study cellular migration into atherosclerotic plaque would be of extreme value in the study of the

469

pathophysiology of coronary disease. Important information may also be obtained by studying the migration of cells into and out of an organ prior to and during the rejection process after organ transplants.

B.   Current Methodology

It has been difficult to use radioactive probes for lymphocytic cell tracking because of problems such as carrier toxicity, internal irradiation effects, as well as poor uptake and rapid elution of the radioactive probe from the cell of interest.

A number of fluorescent probes have been used for these purposes; however, each has exhibited significant limitations. Fluorescein isothiocyanate (FITC) and rhodamine isothiocyanate (RITC) have been used by Butcher *et al.* (1980) and Weissman and Butcher (1980) to label lymphocyte surface proteins. These probes have limited applications because of their rapid elution from cells, leading to reduction of the signal/noise ratio during the first 5 hours of tracking (Olszewski, 1987). In addition, cells that have been labeled with the FITC demonstrate some alterations in functional capability. Other fluorescent probes that have been used by Keller *et al.* (1977) to study cell fusion are fluorescein and rhodamine B conjugated with 18-carbon aliphatic tails. These probes have problems with elution from cells, making long-term studies of cell trafficking difficult. In a later development, Honig and Hume (1986) have used an indocarbocyanine dye for labeling of neurons. Using the procedures described by Honig, one finds the uniform labeling of cells to be a major problem.

We describe here the use of PKH2 cell linker kits. The label is an aliphatic fluorescent chromophore. The staining methodology allows for labeling in the lipids of the cell membrane rather than surface protein labeling. It is made available in two forms: PKH2-GL for general cell labeling and PKH2-PCL for endocytic labeling in phagocytic cells (Zynaxis Cell Science, Inc., Malvern, PA).

II.   Applications

Cell types that have been efficiently labeled with these cell linker kits (PKH2-GL and PKH2-PCL) include tissue culture cells of both adherent and suspension types, and primary cells such as T cells, B cells, macrophages (MΦ), platelets, red blood cells (RBC), bone marrow cells, glial cells, and neurons.

An important feature of these dyes is their stability in cells, which has made it possible to follow the division history of a cell based on fluorescence intensity *in vitro*. If a cell is stained at time zero with a fluorescence intensity of X, upon division, the fluorescence intensity of each daughter cell will be approximately $X/2$ (Horan *et al.*, 1988).

Another application of fluorescent tagging that has been developed is lifetime measurements of cells in animals. For example, one can label RBC with the fluorescent cell linkers *in vitro* and inject the labeled cells back into the animal. Since the RBC does not divide, and the chromophore has a slow elution profile, it is possible to track the cell for its lifetime in circulation. This kind of lifetime measurement could also be performed for other cell types such as platelets. Measurement of blood volume is possible by taking a cohort of cells via venipuncture and uniformly labeling 100% of these cells. After determining the total number of labeled cells in a measured volume, they are then injected into the animal and the dilution factor of fluorescence-positive cells gives a measure of the total blood volume.

The labeling methodology described here has also been applied to a flow cytometric (FCM) cell-mediated cytotoxicity assay (Slezak and Horan, 1989a). In this assay it is imperative that the target cells can be distinguished from the effector cells based on a fluorescent signal. Using the PKH2 fluorescent tracking dyes (emission at 504 nm; see Fig. 1), it is possible to label a target cell population (YAC-1) and identify this population from the effector cell population on the basis of green-fluorescence intensity. Using propidium iodide (PI) as a viability marker (emission at

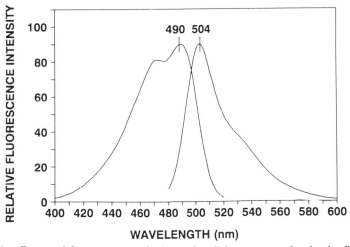

FIG. 1.   Corrected fluorescence excitation and emission spectrum for the dye PKH2.

570 nm), one can identify dead-cell populations on the basis of red fluorescence. Using two-parameter analysis it is possible to distinguish four populations: live targets, dead targets, live effectors, and dead effectors. It should be noted that if the assay is carried out for a sufficient period of time to detect target cell growth, by reduction in fluorescence intensity, a measurement of cytostasis is also possible.

This technology could also be applied to radiation biology studies permitting identification of growing cells after radiation exposure. These cells could be detected by a decrease in fluorescence intensity due to cell growth. Evaluation of the surviving cells might then provide information concerning the cell's protective mechanisms.

The study of *in vivo* differentiation processes may also be possible by isolating and labeling various cell precursors. These cells could then be introduced back into the animal via iv injection, and periodic bone marrow taps could be used to recover a percentage of the cells injected. Using a combination of a green tracking dye to monitor the growth of labeled precursors and phycoerythrin (PE)-conjugated monoclonal antibodies (mAb) to identify cells of a specific lineage, one can characterize the cell differentiation process.

Another potential application of this fluorescent cell-linking technology would be to identify the migrational patterns of cells used in adoptive transfers. Using the adjuvant rat model of arthritis, one could isolate immune cells from the affected joint, uniformly label them with a tracking dye, inject them into naive recipients, and determine whether the fluorescent cells track back to the putative disease site.

As can be seen, there are numerous existing and potential applications for this technology in understanding cell migrational patterns. This understanding may allow for subsequent evaluation of the molecular mechanisms employed and possibly manipulation of cell trafficking patterns.

III.   Materials, Methods, and Critical Aspects of General Cell Labeling

A.   Materials

1. A uniform single-cell suspension in tissue culture medium
2. Culture medium with serum
3. Culture medium without serum or Dulbecco's Ca- and Mg-free phosphate-buffered saline (PBS)

4. PKH2-GL fluorescent cell linker kit (composed of PKH2 dye stock and diluent A)
5. Serum albumin or other system-compatible protein source
6. Polypropylene conical centrifuge tubes
7. Temperature-controlled centrifuge ($0-1000$ g)
8. Instrument(s) for analysis of fluorescence (fluorometer, fluorescence microscope, flow cytometer, or fluorescence image analysis instrumentation)
9. Sterile hood
10. Hemocytometer or Coulter counter
11. Slides and coverslips

B. Methods for General Cell Labeling

The following procedure gives cell concentrations and dye concentrations found to be optimal for tissue culture lines used in our laboratory. You must determine optimum conditions for your cell type(s) and experimental purposes; we suggest an initial experiment similar to that shown in Fig. 2. The range of cell concentrations chosen will be a function of (1) total number of cells available or desired and (2) maximum desired volume of staining solution. Evaluate cell viability (e.g., PI exclusion),

RED BLOOD CELL NUMBER x 10^7

FINAL PKH2 CONCENTRATION (MICROMOLAR)	1	10	100	200	400
1	293 80.2%	261 99.3%	48 99.8%	17 99.9%	4 99.9%
2	627 55.1%	654 97.4%	125 99.7%	54 99.8%	7 99.9%
5	----- 7.4%	750 67.0%	431 99.6%	208 99.8%	26 99.9%
10	----- 6.9%	608 21.8%	632 99.5%	390 99.8%	54 99.9%
20	----- 0.0%	733 28.4%	799 99.5%	672 99.7%	211 99.8%

FIG. 2.   Matrix analysis optimizing labeling of rabbit red cells with PKH2-GL. Fluorescence intensity was measured with an EPICS profile; recovery was measured by counting cells using a Coulter counter. Note that at 10^7 cells/ml optimal staining is not achieved without lysing the cells. At 10^9 cells/ml excellent recoveries were seen at all dye concentrations. Blocked-in area represents samples with a fluorescence per cell above channel 100 (top number in each box) and sample recovery >95% (bottom number in each box).

recovery, stain intensity (mean fluorescence), and stain uniformity (coefficient of variation, CV) to identify a working range of cell and dye concentrations. Within that working range, optimum staining conditions may be selected by testing the effect of staining on pertinent cell functions.

1. If cells are adherent, they must first be removed from culture vessels using proteolytic enzymes (e.g., trypsin–EDTA) and put into a single-cell suspension.
2. Perform all steps at 25°C. Wash the cells once using media without serum.
3. Centrifuge the cells (400 g) for 5 minutes into a loose pellet. The total number of cells should be ~2×10^7 per sample.

For a final concentration of 2×10^{-6} M PKH2 and 10^7 cells/ml of diluent:

4. Prepare 4×10^{-6} M PKH2-GL (2×) in polypropylene tubes. The amount of dye added should be <1% of the individual sample volume. If a greater dilution of the dye stock is necessary, make an intermediate stock by diluting into 100% ethanol. Prepare 1 ml of dye using diluent supplied with kit. The preparation should remain at room temperature (RT; 25°C).
5. After centrifuging cells, carefully aspirate the supernatant leaving no more than 20–30 μl of supernatant on the pellet.
6. Tap the button of cells in the centrifuge tubes so that they are resuspended in the residual media in the tube and add 1 ml of diluent to resuspend cells. Do not vortex.
7. Add the 1 ml of the 2× dye to 1 ml of the 2× cell suspension. Immediately mix the sample by gentle pipeting or inversion of the capped tube.
8. Periodically, gently invert the tube to assure mixing, during the 1- to 10-minute staining period at 25°C.
9. Add an equal volume of serum to stop the staining reaction, and then further dilute with an equal volume of complete media.
10. Underlayer the cells with 5 ml of serum, albumin, or protein solution if compatible with your experimental system.
11. Centrifuge the cells out of the staining solution and into the protein solution underlayer by spinning at 400 g for 10 minutes at 25°C.
12. Remove the supernatant composed of stain and medium/protein.
13. Add complete medium to resuspend the cells to the desired concentration. Wash twice in complete medium.
14. Examine the cells using fluorescence microscopy, FCM, or fluorescence-based image analysis. (See Fig. 1 for excitation emis-

sion spectrum of PKH2.) The stained sample should be checked for cell recovery, cell viability, and fluorescence intensity, which should be uniform and distinguishable from background autofluorescence.

C. Critical Aspects of General Cell Labeling

1. Dye stocks should be stored at RT, and must be examined for crystals prior to use. If crystals are noted in the dye stock, it should be warmed slightly in a 37°C water bath, and/or sonicated to allow crystals to be redissolved into solution. Dye stocks should also be kept tightly capped.
2. The diluent can be stored at RT or refrigerated. Note, however, that it does not contain any preservatives or antibiotics and should be kept sterile.
3. Concentrations of dye should be made immediately or prior to use. Do not store the dye in diluent A.
4. No azide or metabolic poisons should be present at the time of PKH2 staining.
5. It is imperative that single-cell suspensions be used, to obtain uniform staining.
6. Prior to staining, remove all serum proteins and lipids, which will reduce the effective concentration of the dye.
7. During the staining procedure, the cells are pelleted prior to addition of the dye. It is important that the cell button be resuspended in a volume of supplied diluent A prior to the addition of dye. Do not use a vortex to mix the cells as it tends to result in poor cell recoveries. Cells and dye should be prepared separately at 2× concentrations in diluent and mixed in equal volumes for staining. Dye stock should not be added directly into the cell suspension, nor should dye solution be added directly to the cell pellet.
8. The dye and diluent should be applied to the cells for as short a time as possible. If saturation of staining is complete in 1 minute, then the reaction should be stopped at that point in time. Additional exposure to the diluent and dye may have some toxicity toward specific cell types. To evaluate the diluent effects, it is important that the cells be exposed to the diluent alone using the same procedures as those used for staining. The resulting washed cells must then be evaluated for functional impairment. If there is an effect on functional capabilities, contact the manufacturer (Zynaxis Cell Science, Inc., Malvern, PA).
9. The staining reaction should be stopped by adding an equal volume of serum, if possible. Serum lipids absorb the free lipophilic dye and

prevent additional staining of the cells. The addition of complete medium to the serum and dye solution serves to cause any remaining free dye to form micelles and further retards any staining reaction. Do not centrifuge the cells in diluent A prior to the addition of serum. Washing efficiency can be increased if it is possible to use serum proteins or albumin in the stop and washing solutions.

10. After removing the cells from diluent A and stain, they should be washed three to five times using medium. Do not use diluent A.

11. The general cell-labeling procedure can be used for monocytes, MΦ, lymphocytes, or any other cell type where general membrane labeling is desired. It is an *in vitro* procedure and has not been fully characterized for *in vivo* labeling of cells. The PKH2-PCL labeling kit may also be used for selective *in vitro* or *in vivo* labeling of phagocytic cells by endocytosis.

12. General cell labeling should be performed prior to mAb staining. The cell-tracking probes will remain stable during the monoclonal staining at 4°C; however, capping of the mAb is highly probable if the general cell labeling (25°C) is carried out subsequent to antibody labeling.

13. Samples may be fixed using 2% paraformaldehyde and have been found to be stable for up to 3 weeks.

IV. Materials, Methods and Critical Aspects of *in Vivo* Phagocytic Cell Labeling

A. Materials for Cell Labeling

1. Mice or other animals whose phagocytic cells are to be labeled
2. PKH2-PCL fluorescent cell linker kit (composed of PKH2 dye stock and diluent B)
3. Sterile 1-cc syringes with 26-gauge needles (one per animal, plus one extra)
4. Dulbecco's PBS
5. Bunsen burner or alcohol burner in sterile hood
6. Absolute ethanol
7. Sterile gauze squares
8. 70% ethanol in wash bottle
9. Instrument for analysis of fluorescence (e.g., fluorescence microscope, flow cytometer); fluorescein filters to detect PKH2

B.   Procedure for *in Vivo* Labeling of Resident Peritoneal MΦ

Note that the procedure described here is designed to label selectively resident peritoneal MΦ of mice. The resident MΦ (in the peritoneum at the time of injection) can be distinguished from subsequently recruited MΦ, which are not labeled by the green fluorescent tracking dye, (see Fig. 3). This procedure can be adapted to label other phagocytic cells (e.g., neutrophils, resident alveolar MΦ, or monocytes), either *in vivo* or *in vitro*. The reagents in the Zynaxis PCL cell linker kit were formulated to favor selective labeling of phagocytic cells over nonphagocytic cells (e.g., lymphocytes) present in the population.

This procedure has been optimized for labeling the resident peritoneal MΦ of mice weighing ~20 g. Labeling conditions, including dye concentration and volume to be injected ip, should be optimized for larger mice or other species.

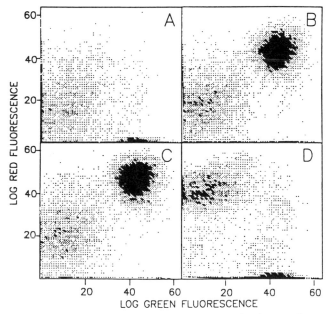

FIG. 3.   Two-color histograms of labeled peritoneal cells after immunofluorescence labeling. (A) PE–goat anti-rat Ig control; no significant red population is seen. (B) F4/80, the red- and green-stained cells (40.9% of total cells) represent MΦ. (C) Mac 1, a MΦ and polymorphonuclear cell marker, also labels the green cells (42%). (D) mAb 14.8, a B-cell marker, labels 39% of the peritoneal cells, which includes the population of cells dimly stained green. Reprinted from Melnicoff *et al.* (1988a) by permission of Wiley/Liss.

1. Remove 0.05 ml PKH2 dye ($1 \times 10^{-3}M$) from kit and mix thoroughly with 0.95 ml absolute ethanol. This will provide 50 μM working stock dye solution.
2. Lift up the center of the metal tab of one diluent bottle under a sterile hood. Do not remove the metal cap or stopper.
3. Remove 0.1–0.2 ml of the working stock (50 μM) of PKH2 with a sterile tuberculin syringe and needle. Sterilize the diluent stopper by passing through the flame. Inject 0.05 ml of the dye into the diluent bottle. It is essential that the contents of the diluent vial remain sterile, as any contamination will induce inflammation into the peritoneal cavity. The dye concentration in the diluent will be 0.2–0.25 μM.
4. Vigorously shake the bottle to mix the dye and diluent. Shake again before removing each sample for injection.
5. Withdraw 0.5 ml of the diluted dye into a fresh sterile syringe for each animal to be injected. Flame the stopper before inserting each needle. Flame the needle again after withdrawing the dye, then flame the needle cap and place it over the needle.
6. Repeat steps 4 and 5 for each syringe to be filled. Use a fresh syringe and needle for each animal.
7. Inject the contents of each syringe ip. Before each injection, swab the abdominal fur with a gauze pad soaked in 70% ethanol. This procedure will minimize the incidence of inflammation from the ip injection.
8. Labeled peritoneal MΦ may be harvested by peritoneal lavage at any time from 2 hours to 21 days after the ip injection of PKH2 dye.
9. Procedures for immunofluorescence staining and functional analysis (including cytotoxicity and ectoenzyme assays) of PKH2-labeled resident MΦ (7) are described in Melnicoff *et al.* (1988a). If the cells are to be evaluated by immunofluorescence, do not pool the cells recovered from the individual animals.

C. Critical Aspects of Phagocytic Cell Labeling

1. This procedure is designed to selectively label resident MΦ. If monocytes/MΦ are recruited to the peritoneum as the result of inflammation induced by the ip injection, they will not be labeled by the dye. The peritoneal cavity is very sensitive to inflammation, and the procedure just described utilizes several precautions to minimize this inflammation, including (1) keeping the injection vial and syringes

absolutely sterile, (2) using a fresh disposable syringe and 26-gauge needle for each animal, and (3) wiping the abdominal fur with 70% ethanol before each injection.

2. In spite of these precautions, we have routinely observed some injection-induced inflammation in as many as 50% of the mice. This figure will decline with practice. Injection-induced inflammation is detected, following immunofluorescence labeling and FCM analysis, as either two populations of MΦ (dye-labeled and unlabeled), one broad population of MΦ with variable stain intensity, or very few to no labeled MΦ in extreme inflammation (Melnicoff *et al.*, 1988b, 1989). At 24 hours after injection, animals with inflammation can be identified by elevated peritoneal cell number and elevated neutrophil counts, relative to untreated animals.

3. We routinely inject extra animals for each study and then, after analyzing the immunofluorescence pattern and differential cell counts of each animal, exclude data from the animals with inflammation.

4. In the absence of inflammation, resident peritoneal MΦ can be distinguished from recruited MΦ by their green-fluorescence intensity for up to 28 days *in vivo*.

5. The diluent in the Zynaxis PCL kit has not been tested for lipopolysaccharide (LPS) content. Contact Zynaxis if you require LPS information.

6. The staining conditions should be optimized for each species (and strain) of animal tested. Contact Zynaxis Cell Science, Inc. (Malvern, PA) if you need advice for optimizing this procedure in your animal model system.

7. We recommend preparing the final dilution of dye in the diluent bottle supplied by Zynaxis to minimize handling (and possible contamination) of the diluent. Any additional handling of the diluent may increase the incidence of inflammation induced from the ip injection.

8. PKH2 diluted in the Zynaxis diluent should be discarded after each day's use. Each bottle of diluent is sufficient to label the peritoneal cells of 20 individual mice (assuming 0.5 ml injected per mouse).

9. The PKH2-PCL staining kit can also be used to label phagocytic cells *in vitro*. The concentration of PKH2 for *in vitro* labeling should be much higher than that used for *in vivo* labeling. The procedure described for general cell labeling can be adapted for labeling phagocytic cells *in vitro*, but using the PCL diluent in place of the GL diluent to enhance labeling of phagocytic cells in the population.

V.   Materials, Methods, and Critical Aspects of Platelet Labeling

Because of the fragility of platelets, a special protocol has been developed to label platelets with the cell-tracking compounds. It differs significantly from the general cell labeling procedure. Platelets are extremely sensitive to centrifugation in the dye–diluent mixture. Even the addition of medium or a protein source has not been effective in reducing this sensitivity, and therefore a column purification step has been developed to remove the labeled platelets from the dye–diluent mixture.

A.   Materials for Preparation of Column for Platelet–Dye Separation

1. Sephadex G-100 (particle size 40–120 μm) chromotography filtration powder
2. PBS
3. Hotplate–stirplate apparatus
4. Polypropylene centrifuge tubes (50 ml or larger)
5. Acetone
6. Centrifuge
7. Hanks' balanced salt solution (HBSS)
8. 30 μm nylon mesh (Small Parts, Inc.)
9. 10 ml Polypropylene syringe
10. O-ring (no. 8)
11. Plastic pipets
12. Ringstand with clamps
13. Tubing to fit hub of syringe
14. Tubing clamp

B.   Materials for Platelet Staining

1. Prepared purification column (see Section V,C)
2. Vacutainer and blood-drawing equipment using citrate as anticoagulant
3. Source for platelets (animals, humans, platelet packs)
4. PKH2-GL fluorescent cell linker kits (composed of PKH2 dye stock and diluent A; Zynaxis Cell Science, Inc., Malvern, PA)
5. Conical centrifuge tubes (15 ml)
6. Acid citrate–dextrose anticoagulant (14.6 mg trisodium citrate, 5.3 mg citric acid, and 16 mg dextrose per 1 ml)

7. Centrifuge
8. Plastic pipets (must *not* be glass!)
9. Cell-counting apparatus (hemocytometer or Coulter counter)
10. Analytical instrument to measure fluorescence, such as spectro-fluorometer, fluorescence microscope, flow cytometer, or fluorescence image analysis equipment.

C. Method for Preparation of Column for Platelet–Dye Separation

1. Weigh out 5 g of Sephadex G-100 powder and place it into a beaker containing 400 ml PBS.
2. Cover the beaker with a watchglass or parafilm and place the beaker on a hotplate–stirplate. With stirring, bring the mixture to a boil and continue boiling for 4–5 hours to rehydrate the gel filtration material. Water may be added as necessary to maintain the volume in the beaker.
3. After 4–5 hours, place the Sephadex mixture into 50-ml or larger conical centrifuge tubes. Fill each tube only one-half full with the Sephadex mixture. Add acetone to fill the centrifuge tubes, and cap. Invert the tubes to mix, and centrifuge the tubes at 500 g for 10 minutes to pellet the Sephadex gel.
4. Remove the supernatant from the Sephadex pellet and resuspend the gel in HBSS for washing.
5. Wash the Sephadex three times using HBSS by resuspending the pellet and centrifuging at 500 g for 10 minutes to pellet the gel. Resuspend the gel in 100 ml HBSS.
6. To prepare the column, cut a circle of nylon mesh to fit the bottom of the 10-ml syringe. Remove the plunger and place the mesh into the bottom of the syringe. Carefully place an O-ring over the mesh to hold it in place. Attach a small piece of tubing to the hub of the syringe and place a tubing clamp on the tubing. Position the syringe vertically using a clamp and the ringstand. Place a flask or beaker under the syringe to collect the eluant from the attached tubing.
7. Stir the Sephadex–HBSS solution to obtain a uniform slurry. Pipet the mixture into the syringe. Fill the syringe with the slurry and slowly release the tubing clamp so that small drops elute from the tubing. Too fast a flow rate will result in uneven packing of the Sephadex column. Continue adding Sephadex until the packed volume of the Sephadex is about four-fifths the volume of the syringe. At no time should the column be allowed to run dry; to prevent this, close the clamp or add more slurry or HBSS.

8. Once the column is packed, HBSS may be allowed to flow through the column to wash the gel thoroughly. Columns may be used immediately or may be covered with parafilm and placed at 4°C for storage. If long-term storage is anticipated, the column should be washed through and stored with HBSS containing 1% sodium azide to prevent bacterial contamination.

D. Method for Platelet Labeling

1. Prepare column for separation of labeled platelets from dye–diluent mixture. Be sure the column is at RT. If the column has been stored in azide, wash the column thoroughly with HBSS to remove the azide.
2. Obtain platelets. If platelets are to be prepared from whole blood, collect blood in a syringe containing a volume of citrate anticoagulant equal to one-tenth the volume of the syringe.
3. Use plastic pipets and centrifuge tubes for the entire procedure to prevent platelet activation. Centrifuge the whole blood at 100 g for 10 minutes at 25°C to obtain platelet-rich plasma (PRP). Remove the cloudy serum supernatant, being careful not to disturb the RBC pellet, and place into a 15-ml polypropylene centrifuge tube.
4. Wash the PRP using an equal volume of citrate anticoagulant. Remove 100 μl of this solution and place into a Coulter vial containing 9.9 ml HBSS for a platelet count. Centrifuge the PRP–citrate mixture at 1000 g for 10 minutes at 25°C to pellet the platelets.
5. Remove the supernatant from the platelet pellet and tap the tube to disperse the pellet gently into the residual fluid. Resuspend the pellet gently in diluent A solution to a concentration of 8×10^8 cells/ml.
6. If a control preparation is required, separate the resuspended platelets into two centrifuge tubes at this point.
7. Prepare a volume of 2× dye solution in diluent A equal to the volume of resuspended platelets. Add 4 μl of dye per 1 ml of diluent A to obtain a concentration of 4 μM.
8. Add the 2× platelet solution to the 2× dye solution and gently mix the sample. The final concentration is 4×10^8 platelets/ml in 2 μM dye. Stain the platelets for 10 minutes at RT. Gently invert the sample to ensure even staining.
9. During the 10-minute staining period, open the tubing clamp and slowly allow the fluid in the column to drop to the head of the packed Sephadex bed. Tighten the clamp to stop the flow through the column.

10. After 10 minutes, pipet the platelet–dye–diluent mixture gently onto the column, being careful not to disturb the packed gel. Open the clamp and allow fluid to run slowly through the column. When all of the platelet–dye–diluent sample has been loaded onto the column, slow the flow rate down and clamp the tubing as the last of the sample enters the packed bed. Gently add 1–2 ml of HBSS to the top of the column and slowly open the tubing clamp. Begin collecting 1-ml fractions off of the column. Continue adding HBSS to the column until the cloudy platelet eluant is collected. Continue collecting fractions from the column until the eluant is clear. At this point, the labeled platelets have passed through the column and the residual dye is retained in the gel.

11. Pool the platelet-containing fractions into one polypropylene centrifuge tube and centrifuge at 1000 *g* for 10 minutes at 25°C. Remove the supernatant and resuspend the platelet pellet in a medium of choice for further experimentation or evaluation.

E. Critical Aspects of Platelet Labeling

1. Use only plastic labware for platelet preparation. Glass pipets tend to enhance platelet activation and aggregation.

2. All serum proteins must be removed before platelets can be labeled with PKH2 tracking dyes. The presence of serum proteins will lower the effective dye concentration.

3. Optimal dye concentration for platelets from different species should be determined on an individual basis. Some type of functional evaluation should be carried out to ensure that the concentration used does not affect platelet function.

4. Labeling of platelets must take place in diluent A. Do not use normal physiological saline solutions.

5. The platelets must be separated from the dye–diluent mixture by use of the prepared column. Centrifugation of the platelets in the dye solution will lead to activation and aggregation.

6. Concentrations of dye should be made immediately prior to use. Do not store the dye in diluent A.

7. If crystals are noted in the dye stock, it should be warmed slightly in a water bath and/or sonicated to allow crystals to be redissolved.

8. Columns that have been prepared and stored at 4°C should be allowed to warm to RT before using. Any columns stored with azide should be thoroughly washed with HBSS before loading platelets.

9. Dye and diluent solutions should be maintained sterile as no antibiotics or preservatives are present in these solutions.

VI. Additional Information

A. Fixation

No organic fixatives can be used (alcohol, acetone, xylene), as they will remove the dye from the tissue. Only water-based fixatives can be used. Thus paraformaldehyde (2% final) or formalin are acceptable fixatives, and should be added in an equal volume at a 2× concentration to cells resuspended in protein-free PBS or medium.

B. Tissue Sections

Because of the nature of these dyes, paraffin embedding or alcohol fixation of tissues is unacceptable; frozen sections, however, are excellent for these compounds. For more complete information; contact the manufacturer (Zynaxis Cell Science, Inc., Malvern, PA).

C. Qualitative Cell Tracking

Since the tracking dye is extremely lipophilic, it is possible to make bulk determinations on the whereabouts of labeled cells. After injecting labeled cells, specific organs can be removed from sacrificed animals, weighed, and homogenized in distilled water. An equal volume of butanol or 2-octanone is added to the homogenate in a closed vessel and vortexed thoroughly. [This step should be carried out in a fume hood.] The samples are then centrifuged at 2,000–4,000 g to permit separation of the emulsion. The organic layer is then removed and analyzed fluorometrically to determine the presence of dye (and therefore cells) in the organ.

VII. Controls and Standards

A. *In Vitro* Growth Experimentation

When using these dyes to monitor cell division *in vitro*, the experiment may continue over a several-day period. It becomes important to control both the biological and instrument variation over this several-day period.

Instrument variation can be monitored by the use of fluorescent microspheres. On a daily basis, the instrument is set up using the same filters and laser power, and adjusting the high voltage on the photomultiplier tube until the modal intensity is identical to the previous day's values.

It is important to establish the fluorescent intensity at time zero and, if possible, to harvest at each time point a sample that has not exhibited any cell growth. Several approaches are potentially applicable. One could fix cells in paraformaldehyde (2% final concentration in PBS) at time zero, and place them at 4°C. These fixed cells could then be sampled concurrently with test samples. Another method is to use cells that have been irradiated or metabolically blocked to ensure a nondividing population. If the cell line being studied can tolerate storage at RT, it is possible to label them at time zero and store them at RT for subsequent sampling. Finally, it is important to ascertain the degree of dye transfer between cells, as each cell type exhibits slightly different dye retention properties. For this reason it is advisable to use a mixture of 50% labeled cells with 50% nonlabeled cells in the same culture. Monitoring the intensity of both the labeled and nonlabeled populations over time permits the determination of the amount of dye transfer between populations.

B.   *In Vivo* Cell Tracking Experimentation

Instrumentation variation can be controlled by the use of fluorescent beads as previously described. The major concerns with *in vivo* cell tracking are related to dye retention by the labeled cells and the potential for alteration of migration patterns resulting from the labeling.

To control for dye elution, it is advisable to use a cell that does not divide and has a long *in vivo* lifetime. For this reason, autologous RBC should be labeled and used as a monitor for approximating *in vivo* dye elution.

To evaluate the effect of labeling on cell tracking patterns, it is advisable to perform preliminary studies to compare results using PKH2 and more traditional labeling methods such as FITC or RITC, or radioisotopic labeling. It is also possible to use genetic markers to evaluate effects of cell labeling on migrational patterns. Cells from congenic animals can be injected into recipient animals with or without the cell-tracking dyes. Their biodistribution can then be determined and the effects of cell labeling ascertained.

If it is desirable to monitor cell growth rates *in vivo*, the dye elution rate and initial fluorescence intensity must be determined. The methods for evaluating fluorescence at time zero have been described already.

VIII.   Instruments

The tracking dyes have been used with several different fluorescence microscopes, flow cytometers (all types), and a Meridian ACAS 470 fluorescence image analyzer (Meridian Instruments, Okemus, MI).

IX.   Results

YAC-1 cells were labeled according to the methods described for the general labeling procedure. Their fluorescence intensity was measured using a Coulter EPICS 753 (488-nm excitation, 525–20 nm emission) flow cytometer. Figure 4 displays the relative fluorescence intensity for positively stained YAC-1 cells and nonstained cells. Note that the fluorescence intensity of the positive cells is more than 3 orders of magnitude higher than that of the unstained cells.

In order to be useful for tracking cells, these compounds must not have any affect on the function of the cells to be studied. One such function is the doubling time. A culture of YAC-1 cells was split, and one-half was labeled using the general labeling procedure described already. Each aliquot was placed into the incubator, and periodically a sample was removed to measure the cell number. Figure 5 displays the Coulter count for cells that are labeled and those that are not labeled. Note that the growth rate of YAC-1 cells is unaffected by the labeling of the cell membrane. Furthermore, this experiment has been performed on many cell types and the same result has been observed. It should be noted that it is possible to add enough dye to affect the growth rate. In tissue culture the upper limit is ~10 μM dye and 10^7 cells/ml. For primary cells the dye concentrations

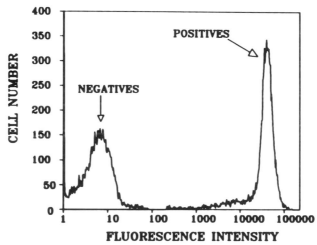

FIG. 4.   Fluorescence intensity distribution for YAC-1 cells labeled with PKH2-GL. Logarithmically growing YAC-1 cells were stained in 10^{-5} M dye at a concentration of 10^6–10^7 cells/ml for 5 minutes.

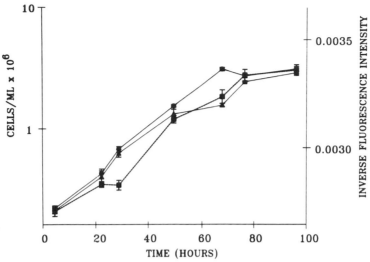

FIG. 5.   Cell number (●, stained; ▲, unstained) and inverse fluorescence intensity (■) as a function of time. Separate flasks of labeled and nonlabeled control cultures (YAC-1) were set at 2×10^5 cells/ml at time zero. Cell counts were obtained using a Coulter ZBI cell counter (Coulter Electronics, Inc., Hialeah, Fl). Mean log fluorescence intensities were determined for each time point, and the inverse was calculated and plotted.

tend to be lower than those used for tissue culture cells. However, each cell type must be tested to define its own limits.

Looking at Fig. 4, it is obvious that if the dye does not leak out of the cells, then the fluorescence intensity should diminish as the cells divide, each daughter cell receiving half the dye in the parent cell. The experiment performed in Fig. 5 was also monitored for fluorescence intensity using the EPICS 753. The mean fluorescence was determined for the stained cells and the inverse of the fluorescence intensity is plotted in Figure 5. Note that the inverse of the fluorescence intensity and the cell number correlate, indicating that fluorescence intensity can be used to determine the number of times a cell population has divided.

One major question is the length of time the dye remains in the membrane of the cell. The RBC was chosen as the test system because the cell does not divide and any loss of intensity would not be confused with division. In addition, the RBC exists in a highly proteinaceous environment in the circulation, and if the dye is stably bound to the RBC in the presence of lipids and albumin, then it should be stable in nearly any other cell type *in vivo*. To test that question, we removed 30 ml of blood from a rabbit and labeled the cells with PKH3, which is a red dye that excites at 560 nm. The procedure used is the same as described in the section for

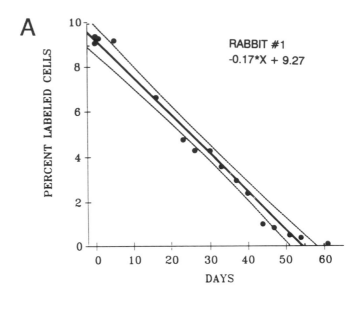

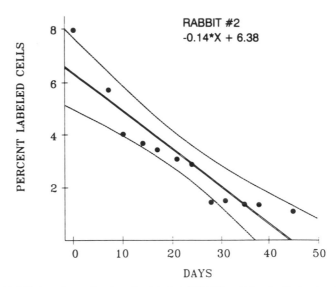

FIG. 6. (A) Lifetime determinations for labeled (PKH3) red blood cells *in vivo*. Percentage positive determined by FCM and plotted as a function of time. Regression analysis showed a starting percentage of 9.3% and an average lifetime of 54 days. (B) Mean fluorescence intensity of labeled red cells *in vivo*. Mean log intensity of the circulating labeled cells was determined by periodic venipuncture and FCM analysis.

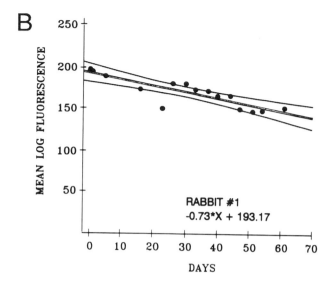

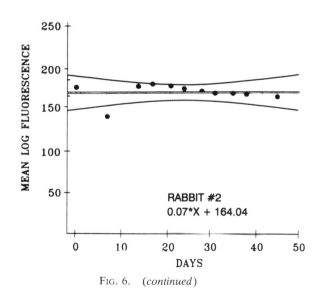

FIG. 6. (*continued*)

general cell labeling. The cells were washed extensively and reinjected into the rabbit. A small aliquot was removed periodically by venipuncture and analyzed by FCM (Coulter EPICS 753). Using fluorescence standards, the fluorescence intensity was measured and the percentage positive cells calculated. It can be seen from Figure 6 that even as the number of fluorescence-positive cells is declining, the fluorescence per cell is relatively constant (Slezak and Horan, 1989b).

Macrophages were labeled with the green tracking dye as described in Section IV on phagocytic cell labeling. In addition, they were labeled with mAb using an indirect procedure and PE–goat anti-rat mAb as the secondary reagent (Melnicoff *et al.*, 1988a). The fluorescence intensity was measured using an EPICS V flow cytometer with the argon laser set at 500 mW power at 488 nm. The forward- and right-angle scatter were used to gate out debris and electronic noise. Green fluorescence of the dye was collected through a 525–20 nm bandpass filter and the orange–red fluorescence of the PE was collected through a 570-nm interference-type filter. In Fig. 3 we see the two-color histograms of labeled peritoneal cells after immunofluorescence labeling. This demonstrates that the green tracking dyes can be used effectively with PE antibody staining.

References

Butcher, E. C., Scullay, R. G., and Weissman I. L. (1980). *J. Immunol. Methods* **37,** 109.

Honig, M. G., and Hume, R. I. (1986). *J. Cell Biol.* **103,** 171–187.

Horan, P. K., Slezak, S. E., and Jensen, B. D. (1988). *Cytometry Suppl.* **2,** 38.

Keller, P. M., Person, S., and Snipes, W. (1977). A *J. Cell Sci.* **28,** 167–177.

Melnicoff, M. J., Morahan, P. S., Jensen, B. D., Breslin, E. W., and Horan, P. K. (1988a). *J. Leuk. Biol.* **43,** 387–397.

Melnicoff, M. J., Horan, P. K., Breslin, E. W., and Morahan, P. S. (1988b). *J. Leuk. Biol.* **44,** 367–375.

Melnicoff, M. J., Horan, P. K., and Morahan, P. S. (1989). *Cell. Immunol.* **118,** 178–191.

Olszewski, W. L. (1987). "IN VIVO Migration of Immune Cells." C. R. C. Press, Boca Raton, Florida.

Slezak, S. E. and Horan, P. K. (1989a). *J. Immunol. Methods* **177,** 205–214.

Slezak, S. E. and Horan, P. K. (1989b). *Blood* **74,** 2172–2177.

Weissman, I. L., and Butcher, E. C. (1980). *J. Immunol. Methods* **37,** 97.

Chapter 43

Rapid Determination of Cellular Resistance-Related Drug Efflux in Tumor Cells

AWTAR KRISHAN

Department of Oncology
University of Miami School of Medicine
Miami, Florida 33101

I. Introduction

Several studies suggest that cellular resistance to cancer chemotherapeutic agents such as alkaloids and antibiotics (multiple drug resistance, MDR) is related to their rapid efflux from the intracellular environment (Kessel *et al.*, 1968; Dano, 1973; Skovsgaard, 1978). Analytical methods such as high-pressure liquid chromatography and spectrofluorometry can be used for monitoring of cellular drug retention and efflux, but of necessity these methods are slow, need large samples, and cannot measure drug retention in single cells. Laser flow cytometry (FCM) offers a unique tool for monitoring of fluorescent antitumor drug retention and its modulation in tumor cells (Krishan and Ganapathi, 1979, 1980). Besides its rapidity, the laser FCM method can identify heterogeneity of drug retention as well as allow for sorting of subpopulations for further biochemical or morphological characterization.

Anthracyclines such as doxorubicin and daunomycin are important cancer chemotherapeutic agents. Most of the anthracyclines are fluorescent and can be excited with the 488-nm laser line from an argon ion laser (Krishan, 1986). Thus, this method can be used for rapid monitoring of anthracycline transport and retention in drug-resistant and sensitive tumor cells.

Cellular resistance to some of the clinically important anthracyclines has been suggested to be due to rapid drug efflux. Thus, drugs that block

491

anthracycline efflux and thereby enhance retention can reduce cellular resistance to anthracyclines (Ganapathi and Grabowski, 1983; Tsuruo *et al.*, 1981). Several studies have shown that certain unrelated drugs, such as the calcium channel blocker (e.g., verapamil) and phenothiazine (e.g., trifluoperazine), will inhibit anthracycline efflux from resistant cells and thereby render them drug-sensitive. Similarly, reduced cellular retention of the vital DNA dye Hoechst 33342 and the calcium indicator dye, Indo-AM, in certain refractory cells may also be related to rapid efflux. Drugs such as phenothiazines or calcium channel blockers can block Hoechst 33342 efflux and make it possible to generate DNA distribution histograms from living cells, which were heretofore difficult to stain with this vital dye (Krishan, 1987).

The use of efflux blockers has been advocated for chemotherapy of human malignancies. It would be useful if before the administration of anthracyclines and agents that affect their cellular efflux *in vivo*, tumor cells could be screened *in vitro* for their anthracycline retention and efflux characteristics. As shown in the following sections, we have used laser FCM to monitor anthracycline fluorescence in tumor cells and to monitor the effect of drug efflux-blocking agents on drug retention.

We have used this method to monitor anthracycline retention and its modulation by phenothiazines and amphotericin-B in P388 doxorubicin-sensitive and -resistant cells (Krishan *et al.*, 1985a,b, 1986). Some of our data show that the effect of efflux blockers on doxorubicin retention is cell cycle- and cell proliferation-related, and often one can identify subpopulations based on their differential response to efflux blockers. From these studies it follows that modulation of drug transport and thereby cellular resistance by phenothiazines or calcium channel blockers may not be uniform in a population but may be selective and confined to only certain types of cells and subpopulations.

II. Applications

We initially used the laser FCM method for monitoring of anthracycline–doxorubicin (adriamycin) transport and retention in tumor cells. In subsequent studies, this method was used to monitor the effect of drugs that enhance influx (e.g., amphotericin-B) and efflux blockers (such as verapamil and phenothiazine). Subsequently, we used this method to determine heterogeneity of retention and response to efflux blockers in human tumor cells (Krishan *et al.*, 1987). We have recently used this method to monitor the effect of efflux blockers on retention of the DNA-binding drug Hoechst 33342 and the calcium indicator dye Indo-AM in tumor cells (Krishan, 1987). Thus this rapid laser FCM method can be used in several ways.

1. Monitoring of fluorescent drug transport (influx, efflux, retention)
2. Effect of drugs that either increase influx or enhance retention by blocking efflux
3. Determination of heterogeneity in retention and response to efflux blockers of tumor cells
4. Selection of ideal efflux blockers for possible clinical use
5. Rapid identification and sorting of cells that have efflux as a major mechanism of drug resistance.

III. Materials

1. Suspension cultures of murine leukemic P388 and its adriamycin-resistant cell line P388/R84 or P388/ADR
2. Single-cell suspension of tumor cells from bone marrow aspirate, ascites, pleural fluid, or after enzymatic digestion of solid-tumor biopsy samples.
3. Doxorubicin (Adriamycin, NSC-123127, Adria Labs, Columbus, OH); daunomycin (Cerubidine, NSC 821151, Ives Labs, New York NY; or Hoechst 33342 (Calibiochem Inc., San Diego, CA).
4. Efflux blockers: trifluoperazine (Stelazine, Smith, Kline, and French Labs, Carolina, Puerto Rico); verapamil (Calan, Searle Pharmaceutical Inc., Chicago, IL).

Suspension cultures of P388 cell line and its doxorubicin-resistant subline (P388/R84) grown in RPMI-1640 medium supplemented with 10% heat-inactivated fetal bovine serum (FBS), penicillin, and streptomycin are used for calibration and as standards. In soft-agar assays, the ID_{50} for the P388 and P388/R84 cells are 0.05 and 4.8 μg/ml of doxorubicin, respectively.

Human tumor cells are recovered by centrifugation of pleural fluid (lung cancer), bone marrow or peripheral blood (leukemia), or ascites (ovarian), or other solid-tumor material after enzymatic digestion. Filtration through nylon mesh (10–40μm) is used to remove clumps. The supernatant fluid is aspirated (after centrifugation) and the cell pellets resuspended and washed in Ca^{2+} and Mg^{2+}-free phosphate-buffered saline (PBS). After centrifugation, the pelleted cells are resuspended in fresh tissue culture medium supplemented with 10% heat-inactivated FBS. To remove erythrocytes from bone marrow aspirates and peripheral blood samples, specimens are diluted with Ca^{2+}- and Mg^{2+}-free PBS and centrifuged over a 70% preformed Percoll gradient (Pharmacia, Piscataway, NJ). Mononuclear cells are recovered and washed before incubation with the drugs at 37°C. Cytospin is used for analysis of morphological heterogeneity and for differential counting purpose.

Stock solutions of drugs are prepared in Ca^{2+}- and Mg^{2+}-free Hanks' balanced salt solution (HBSS). Fresh dilutions of the drugs are prepared in normal saline before each experiment.

IV. Staining Procedure

For generation of two-parameter histograms based on cellular drug fluorescence and length of incubation (time), cell suspensions are directly mixed with the drug-containing medium in the sampling cuvet of the flow cytometer (Coulter EPICS 753), maintained at 37°C with constant stirring. Final drug concentrations used are doxorubicin and daunomycin (1–3 μM) and Hoechst 33342 ((10 μM).

For efflux-blocking experiments, cells are incubated with or without the addition of trifluoperazine (5–100 μM) or verapamil (2–100 μM). Samples are run after 30–60 minutes of incubation.

V. Critical Aspects

Several parameters related to specimen preparation and instrumentation can cause artifacts and lead to generation of erroneous data. Special consideration should be given to the following factors.

1. Several anthracyclines quench their fluorescence on binding to DNA and other target molecules. It is important to keep this in mind and use nonquenching anthracyclines (e.g., AD-32: Krishan et al., 1978) or Hoechst 33342 for critical experiments.
2. pH can have a major effect on drug fluorescence (Albaster et al., 1989), either by shifting the excitation maxima or by altering drug transport and retention.
3. Some efflux blockers may precipitate or bind to glass. Proper precautions should be taken to avoid these artifacts.

4. High anthracycline concentrations can overcome efflux, and high levels of phenothiazines can damage cell membrane and thus increase cellular drug fluorescence. However, the cell membrane-damaged cells can be readily recognized on the basis of their reduced light scatter signal.

5. Coated filters used in the flow cytometer can often, with age, develop pinholes that may result in light leaks. Similarly, certain commercially available filters are notorious for generating autofluorescence when excited with high laser excitation.

6. Dead cells (cells with damaged cell membrane) will rapidly stain with the dye and give erroneous results. Forward- and/or right-angle light scatter should be used to isolate and identify these cells in multiparameter histograms. Furthermore, dead cells can be selectively removed by treatment with DNase I, as described by Frankfurt (Chapter 2, this volume).

VI. Control, Standards

We regularly use adriamycin-sensitive and -resistant P388 cells coincubated with similar drug concentrations under identical conditions as controls for each experiment. Once the photomultiplier (PMT) high voltage, laser power, and amplifications are optimized and set, all samples are analyzed without altering any of these parameters. In general, drug-sensitive P388 cells (ID_{50} 0.05 μg/ml), after incubation with 1–2 μg/ml of doxorubicin for 30–60 minutes, are 20 times less fluorescent than human diploid nuclei stained with the propidium iodide (PI)–hypotonic citrate method.

VII. Instruments

For FCM analysis, cell suspensions incubated *in vitro* with anthracyclines with or without the addition of phenothiazines or verapamil, are analyzed for their total cellular fluorescence in a Coulter Electronics EPICS 753 cell sorter interfaced to a MDADS II data acquisition and analysis system. Fluorescence emission (>530 nm) and forward-angle light scatter are collected, amplified, and scaled to generate multiparameter histograms. A minimum of 10,000 cells are analyzed for each histogram.

For generation of two-parameter histograms based on cellular drug fluorescence and length of incubation (time), cell suspensions are directly mixed with the drug-containing medium in the sampling cuvet of the flow cytometer (Coulter EPICS 753), maintained at 37°C with constant stirring. Time as a parameter is available in software for the Coulter MDADS system.

For efflux-blocking experiments, P388 and R84 cells are incubated with doxorubicin (1–2 μM), for 30–60 minutes at 37°C, with or without the

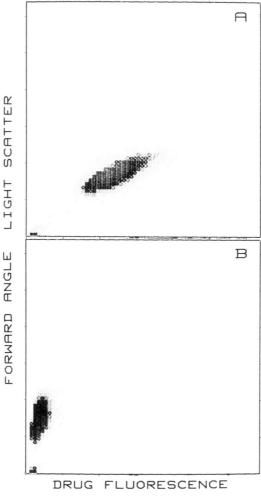

FIG. 1. Scattergrams of (A) adriamycin-sensitive and (B) -resistant P388 cells, respectively. Forward-angle light scatter (ordinate) and peak fluorescence (abscissa) in linear scale are shown. Note that by the fluorometric method, P388/R cells have 4- to 6-fold less drug content (due to rapid drug efflux) than the parental drug-sensitive cells.

addition of chlorpromazine (5–100 μM), trifluoperazine (5–100 μM), or verapamil (2–100 μM).

An EPICS V cell sorter (Coulter Electronics, Hialeah, FL) equipped with an argon ion laser is used to measure forward-angle light scatter and fluorescence from cells excited with the 488-nm line for anthracyclines or with ultraviolet (UV) excitation (for Hoechst 33342).

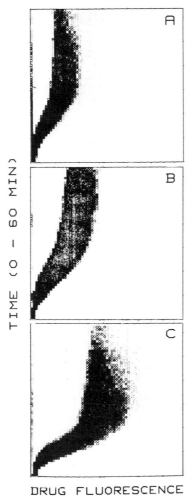

DRUG FLUORESCENCE

FIG. 2. Scattergrams showing the effect of pH on adriamycin fluorescence of P388 cells: (A) pH 6; (B) pH 7; (C) pH 8. Abscissa measures linear peak fluorescence whereas ordinate records time (60 minutes). Note that cells incubated at the alkaline pH of 8.0 (C) have much higher fluorescence than cells incubated in acidic medium (pH 6.0, A).

Appropriate filters are used to collect emitted light. Data are collected and scaled in a Coulter Electronics MDADS data acquisition and analysis system.

VIII.   Results

Figures 1A and 1B are scattergrams of adriamycin-sensitive and -resistant P388 cells, respectively. Forward-angle light scatter (ordinate) and peak fluorescence (abscissa) in linear scale are shown. Note that by the fluorometric method, P388/R cells have 4- to 6-fold less drug content (due to rapid efflux) than the parental drug-sensitive cells.

Figures 2A, 2B, and 2C are scattergrams that show the effect of pH 6, 7, and 8, respectively, on adriamycin fluorescence of P388 cells. Abscissa

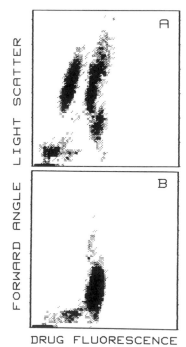

Fig. 3.   Scattergrams of cells from human solid-tumor ascites incubated with doxorubicin. (A) Note the appearance of four subsets. In cells coincubated with the efflux blocker chlorpromazine, a homogeneous population with high drug fluorescence emerges, suggesting that the subpopulations with the lower amount of drug fluorescence (in A) were possibly effluxing the drug.

measures linear peak fluorescence, whereas ordinate records time (60 minutes). Note that cells incubated at alkaline pH of 8.0 (Fig. 2C) have much higher fluorescence than cells incubated in acidic medium (pH 6.0, Fig. 2A).

The scattergrams shown in Fig. 3 are of cells from human solid-tumor ascites incubated with doxorubicin. Note the appearance of four subsets in Fig. 3A. In cells coincubated with the efflux blocker chlorpromazine, a homogeneous population with high drug fluorescence emerges, suggesting that the subpopulations with the lower amount of drug fluorescence (in Fig. 3A) were possibly effluxing the drug.

The scattergrams shown in Fig. 4 are of P388 (Fig. 4A–C) and P388/R (Fig. 4D–F) cells incubated with the calcium indicator dye, Indo-AM. Note that low fluorescence in P388/R cells (Fig. 4D) is significantly enhanced in the presence of efflux blockers trifluoperazine (Fig. 4E) or verapamil (Fig. 4F).

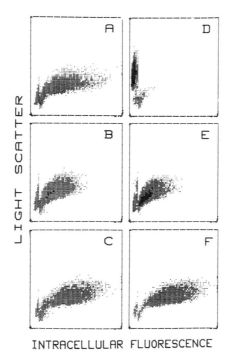

INTRACELLULAR FLUORESCENCE
INDO-1

FIG. 4. Scattergrams of P388 (A–C) and P388/R (D–F) cells incubated with the calcium indicator dye, Indo-AM. The efflux blockers trifluoperazine (B, E) and verapamil (C, F) were present as indicated. Note that low fluorescence in P388/R cells (D) is significantly enhanced in the presence of these efflux blockers (E, F).

References

Albaster, *et al.* (1989). *Cancer Res.* **49,** 5638–5643.

Dano, K. (1973). *Biochim. Biophys. Acta* **323,** 1466–1483.

Ganapathi, R., and Grabowski, D. (1983). *Cancer Res.* **43,** 3696–3699.

Kessel, D., Botterill, V., and Wodinsky, I. (1968). *Cancer Res.* **28,** 938–941.

Krishan, A. (1986). *In* "Techniques in Cell Cycle Analysis" (J. E. Gray and Z. Darzynkiewicz, eds.), pp. 337–366. Humana Press, Clifton, New Jersey.

Krishan, A. (1987). *Cytometry* **8,** 642–645.

Krishan, A., and Ganapathi, R. (1979). *J. Histochem. Cytochem.* **27,** 1655–1656.

Krishan, A., and Ganapathi, R. (1980). *Cancer Res.* **40,** 3895–3900.

Krishan, A., Ganapathi, R. N., and Israel, M. (1978). *Cancer Res.* **38,** 3656–3662.

Krishan, A., Sauerteig, A., and Gordon, K. (1985a). *Cancer Res.* **45,** 4097–4102.

Krishan, A., Sauerteig, A., and Wellham, L. (1985b). *Cancer Res.* **45,** 1046–1051.

Krishan, A., Sauerteig, A., Gordon, K., and Swinkin, C. (1986). *Cancer Res.* **46,** 1768–1773.

Krishan, A., Sridhar, K. S., Davila, E., Vogel, C., and Sternheim, W. (1987). *Cytometry* **8,** 306–314.

Skovsgaard, T. (1978). *Cancer Res.* **38,** 1785–1791.

Tsuruo, T., Lida, H., Tsukagoshi, S., and Sakurai, Y. (1981). *Cancer Res.* **41,** 1967–1972.

Chapter 44

Slit-Scan Flow Analysis of Cytologic Specimens from the Female Genital Tract

LEON L. WHEELESS, JR., JAY E. REEDER, AND
MARY J. O'CONNELL

Analytical Cytology Unit
Department of Pathology
University of Rochester Medical Center
Rochester, New York 14642

I. Introduction

Automated detection of precancerous and cancerous lesions of the uterine cervix is possible employing methods and instrumentation exploiting differential staining of normal and malignant cells. The procedure employs acridine orange (AO) and specialized flow cytometry (FCM) instrumentation. Acridine orange is an extremely useful marker in the recognition of abnormal cells in that it provides information on both cellular DNA content and configuration (accessibility).

Accessibility of nucleic acids *in situ* to AO has been demonstrated to be dependent on cell type and differentiation (Darzynkiewicz *et al.*, 1984; Gill *et al.*, 1978, 1979). Utilizing slide-based, quantitative fluorescence instrumentation, it has been documented that abnormal cells found in cytologic specimens from the uterine cervix show elevated nuclear fluorescence when compared to normal cells from the same specimen (Wheeless and Patten, 1973; Wheeless *et al.*, 1975). These measurements of nuclear fluorescence were made in the green range from 520 to 570 nm. This differential in green nuclear fluorescence intensity reflects both quantitative and qualitative differences in nucleic acid content. In contrast to techniques used for strictly stoichiometric DNA and RNA content measurements (Darzynkiewicz *et al.*, 1984), this staining technique does

METHODS IN CELL BIOLOGY, VOL. 33

not employ extraction or enzyme digestion steps. Therefore, a measurement of native chromatin configuration is obtained by assessing DNA accessibility to the fluorochrome.

Using the staining conditions described here, cellular constituents that bind AO and thus contribute to the measured green fluorescence include DNA, RNA, and glycosaminoglycans. A potential problem in the general use of AO is cytoplasmic fluorescence. Acridine orange stains both the nucleus and cytoplasm of cells. Normal intermediate squamous cells show a pale-green cytoplasmic fluorescence with a more intense nuclear fluorescence. Because of this cytoplasmic fluorescence and its variability, it is necessary to employ specialized FCM instrumentation that permits discrimination of nuclear from cytoplasmic fluorescence. Otherwise, differences in cytoplasmic area and staining will mask the differences in nuclear fluorescence between normal and abnormal cells. Slit-scan flow systems permit this separation of nuclear from cytoplasmic fluorescence while providing additional morphologic information.

II.   Slit-Scan Flow Instrumentation

Slit-scan systems provide quantitative fluorescence together with low-resolution morphologic information on cells in flow (Wheeless, 1979, 1990; Wheeless and Kay, 1985). A beam of laser excitation illumination is focused to a thin ribbon or slit of light. The extent to which the beam may be reduced is dependent on the size of the cells to be measured. For cells from the female genital tract, the beam is focused to a 4-μm slit. Cells are sequentially illuminated as they flow through the slit of excitation illumination. Measurement of fluorescence from a cell as it passes through the excitation slit yields a one-dimensional fluorescence intensity distribution or slit-scan contour. Fluorescence is typically recorded 256 times as the cell passes through the excitation slit.

Slit-scan contours are computer-processed in real time for extraction of features (Wheeless, 1979, 1990). Features available from a one-dimensional slit-scan contour include nuclear fluorescence, cytoplasmic fluorescence, cell diameter, nuclear diameter, nuclear/cell diameter ratio, and other low-resolution morphologic information providing identification of most binucleate, clumped, and overlapping cells in flow.

Specimens are classified as normal or abnormal based on the total number of alarms detected. An alarm is a cell having a nuclear fluorescence value >2.5 times the mean nuclear fluorescence of an identifiable

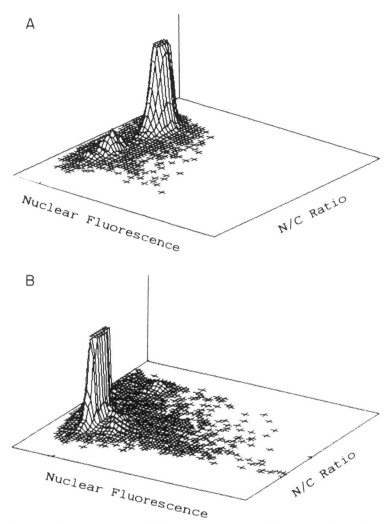

FIG. 1.   (A) Isometric bivariate display for a normal specimen from the uterine cervix. The major population is composed of normal intermediate and superficial cells. These cells have low N/C ratio and normal values of nuclear fluorescence. (B) Isometric bivariate display for a specimen from a patient with carcinoma of the cervix. the major population is composed of polymorphonuclear leukocytes associated with an inflammatory response. Abnormal cells exhibit elevated nuclear fluorescence.

normal population in the specimen. For specimens from the female genital tract, the normal intermediate squamous cell population is used as the "on-board" staining reference, as its position in multiparameter feature space is well known. The value of the mean nuclear fluorescence of this normal population is established during the initial stages of flow analysis to permit the identification of alarm cells in real time.

A multidimensional slit-scan flow system was fabricated for processing of specimens from the female genital tract (J. L. Cambier *et al.*, 1979; Kay *et al.*, 1982). This instrument provides three orthogonal one-dimensional slit-scan contours on each cell in flow with real-time processing of all contours at rates exceeding 100 cells/second. In a single-blind study, this system provided results close to established laboratory standards for the segregation of normal and abnormal specimens (Wheeless *et al.*, 1984). For this study, specimens were classified abnormal if they had an alarm rate >0.18%. Given the false-alarm rate in normal specimens to be ~0.12%, this represents a sensitivity to abnormal cells of ~1 in 1000.

Representative isometric bivariate displays for a normal specimen from the uterine cervix and a specimen from a patient with carcinoma of the cervix appear in Fig. 1. The plots depict nuclear fluorescence versus nuclear/cytoplasmic diameter ratio (N/C ratio). In Fig. 1A, the major population is composed of cells having low N/C ratio and normal values for nuclear fluorescence, the normal intermediate and superficial squamous cells. The small population with high N/C ratio represents polymorphonuclear leukocytes (PMN). The false-alarm rate (cells having elevated nuclear fluorescence) is <0.1%. In contrast to this normal pattern, the specimen from the patient with carcinoma of the cervix (Fig. 1B) has few normal squamous cells. The major cell population is composed of PMN associated with an inflammatory response. A population of cells with elevated nuclear fluorescence appears as a shoulder on the PMN population. Additionally, there are numerous other cells with elevated nuclear fluorescence. The alarm rate for this abnormal specimen is 0.7%.

III. Methods

Method development studies have been reported in a series of publications from this laboratory. These studies described the differential stainability of abnormal nuclei (Cambier *et al.*, 1976; Gill *et al.*, 1978; 1979; Wheeless and Patten, 1973; Wheeless *et al.*, 1975), the development of an automated syringing device for tissue disaggregation (Mead *et al.*, 1978;

Lopez *et al.*, 1981), a poststaining fixation technique for AO (M. A. Cambier *et al.*, 1977, 1979;), and the sensitivity and specificity of the slit-scan technique for screening cytologic specimens from the female genital tract (Wheeless *et al.*, 1984) and the urinary bladder (Wheeless *et al.*, 1986).

Specimen preparation methods are described here, beginning with a list of reagents and their preparation. All solutions are filtered through a 0.2-μm filter prior to use.

The procedure has two parts. Part 1 describes the collection, staining, fixation, and storage of the sample. Staining and fixation are accomplished within 1 week of sample collection. Following these steps, the sample can be stored refrigerated up to 1 year.

Part 2 of the procedure consists of a wash to remove excess stain and fixative, and adjustment of cell concentration for flow analysis.

IV. Materials

A. Reagents

1. Phosphate buffer: 0.2 M Phosphate buffer, pH 7.0, prepared from sodium phosphate, monobasic and dibasic
2. Phosphate-buffered saline (PBS): Prepared from FTA hemagglutination buffer (Fisher catalog no. B11248)
3. Glutaraldehyde fixative: Mix in the following order and proportions:
   0.2 M Sodium phosphate monobasic, 70.3%
   0.63 M NaOH, 14.4%
   25% Glutaraldehyde (Baker 2127-03), 15.3%
4. AO stock: 0.1% Purified AO (Polysciences, Inc., catalog no. 4539) in distilled water
5. AO staining solution: 0.005% AO in phosphate buffer, prepared by dilution of 5 ml AO stock to 100 ml with phosphate buffer
   Dilution is stable for 1 week at 4°C.
6. Collection medium:
   Acetamide (Sigma A0500), 22.91%
   Lithium chloride (Sigma L0505), 0.53%
   Water, 70.29%
   Polyethylene glycol (PEG) 1450 (Sigma P-5402), 6.12%
   Methyl paraben (Tri-K Industries), 0.1%
   Propyl paraben (Tri-K Industries), 0.05%

   a. Liquefy PEG with gentle warming in a water bath.

b. With continuous stirring, dissolve acetamide and lithium chloride in approximately three-fourths of total volume of water.
c. Add PEG.
d. Dissolve methyl and propyl paraben in remaining 25% of water and add to acetamide solution.
e. At intervals of 15 minutes, measure pH, until two consecutive measures agree to 0.1 pH units. The target pH is between 4.7 and 6.7. Stability is routinely obtained within 1 hour.
f. A series of suction filtration steps are necessary prior to use. Start with 8-μm polycarbonate membrane filtration and follow by 3- and 1-μm pore sizes.

B. Preparation: Part 1

1. Obtain cervical scrape using a plastic Auer-type spatula and immediately rinse in 15 ml of collection medium, seal, and refrigerate. Care should be exercised to prevent contamination of the sample by glove powders and lubricants.
2. Autosyringe 35 strokes, 22 psi using a 10-ml plastic syringe fitted with a $1\frac{1}{2}$-in. 18-gauge needle.
3. Transfer sample to a 15-ml plastic tube and centrifuge for 10 minutes at 10°C, 2000 rpm (725 g).
4. Aspirate and discard supernatant. Carefully vortex tube to resuspend cells in volume left in tube.
5. Add 5 ml AO stain; manually syringe 10 times with a 5-ml syringe fitted with an 18-gauge needle. Observe tubes and continue syringing if large clumps persist. Add 5 ml of AO stain for 10 ml total.
6. Incubate 20 minutes at room temperature (RT) in the dark.
7. Centrifuge for 10 minutes at 10°C, 2000 rpm.
8. Aspirate supernatant and discard.
9. Vortex tube to resuspend cells.
10. Add 5 ml glutaraldehyde fixative solution, syringe to break up any clumps, then add 10 ml more fixative and cap tubes.
11. Store at 4°C.

C. Preparation: Part 2

12. Centrifuge samples for 10 minutes at 10°C, 2000 rpm.
13. Aspirate and discard supernatant.
14. Resuspend pellet in 5 ml PBS. Count aliquot in hemocytometer with fluorescence microscope.
15. Centrifuge samples for 10 minutes at 10°C, 2000 rpm.

16. Aspirate enough supernatant to achieve 1×10^6 cells/ml.
17. Store at 4°C.
18. Filter through 105-μm nylon mesh prior to flow analysis.

ACKNOWLEDGMENTS

This work was supported by National Institutes of Health grants CA43161 and CA33148.

REFERENCES

Cambier, J. L., Kay, D. B., and Wheeless, L. L. (1979). *J. Histochem. Cytochem.* **27,** 321–324.

Cambier, M. A., Christy, W. J., Wheeless, L. L., and Frank, I. N. (1976). *J. Histochem. Cytochem.* **24,** 305–307.

Cambier, M. A., Wheeless, L. L., and Patten, S. F. (1977). *Acta Cytol.* **21,** 477–480.

Cambier, M. A., Wheeless, L. L., and Pattern, S. F. (1979). *Anal. Quant. Cytol.* **1,** 57–60.

Darzynkiewicz, Z., Traganos, F., Kapuscinski, J., Staiano-Coico, L., and Melamed, M. R. (1984). *Cytometry* **5,** 355–363.

Gill, J. E., Wheeless, L. L., Hanna-Madden, C., Marisa, R. J., and Horan, P. K. (1978). *Cancer Res.* **38,** 1893–1898.

Gill, J. E., Wheeless, L. L., Hanna-Madden, C., and Marisa, R. J. (1979). *J. Histochem. Cytochem.* **27,** 591–595.

Kay, D. B., Wheeless, L. L., and Brooks, C. L. (1982). *IEEE Trans. Biomed. Eng.* **29,** 106–111.

Lopez, P. A., Cambier, M. A., and Wheeless, L. L. (1981). *Anal. Quant. Cytol.* **3,** 235–238.

Mead, J. S., Horan, P. K., and Wheeless, L. L. (1978). *Acta Cytol.* **22,** 86–90.

Wheeless, L. L. (1979). *In* "Flow Cytometry and Sorting" (M. E. Melamed, T. Lindmo, and M. L. Mendelsohn, eds.), pp. 125–135. Wiley, New York.

Wheeless, L. L. (1990). *In* "Flow Cytometry and Sorting" (M. R. Melamed, T. Lindmo, and M. L. Mendelsohn, eds.), Vol. II, pp. 109–125. Liss, New York.

Wheeless, L. L., and Kay, D. B. (1985). *In* "Flow Cytometry: Instrumentation and Data Analysis" (M. A. Van Dilla, P. N. Dean, O. D. Laerum, and M. R. Melamed, eds.), pp. 21–76. Academic Press, London.

Wheeless, L. L., and Patten, S. F. (1973). *Acta Cytol.* **17,** 391–394.

Wheeless, L. L., Patten, S. F., and Onderdonk, M. A. (1975). *Acta Cytol.* **19,** 460–464.

Wheeless, L. L., Lopez, P. A., Berkan, T. K., Wood, J. C. S., and Patten, S. F. (1984). *Cytometry* **5,** 1–8.

Wheeless, L. L., Berkan, T. K., Patten, S. F., Eldidi, M. M., Hulbert, W. C., and Frank, I. N. (1986). *Cytometry* **7,** 212–216.

Chapter 45

Cell Sorting with Hoechst or Carbocyanine Dyes as Perfusion Probes in Spheroids and Tumors

RALPH E. DURAND, DAVID J. CHAPLIN, AND PEGGY L. OLIVE

Medical Biophysics Unit
B. C. Cancer Research Centre
Vancouver, British Columbia V5Z 1L3
Canada

I. Introduction

The widespread availability of fluorescence-activated cell sorters (FACS) currently allows not only the quantitative demonstration of biological heterogeneity among cells of tissues, but also, the ability to selectively recover cells with particular characteristics. Fluorescent probes that recognize variations in expression of surface molecules, proteins, and DNA or RNA have become fairly commonplace; a number of functional stains have subsequently been developed that are responsive to membrane potential, respiratory activity, intracellular pH, and a growing list of other parameters.

In many aspects of biology, and particularly in cancer research, cellular location within a tissue or tissue mass can have considerable impact on the characteristics of the cell and its response to cytotoxic or mutagenic agents. For this reason, it is of key importance to be able to identify and collect cells from different regions within solid tumors. This has been approached historically by techniques ranging from gross dissection to sedimentation on velocity or density gradients. Subsequently, separation of cells based on their access to fluorescent perfusion probes has been introduced, starting with *in vitro* work with "spheroids," a three-dimensional cell culture

509

system in which gradients of metabolite supply produce conditions very similar to those found in solid tumors outgrowing the blood supply (Sutherland and Durand, 1976). Recognition of the microenvironmental heterogeneities in growing spheroids prompted the development of techniques for isolation of specific, defined cell subpopulations based on cellular position within the spheroid mass (Durand, 1982).

Cell sorting techniques with Hoechst or carbocyanines as perfusion probes utilize the rapid binding of these agents as they diffuse from the medium into a spheroid, or from the blood vessels into a tumor. Both agents remain localized in undamaged target cells after the multicell structure is disaggregated into a single-cell population. From this, it follows that sorting cells of given fluorescent intensities permits recovery of subpopulations from equivalently perfused regions of the spheroid or tumor system (presumably, from geometrically or architecturally similar regions).

These techniques, in general, require the following implicit assumptions concerning the stains:

1. Nontoxic at usable concentrations
2. Capable of passive diffusion through cells
3. Rapid, irreversible, and nonsaturable binding
4. Uniform binding within a cell population
5. No influence of metabolic or physiological processes
6. High fluorescence quantum efficiency

Since none of these criteria is entirely met by any dye examined to date, the following discussion focuses on the practical question of how to best use available stains.

II. Applications

Most studies have used these techniques with the goal of selecting cellular subpopulations that respond differently to antineoplastic agents. In particular, a large data base exists for drug and radiation effects in spheroids (Durand, 1982, 1986a,b; Durand and Olive, 1987; Olive and Durand, 1987), and a similar data base is developing for transplantable tumors in murine systems (Chaplin *et al.*, 1985, 1986, 1987). Less work has been reported using normal tissue end points, or in cell systems composed of heterogeneous cell populations. Nonetheless, there is a suggestion that under carefully controlled conditions, such end points can be used (Loeffler *et al.*, 1987).

Once the separated cells have been sorted by the FACS, other traditional studies can be undertaken. The non-toxic nature of these stains permits studies of cellular function or viability studies. Shorter term assays—including, for example, nucleic acid or protein precursor uptake, or other functional staining—can also lead to new information. These perfusion stains can also identify hypoxic cells as a result of their relatively slow penetration characteristics (Olive and Durand, 1989), and can be used as an analytical tool to demonstrate the vasculature (Reinhold and Visser, 1983), and fluctuations in blood flow in animal tumor systems (Trotter *et al.*, 1989).

III.   Materials and Methods

Hoechst 33342, originally supplied in North America only by American Hoechst in New Jersey, is now widely available from many commercial chemical companies (Sigma, Aldrich, Calbiochem, etc.). Another valuable source for potential perfusion probes is Molecular Probes, Inc. (Eugene, OR). Several Hoechst stains with relatively similar characteristics are available; we have found that 33342 offers the best characteristics of uptake, fluorescence intensity, and low toxicity. Of the carbocyanines, $DiOC_7(3)$ is the stain of choice, again for the same reasons. Other probes, including acridine orange (AO), fluorescein derivatives, and several rhodamine compounds can also be used; with these compounds, staining is often much more affected by stain concentration and pH gradients than for the Hoechst or carbocyanine probes, and the gradient for sorting is typically shallower.

Staining is easily accomplished *in vitro* (dye can be added directly to the culture medium); *in vivo*, stains can be administered by iv injection or infusion, or more easily by ip injection (the latter, however, produces less intensity and poorer gradients). In tumors, however, blood flow can be irregular. As a result, a more representative picture of the "average perfusion" of the tumor can be gained by relatively long-term infusion of the agents. We use a Harvard infusion pump, with a needle placed in the lateral tail vein of the mouse, and with a total volume of 0.1–0.3 ml infused into the animal over a 20- to 60-minute period. Following infusion, the cells to be studied must be isolated. To ensure stain retention and lack of exchange between cells, the tissue should be cooled to ice bath temperature as quickly as possible. A number of different disaggregation techniques have been used; most are acceptable if they result in a good single-cell suspension. Two problems must be remembered during the disaggregation process: if cells are damaged by cutting or chopping the tissues, stain can be released and reutilized by neighboring cells. This is particularly true for the carbocyanine dye $DiOC_7(3)$. Additionally, enzymatic damage can result in cell lysis and provide another source of extraneous stain. For these reasons, optimization procedures *must* be developed for each tissue or cell type to be observed, to minimize cell damage and dye exchange during the single-cell preparation step (Olive *et al.*, 1985). An additional factor that should be considered is that while these stains are relatively nontoxic by themselves, their toxicity tends to increase with time as cell suspensions are held at ice bath temperatures.

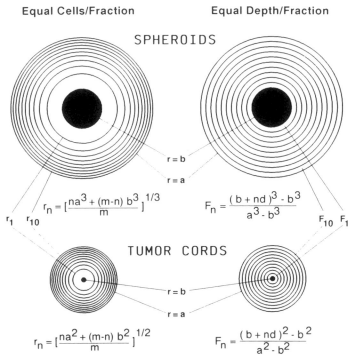

Equal Cells/Fraction　　　　　　**Equal Depth/Fraction**

SPHEROIDS

$r = b$

$r = a$

$$r_n = [\frac{na^3 + (m-n)\,b^3}{m}]^{1/3}$$

$$F_n = \frac{(b + nd)^3 - b^3}{a^3 - b^3}$$

r_1　r_{10}　　　　　　　　　　　　　F_{10}　F_1

TUMOR CORDS

$r = b$

$r = a$

$$r_n = [\frac{na^2 + (m-n)\,b^2}{m}]^{1/2}$$

$$F_n = \frac{(b + nd)^2 - b^2}{a^2 - b^2}$$

Fɪɢ. 1.　Idealized "sections" through spheroids or tumor cords, with layers of viable cells surrounding the central necrosis of the spheroid (black region), or individual blood vessels of tumors (central black region). Two methods of cell sorting are compared: equal cells recovered per fraction sorted (left figures), or cells recovered from "shells" of equal thickness (right). When sorting equal numbers of cells per fraction, if m fractions are to be recovered, one simply integrates the cell distribution profile and sets the m windows so that $1/m$ of the cells are recovered per window. The radius from which fraction n of the cells is recovered is then calculated as shown (left figures). When sorting "shells" of constant depth, each shell constitutes a different fraction. The fraction of cells to be sorted for shell n is then calculated using the expression shown (right) for spherical or cylindrical geometry. See also Table I.

We use a number of different sorting techniques implemented on a
FACS 440 system. Software has been developed to control the instrument
completely; sorting windows can be set on the basis of dye intensity, or on
the basis of dye concentration where concentration is estimated by the
ratio of the probe intensity to cell size (forward or peripheral light scatter).
With computer-controlled sorting, it is quite easy to integrate a cell
fluorescence profile, and to determine where windows should be set for any
particular percentage of the cells to be sorted. We have adopted two
different sorting philosophies that are slightly different in their implementation for spheroids versus tumors. These are illustrated in the
schematic cross sections through spheroids or tumors shown in Fig. 1. The

TABLE I

VALUES OF SORTING PARAMETERS DERIVED FOR SORTS
PERFORMED AS IN FIG. 1

	Equal cells fraction			Equal depth fraction	
n	r_n	F_n	n	r_n	F_n
		Spheroids[a]			
0	70.0	0.0	0	70	0.0000
1	123.2	0.1	1	88	0.0221
2	150.4	0.2	2	106	0.0555
3	170.2	0.3	3	124	0.1023
4	186.2	0.4	4	142	0.1649
5	199.9	0.5	5	160	0.2456
6	211.9	0.6	6	178	0.3466
7	222.7	0.7	7	196	0.4703
8	232.5	0.8	8	214	0.6189
9	241.6	0.9	9	232	0.7947
10	250.0	1.0	10	250	1.0000
		Tumors[b]			
0	8.0	0.0	0	8	0.0000
1	38.1	0.1	1	19	0.0214
2	53.3	0.2	2	30	0.0603
3	65.0	0.3	3	41	0.1167
4	74.9	0.4	4	52	0.1905
5	83.6	0.5	5	63	0.2817
6	91.5	0.6	6	74	0.3905
7	98.8	0.7	7	85	0.5167
8	105.6	0.8	8	96	0.6603
9	112.0	0.9	9	107	0.8214
10	118.0	1.0	10	118	1.0000

[a] $a = 250$ μm, $b = 70$ μm, $m = 10$.
[b] $a = 118$ μm, $b = 8$ μm, $m = 10$.

upper two panels show the spheroid configuration (spherical symmetry where the viable cells surround a central necrotic region); in the lower two panels, the *in vivo* situation (where viable cells surround a blood vessel, and necrosis appears at large distances from the blood vessels) is depicted.

The difference in the sorting philosophy is shown in the left panels versus the right panels: the left panels show the geometric consequence of sorting constant numbers of cells in each sort fraction; the right panels show an approach in which "shells" of constant thickness can be sorted from the two systems. Mathematically, it is quite easy to set up the sorting criteria for these two cases: for the spheroid system, the spherical geometry permits rapid solution of the spheroid volume equation; and for tumors, cylindrical geometry around the blood vessel is used.

As can be appreciated from Table I, the most efficient usage of sorter time is achieved by sorting constant numbers of cells per fraction. This does, however, result in progressively "narrower" shells of cells being recovered from the spheroid as one progresses outward from the necrotic center, and similarly, progressively thinner shells sorted from the tumor cord as one moves away from the blood vessel. Consequently, relatively poor spatial resolution (thicker shells) is achieved in the "inner" regions of each system.

IV. Critical Considerations

Each of the characteristics previously listed for an "ideal" stain indicates limitations of available stains. In particular, the use of the Hoechst and carbocyanine dyes is limited by the heterogeneity of cellular dye uptake, dye toxicity, potential problems of dye exchange, stability, and the influence of sample preparation techniques.

Uptake heterogeneity can be demonstrated fairly easily for isolated cells of different types; tumor cells are typically less amenable to staining at a given dye concentration than are normal cells. As a consequence of this, attempting to determine the location of cells *of different types* within a tumor on the basis of these stains alone is virtually impossible; if, however, the cells can be distinguished by other criteria, the distribution profiles for the perfusion stains can be of considerable value. An additional problem is the influence of cell size; since a flow cytometer gives an integrated signal (stain concentration × cell size), it follows that two cells containing the same concentration of stain will appear to have a different fluorescence intensity unless they are of equal volume. The corollary to this point is that for two cells of different sizes to be sorted as though they had the same intensity, they must have markedly different intracellular concentrations of the stain.

Numerically, the influence of these parameters can have marked implications for sorting. If one is dealing with cells that vary, for example, by a factor of 2 to 3 in volume, and if one wishes to sort 10 fractions of such cells, then the stain gradient required in order to lead to unambiguous separation of such cells on the basis of their stain uptake would be of the order of the difference in size raised to the power of the number of fractions to be sorted (in our hypothetical example with a cell size differential of 2, a stain intensity gradient of 2^{10} or > 1000-fold difference between the dimmest and brightest cells would be required for unequivocal separation of cells based on staining).

Since dye uptake *in vivo* is dependent on delivery, which is in turn dependent on blood flow rate, volume, and exposure time, uptake can be quantitatively compared between adjacent blood vessels only if the integrated flow through those vessels is constant in time. If blood flow rate varies (as it does for most tumor and normal tissues), it follows that the staining pattern at best reflects net tissue perfusion during a finite time interval, and thus the "functional" rather than "physical" position of the cells relative to the blood supply.

Technical problems can also markedly influence the quality of data that can be obtained. Intravenous injections in mice can be difficult to perform; to place a needle in the tail vein and maintain it for long perfusions is even more difficult. Thus, differences in staining intensity may well be as much an indication of stain delivery technique as of the biology of the system *in vivo*. We have evaluated several methods of normalizing for such variations; an easy and useful approach is to use a second tissue (e.g., spleen cells) as a "standard." Thus, for each animal, the tumor/spleen cell staining ratio can be used to "normalize" the tumor cell staining profile on the basis of that of the spleen cells. This approach, of course, presupposes that the blood supply to the spleen is relatively constant among animals. Parenthetically, it should also be noted that cell populations from different animals cannot be mixed together prior to sorting or analyses; the dimmest (or brightest) cells will undoubtedly all come from one animal.

Toxicity is also a major consideration for these perfusion stains (e.g., Arndt-Jovin and Jovin, 1977; Durand and Olive, 1987; Siemann and Keng, 1986). Toxicity is typically defined in terms of interference with reproductive integrity; within the context of that definition, usable concentrations of stains can truly be nontoxic. However, as with all biological materials, a dose response is inherent, and differential sensitivity of various target cells is expected. It is thus essential to determine toxicity parameters independently for each cell type to be studied. In addition, one should extend toxicity beyond the definition of reproductive integrity; most of these stains are quite efficient at inhibiting other metabolic processes (e.g., DNA synthesis) and in some cases, can be quite vasoactive as well. Any type of differential toxicity of this sort will, of course, lead to potential artifacts in reported results.

In summary, these stains can be used effectively to select cells of a particular cell type from areas of tissue that are equivalently perfused by the dyes. However, great care is indicated when processing populations of differing cell types, or under conditions where blood flow or other physiological factors are altered. Because of the problems with differential stain uptake and cell size, it is difficult if not impossible to be confident that adjacent cells of different types really do stain with comparable intracellular dye *intensity*. Interestingly, neither technique of sorting (on the basis of dye intensity, or on the basis of dye concentration) leads to good results with mixed-cell types: in the first case, intensity is disproportionately affected by cell size; in the second case, intracellular concentration differences due to varying avidity of drug uptake may override stain delivery. Thus, while dye uptake is a very good parameter indicating the net perfusion of a *single* cell type within an organ, this technique alone generally cannot be extended to separating cells of different types on the basis of location within the tissue.

V.   Controls and Standards

As might be inferred from the preceding discussion, absolute standards do not exist for these techniques. Probably one of the better standards available is historical experience; since the sorting techniques can only be performed on the basis of numerical analyses of the stained cell populations, it necessarily follows that one can develop a data base relating staining gradients to administered stain. We highly recommend quantitative reanalysis of the staining gradient. For example, if one is sorting 10 fractions of cells either on the basis of equal cell numbers or equal depth increments, one can calculate the mean fluorescence intensity of each cell population sorted, and use this data to ensure that comparable gradients are achieved from experiment to experiment. With spheroids, the brightest 10% of the cells are typically 200-fold more fluorescent than the dimmest 10%; in the murine tumors studied to date, a 20-fold differential in stain concentration can usually be seen. Lack of a "reasonable" gradient, based on previous experience, is certainly sufficient cause to reject a particular cell population as being nonrepresentative.

Many of the technical problems of stain delivery can be overcome by introducing a biological standard into the experiment. In the case of *in vitro* staining of spheroids, problems caused by such variables as differing cell densities or differing spheroid sizes can be addressed by adding single cells at a known cell density to the spheroid flask. Following exposure, if the flask is removed from the magnetic stirrer, the spheroids will sediment rapidly to the bottom of the flask and can be recovered independently from the suspended cells. Thus, stain uptake profiles in spheroids can be normalized on the basis of the stain uptake by the free cells. Similarly, *in vivo*, spleen or another organ (spleen is an obvious choice because of the ease of preparing a single-cell suspension) can be utilized to control for differences in injection/infusion techniques.

We also recommend the practice of establishing the stain uptake characteristics of a particular tissue by restaining separated cells. Differential uptake at a particular stain concentration can be identified, and can then be used to interpret further the response of different cell types within the population. Differential stain uptake can also occur within a single type of cell, based on (for example) cell cycle position, pH, and, quite interestingly, temperature. Thus, while staining intensity *in vivo* is qualitatively related to blood flow, rigorously quantifying blood flow with these techniques may prove impossible.

An additional "control" is to evaluate the staining technique in the context of other end points to be studied. Concentrations of stain can be selected that are minimally perturbing to the system in the absence of other treatment; it does not necessarily follow, however, that the stain will not interact with other drugs or agents to be evaluated. We believe that the most useful method of evaluating potential interactions is to evaluate escalating doses of the stain, and determine the stain concentration at which interactions begin to be evident. When meaningful results can be obtained with much lower stain concentrations, one can be relatively sure that interactions are minimal.

VI.   Conclusion

Despite the uncertainties discussed here, we remain convinced that cell sorting, utilizing these perfusion probes, can provide new opportunities for addressing important questions in tumor biology. Functional differences in cellular response to antineoplastic agents can be documented, with the result that treatments can now be chosen with the aim of targeting specific tumor cell subpopulations. In addition to "applied" studies, use of these techniques is also permitting reassessment of the fundamental question of the "physiology" of tumor blood flow (Chaplin *et al.*, 1987).

ACKNOWLEDGMENT

This work was supported by USPHS grant CA-37775.

REFERENCES

Arndt-Jovin, D. J., and Jovin, T. M. (1977). *J. Histochem. Cytochem.* **25**, 585–593.

Chaplin, D. J., Durand, R. E., and Olive, P. L. (1985). *Br. J. Cancer* **51**, 569–572.

Chaplin, D. J., Durand, R. E., Stratford, I. J., and Jenkins, T. C. (1986). *Int. J. Radiat. Oncol. Biol. Phys.* **12**, 1091–1095.

Chaplin, D. J., Olive, P. L., and Durand, R. E. (1987). *Cancer Res.* **47**, 597–601.

Durand, R. E. (1982). *J. Histochem. Cytochem.* **30**, 117–122.

Durand, R. E. (1986a). *J. Natl. Cancer Inst.* **77**, 247–252.

Durand, R. E. (1986b). *Cancer Res.* **46**, 2775–2778.

Durand, R. E., and Olive, P. L. (1987). *Cancer Res.* **47**, 5303–5309.

Loeffler, D. A., Keng, P. C., Wilson, K. M., and Lord, E. M. (1987). *Br. J. Cancer* **56**, 571–576.

Olive, P. L., and Durand, R. E. (1987). *Cytometry* **8**, 571–575.

Olive, P. L., and Durand, R. E. (1989). *Int. J. Radiat. Oncol. Biol. Phys.* **16**, 755–761.

Olive, P. L., Chaplin, D. J., and Durand, R. E. (1985). *Br. J. Cancer* **52**, 739–746.

Reinhold, H. S., and Visser, J. W. M. (1983). *Int. J. Microcirc. Clin. Exp.* **2**, 143–146.

Siemann, D. W., and Keng, P. C. (1986). *Cancer Res.* **46**, 3556–3559.

Sutherland, R. M., and Durand, R. E. (1976). *Curr. Topics Radiat. Res. Q.* **11**, 87–139.

Trotter, J. M., Chaplin, D. J., Durand, R. E., and Olive, P. L. (1989). *Int. J. Radiat. Oncol. Biol. Phys.* **16**, 931–934.

Chapter 46

DNA Measurements of Bacteria

HARALD B. STEEN, KIRSTEN SKARSTAD, AND ERIK BOYE

Department of Biophysics
Institute for Cancer Research
Montebello, 0310 Oslo
Norway

I. Introduction

Whereas flow cytometry (FCM) has had a major impact on cell cycle studies of mammalian cells, it has not as yet been applied much to such measurements of bacteria (Steen, 1990b). This is in spite of the fact that the cell cycle of bacteria presents a number of essential problems that are difficult to resolve without a method that facilitates precise determination of the DNA content of large numbers of individual cells. The cell cycle of bacteria is basically different from that of mammalian cells in that more than one replication cycle can be in progress in the same chromosome at the same time. This makes DNA histograms of bacteria look very different from what we are used to seeing for mammalian cells and complicates cell cycle analysis quite significantly. Furthermore, the bacterial cell cycle varies much more with growth conditions than that of mammalian cells.

From an experimental, FCM point of view, bacteria differ from mammalian cells in several important respects:

1. The DNA content of bacteria is generally much lower than that of mammalian cells. For example, the DNA content of the *Escherichia coli* chromosome is about 1400 times less than that of a diploid human cell. This means that measurement of the bacterial DNA content with sufficient precision for applications like determination of cell cycle distribution— that is, with coefficients of variation (CV) of the order of a few percent— requires a combination of highly fluorescent staining and a sensitive instrument.

2. On the other hand, bacteria in some situations (e.g., when grown under optimal conditions) have a relatively much higher RNA content

519

(RNA/DNA ratio) than typical mammalian cells. This means that dyes with some affinity for RNA, like ethidium bromide (EB) and propidium iodide (PI) are not suitable unless RNA has been removed (e.g., by treating the cells with RNase). However, we have not been able to obtain consistent results subsequent to RNase treatment of bacteria. Hence, dyes with higher DNA specificity such as DAPI (4',6-diamidino-2-phenylindole), Hoechst 33342, Hoechst 33258, chromomycin, mithramycin (MI) or 7-AMD (7-aminoactinomycin D) are desirable. The Hoechst dyes and DAPI require ultraviolet (UV) excitation, which means that the flow cytometer light source must be either a high-power ion laser or a mercury high-pressure arc lamp; the latter has a strong emission line at 366 nm, which coincides with the peak absorption of these dyes. In laser instruments MI and its analog chromomycin can be excited only marginally by means of the 458-nm line of a tunable high-power argon laser. Furthermore, these dyes are not very bright due to a relatively low fluorescence quantum yield. The fluorescence yield of 7-AMD is too low to facilitate measurements on bacteria.

3. Bacteria differ from eukaryotic cells in that the chromosome does not contain histones and other proteins that inhibit the binding of many DNA-specific dyes and thereby destroys the stoichiometry of the staining. Presumably the bacterial DNA is more loosely packed, so that "chromatin structure" may not be expected to affect the staining of such cells.

4. Many bacteria species, such as *E. coli*, are rods rather than spheres and may therefore create orientation artifacts in some flow cytometers, especially laser-based instruments with near parallel excitation light.

5. In contrast to mammalian cells, which—with very few exceptions—are spherical in suspension and with a nucleus that is concentric with the outer cell wall and with a size roughly constant relative to that of the cell, the distribution of the DNA of bacteria may vary greatly with growth conditions and other factors. Thus, while under certain conditions the DNA appears to be evenly distributed in all of the cytoplasm, it may be concentrated into a minor portion of the cell volume in other cases. Again, geometry of DNA distribution may cause artifacts in some instruments.

6. The volume of bacteria is typically three orders of magnitude lower than that of mammalian cells. In some instruments this creates problems with the light-scattering measurement; the large-angle (90°) detection especially lacks sufficient sensitivity.

It may appear that the main reason why FCM has been so little applied to bacterial studies can be found in a combination of some of the aforementioned problems. In particular, the poor overlap between the emission lines of the argon laser, which is the standard excitation light source in most flow cytometers, and the absorption spectra of the most DNA-specific dyes may have caused weak fluorescence yields and discouraged some investigators from taking on a challenge that, in our opinion, is going to prove most rewarding.

Using an arc lamp-based flow cytometer, we have been able to carry out DNA measurements of bacteria with sufficient sensitivity and precision to allow asessment of essential cell cycle parameters (Boye *et al.*, 1983; Skarstad *et al.*, 1983, 1985; Steen, 1990b) and to study the mechanisms of the replication of the chromosome in more detail than was previously possible (Kogoma *et al.*, 1985; Skarstad *et al.*, 1986, 1988). Although bacteria are very easy to grow and process for measurement, we have found that the results are generally much more sensitive to growth conditions and other experimental parameters than in the case of mammalian cells. Strict reproducibility of culture conditions are essential to obtain consistent data. The stain also appears more critical than with mammalian cells. In our hands the dye giving the best results is MI. Although DAPI and the Hoechst dyes are known to be highly DNA-specific, we have not been able to achieve as good results with these dyes, in terms of sharp peaks and reproducibility, as we do with MI, although the fluorescence signals are of the same magnitude. The fluorescence quantum yield of MI is quite low (i.e., $\sim 0.05\%$ (Langlois and Jensen, 1979), thus putting great demands on instrument sensitivity. We therefore use it in combination with EB in order to obtain a higher fluorescence yield. Ethidium bromide has negligible absorption at the excitation wavelength being used (mainly the strong 436-nm line of the mercury arc lamp), and is therefore excited primarily by resonance energy transfer from MI molecules bound nearby (< 5 nm away) to DNA (Langlois and Jensen, 1979). Because of the higher fluorescence quantum yield of EB, excitations transferred to this dye produce more fluorescence than does MI alone. Hence, the DNA specificity of MI is combined with the higher fluorescence yield of EB. The net result is an increase in fluorescence intensity by a factor of ~ 2. It is interesting that this increase is significantly smaller than the value reported for mammalian cells—that is, 3.4 (Zante *et al.*, 1976; Langlois and Jensen, 1979). Fluorescence from RNA-bound EB appears to be negligible with this staining and excitation wavelength.

II. Materials

The following solutions are required:

1. Fixing solution, consisting of 77% ethanol in water. (If 96% ethanol is used, mix 80.2 parts of 96% ethanol with 19.8 parts of water.)

2. Staining buffer, consisting of 10 mM Tris, pH 7.4 and 10 mM MgCl$_2$. Prepare by adding 1.21 g Trizma base (Sigma Chemical, St. Louis, MO) and 2.03 g MgCl$_2 \cdot 6$ H$_2$O to 1 liter of distilled water (dH$_2$O) and adjust pH with HCl.
3. Ethidium bromide stock solution, consisting of 1000 μg/ml EB in staining buffer.
4. Staining solution, consisting of 180 μg/ml MI and 40 μg/ml EB in staining buffer. Prepare by dissolving the contents of one 2.5-mg ampule of Mithracin (Pfizer, New York, NY) into 13.3 ml of the staining buffer and add 0.56 ml of the EB stock solution.

The EB stock solution is stable for many months when stored in the dark at 4°C. The staining solution should be stored in the dark at 4°C and be used within a week of being prepared.

Note that dH$_2$O may contain significant amounts of microscopic particles (presumably mainly silicates) with sizes that produce light-scattering signals of the same order of magnitude as do many bacteria. The water may therefore produce a sizable background in the light-scattering channel. In order to avoid this artifact, filter both sheath water and reagent water, preferably with a somewhat finer filter than the standard 0.22-μm pore size, which transmits significant numbers of particles > 0.5 μm.

III. Fixation and Staining

Wash 1 ml of bacteria suspension once by centrifugation in staining buffer, resuspend in the same volume of buffer, and draw into a hypodermic syringe fitted with a 27-gauge needle. Squirt into 10 ml of ice-cold fixing solution while vigorously stirring (vortexing).

Fixation is completed within a few minutes. The cells may be stored at 4°C in the fixing solution for a few months, and at -20°C for much longer.

Before staining, wash the fixed cells once in staining buffer. Centrifugation may be carried out for 10 minutes at ~ 4000 g in a standard refrigerated centrifuge or for 2–4 minutes in a microcentrifuge at ~ 13000 rpm.

Resuspend the cell pellet in ice-cold staining buffer to a density that should not exceed 2×10^9 cells/ml. Mix the cell suspension with an equal volume of staining solution. This procedure ensures a reproducible dye concentration in the samples. The final sample volume is normally ≤ 1 ml. Keep the stained cells at subdued light on ice until a few minutes before measurement; capped Eppendorf tubes are suitable containers. The cells may be measured after a few minutes of staining; however, to obtain perfectly reproducible results they should be left in the staining solution for at least 1 hour before measurement.

Stained samples may be stored on ice for at least 8 hours before any deterioration—in terms of reduced quality of the measuring data—can be observed. In our experience the storage time that can be allowed appear to vary considerably from one experiment to the next. In some cases samples have been stored for several days without detrimental effects. We suspect that the main reason for the degradation of the samples is the presence of trace amounts of DNase. Hence, efficient washing of the fixed cells and careful cleaning of labware is important.

IV. Flow Cytometric Measurement

The small size and low DNA content of bacteria necessitate high sensitivity of the flow cytometer with regard to light scattering as well as fluorescence detection. We have obtained data with adequate quality using an arc lamp-based instrument (Steen, 1990a). A commercial version of this instrument is available from Skatron A/S (Lier, Norway). (Our laser-based instrument, the Coulter EPICS V, has been found insufficient with regard to sensitivity for detection of both fluorescence and light-scattering signals from bacteria.) The optical configuration of the arc lamp-based instrument is essentially similar to that of the epifluorescence microscope (Steen and Lindmo, 1979). In addition, it comprises a dark-field configuration that facilitates measurement of light scattering at small and large scattering angles in separate detectors (Steen and Lindmo, 1985). For the present type of application we use a 100-W high-pressure mercury arc lamp as the excitation light source and a "B1" filter block, which contains an interference excitation filter with a transmission band between 390 and 440 nm, a dichroic beam splitter having a characteristic wavelength of 460 nm, a long-pass emission filter with transmission from 470 nm, and an additional short-pass emission filter with transmission below 720 nm. The excitation filter transmits primarily the 405- and 436-nm mercury emission lines. Especially the 436-nm line is quite intense and coincides with the absorption peak of DNA-bound. The 720-nm short-pass filter is used to eliminate red and infrared background light, which would otherwise increase the shot noise on the signal and thereby reduce the effective sensitivity.

Instrument sensitivity is the major concern in measurement of bacteria. It is a function of the background of shot noise on the signal as much as of the signal itself. Thus, the signal/noise ratio is the essential parameter in this regard (Steen 1990a). In order to reduce noise, use the flow cytometer with the smallest permissible excitation and emission slits. (The latter is equivalent to the "pinhole" in other types of flow cytometers). To enhance signal, reduce the flow velocity by lowering the sheath pressure. (In the Argus instrument a pressure around 0.5 kg/cm^2 is suitable.) This can be done without affecting measuring precision, since the cell density of samples from bacteria cultures is typically high (i.e., 10^8–10^9 cells/ml) and the sample flow rate can therefore be kept correspondingly low (i.e., 0.1–1 μl/minute.) An instrument with volumetric sample injection is preferable to achieve a stable sample flow at this level.

Simultaneous light-scattering detection to gate the fluorescence measurement is always carried out in order to distinguish degraded cells and debris from intact cells. Thus, fluorescence and light scattering are recorded in dual-parameter mode. With samples having a high content of debris, light scattering can be a prerequisite for adequate data quality. Dual-parameter low-angle and large-angle light-scattering measurement can be used to distinguish dead cells from viable ones (Steen, 1986).

V.   Standards and Controls

Monodisperse fluorescent particles are always run when the instrument is set up to check the nozzle, that is, to see that the flow is perfectly laminar, and to optimize the instrument with regard to sensitivity and resolution (i.e., narrow histogram peaks). We use 1.5-μm fluorescent particles (Skatron A/S, Lier, Norway). The CV of the fluorescence peak should not exceed 2%, provided the particles are sufficiently monodisperse with regard to fluorescence.

DNA histograms of bacteria in many cases have no sharp peaks that can be used to check the preparation of the sample and the instrument function. A standard cell sample is therefore almost indispensable. For this purpose we use *E. coli* cells treated with the antibiotic rifampicin. This drug produces cells giving DNA histograms with several prominent and narrow peaks, each of which represents an integral number of chromosomes (Fig. 1), which may thus be used both to calibrate the fluorescence scale versus DNA content and to check the resolution of the measurement.

The procedure for preparing this standard sample is as follows: *E. coli* K-12 cells, strain CM735 (any wild-type strain that is not resistant to rifampicin can be used) is grown in LB medium with 0.2% glucose at 37°C

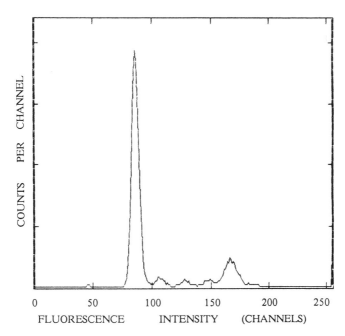

FIG. 1.   Fluorescence histogram of *E. coli* K-12 cells, strain CM735, harvested from a culture grown for 3 hours in the presence of the antibiotic rifampicin, which was given while the cells were in rapid exponential growth. The cells were fixed and stained with a combination of mithramycin and ethidium bromide according to the present procedure. The two prominent peaks represent cells with four and eight chromosomes, respectively. Minor peaks due to cells with two, five, six, and seven chromosomes are also evident. The histogram can thus be used to calibrate the fluorescence axis with regard to cellular DNA content. The CV of the main histogram peak is 3.5%.

to an optical density (OD) of 0.1 at 600 nm. At this point, when the cells are in exponential growth, rifampicin (Ciba-Geigy, Basel, Switzerland; 15 mg/ml in methanol) is added to the culture to a final concentration of 150 μg/ml. After a further 3 hours of culture, cells are harvested and fixed according to the procedure given already. The DNA histogram of such samples typically has prominent peaks representing cells with two, four, and eight chromosomes, respectively (Fig. 1). Provided that the instrument has sufficient sensitivity and is properly tuned, the CV of these peaks should be of the order of ≤5%.

Another sample that may serve as an absolute standard for DNA content is *E. coli* from a culture that is in stationary-growth phase. A wild-type *E. coli* (e.g., the aforementioned K-12 strain) is grown as described before until it has reached stationary-growth phase, namely until the OD of the culture has been constant to within 10% for at least 2 hours. Cells harvested from such cultures produce DNA histograms with two prominent peaks that represent cells with one and two chromosomes, respectively. The CV of these peaks is usually not as low as that found for rifampicin-treated cells, but should be well below 10%.

REFERENCES

Boye, E., Steen, H. B., and Skarstad, K. (1983). *J. Gen. Microbiol.* **129,** 973–980.
Kogoma, T., Skarstad, K., Boye, E., von Meyenburg, K., and Steen, H. B. (1985). *J. Bacteriol.* **163,** 439–444.
Langlois, R. G., and Jensen, R. H. (1979). *J. Histochem. Cytochem.* **27,** 72–78.
Skarstad, K., and Boye, E. (1988). *J. Bacteriol.* **170,** 2549–2554.
Skarstad, K., Steen, H. B., and Boye, E. (1983). *J. Bacteriol.* **154,** 656–662.
Skarstad, K., Steen, H. B., and Boye, E. (1985). *J. Bacteriol.* **163,** 661–668.
Skarstad, K., Boye, E., and Steen, H. B. (1986). *EMBO J.* **5,** 1711–1717.
Steen, H. B. (1986). *Cytometry* **7,** 445–449.
Steen, H. B. (1990a). *In* "Flow Cytometry and Sorting" (M. R. Melamed, T. Lindmo, and M. L. Mendelsohn, eds.), pp. 11–25. Liss, New York.
Steen, H. B. (1990b). *In* "Flow Cytometry and Sorting" (M. R. Melamed, T. Lindmo, and M. L. Mendelsohn, eds.), pp. 605–622. Liss, New York.
Steen, H. B., and Lindmo, T. (1979). *Science* **204,** 403–404.
Steen, H. B., and Lindmo, T. (1985). *Cytometry* **6,** 281–285.
Zante, J., Schumann, J., Barlogie, B., Goehde, W., and Buchner, Th. (1976). *In* "Pulsecytophotometry. Second International Symposium" (W. Goehde, J. Schumann, and Th. Buchner, eds.), pp. 97–106. European Press, Gent, Belgium.

Chapter 47

Isolation and Flow Cytometric Characterization of Plant Protoplasts

DAVID W. GALBRAITH

Department of Plant Sciences
University of Arizona
Tucson, Arizona 85721

I. Introduction

In order to apply techniques of flow cytometric (FCM) analysis to eukaryotic cells, we need single-cell suspensions. However, the somatic cells of higher plants typically are found in the form of complex three-dimensional organs, comprising a variety of tissues. Tissues are formed as a direct consequence of the means whereby cell division is achieved in plants. Cytokinesis occurs through the coalescence of Golgi-derived vesicles to form a cell plate, which progressively extends from the mother cell wall and eventually divides the cytoplasms of the sister cells. A continuous wall structure therefore links the cells within plant tissues, and individual cells cannot separate following completion of cell division. In order to convert plant tissues into cell suspensions, and thereby render them amenable to FCM techniques, we must remove the cell wall; such wall-less cells are termed protoplasts.

Protoplasts are prepared through the use of commercial preparations of hydrolases (cellulases, hemicellulases, and pectinases). These enzymes solubilize the polymeric components of the plant cell wall and, in the presence of suitable osmoprotectants, allow the release of intact protoplasts as a single-cell suspension. Removal of the cell wall, coupled to an absence of an elaborate glycocalyx, results in the adoption of a spherical shape by the protoplasts, even if they are derived from nonspherical cells. From an optical point of view, protoplasts are well suited for FCM analysis

METHODS IN CELL BIOLOGY, VOL. 33

and sorting, although some modifications to instrumentation and methodologies are essential (Harkins and Galbraith, 1984, 1987). These modifications are required primarily to accommodate the distinctive physical
characteristics of protoplasts. They are exceptionally fragile, in the absence of the cell wall, and are almost always larger than the animal cells
commonly used in FCM. Furthermore, their sizes (typically 20–60 μm in
diameter and sometimes larger) approximate and occasionally exceed that
of the flow cell tips employed in FCM. This means that for successful
analysis and sorting of protoplasts, use of large flow tips (100–200 μm in
diameter) becomes essential (Harkins and Galbraith, 1984, 1987; Sundberg
et al., 1987). Although traditional procedures of FCM and cell sorting
and the designs of the instrumentation have been optimized using small
biological cells and artificial microspheres (i.e., those falling in the range of
3–20 μm in diameter), there are no theoretical limitations either on the
sizes of the cells that can be analyzed and sorted using FCM or on the sizes
of the flow cells that can be employed (Kachel et al., 1990). Nevertheless,
a variety of practical considerations enter into the picture. This article outlines some of the problems and their resolution in the sorting of biological
cells as large as 100 μm using commercially available FCM instrumentation. In principle, the procedures are applicable to all biological cells and
cell aggregates within this size range, although they are illustrated for
plant protoplasts and pollen.

Recent developments in the application of FCM to plant protoplasts
include methods for the quantitative examination of the cellular sizes and
chlorophyll contents of populations derived from leaf tissues (Galbraith
et al., 1988). Size analysis is achieved through measurement of the time
taken for the protoplasts to pass through the laser beam. This parameter,
termed the time-of-flight (TOF) parameter, is obtained by electronic-pulse
shape analysis. A measurement is made of the time widths of the fluorescence signals produced by the protoplasts, based either on chlorophyll autofluorescence or on fluorescein diacetate (FDA) fluorochromasia. Under
appropriate conditions, these fluorochromes are evenly dispersed throughout the body of the cell (in the former case within chloroplasts and in the
latter case within the cytoplasm). Thus the time width of the signal contains
information about the sizes of the protoplasts. This can be processed to
provide information about protoplast diameters. As will be illustrated,
simple circuitry can then be employed to provide an accurate estimate of
cellular surface areas and volumes. The autofluorescence signals produced
from the chloroplasts also can be employed to provide information about
the chlorophyll contents of protoplasts (Galbraith et al., 1988). The combination of size and chlorophyll measurements allows a discrimination

between different subsets of protoplasts produced from plant organs such as leaves and for the first time permits an analysis at the single-cell level of biochemical functions associated with these protoplast subpopulations (Harkins *et al.*, 1990).

II. Application

The preparation of plant protoplasts is not restricted to particular plant species or tissues; however, experimental conditions that give rise to the release of viable protoplasts must be tailored to the particular plant tissues under study. Details of these conditions are beyond the scope of this article but can be found elsewhere (see, for example, Fowke and Constabel, 1985; Reinert and Binding, 1986; Puite *et al.*, 1988). In principle, since all protoplasts comprise spherical structures, the methods described for measurement of cell size and chlorophyll contents should be universally applicable, although some possible limitations are discussed in Section V. This article outlines methods for the analysis and sorting of a variety of different protoplast and cell types. It also describes a simple analog circuit that permits the conversion in real time of cell size measurements into measurements of surface area and volume.

III. Materials

A. Plant Materials

Tobacco (*Nicotiana tabacum* cv Xanthi) seed was obtained from the USDA Tobacco Research Laboratory (Oxford, NC) and was maintained by selfing under normal greenhouse conditions. Pollen (*Zea mays, Carya illinoiensis*, paper mulberry and ragweed) and *Lycopodium* spores were obtained from Polysciences, Inc. (Warrington, PA). A cell suspension culture of *Nicotiana sylvestris* was kindly provided by Dr. Roy Jensen. It can be maintained under sterile conditions as 100-ml aliquots in basal MS medium at pH 5.7 (Murashige and Skoog, 1962) containing 3% (w/v) sucrose and 1 mg per 1 liter of 2, 4-dichlorophenoxyacetic acid. The cell cultures are contained in 500-ml Erlenmeyer flasks at 25°C with constant orbital agitation (100 rpm). Subculture occurs at 5-day intervals by transfer of cells into 100 ml fresh medium.

B.   Chemicals

We have obtained all chemicals from the Sigma Chemical Co. (Saint Louis, MO), unless otherwise noted. Macerase and Cellulysin are from Calbiochem, Inc. (La Jolla, CA), cellulase from Worthington, aniline blue from the Fisher Scientific Co. (Pittsburgh, PA), and fluorescent microspheres from Polysciences, Inc. (Warrington, PA).

IV.   Procedures

A.   Protoplast Preparation

1.   TOBACCO LEAF TISSUES

Protoplasts are prepared from leaves selected from vegetative plants growing vigorously in Magenta boxes in basal MS medium (Murashige and Skoog, 1962) containing 3% (w/v) sucrose, solidified with 0.8% agar. Leaf tissue (0.5 g) is excised under sterile conditions and sliced into 1×10 mm segments, which are incubated within an 85-mm-diameter sterile plastic Petri dish in 10 ml of a filter-sterilized osmoticum containing 10 mM $CaCl_2$, buffered with 3 mM 2- [N-morpholino]ethane sulfonic acid (MES), pH 5.7; the osmotic pressure depends on the particular application. For isolation of mesophyll protoplasts either 0.7 M mannitol or 0.35 M KCl can be employed; for applications involving epidermal protoplasts an ionic osmoticum is essential, since these protoplasts have a lower buoyant density than isotonic mannitol. The cell walls are removed by including in the osmotica 0.1% (w/v) driselase, 0.1% (w/v) macerase, and 0.1% (w/v) cellulysin. Incubation is continued for 12–15 hours at room temperature (RT). The protoplast digest is filtered through sterile cheesecloth into plastic conical centrifuge tubes and is centrifuged at 50 g for 10 minutes. The protoplast pellet is gently resuspended in 5 ml of a solution containing 25% (w/w) sucrose dissolved in 3 mM MES and 10 mM $CaCl_2$, pH 5.7, and is overlaid with 5 ml of osmoticum. The protoplasts are centrifuged at 50 g for 2 minutes. The interface, which contains the viable protoplasts, is collected, diluted with 10 ml KCl osmoticum, and the protoplasts are pelleted by centrifugation at 50 g for 5 minutes. The protoplasts are resuspended in osmoticum to a final concentration of $\sim 10^5$ per 1 ml. The total protoplast yield is determined through hemocytometry and the proportion that are viable is found by resuspension of the protoplasts in osmoticum containing 0.1% v/v of a solution of fluorescein diacetate (FDA; 1 mg/ml in acetone). Viable protoplasts accumulate fluorescein, and the proportions of these can be determined by fluorescence microscopy (Harkins and Galbraith, 1984).

2. *NICOTIANA SYLVESTRIS* CELL SUSPENSION CULTURES

Protoplasts are prepared from actively growing cell cultures 3 days after subculture. The cells contained within 150 ml of the cultures are allowed to settle under gravity for ~15 minutes. The supernatant is removed and the cells (~14 ml) are resuspended in five volumes of an osmoticum (NTTO; Galbraith and Mauch, 1980) containing 0.5% (w/v) cellulase, 0.1% (w/v) cellulysin, 0.05% (w/v) driselase, and 0.02% (w/v) macerase. After incubation with gentle orbital shaking at 25°C for 20 hours, the protoplasts are filtered through five layers of cheesecloth into 50-ml conical tubes. The protoplasts are sedimented by centrifugation at 50 g for 5 minutes, are resuspended in 5 ml 25% (w/v) sucrose, and are overlaid with 5 ml of osmoticum. After centrifugation at 50 g, the intact protoplasts are collected from the interface (~2 ml), are diluted with five volumes of osmoticum, and are collected by centrifugation at 50 g for 5 minutes.

B.   Fluorescent Standards

Protoplasts are fixed by incubation for 60 minutes at 20°C in 2% (w/v) freshly prepared paraformaldehyde dissolved in osmoticum. Pollen (5 mg) is stained by resuspension in 2 ml of a solution (0.1%, w/v) of aniline blue in phosphate-buffered saline (PBS) pH 9.0.

C.   Flow Cytometry

The EPICS V flow cytometer/cell sorter (Coulter Electronics, Hialeah, FL) is operated as detailed here for the various applications.

1.   ANALYSIS OF INTACT MESOPHYLL PROTOPLASTS

The laser is tuned to 457 nm with a power output of 100 mW. We employ two barrier filters to exclude scattered light; these have half-maximal transmittance at 510 and 515 nm (termed LP510 and LP515). A further barrier filter (LP610) screens the red-channel photomultiplier (PMT). Most chlorophyll autofluorescence emission occurs above 620 nm; the first two filters are routinely present in the light path but may not be needed for this application. Integral or log-integral red fluorescence, one-parameter frequency distributions are typically accumulated to a total count of 20,000.

2. ANALYSIS OF VIABLE MESOPHYLL PROTOPLASTS

Protoplasts are stained with FDA. The laser is tuned to 457 nm with a power output of 100 mW. Fluorescence emission is detected using the green-channel PMT. Barrier filters LP510 and LP515 are used to eliminate scattered light. Light is reflected into the PMT using a dichroic mirror (DC590) that splits at a wavelength of 590 nm (for this application, a fully-silvered mirror would also be appropriate; the dichroic is used for convenience in other applications). Two blue-glass BG38 filters (Optical Instrument Laboratory, Houston, TX) are used to screen the PMT. Integral or log-integral green fluorescence, one-parameter frequency distributions are accumulated to a total count of 20,000.

3. ANALYSIS OF VIABLE PROTOPLASTS FROM SUSPENSION CULTURES

Protoplasts are stained with FDA. The flow cytometer is operated as described in Section IV, C, 2, with the exception that the BG38 barrier filters can be omitted.

4. SORTING POLLEN

Pollen is stained with aniline blue. The laser is tuned to 514 nm with an output of 200 mW. Barrier filters LP530 and LP540 are used to screen the green-channel PMT. Pulse-width time-of-flight (PW-TOF) analysis is performed based on peak green fluorescence. The TOF module is set to analyze the time that the pulses of fluorescence remain above thresholds set to 50% of the peak value, although this threshold can be lowered if necessary (Galbraith et al., 1988). One-parameter frequency distributions of protoplast size result from this analysis; further one-parameter frequency distributions can be accumulated based on peak or integral fluorescence and on forward-angle light scatter.

5. SORTING FIXED MESOPHYLL PROTOPLASTS

The fluorescent signal derives mostly from chlorophyll autofluorescence, although some contribution from paraformaldehyde-induced fixation will occur. The laser is tuned to 457 nm with an output of 100 mW. Barrier filters LP510 and LP515 and 590 are used to screen the red-channel

PMT. Log-integral red-fluorescence signals are accumulated to give one-parameter frequency distributions. Sort windows are positioned according to the location of the peak.

6. SEPARATION OF MESOPHYLL AND EPIDERMAL PROTOPLASTS

The protoplasts are prepared in ionic osmoticum. The purified protoplasts are stained with FDA, as described previously. The laser is tuned to 457 nm with a power output of 100 mW. Fluorescence emission is detected using both the red- and green-channel PMT. Barrier filters LP510 and LP515 are used to eliminate scattered light. Light is reflected into the green PMT using a dichroic mirror (DC590) and two BG38 barrier filters to eliminate entry of chlorophyll autofluorescence. Light transmitted through the dichroic mirror enters the red PMT screened by barrier filter LP610. Some signal subtraction may be necessary to eliminate spectral overlap. Cell size is determined through PW-TOF based either on peak green (FDA) fluorescence (for epidermal protoplasts) or on peak red fluorescence (for mesophyll protoplasts). Two-dimensional PW-TOF (FDA) versus integral red fluorescence allows the generation of contour plots permitting the sorting of epidermal from mesophyll protoplasts (Galbraith *et al.*, 1988). Frequency distributions are typically accumulated to a total count of 20,000.

D. Cell Sorting: Instrument Alignment

The system pressures and bimorphic crystal drive frequencies suitable for use with the various flow tip sizes can be found in Harkins and Galbraith (1987). For the 200-μm flow tip, we routinely use a system pressure of 6 psi and a drive frequency of 8 kHz. The sheath fluid comprises mannitol (50.5 g per 1 liter) and glucose (68.4 g per 1 liter), buffered with 3 mM MES, pH 5.7. We employ two 2.5-liter sheath tanks connected in parallel; this provides ~2–3 hours of sorting and analysis between refills.

Sort alignment is achieved as follows:

1. Using the sort test mode, the bimorphic crystal drive amplitude is adjusted to provide a uniform and stable sorted stream.
2. The number of undulations are counted to provide an estimate of the position of droplet break-off and hence the charge delay setting.
3. Analysis is initiated using standard particles that approximate the size of the protoplasts that are to be sorted (either paraformaldehyde-fixed protoplasts or aniline blue-stained pollen). Sort windows are set on the appropriate frequency distributions.

4. A sort matrix analysis is performed, to define precisely the charge delay setting that yields a sort efficiency of 100%. This can be conveniently measured using a single-cell deposition device (the Coulter Autoclone) to sort 10 particles per well in a 96-well culture plate; the actual numbers of particles recovered are then determined using an inverted light microscope.

5. Analysis and sorting of the protoplasts is initiated, using the analysis parameters previously described and the sort conditions defined by use of the standard particles.

For sterile sorting, we utilize a 0.2-μm in-line filter from Pall (Ultipor Type DFA4001ARP) that can be autoclaved. The sample tube, sample pick-up, and sample introduction line are autoclaved. The sheath tanks and sheath lines are thoroughly cleaned with dilute bleach and are rinsed with 70% ethanol. The sheath fluid is sterilized either by autoclaving or by passage through Millipore 0.22-μm filters. Prior to sorting, the sample lines are cleared of residual ethanol by passage of sterile sheath fluid. Sorting is performed into 96-well plates prefilled with 50–100 μl of sterile growth medium. Protoplast density affects further development in culture. Either sufficient protoplasts must be sorted ($\sim$1000 per well) or feeder cells (nonmorphogenic *N. tabacum* cell suspensions) must be included in the wells (Ayres, 1987; Galbraith, 1989a, b).

E. Transformations of Data

1. PROTOPLAST SIZE MEASUREMENT

Previous work has shown that the TOF parameter can be used for the accurate measurement of protoplast size, assuming appropriate calibration (Galbraith *et al.*, 1988). For this calibration, we require particles that approximate the size ranges over which the TOF measurements are to be made; these particles must be fluorescent and must not deform or degrade during passage through the flow cytometer. In general, artificial fluorescent microspheres with diameters similar to those of protoplasts are not readily available. We therefore have employed natural microspheres (pollen) stained with aniline blue for calibrating the TOF parameter (Galbraith *et al.*, 1988). Since pollen is essentially indestructible, it can also be employed for the purpose of optimizing the sort process (see Section VI). The linearity of the relationship between the TOF parameter and actual protoplast diameter can be directly determined through sorting of the protoplasts using a series of nonoverlapping sort windows spaced across

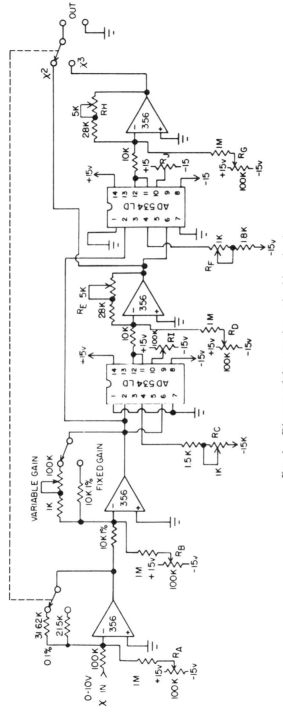

Fig. 1. Diagram of the squaring and cubing circuit.

the TOF frequency distribution, followed by measurement of protoplast diameters from light micrographs (Galbraith *et al.*, 1988).

2. PROTOPLAST SURFACE AREA AND VOLUME MEASUREMENT

Monolithic JFet input low-power operational amplifiers (LF356N) can be purchased from DigiKey Corporation (Thief River Falls, MN). Internally trimmed precision integrated-circuit four-quadrant multiplier/ dividers (AD534LD) are available from Analog Devices Corporation (Norwood, MA). The design of the squaring and cubing (S/C) circuit (Fig. 1) follows the recommendations of the manufacturers. The circuit involves six stages, each of which acts as a voltage inverter. The first operational amplifier scales the input voltage to a range appropriate for subsequent squaring or cubing. The second provides a fixed or variable gain feature. The third squares the input voltage. This is rescaled by the fourth stage prior to output on a 0–10 Volt scale, or is further multiplied by the fifth stage to provide a cubic output, which is then rescaled by the sixth stage to the 0–10 V scale. The response time of the circuit was found to be <10 μsec; this is much smaller than the conversion time employed by the multiple data analysis and display system (MDADs) of the flow cytometer (50 μsec).

3. PROTOPLAST CHLOROPHYLL MEASUREMENT

Correlation between the emission of red autofluorescence and protoplast chlorophyll content can be achieved in a manner analogous to that described for TOF/size measurements. Thus, different, nonoverlapping sort windows are placed on the one-dimensional frequency distributions of red-autofluorescence emission, and defined numbers of protoplasts are sorted. Cellular chlorophyll amounts can then be quantitated through spectrofluorometric analysis (Galbraith *et al.*, 1988).

V.   Critical Aspects of the Procedures

In general, the most critical aspect relates not to the procedures per se but to the biological materials under study. Familiarity with procedures of preparation and culture of protoplasts is essential for their successful use in flow analysis and sorting. Protoplasts represent some of the most fragile types of cells yet subjected to flow techniques. Correspondingly, the researcher must recognize those features of protoplasts that indicate physiological health, if these experiments are to be successful; the light microscope is probably the most important instrument in this regard. A variety of observational characteristics can be listed: for example, viable protoplasts are turgid and spherical, they exhibit FDA fluorochromasia and cytoplasmic streaming, and organelles are evenly dispersed through the cytoplasm. Other, more phenomenological observations include that the physiological status of the donor plant or cell type is critical for the production of viable protoplasts that can be readily manipulated using flow techniques. Variation in light intensity, day length, plant stage, and leaf number can affect protoplast yield and, in general, there appears to be a direct correlation between protoplast yield and subsequent success in experimental manipulation. Controlling for these variables through use of *in vitro* methods of plantlet propagation is strongly recommended. Finally, it is a truism to state that no one set of experimental conditions is applicable to all types of protoplasts. Wide variations in the composition of media are encountered in reports concerning the successful culture of different protoplast types (Puite *et al.*, 1988). An obvious and sensible approach to take when addressing the flow analysis and sorting of different protoplast types would be the use of optimal culture media as the sheath fluid, bearing in mind the conductivity requirements associated with the cell sorting process.

A. Flow Analysis of Protoplasts

1. CHOICE OF FLOW TIPS

As previously noted, plant protoplasts are generally much larger than animal cells, and the sizes of those commonly employed in FCM applications (30–60 μm) approach the diameters of the standard flow tips (60–75 μm). Evidently, it is important to employ flow tips larger than the cells to be analyzed, although use of flow tips much larger in diameter could lead to errors associated with the hydrodynamic process of centering the particles in the fluid stream.

2. COMPETING PIGMENTS

A feature of higher plants is the presence of a variety of intracellular pigments, many of which are associated with photosynthesis. The presence of these pigments complicates flow analysis of protoplasts, depending on the degree of spectral overlap between these pigments and the specific fluorochromes to be used in the particular study. The researcher should be aware of these potential problems and be prepared to deal with them through appropriate selection of fluorochromes, fixation procedures, and excitation and emission wavelengths.

3. SAMPLE PREPARATION

Flow cytometry allows measurements of all objects contained within the sample. Since pulse analysis is typically triggered on light-scatter, and since plant cells contain significant numbers of intracellular organelles, some confusion can be experienced in initial attempts to analyze plant protoplast populations, as broken protoplasts can contribute significant amounts of subcellular debris (Harkins and Galbraith, 1984). Use of techniques for purification of the protoplasts is recommended, such as sucrose gradient centrifugation. Alternatively, the flow cytometer can be adjusted to trigger on fluorescence, cell size (TOF), or other parameters that more precisely define the population of interest. Finally, use of the fluorescence microscope is strongly recommended in order to determine whether appropriate staining of the protoplasts has been achieved.

4. TOF LIMITATIONS

Theoretical considerations indicate that the relationship between protoplast diameter and the PW-TOF parameter will increasingly deviate from a linearity as the size of the protoplast approaches that of the beam (Leary et al., 1979). For the EPICS system using standard optics, deviation becomes noticeable at a protoplast diameter of ~15 μm (Galbraith et al., 1988). Below this point, correct sizing can still be achieved, although it is necessary to use a series of appropriately sized standard particles in order to calibrate these measurements properly. We have not empirically established the upper limit to TOF size analysis that will be dictated by the maximal time domain that can be accommodated by the TOF circuitry. However, a few calculations suggest that this will not be a limitation: for the standard EPICS V hardware, the upper limit on TOF analysis is 40 μsec. For large-particle sorting using the larger flow tips, the system pressure must be lowered in order to obtain a point of droplet break-off above the position of the deflection assembly (Harkins and Galbraith, 1987). For the 200-μm flow tip, the highest fluid stream velocity is typically 10.8 m/second. This means that the effective upper limit for TOF measurements set by the electronics is ~400 μm, which is an order of magnitude larger than tobacco leaf protoplasts.

Biological factors might be expected to influence the linearity of the TOF analysis. For example, deformation of cellular shapes at the laser interrogation point due to hydrodynamic forces experienced within the flow cytometer might be expected to introduce errors in sizing; our data indicate that this is not the case for protoplasts (Galbraith *et al.*, 1988). On the other hand, the presence of chloroplasts within mesophyll protoplasts introduces errors in protoplast sizing when TOF measurements are performed based on FDA fluorescence, but not when they are based on chlorophyll autofluorescence (Galbraith *et al.*, 1988). This is probably due to quenching of FDA fluorescence by the highly absorbent pigments within the chloroplasts.

B. Protoplast Sorting

Critical to the successful sorting of protoplasts are the following elements: the generation of high-quality frequency distributions in which the location of the desired protoplast populations are well defined, the optimization of sort parameters in order to provide accurate and high-efficiency sorting of the desired protoplasts, the provision of sterile conditions, and the establishment of conditions for protoplast growth after sorting. Optimization of sort parameters is most easily achieved with large flow cell tips, coupled to the use of appropriate standard particles. These are detailed in later sections. Sterile conditions are achieved through use of a combination of liquid sterilants and autoclaving. Use of a single-cell deposition device to sort protoplasts into 96-well plates also greatly facilitates maintenance of sterile conditions. Inclusion of certain antibiotics (e.g., β-lactam derivatives at 50–500 mg per 1 liter) can be helpful; these appear innocuous to protoplasts. In terms of optimization of culture conditions after sorting, maintenance of a minimal cell density appears important. This can be achieved either by sorting large numbers of protoplasts (>1000) within the wells of the culture plates, or through the use of conditioned media or feeder cells (Afonso *et al.*, 1985; Ayres, 1987). Feeder cells can be provided as a liquid cell suspension or can be embedded in agarose, with or without a physical barrier. Finally, it should be noted that some types of tissue culture plates appear to be toxic to protoplasts.

VI. Controls, Standards

We routinely employ pollen, *Lycopodium* spores, or fixed protoplasts for setting up and standardizing the sort process (Table I). We have obtained samples of standard fluorescent particles as large as 70 μm. These give a single discrete peak upon one-dimensional frequency distributions of

TABLE I

Biological Particle Size Ranges

Particle type	Diameter (μm; mean $\pm$ SD)	Fluorescence source
Broussonetia papyrifera pollen	13.8 $\pm$ 1.3–20	Callose
Ambrosia elateior pollen	20.9 $\pm$ 1.2–22	Callose
Lycopodium spores	28.6 $\pm$ 2.5–12	Callose
Nicotiana tabacum leaf protoplasts (fixed)	33.7 $\pm$ 6.2–25	Chlorophyll
N. tabacum leaf protoplasts	42.3 $\pm$ 6.9–20	Chlorophyll
N. tabacum suspension culture protoplasts	41.2 $\pm$ 10.4–21	FITC/FDA
Carya illinoiensis pollen	51.1 $\pm$ 3.3–11	Callose
Zea mays pollen	95.3 $\pm$ 4.1–11	Callose

Adapted from Harkins and Galbraith (1987) with permission of Wiley-Liss.

fluorescence, and although the coefficients of variation (CV values) are somewhat high ($\sim$7.2%), they should be suitable for setting up the sort process. It should be emphasized that correct selection of sort parameters for fragile plant protoplasts requires the use of particles that approximate the size of the protoplasts. This is because the forces that drive the formation of droplets appear to interact with the particles in a size-dependent manner. Thus conditions for sorting that are optimal for small particles may not necessarily be optimal for large particles and cells (Harkins and Galbraith, 1987). As a general rule, lowering the crystal drive frequency tends to improve the sort efficiency.

VII.   Instruments

The procedures first described have been worked out for the Coulter EPICS V system. Large-particle sorting has also been successfully achieved using a Becton Dickinson FACStar Plus equipped with a 200-μm flow tip. The procedures should apply in principle to all other types of flow instrumentation, assuming the availability of appropriate hardware.

VIII.   Results

Use of these procedures is illustrated using three general examples.

A. Analysis of the Dimensions of Plant Protoplasts and Cells

1. MEASUREMENT OF DIAMETERS

Characterization of the TOF distributions of populations of tobacco leaf protoplasts is presented in Fig. 2A. This is a near-normal distribution with a mode channel of 98. A correlation between the TOF parameter and actual protoplast size can be obtained by sorting protoplasts using a series of defined, nonoverlapping sort windows. Protoplast sizes are then subsequently measured using light microscopy. Figure 3 illustrates the linear relationship obtained in this manner (Galbraith et al., 1988). Use of standard particles allows the rapid calibration of TOF/size relationship; thus for the data presented in Fig. 2A, the true size of the protoplasts is 41 μm.

2. MEASUREMENT OF SURFACE AREAS AND VOLUMES

The accuracy of the S/C circuit (Fig. 1) was measured using a Hewlett-Packard 8011A pulse generator to produce a series of input square-wave pulses of different defined widths adjusted to span the linear TOF range at 10-channel intervals. One-dimensional TOF histograms were accumulated, either of the linear input, or of the squared or cubed output of the S/C module. Based on regression analysis, we determined that the S/C module operates with an accuracy that is no less than that of the EPICS MDADS. Analysis and processing of real-time distributions of TOF signal produced by protoplasts is illustrated in Fig. 2B and 2C. Since protoplasts are spherically symmetrical, these distributions provide a measure of protoplast surface area and volume, respectively.

B. Analysis of the Chlorophyll Contents of Leaf Protoplasts

Uniparametric analysis of the autofluorescence emission from tobacco leaf protoplasts is presented in Fig. 4. This profile is gated to exclude the contributions from free chloroplasts (Harkins and Galbraith, 1984; Galbraith et al., 1988). In order to obtain a correlation between fluorescence emission and chlorophyll content, this near-normal distribution can be divided into a series of nonoverlapping regions for the purposes of sorting. In this example, we selected five windows that were 15 channels wide and that were spaced by intervals 5 channels wide. Defined numbers of protoplasts were sorted and were subjected to spectrofluorometric analysis for

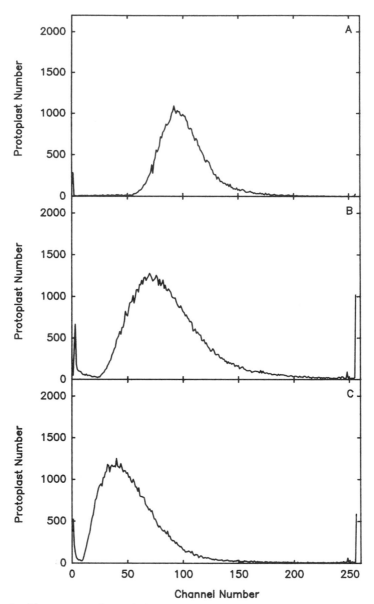

Fig. 2. Flow cytometric analysis of size according to time of flight (TOF). One-dimensional frequency distributions were accumulated to a total count of 20,000, corresponding to (A) TOF, (B) TOF2, and (C) TOF3 signals derived from chlorophyll autofluorescence of leaf protoplasts. The means and CV values for the distributions are 98 (18.4%), 80 (33.3%), and 48 (47.6%), respectively.

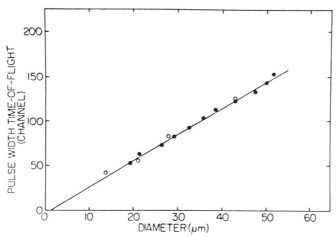

FIG. 3. Correlation between the pulsewidth time-of-flight (PW-TOF) parameter and protoplast or pollen/spore diameters, based on chlorophyll or aniline blue-induced fluorescence, respectively: protoplasts (●); pollen/spores (○). Reproduced from Galbraith *et al.* (1988) with permission of Wiley-Liss.

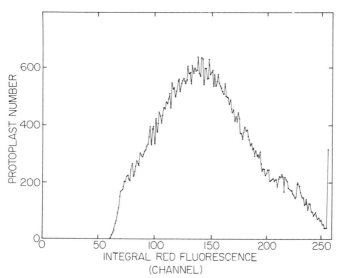

FIG. 4. Uniparametric flow analysis of the emission of chlorophyll autofluorescence from leaf protoplasts, using a linear abscissa gated to exclude free chloroplasts (below channel 60). Reproduced from Galbraith *et al.* (1988) with permission of Wiley-Liss.

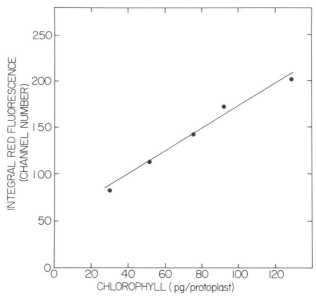

FIG. 5.   Correlation between the chlorophyll autofluorescence emission and the protoplast chlorophyll content ($r^2 = 0.983$). Reproduced from Galbraith *et al.* (1988) with permission of Wiley-Liss.

chlorophyll content (Galbraith *et al.*, 1988). There is a high degree of correlation between protoplast chlorophyll content and the autofluorescence yield (Fig. 5).

C.   Combined Chlorophyll–TOF Analysis of Leaf Protoplasts

The tissues of dicotyledonous leaves comprise a variety of different cell types. Predominant are the photosynthetic tissues of the mesophyll, which contain large numbers of mature chloroplasts, and the nonphotosynthetic tissues of the epidermis and perivascular parenchyma. Biparametric flow analysis of protoplasts prepared from leaves according to TOF and chlorophyll content readily permits an identification of two populations of protoplasts (Fig. 6). As would be expected from the uniparametric analysis (Fig. 5), the mesophyll protoplasts exhibit a broad range of chlorophyll contents, whereas those from the epidermis and perivascular parenchyma are essentially devoid of chlorophyll. It should be noted that for accurate sizing of the mesophyll protoplasts, TOF processing of the chlorophyll autofluorescence signal rather than the FDA fluorochromasia signal is required (Galbraith *et al.*, 1988); obviously, it is not possible to analyze

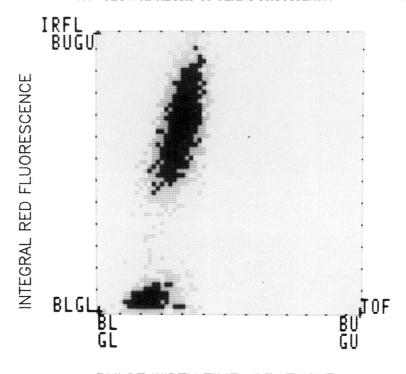

PULSE WIDTH TIME—OF—FLIGHT

FIG. 6. Contour analysis of two-parameter measurements of the chlorophyll autofluorescence versus pulsewidth time-of-flight characteristics of leaf protoplasts. Fluorescein diacetate fluorochromasia was used as the source of signals for the TOF analysis. The contour levels correspond to 5, 15, and 50% of the peak channel. Reproduced from Galbraith *et al.* (1988) with permission of Wiley-Liss.

nonphotosynthetic protoplasts in this way. Thus for combined, accurate TOF/size analyses of these different protoplast populations, two independent TOF modules would be required.

D. Sorting and Culture of Leaf Protoplasts

Under appropriate conditions, protoplasts can be sorted at efficiencies approaching 100% with no loss of viability (Afonso *et al.*, 1985; Harkins and Galbraith, 1984, 1987). A convenient monitor of the integrity of mesophyll protoplasts is the presence of free chloroplasts in the culture plate. Under optimal sorting conditions, the protoplasts remain metabolically active and at appropriate times will resynthesize cell walls and enter into cell division (Fig. 7). We have successfully sorted and cultured protoplasts following induced fusion (Afonso *et al.*, 1985; Ayres, 1987) and

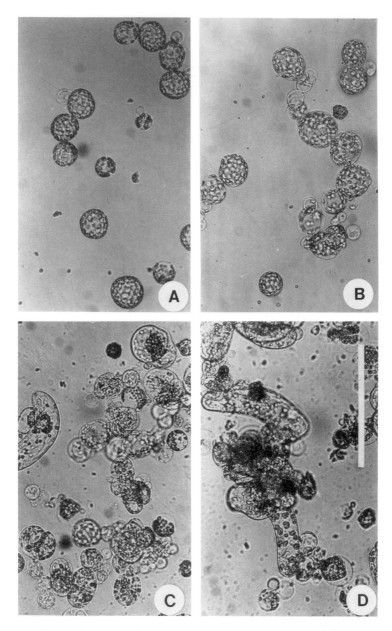

FIG. 7.   Growth in culture of sorted *Nicotiana tabacum* protoplasts (A) after 1 day, (B) after 2 days, (F) after 5 days, and (D) after 12 days. (×200). Bar, 150 μm. From Galbraith (1984).

following transfection (Harkins *et al.*, 1990). Finally, we have been able to sort and culture protoplasts derived both from the photosynthetic and nonphotosynthetic tissues of the leaf with essentially equal efficiencies.

ACKNOWLEDGMENTS

I thank Kristi Harkins for valuable assistance and Walt Hancock for the design, assembly, and testing of the S/C circuit. This work was supported in part by grants from the U. S. Departments of Agriculture and Energy, and the National Science Foundation.

REFERENCES

Afonso, C. L., Harkins, K. R., Thomas-Compton, M., Krejci, A. E., and Galbraith, D. W. (1985). *Biotechnology* **3,** 811–815.
Ayres, N. M. (1987). Ph. D. thesis, University of Nebraska-Lincoln.
Fowke, L. C., and Constabel, F. (1985). "Plant Protoplasts." CRC Press, Boca Raton, Florida.
Galbraith, D. W., and Mauch, T. J. (1980). *Z. Pflanzenphysiol.* **98,** 129–140.
Galbraith, D. W., Harkins, K. R., and Jefferson, R. A. (1988). *Cytometry* **9,** 75–83.
Galbraith, D. W. (1984). *In* "Cell Culture and Somatic Cell Genetics of Plants" (I. K. Vasil, ed.), pp. 433–447. Academic Press, New York.
Galbraith, D. W. (1989a). *Intl. Rev. Cytol.* **116,** 165–227.
Galbraith, D. W. (1989b). *In* "Biotechnology in Agriculture and Forestry", Vol. 9, (Y. P. S. Bajaj, ed.), pp. 304–327. Springer-Verlag, New York.
Harkins, K. R., and Galbraith, D. W. (1984). *Physiol. Plant.* **60,** 43–52.
Harkins, K. R., and Galbraith, D. W. (1987). *Cytometry* **8,** 60–70.
Harkins, K. R., Jefferson, R. A., Kavanagh, T. A., Bevan, M. W., and Galbraith, D. W. (1990). *Proc. Natl. Acad. Sci. U.S.A.* **87,** 816–820.
Kachel, V., Fellner-Feldegg, H., and Menke, E. (1990). *In* "Flow Cytometry and Sorting" (M. R. Melamed, T. Lindmo, and M. L. Mendelsohn, eds.), 2nd Ed., pp. 27–44. Liss, New York.
Leary, J. F., Todd, P., Wood, J. C. S., and Jett, J. H. (1979). *J. Histochem. Cytochem.* **27,** 315–320.
Murashige, T., and Skoog, F. (1962). *Physiol. Plant.* **15,** 473–497.
Puite, K. J., Dons, J. J. M., Huizing, H. J., Kool, A. J., Koornneef, M., and Krens, F. A. (1988). "Progress in Plant Protoplast Research." Kluwer, Dordrecht.
Reinert, J., and Binding, H. (1986). "Differentiation of Protoplasts and of Transformed Plant Cells." Springer-Verlag, Berlin.
Sundberg, E., Landgren, M., and Glimelius, K. (1987). *Theor. Appl. Genet.* **75,** 96–104

Chapter 48

Flow Cytometric Analysis of Plant Genomes

DAVID W. GALBRAITH

Department of Plant Sciences
University of Arizona
Tucson, Arizona 85721

I. Introduction

Compartmentalization of genomes within specific subcellular organelles is a characteristic feature of eukaryotic cells. Higher plants contain three separate genomes, located within the nucleus, the plastids, and the mitochondria, respectively. Correct coordination of gene expression within these genomes underlies the development of the organism from the fertilized egg, which occurs as a consequence of controlled cellular growth and division. As for other eukaryotes, the duplication of cells occurs through the operation of a cell division cycle. This cycle can be charted from the point of mitosis (M phase) and cell division; this is followed by a period of cellular growth (G_1 phase) preceding duplication of the nuclear genome (S phase), which in turn is followed by a second period of cellular growth (G_2 phase) prior to the next point of mitosis. The proportions of cells within the various phases of the cell division cycle can be empirically defined through measurement of the nuclear DNA content and its relationship to the haploid genome size (C), defined as the nuclear DNA content of the gamete. For somatic cells in G_1 phase, the nuclei are defined as having a 2C DNA content. For those in G_2 phase, the nuclei have a 4C DNA content. For those in S phase, the amount of nuclear DNA (which is intermediate between 2C and 4C) acts to chart the progress of the cells through S phase.

In animal cell systems, a variety of flow cytometric (FCM) procedures have been developed for the analysis of nuclear DNA contents. These

METHODS IN CELL BIOLOGY, VOL. 33

involve the staining of the nuclei with specific fluorochromes, under conditions in which the amount of emission of fluorescence is linearly correlated to DNA content. In order to provide access of the fluorochromes to the nuclei, the plasma membrane must be permeabilized, either through fixation or through the use of detergents. The types of fluorochromes that have been employed include molecules that selectively bind to DNA such as the Hoechst dyes, mithramycin (MI) and chromomycin A (CH-A). Alternatively, dyes that intercalate double-stranded nucleic acids, such as ethidium bromide (EB) and propidium iodide (PI), can be employed, in which case an RNase pretreatment becomes necessary. Specialized applications include the use of acridine orange (AO) for combined biparametric analysis of DNA and RNA, and the use of fluorescein-labeled monoclonal antibodies (mAb) directed against bromodeoxyuridine (BrdUrd) for biparametric analysis of the proportion of cells within S phase and their rate of DNA synthesis.

Similar types of approaches can be adapted for measurement of nuclear genome sizes in higher plants (Galbraith and Shields, 1982; Galbraith et al., 1983; Galbraith, 1984). The main consideration concerns the fact that higher plants comprise complex interspersions of interconnected tissues rather than the single-cell suspensions required for FCM analysis. As discussed in Chapters 47 (Galbraith) and 49 (Bergounioux and Brown) of this volume, one solution to this problem involves the production of suspensions of wall-less cells (protoplasts). Since protoplasts are spherically symmetrical, these suspensions are optically suited for flow analyses, if accommodation is made for their large diameters. Two strategies are available for measurement of DNA-specific fluorescence. Because the available fluorochromes cannot penetrate the plant plasma membrane; either the protoplasts can be fixed using organic solvents. This approach conveniently elutes cellular pigments such as chlorophyll (Galbraith and Shields, 1982; Galbraith, 1984). Alternatively, the protoplasts can be lysed through resuspension in the presence of detergents and the nuclei isolated. In both cases, addition of DNA-specific fluorochromes then permits the quantitative staining and subsequent FCM analysis of cellular or nuclear DNA contents.

It has been found that FCM analysis of DNA content can successfully be accomplished based on measurements at isolated nuclei (Galbraith et al., 1983; Sharma et al., 1983; Galbraith, 1984). These can be isolated from protoplasts or can be directly released from most plant tissues through maceration. Under controlled conditions, DNA degradation can be minimized and with the inclusion of suitable internal standards, a rapid and accurate estimate of the nuclear DNA content can be made based on analysis of uniparametric frequency distributions histograms of fluorescence emission.

These frequency distributions also include information about the proportions of nuclei within the various phases of the cell cycle. It should be noted, however, that interpretation of this information is complicated by two factors. The first is that the frequency distributions only represent the status of the nuclei present at the time of assay and therefore cannot by themselves indicate whether these nuclei are from cycling or noncycling cells. The second is a consequence of the fact that most plant species are polysomatic (d'Amato, 1986); this makes it difficult to determine whether (for example) all the nuclei within the "4C" peak are from diploid cells in G_2 or from tetraploid cells in G_0/G_1. The widespread occurrence of endoreduplication within mature plant tissues, possibly related to the process of cytodifferentiation, can further complicate the situation.

II. Application

This chapter outlines a simple series of procedures that permit the FCM measurement of plant nuclear DNA contents and analysis of the cell cycle using *Nicotiana tabacum* as a model. We provide methods for use with intact tissues, protoplasts, and callus in which MI is employed as the DNA-specific fluorochrome. Other fluorochromes that can be used include Hoechst 33258, DAPI (4', 6-diamidino-2-phenylindole), PI, EB, or CH A3 (Puite and Ten Broeke, 1983; Bergounioux *et al.*, 1986, 1988; Sgorbati *et al.*, 1986; Sundberg and Glimelius, 1986; Sundberg *et al.*, 1987; Ulrich and Ulrich, 1986; Ulrich *et al.*, 1988).

III. Materials

A. Biological Materials

Tobacco (*N. tabacum* cv xanthi) seed was obtained from the USDA Tobacco Research Laboratory (Oxford, NC) and was maintained by selfing under normal greenhouse conditions or as sterile shoot cultures on agar-solidified basal MS medium (Murashige and Skoog, 1962) containing 3% (w/v) sucrose. The cell suspension culture of *N. tabacum* cv xanthi was kindly provided by Dr. E. Nester. It is maintained under sterile conditions as 100-ml aliquots in MS3+ medium at pH 5.7 (Meyer *et al.*, 1988) with 3% (w/v) sucrose. The cell cultures are contained in 500-ml Erlenmeyer flasks at 25°C in darkness with constant orbital agitation (100 rpm). Subculture is achieved at 3-day intervals by transfer of ~50 ml of cells into 100 ml fresh medium. Standard chicken whole blood is obtained by venous puncture. The chicken red blood cells (CRBC) are employed fresh or are stored as a 1 : 5 dilution in Alsever's solution at 4°C, in which case they are used within a period of 4 weeks.

B. Chemicals

We have obtained all chemicals from the Sigma Chemical Co. (Saint Louis, MO), unless otherwise noted. Mithramycin is available from the Pfizer Chemical Company (Groton, CT).

IV. Procedures

A. Sample Preparation

1. PLANT ORGANS, CALLUS, AND CELL SUSPENSION CULTURES

The fresh plant samples are excised and weighed. Tissue from plant organs (~0.1 g) is transferred into a 60×15 mm Petri dish containing 0.75 ml of an ice-cold "lysis buffer" (pH 7.0) containing 45 mM MgCl$_2$, 30 mM sodium citrate, 20 mM MOPS [3-(N-morpholino)propanesulfonate], and 1 mg/ml Triton X-100. Tissue homogenization is achieved by chopping for 2 minutes on ice in a walk-in cold room, using a single-edged razor blade, until the tissue is finely minced and, for photosynthetic tissues, appears light green. For callus and cell suspension cultures an identical procedure is followed, except that the proportion of tissue to lysis buffer is increased 2-fold. The cell suspension cultures are collected prior to chopping by centrifugation at 50 g for 5 minutes.

2. PROTOPLASTS

Protoplasts are prepared and purified under conditions appropriate for the species and tissues under consideration. An example, for *N. tabacum* grown as *in vitro* shoot cultures, is given in Chapter 47 (Galbraith, this volume; Galbraith, 1989). In this example, the protoplasts are recovered by centrifugation at 50 g for 5 minutes after purification from nonviable cells and subcellular debris through flotation on a sucrose step-gradient.

For fixation, protoplasts (10^5) are resuspended in 2 ml of an ice-cold fixative comprising acetic acid–ethanol–water (18 : 3 : 2) containing 1%

sorbitol (Galbraith and Shields, 1982; Galbraith, 1984). The protoplasts are then recovered by centrifugation at 50 g for 5 minutes and are washed in 70% ethanol.

B. Sample Staining

Mithramycin is dissolved in lysis buffer to a nominal concentration of 0.1 mg/ml. Since different samples of the commercially available dye differ in purity, we routinely standardize the optical density of the solution to 0.6 at 420 nm.

For staining of chopped lysates prepared from intact plant organs and from callus and cell suspension cultures, an aliquot (0.2 ml) of the homogenate is added to 1.8 ml of ice-cold MI solution. The sample is then filtered through nylon mesh (pore size 60 μm). For lysis and staining of nuclei from viable protoplasts, and for staining of fixed protoplasts, the pelleted protoplasts are resuspended to a final concentration of 5×10^4/ml in the MI solution.

For provision of an internal standard, CRBC (50 μl) are added to the samples prior to analysis. The CRBC suspension comprises 20 μl of the diluted whole blood resuspended in 200 μl lysis buffer, added to 1.8 ml of the MI solution.

C. Flow Cytometry

Prior to flow analysis, all samples are filtered through 15-μm nylon mesh. Flow cytometric analysis is performed at 457 nm using a laser output of 100 mW. Blocking filters with half-maximal (cut-on) transmittance at 510/515 nm (LP510, LP515) exclude scattered laser light, and two 1-mm-thick BG38 glass filters eliminate chlorophyll autofluorescence. The sheath fluid comprises lysis buffer lacking Triton X-100. One-parameter histograms are accumulated at a flow rate of 50 per second typically to a total count of 5000–10,000.

D. Data Analysis: Measurement of Nuclear of Cellular DNA Contents

Comparison of the relative positions of the peaks of fluorescence corresponding to the plant cells or nuclei and of the standard RBC permits an accurate determination of the "unknown" DNA content, by division of the respective mode positions and multiplication of this amount by 2.33 pg (the genomic DNA content of CRBC). This calculation assumes linearity of the analog-to-digital converters (ADC) as well as a constant zero-level adjustment. Use of dual internal standards can account for variations in this latter adjustment (Vindeløv et al., 1983). The accuracy of measurement of genomic DNA contents also is a function of the difference in amounts between the unknown and the internal standard. Obviously, this accuracy is greatest if the "unknown" and the standard differ only enough to be distinguished. [They cannot be nearly identical; otherwise the two peaks of fluorescence will overlap and, if the mean values are closer than twice the coefficient of variation (CV) of the peaks, a single peak of fluorescence will be observed.] Plant species have unusually large ranges of nuclear DNA contents (Bennett et al., 1982; see Table I) and this means that it may be important to establish internal standards alternative to CRBC for use with some species.

V.   Critical Aspects of the Procedures

A.   Experimental Design, Implementation, and Standardization

Both the isolation of intact nuclei and the quantitative staining of DNA by MI require the presence of Mg^{2+} ions (Ward et al., 1965). Deoxyribonucleases typically are activated under similar conditions (Maniatis et al., 1982). It is critical that the samples be maintained at ice-cold temperatures following tissue homogenization or protoplast lysis. For this reason, use of a walk-in cold room and performance of all procedures on ice is recommended.

Cell cycle parameters are usually obtained from DNA frequency distributions through use of various computer programs. The accuracy with which cell cycle parameters can be determined is a function both of the numbers of cells or nuclei that are analyzed and of the accuracy inherent in the FCM measurements. Since in plant cell systems the proportions of cells in S phase often appear unusually low (Galbraith, 1989), accumulation of total counts to larger totals may become necessary. The overall accuracy of FCM measurements reflects factors intrinsic to the instrumentation (such as alignment and ADC linearity) and factors associated with the biological material and the staining procedure. The use of two sets of standard fluorescent microspheres is recommended for purposes of instrument standardization and calibration, in which case the positions of the peaks corresponding to the microspheres and their CV values should be monitored. An increase in the photomultiplier (PMT) voltage should not affect the ratio of the mean channel positions of the microspheres nor the CV values. The CV values of the DNA distributions should also be monitored. In general, these CV values should be as low as possible. Under most circumstances, we have observed values in the range of 3–6%, with fixation appearing to increase the CV values. Other workers have reported lower values, using different fluorochromes, hardware, and plant species (Ulrich *et al.*, 1988).

B. Interpretation of Data

The use of FCM data for the establishment of the nuclear genome size of a particular plant species requires an unambiguous definition of the haploid and diploid state within that species, both in terms of the chromosome number (n and 2n) and in terms of the DNA content of haploid and diploid cells (C and 2C). Since cells within somatic tissues can arrest within both G_0/G_1 and G_2, a precise definition of the diploid state can only be achieved through chromosome counts coupled to spectrophotometric analysis of the DNA contents of mitotic cells, followed by a statistical survey of the proportions of cells within the tissue correlated with FCM data from these tissues. This approach is considerably time-consuming, and strong circumstantial evidence can be obtained more simply through FCM. For example, analysis of nuclei released from chopped tobacco leaves reveals a population in which the predominant peak has a DNA content of 9.67 pg (Figs. 1 and 4, and Table I). Similar DNA contents are observed for protoplasts isolated from leaf tissues (Galbraith, 1984). These protoplasts during incubation in culture develop toward a conventional cell cycle in which the majority of the cells presumptively have a 2C (G_1) DNA content (Galbraith *et al.*, 1981). Correspondingly, chopping of tobacco anther tissues reveals a peak of fluorescence at one-half of the somatic cell DNA content, presumably the contribution of nuclei derived from haploid cells of the developing male gametophyte (Fig. 2). Finally, haploid plants, derived from anther culture, display half the somatic nuclear DNA content upon FCM analysis (Sharma *et al.*, 1983). Taken together, we may conclude with almost complete certainty that the 2C DNA content of tobacco is 9.67 pg.

VI. Instruments

The techniques just described, though originally developed for use with the Coulter EPICS V system (Coulter Electronics, Hialeah, FL), are applicable to all commercial flow cytometers.

VII.   Results

A.   Analysis of Plant Organs

When nuclei released from mature tobacco organs by the chopping procedure are stained with MI, the resultant one-dimensional frequency distributions display two discrete peaks (Fig. 1). The lower peak contains those nuclei from cells in G_0/G_1 and the upper peak those nuclei from cells in G_2. Nuclei from cells in S phase are found between these two peaks. We presume that mitotic nuclei are lost. The CV values of the two peaks are comparable to those obtained with animal cell lines.

Application of the chopping technique for the analysis of tobacco anther tissue is illustrated in Fig. 2. In this case, a third peak of fluorescence corresponding to a nuclear DNA content of 1C (channels 52–55) is apparent within the frequency distribution. It presumably represents the contribution of haploid cells located in the developing anther.

B.   Analysis of Cell Suspension Cultures

The chopping technique can also be successfully employed for the release of nuclei from callus and cell suspension culture sources in a condition suitable for FCM analysis. This type of analysis is illustrated for cell suspension cultures in Fig. 3. The CV of the G_0/G_1 peak is 5.4% and the cell cycle distribution is 57.8% G_0/G_1, 10.4% S, and 31.8% G_2. The interpretation of these phase distributions depends on the assumption that the cell suspension culture comprises a homogeneous cell population in which all cells are actively cycling. Evidently, the presence of noncycling cells, arrested either in G_0/G_1 or G_2, will alter this interpretation.

C.   Analysis of Protoplasts.

Lysis of protoplasts in the presence of MI releases nuclei that provide very high-quality DNA distributions (Fig. 4). In this case, CRBC have been included to permit an analysis of genome sizes. The CRBC appear as a discrete peak with a mode of channel 43 and a CV of 3.1%. The tobacco nuclear peak is found at a mode channel of 169 and has a CV of 2.5%. Fixation of the protoplasts prior to staining has two major effects (Fig. 5). First, it means that the measured amount of DNA represents that of a complete protoplast, rather than simply that of the nucleus; this includes a contribution from plastids and from mitochondria. Second, it tends to increase the CV values, for both the CRBC (6.2%) and the protoplasts (8.3%). In the latter case, the rather disproportionate increase may be a

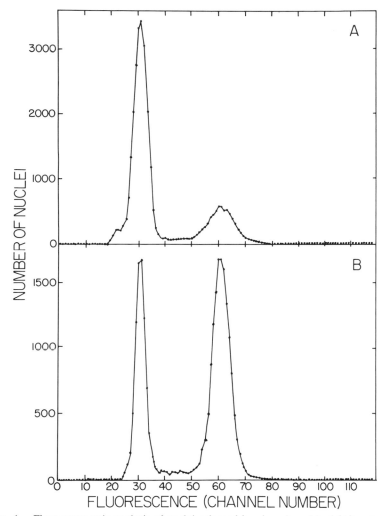

FIG. 1. Flow cytometric analysis of nuclei released by chopping of leaf (A) or terminal root (B) tissue of *Nicotiana tabacum*. The nuclei were stained with mithramycin before analysis. The CV values for the G_1 peaks were 8.1% and 6.0%, respectively, and the profiles represent (A) phase G_1 71.1%, S 6.6%, and G_2 22.3%; and (B) G_1 30.5%, S 13.6%, and G_2 55.9%. Reproduced from Galbraith *et al.* (1983) with permission of the AAAS.

consequence of between-cell variation in cytoplasmic DNA levels (from the plastids and mitochondria). It may also be a consequence of geometric considerations: fixation of protoplasts usually results in a localization of the nucleus at one side of the cell within a thin layer of peripheral cytoplasm, just underneath the plasma membrane. The intensity of illumination of this

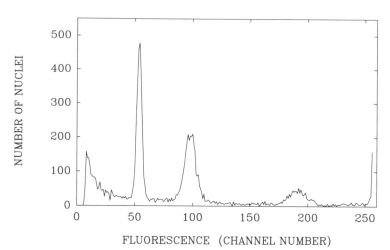

FIG. 2.   Flow cytometric analysis of nuclei released by chopping of anthers of *Nicotiana tabacum*. The peak located in channels 50–55 corresponds to haploid nuclei from cells in G_0/G_1 (D. P. Sharma, K. R. Harkins, and D. W. Galbraith, unpublished data).

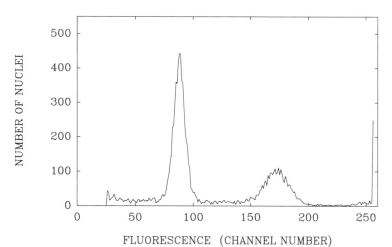

FIG. 3.   Flow cytometric analysis of the DNA contents of nuclei, prepared from cell suspension cultures of *Nicotiana tabacum*. Nuclei were released using the chopping procedure prior to staining with mithramycin.

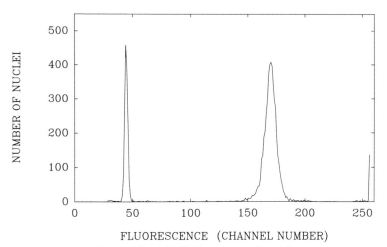

FIG. 4. Flow cytometric analysis of the DNA contents of nuclei, prepared by lysis of leaf protoplasts of *Nicotiana tabacum* in the presence of mithramycin.

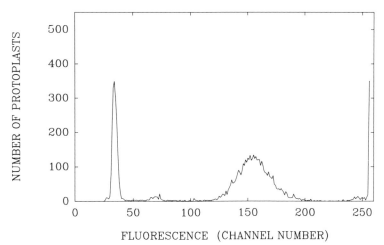

FIG. 5. Flow cytometric analysis of the DNA contents of fixed tobacco leaf protoplasts.

TABLE I

NUCLEAR DNA CONTENTS OF A VARIETY OF DIFFERENT
PLANT SPECIES AS DETERMINED THROUGH
FLOW CYTOMETRY[a]

Species	2C Nuclear DNA content (pg)
Agropyron smithii	0.81
Andropogon gerardii	7.46 (2n = 60)
A. gerardii	8.49 (2n = 70)
A. gerardii	9.79 (2n = 80)
Antirrhinum majus	1.03
Bouteloua gracilis	39.38
Brassica napus	2.47
Capsicum annuum	5.52
Catharanthus roseus	4.84
Coleus blumei	3.43
Crepis capillaris	3.55
Dianthus caryophyllus	5.05
Elymus canadensis	21.60
Euphorbia pulcherrima	2.62
Fragaria xanassa	1.16
Helianthus annuus	3.57
Ipomoea purpurea	1.08
Lycopersicon esculentum	1.48
Mesembryanthemum crystallinum	0.83
Nicotiana alata	4.57
N. glauca	6.91
N. glutinosa	4.08
N. knightiana	6.04
N. nesophila	10.13
N. sylvestris	5.43
N. tabacum	9.67
N. stocktonii	9.45
Panicum maximum	2.48
P. thermale	2.12
Pelargonium × hortorum	4.33
Pennisetum purpureum	5.58
Pisum sativum	7.69
Portulaca grandiflora	5.11
Saintpaulia onantha	1.35
Solidago canadensis	3.13
Solanum melongena	2.33
Zea mays	5.99

[a] Extended from Galbraith *et al.* (1983) with permission of the AAAS.

asymmetric spheroid is probably a function of the position of the nucleus as the protoplast passes through the laser intersection point within the flow cytometer.

D. Plant Genome Sizes

A compilation of plant 2C nuclear DNA contents obtained through the use of FCM is given in Table I. The plants that were analyzed represent a broad spectrum of monocotyledonous and dicotyledonous species. In most cases, these values were obtained through use of the chopping procedure, although we have observed identical values in those cases in which protoplasts were analyzed. Broad correspondence exists between these results and those obtained using Feulgen microspectrophotometry (Galbraith *et al.*, 1983).

ACKNOWLEDGMENTS

I thank Kristi Harkins for valuable assistance. This work was supported in part by grants from the U. S. Departments of Agriculture and Energy, and the National Science Foundation.

REFERENCES

Bennett, M. D., Smith, J. B., and Heslop-Harrison, J. S. (1982). *Proc. R. Soc. London (Biol.)* **216**, 179–199.
Bergounioux, C., Perennes, C., Miege, C., and Gadal, P. (1986). *Protoplasma* **130**, 138–144.
Bergounioux, C., Perennes, C., Brown, S. C., Sarda, C., and Gadal, P. (1988). *Protoplasma* **142**, 127–136.
D'Amato, F. (1986). *C. R. C. Crit. Rev. Plant Sci.* **3**, 73–112.
Fowke, L. C., and Constabel, F. (1985). "Plant Protoplasts." CRC Press, Boca Raton, Florida.
Galbraith, D. W. (1984). *In* "Cell Culture and Somatic Cell Genetics of Plants" (I. K. Vasil, ed.), pp. 765–777. Academic Press, Orlando, Florida.
Galbraith, D. W. (1989). *Int. Rev. Cytol.* **116**, 165–228.
Galbraith, D. W., and Shields, B. A. ((1982). *Physiol. Plant.* **55**, 25–30.
Galbraith, D. W., Mauch, T. J., and Shields, B. A. (1981). *Physiol. Plant.* **51**, 380–386.
Galbraith, D. W., Harkins, K. R., Maddox, J. R., Ayres, N. M., Sharma, D. P., and Firoozabady, E. (1983). *Science* **220**, 1049–1051.
Harkins, K. R., and Galbraith, D. W. (1984). *Physiol. Plant.* **60**, 43–52.
Maniatis, T., Frisch, E. F., and Sambrooke, J. (1982). "Molecular Cloning, a Laboratory Manual." Cold Spring Harbor Laboratory, Cold Spring Harbor, New York.
Meyer, D. J., Afonso, C. L., and Galbraith, D. W. (1988). *J. Cell Biol.* **107**, 163–174.
Murashige, T., and Skoog, F. (1962). *Physiol. Plant.* **15**, 473–497.
Puite, K. J., and Ten Broeke, W. R. R. (1983). *Plant Sci. Lett.* **32**, 79–88.
Sgorbati, S., Levi, M., Sparvoli, E., Trezzi, F., and Lucchini, G. (1986). *Physiol. Plant* **68**, 471–476.

Sharma, D. P., Firoozabady, E., Ayres, N. M., and Galbraith, D. W. (1983). *Z. Pflanzenphy-siol.* **111,** 441–451.

Sundberg, E., and Glimelius, K. (1986). *Plant Sci.* **43,** 155–162.

Sundberg, E., Landgren, M., and Glimelius, K. (1987). *Theor. Appl. Genet.* **75,** 96–104.

Ulrich, I., and Ulrich, W. (1986). *Z. Naturforsch.* **41c,** 1052–1056.

Ulrich, I., Fritz, B., and Ulrich, W. (1988). *Plant Sci.* **55,** 151–158.

Vindeløv, L. L., Christensen,, I. J., and Nissen, N. I. (1983). *Cytometry* **3,** 328–331.

Ward, D. C., Reich, E., and Goldberg, I. H. (1965). *Science* **149,** 1259–1263.

Chapter 49

Plant Cell Cycle Analysis with Isolated Nuclei

CATHERINE BERGOUNIOUX

Laboratoire de Physiologie Végétale Moléculaire
Faculté des Sciences
Université d'Orsay
91405 Orsay
France

SPENCER C. BROWN

Cytométrie
Institut des Sciences Végétales
Centre National de la Recherche Scientifique
91198 Gif-sur-Yvette
France

I. Introduction

The complex three-dimensional structures of plant tissues and the walls of whole cells are generally incompatible with direct-flow cell cycle analysis. To obtain single-cell suspensions calls for isolation of protoplasts, but fluorescent natural pigments and the numerous organelles still hinder good resolution. Moreover, extranuclear DNA is significant in plant cells: chloroplast DNA typically accounts for 5–8% of total DNA, while mitochondrial DNA constitutes $< 1\%$. Thus the coefficient of variation (CV) of the mean DNA content of the G_1 cell population, even after protoplast fixation, is ~6% (Galbraith *et al.*, 1983). But using isolated nuclei obtained via protoplast lysis or directly by chopping the organs, nuclear DNA may be assessed with a CV of 3%.

II. Application

The method has been applied to a large variety of plant species and organs:

1. To characterize differentiated tissues according to the relative frequencies of nuclei having G_0/G_1, S and G_2 levels of DNA (Galbraith *et al.*, 1983; Bergounioux *et al.*, 1988b).
2. To quantify the DNA level of G_0/G_1 cells so as to establish the ploidy (haploid, dihaploid, etc.) or the DNA variability between lines or differentiated organs (Petit *et al.*, 1986; de Laat *et al.*, 1987)
3. To study nucleic acid synthesis during the regeneration of protoplasts back to whole cells and their subsequent division under hormonal control, notably using acridine orange (AO; Bergounioux *et al.*, 1986, 1988a,c).

III. Material

A. Plant Material

Petunia hybrida or *Nicotiana plumbaginifolia* plants were grown in the greenhouse of the Phytotron in Gif-sur-Yvette at 17°C night and 24°C day with a 16-hour photoperiod.

Sorghum bicolor or *Zea mays* seeds were surface-sterilized for 5 minutes in 70% ethanol and for 1 hour in 6% calcium hypochlorite and germinated on a synthetic medium (Bergounioux *et al.*, 1988b) supplemented with 7% agar; the final pH was adjusted to 5.8. The photoperiod was 16 hours and the temperature 27°C.

B. Protoplast Isolation

To obtain *Petunia* and *Nicotiana* mesophyll protoplasts:

1. Open and expanding leaves from plants grown under greenhouse conditions are washed in tap water with 1% Teepol or Tween 20, sterilized by immersion in 6% calcium hypochlorite for 10 minutes, and rinsed with sterile water.
2. All further operations are carried out in a laminar sterile airflow hood.
3. The lower epidermis is peeled off and the leaf fragments floated in Petri dishes on a 20-ml enzyme solution composed of cellulase R10

0.5% (Onozuka), macerozyme 0.25% (Onozuka), 0.4 M mannitol (pH 5.6).

The enzyme solution can be made as follows:

0.5 g of cellulase, 0.25 g macerozyme, and 7.2 g mannitol are gently stirred until dissolved in 100 ml distilled water (dH$_2$O).

Adjust the pH to 5.6 with HCl N/100.

Sterilize by filtering through a 0.45 μm Millipore filter.

Store as frozen aliquots.

4. Incubate at 25°C during 3 hours in the light.
5. Protoplasts are separated from large pieces of debris (vessels, upper epidermis) by filtration through a 60-μm mesh stainless-steel filter.
6. The protoplast suspension is centrifuged at 54 g for 5 minutes.
7. The pellet is promptly resuspended in 0.5 M P medium and twice rinsed by centrifugation to remove all cellulytic enzymes.

P medium "0.5 M pH 5.8" (before sterilization) consists of:
  a. Macro elements (grams per liter)
     NH$_4$NO$_3$, 1.65
     KNO$_3$, 1.9
     CaCl$_2 \cdot$ 2 H$_2$O, 0.44
     MgSO$_4 \cdot$ 7 H$_2$O, 0.37
     KH$_2$PO$_4$, 0.17
  b. Micro elements (milligrams per liter)
     ZnSO$_4$, 1
     H$_3$BO$_3$, 1
     MnSO$_4 \cdot$ H$_2$O, 0.076
     CuSO$_4 \cdot$ 5 H$_2$O, 0.03
     AlCl$_3 \cdot$ 6 H$_2$O, 0.05
     KI, 0.01
     NiCl$_2 \cdot$ 6 H$_2$O, 0.03
  c. Mannitol, 72.2 g/liter
     Glucose, 18 g/liter

0.7 M P medium = (a) + (b) + sorbitol 126 g/liter.

8. Protoplasts in P medium can be transferred into sterile Petri dishes to be cultivated in low light until their division (leading possibly to formation of callus and plantlets) or used directly to obtain isolated nuclei.

Maize or Sorghum Leaf Protoplasts

1. Leaves from 6-, 7-, and 9-day-old *Sorghum* seedlings are chopped into 10 ml 0.35 M KCl solution.

2. This medium is then replaced by 10 ml 0.5% cellulase R10, 0.5% Cellulysin, and 0.6 M mannitol, pH 5.6.
3. Cell walls are hydrolyzed at 21°C for 15 hours in darkness.
4. The protoplast suspension is filtered through a 63 μm mesh steel filter, and pelleted twice at 100 g for 1 minute in 0.5 M P medium (all the media are sterilized by autoclaving or microfiltration).

Sorghum Roots

1. Roots from 6-day-old *Sorghum* seedlings are chopped into 10 ml 0.35 M KCl solution.
2. The medium is then replaced by 2% cellulase R10, 0.2% pectolyase Y23 (Yakult), 0.05% sodium citrate, and 0.6 M sorbitol, pH 5.6.
3. Protoplasts are then washed as described before.

C.   Nuclei Isolation

Two distinct methodologies may be applied:

1. The protoplast lysis method (e.g., to analyze RNA vs DNA during *Petunia* protoplast regeneration and division).
2. Direct isolation of nuclei by chopping organs in an extraction buffer (e.g., to evaluate nuclear DNA in different *Petunia* species, in polyploids and trisomic lines, or in different organs)

Protoplast Lysis Method

1. Triton X-100 (final concentration 1%) is added to 10^6 protoplasts per 1 ml in 0.5 M P medium.
2. The cell wall material is removed with a 10-μm mesh nylon filter, and further organellar debris eliminated by pelleting the nuclei with a 110-g centrifugation for 5 minutes (repeat twice more).
3. Immediately the pellet is resuspended in 0.7 M P medium (pH 6.6 after autoclaving).

Mechanical Chopping Method
Leaves

1. Avoiding major vascularization, 0.3 g is chopped by hand with a single-edge razor blade directly into 1 ml of 0.7 M P medium.
2. 0.1 ml of 10% (w/v) aqueous Triton X-100 (final concentration, 1%) is added.
3. Nuclei are filtered through 40-μm stainless-steel mesh.

Roots

1. (0.1–0.3 g) Root tips are chopped in 1 ml of 0.7 M P medium.
2. After 30 minutes of maceration in 1% Triton X-100, nuclei are released with gentle grinding.
3. The subsequent steps are as described for leaves.

Rat Liver

1. Half a liver is chopped by hand with a razor blade with 3 ml of 0.7 M P buffer, ground in a mortar, homogenized in a potter, and filtered sequentially through 100-μm and 40-μm filters.
2. Centrifuge 5 minutes at 120 g.
3. 1% Triton X-100 final is added to the pellet; this is rapidly homogenized, and P medium is added and centrifuged 5 minutes at 120 g.
4. The pellet is resuspended in 1 ml 0.7 M P medium.
5. 100 μl are resuspended in 1 ml P medium for staining and used as stock solution ($\sim 10^8$ nuclei/per 1 ml).

IV.   Staining Procedure

A.   DNA

1. 20 μl of Hoechst 33342 (HO) are added to 10^6 nuclei/ml 5 minutes before analysis (1.2 μg/ml final). HO (Polysciences, Inc., Warrington, PA) stock solution is prepared with 6 mg/100 ml sterile dH$_2$O and may be stored for several months at 4°C in the dark.
2. 5 μl ethidium bromide (EB) are added to 10^6 nuclei/ml 15 minutes before analysis (5 μg/ml final). EB (Polysciences) stock solution is prepared with 1 mg/ml of sterile dH$_2$O and stored at 4°C in the dark.

B.   RNA versus DNA

Acridine orange (AO) is a stoichiometric dye for simultaneous assessment of DNA (green fluorescence) and RNA (red luminescence). Staining with AO (8 μg/ml final) is performed according to Darzynkiewicz et al. (1980) and Crissman et al. (1985), but on isolated nuclei instead of entire cells (Bergounioux et al., 1988b). The principle of this method is detailed by Darzynkiewicz in Chapter 27, this volume. This staining is done at 4°C, immediately after release of nuclei with Triton X-100.

1. 10^6 Nuclei are resuspended in 100 μl of sterile 0.7 M P medium pH 6.6, 1 mM EDTA supplemented with 25 units/ml RNase inhibitor (Ribonucleasin, BRL) at 4°C.
2. 400 μl of cold AO stain buffer is added.
3. 5 μl Formaldehyde (37% w/v, Merck) is added.
4. Nuclei are kept on ice until analysis.

Modified Darzynkiewicz AO buffer prepared daily:
   a. 37 ml of 0.1 M citric acid with 63 ml of 0.2 M Na$_2$HPO$_4$.
   b. 12.7 g sorbitol and 34 mg of EDTA disodium salt.
   c. Add 1 ml of a stock aqueous solution of AO (1 mg/ml; Polysciences). Store at 4°C.

V.   Critical Aspects

1. Triton X-100 treatment simultaneously disrupts plasmalemma, organelles, and particularly chloroplasts (natural fluorescence is thus discarded), permeabilizes the nuclear membrane, and reduces nonspecific Hoechst fixation. The use of Triton produces better CV values and uptake of dye by nuclei. Consequently the fluorescence amplitude may only be compared for nuclear samples obtained under the same conditions.

2. HO dye may be used without purification of nuclei, immediately after chopping the tissues.

3. HO stock must not be sterilized by filtration through Millipore membrane because the membrane retains the dye.

4. HO is preferentially added after nuclear release rather than in the chopping buffer, since it can bind tissue and debris, modifying the effective concentration.

5. The recovery of nuclei from protoplasts is virtually complete, while the recovery rate is very low with direct organ chopping. As the ploidy of cells may be different in different organs of the same plant, tissue sampling must be strictly controlled when comparing the DNA content of different plants.

6. The biparametric analysis of fluorescence pulse peak against fluorescence pulse integral enables one to eliminate doublets and to determine the real frequency of 4C nuclei (Brown and Bergounioux, 1988).

7. The biparametric analysis time-of-flight (TOF; on fluorescence pulse peak) against fluorescence pulse integral enables one to assess size (length of the longer axis) of 2C and 4C nuclei, and to contrast these with the relevant forward-angle light scatter (FALS) values.

8. Various telo-trisomic plant clones for six different chromosomes (2n + 1 fragment) have been compared with strictly 2n (= 14) or n *Petunia* lines using leaf chopping and HO staining: despite the known cytogenetic differences, the measured nuclear DNA of these lines was not significantly higher than that of the reference plants. However, whole-chromosome addition line (2n + 1) did always show higher nuclear DNA than the diploid plant. Therefore, the least significant difference with this technique is ∼ 10% of nuclear DNA (results not shown).

VI. Instruments

Flow analyses were conducted on an EPICS V cytofluorometer (Coulter, Hialeah, FL) with an argon laser (Spectra-Physics 2025-05). A 100-μm nozzle, water as sheath fluid, and the standard optics were used. 10^4 Nuclei per sample are analyzed. Forward angle light scatter representing the light dispersion at low angles (11–20°) is widely used to estimate object volume; however, light scattering and cell volume can vary independently in living cells (McGann *et al.*, 1988).

Time of flight (TOF), the duration at half-height of the fluorescence pulse as the object crosses the focal point, is a measure of the size of each object (Galbraith *et al.*, 1988).

A. DNA

The blue emission of the DNA–HO dye complex was collected through 418 long-wave pass and 530 short-wave pass filters. The excitation was 351 + 364 nm using 60 mW. The fluorescence histograms were resolved into $G_0 + G_1$, S, and G_2 cell cycle compartments with a peak–reflect algorithm using two Gaussian curves (PARA 1, Coulter, Inc.) or a DNA-fit algorithm (Cytologic, Coulter).

B. RNA versus DNA

Luminescence of individual cells was measured at two wavelength bands: green (between 515 and 540 nm) and red (above 610 nm). The excitation was 488 nm using 300 mW. The analyses were done on 10^4 nuclei at ice temperature. The cytogram of RNA red luminescence against DNA green fluorescence was used to evaluate the relative RNA specific to each cell cycle phase (2C, S, or 4C DNA quantity). Routinely, the DNA distribution was adjusted with G_1 at the channel 80 (256-channel analyzer) and the red photomultiplier (PMT) was set such that freshly isolated protoplasts had a mean RNA around 55 (arbitrary units, 256-channel analyzer).

VII.   Results

Figure 1 illustrates the variability of nuclear DNA distributions derived from various organs. In different plant organs differentiation may occur in G_1 and G_2 cell cycle states. Nuclei are obtained by chopping melon first leaf (Fig. 1A) and cotyledons (Fig. 1B). However, in the case of sorghum nuclei released from protoplasts (Fig. 1C,D), difference between leaf and root apex (0–5 mm) may be also attributed to a degree of cell selection during the enzyme digestion, as we have shown in parallel studies using direct nuclear extraction by chopping (Bergounioux et al., 1988b).

Figure 2 illustrates the application of direct nuclear release to the determination of plant ploidy (Petit et al., 1986; de Laat et al., 1987) and DNA quantification (Galbraith et al., 1983; Ulrich and Ulrich, 1988) with the chopping method. A logarithmic amplifier was used to allow widely disparate samples to be represented on a single scale. Moreover, this scale can be kept constant and independent of linear amplifier gain, which may have to be changed when analyzing diverse species. The calculations from an internal standard, however, are done using the linear scale which has superior resolution and which is simultaneously obtained. Figure 2A shows an analysis of a mixture of nuclei from haploid and diploid Petunia hybrida, the internal rat nucleus standard being stained simultaneously with the Petunia nuclei. The haploid plant contains nuclei in G_0/G_1, seen as peak 1; its G_2 nuclei are confounded in peak 2 with those from G_0/G_1 of the diploid plant. Peak 4 comprises uniquely nuclei from G_2 diploid cells; peaks 3 and 5 correspond, respectively, to G_0/G_1 and G_2 nuclei of the rat standard. Figure 2B illustrates quantification of maize nuclear DNA with the same standard.

Figure 3 illustrates an application of the RNA/DNA-staining technique to Petunia nuclei released after 18 hours of protoplast culture. The protoplasts from two quite homogeneous groups having RNA and DNA levels (X,Y) correspond to inactive G_0/G_1 and G_2 cells: these positions are marked as dark spots and constitute particularly homogeneous groups at 0 hours. During 18 hours of culture, many protoplasts synthesize RNA; a few also synthesize DNA and enter the S compartment. Protoplasts

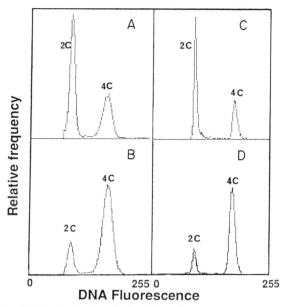

FIG. 1. Nuclear DNA distributions derived from (A) melon first leaf, (B) melon cotyledons, (C) nuclei from sorghum leaves, and (D) nuclei from sorghum root apex.

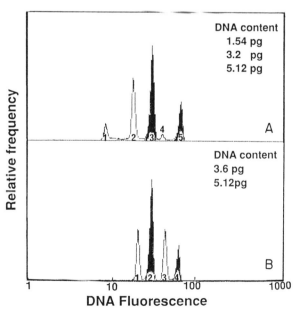

FIG. 2. Logarithmic amplifier analysis of (A) a mixture of nuclei from haploid and diploid *Petunia hybrida* [peak 1, n(1C DNA); peak 2, 1n(2C DNA) + 2n(2C DNA); peak 4, 2n (4C DNA)] and an internal rat nucleus standard (peak 3, 2C DNA; peak 5, 4C DNA); (B) maize nuclear DNA (peak 1, 2C; peak 3, 4C) with the same rat internal standard (peak 2, 2C; peak 4, 4C).

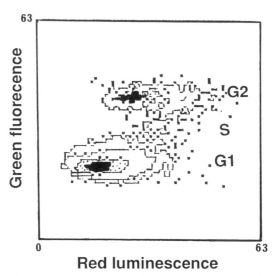

FIG. 3.   Red and green fluorescence distribution using RNA/DNA-staining technique with *Petunia* nuclei released after 18 hours of protoplast culture.

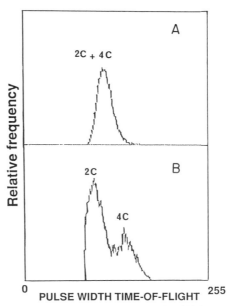

FIG. 4.   Fluorescence histograms of *Petunia* nuclear size (A) immediately after protoplast release and (B) after 48 hours of protoplast culture.

originally with 4C DNA have also synthesized RNA, but microscopic observation reveals no divisions. (Discontinuous subcompartments of G_0 and G_1 are not revealed; the synthesis of RNA forms a continuum between these putative cell states.) The effects of hormones (auxins, cytokinins) on RNA and DNA synthesis have been reported (Bergounioux *et al.*, 1988a,b).

Figure 4 illustrates a measure of nuclear length with TOF fluorescence pulse peak. Note that G_1 nuclear size decreases during protoplast culture. Histograms are presented for nuclear size (Fig. 4A) immediately after protoplast release and (Fig. 4B) after 48 hours of protoplast culture (10^4 nuclei). The peaks corresponding to 2C and 4C could be ascertained from the cytogram (not shown) of TOF versus DNA (fluorescence pulse integral).

ACKNOWLEDGMENTS

We thank C. Perennes for her kind and excellent assistance, and Drs. C. Raquin, D. Maizonnier, and A. Borgel for generous gifts of haploid *Petunia*, trisomic and telotrisomic *Petunia* clones, and melon plantlets.

REFERENCES

Bergounioux, C., Perennes, C., Miège, C., and Gadal, P. (1986). *Protoplasma* **130**, 138.
Bergounioux, C., Perennes, C., Brown, S. C., and Gadal, P. (1988a). *Cytometry* **9**, 1.
Bergounioux, C., Perennes, C., Brown, S. C., Sarda, C., and Gadal, P. (1988b). *Protoplasma* **142**, 127.
Bergounioux, C., Perennes, C., Brown, S. C., and Gadal, P. (1988c). *Planta,* **175**, 500.
Brown S. C., and Bergounioux, C. (1988). *In* "Flow Cytometry: Advanced Experimental and Clinical Applications" (A. Yen, ed.) Vol. II, p 195. CRC Press, Boca Raton, Florida.
Crissman, H. A., Darzynkiewicz, Z., and Tobey, R. A. (1985). *Science* **228**, 1321.
Darzynkiewicz, Z., Sharpless, T., Staino-Coico, L., and Melamed, M. R. (1980). *Proc. Natl. Acad. Sci. U.S.A.* **77**, 6696.
De Laat, A. M. M., Gohde, W., and Vogelzang M. J. (1987). *Plant Breed.* **99**, 307.
Galbraith, D. W., Harkins, K. R., Maddox, J. M., Ayres, N. M., Sharma, P., and Firoozabady, E. (1983). *Science* **220**, 1049.
Galbraith W. D., Harkins K. R., and Jefferson R. A. (1988). *Cytometry* **9**, 75.
McGann, L. E., Walterson, M. L., and Hogg, L. M. (1988). *Cytometry* **9**, 33.
Petit, P., Conia, J., Brown, S., and Bergounioux, C., (1986). *Biofutur* **51**, 28.
Ulrich, B., and Ulrich, W. (1988). *Plant Sci.* **55**, 151.

Chapter 50

Environmental Health: Flow Cytometric Methods to Assess Our Water World

CLARICE M. YENTSCH

J. J. MacIsaac Flow Cytometry/Sorting Facility
Bigelow Laboratory for Ocean Sciences
West Boothbay Harbor, Maine 04575

I. Introduction

One visual impression of environmental scientists, of which the aquatic scientists limnologists and oceanographers are a subset, is the field biologist with hip boots and a silk butterfly or plankton net. Although such an impression stirs nostalgic memories of our roots in natural history, it is not an accurate representation of the present situation. Today, laser-based flow cytometers/sorters are taken on shipboard and used round-the-clock on research expeditions. Their performance under such rigorous conditions has helped to demonstrate their durability, and thus it is proposed that flow cytometers be placed in spacecraft (Winfield, 1987) to study the gravity-free cytology of plant and animal cells. Such experiments will be akin to oceanographic remote-sensing devices that have been developed over the past decade. These sensors transmit information to ground terminals about water transparency/color, water temperature, wave height, and salinity. At ground stations, it is just as likely that statisticians will be processing multiparametric analyses of satellite remote-sensing imagery of ocean color, as flow cytometry (FCM) data sets on individual phytoplankton. The spatial scale of signal detection of these 0.5–100 μm diameter cells spans from submicron detection by flow cytometers to thousands of kilometers by satellites (Fig. 1).

METHODS IN CELL BIOLOGY, VOL. 33

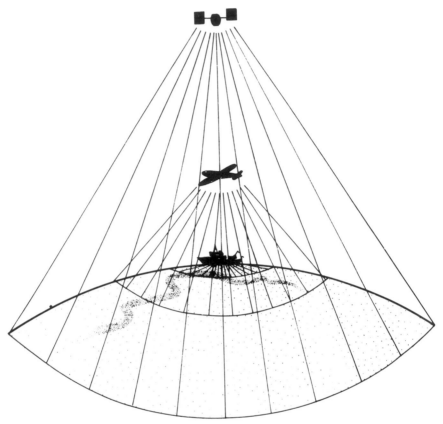

FIG. 1. Perspectives. Satellite and aircraft remote sensing of particles plus shipboard individual particle analysis.

A. Signal Packaging and Processing

Curiously enough, the source of the primary signal of water color is the same for the satellite as it is for the flow cytometer. That source is photosynthetic membrane-bound pigments packaged in microscopic drifting plants called phytoplankton. This chapter presents useful methods to study signal packaging and processing, with special reference to the water world. The goal of aquatic scientists is to understand this signal packaging and processing, to understand the underlying biological fundamentals, and thus to be able to deconvolute and reconstruct the light budget in the world's oceans, rivers and lakes. There is a search for knowledge of both light and life—as they are inseparably linked.

The advent of flow cytometry in oceanography is a development of great significance. It is not merely a matter of increased speed and convenience: the benefits extend to the most fundamental bases of both physical (optical) and biological oceanography. The essential advantage is that, for any sample of seawater, flow cytometry permits the optical characterization of large numbers of individual particles contained therein. For physical applications, the value is that it becomes possible to attempt the recovery of the bulk optical properties of seawater from the details of the properties of the constituents. For biological applications, where the focus is on the unicellular biota, the value is that it permits the analysis of population and community structure, that is of the variance between individuals. Platt (1989, p. 500).

B. Environmental Health

Aquatic scientists, just like their biomedical counterparts, are interested in health. However, it is environmental or planetary health that is their major concern. The urgent need to be knowledgable about our home planet is now recognized. The earth is changing rapidly (Kornberg, 1988) because the natural rate of change has been altered by human activities. Knowledge of the optical properties of water and phytoplankton permit calculations on the natural processes of solar heating and on the light that is available for photosynthesis. Both general and specific knowledge of the biota helps scientists to recognize the vital signs, as well as any symptoms of ill health of the ocean–river–lake system that is now viewed as one "living organism."

II. Light and Life

Visible light and life are closely related. In water, photosynthetic organisms must deal with variable light intensity as well as spectral change due to the unequal attenuation of different wavelengths. As water depth increases, the ultraviolet (UV) and infrared (IR) wavelengths disappear, and the blue and red regions of the spectrum phase out until eventually only blue-green light remains, which finally disappears as well. Phytoplankton occur as suspensions in this upper sunlit layer. All of these organisms contain chlorophyll a (Chl$a$) plus accessory pigments that span the visible spectrum, and can be exploited as molecular probes in cytology (Fig. 2).

Most living organisms in aquatic systems exist as independent, single-cell entities in a fluid medium. In even the most nutrient-impoverished "oligotrophic" waters of the world, small cells (<1 μm in cell diameter) occur at

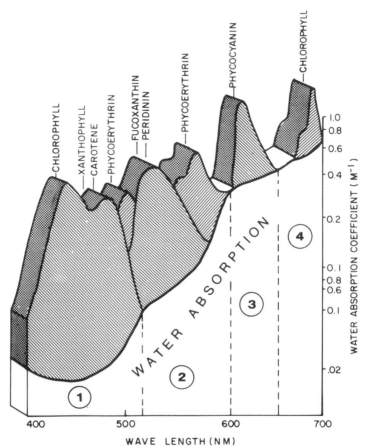

FIG. 2. Absorption of light by water and by different algal pigments in the windows of "clarity" of water; the spectra for the pigments approximate those measured *in vivo*; fucoxanthin and peridinin are superimposed. Note that not all pigments are in all algae. From Yentsch and Yentsch (1979).

concentrations of nearly one million cells per liter. In more nutrient-rich, "eutrophic" waters, numbers per unit volume are several orders of magnitude higher. Polluted waters can be far in excess of this. It is no wonder that recording of the number and size, and the characterization of individual cells was a preoccupation of early naturalists interested in aquatic ecosystems. For the most part, cells are in the 0.5–100 μm diameter size range. Grouping organisms by size is termed *allometry* and is an expedient clustering technique used to test various hypotheses (Silvert and Platt, 1978, 1980; Legendre and LeFèvre, 1989). Generally, in open, nonpolluted waters, the smaller the cells the more numerous they are (Fig. 3); anomalies are significant. Counting and tracking cells from various water masses is important in determining how cells are affected by the physical and chemical environment expressed in three-dimensional fields (latitude, longitude, and depth). Flow cytometers with simultaneous volume sensing are extremely valuable tools for rapid allometric analysis.

Taxonomic genera of phytoplankton have been clustered into groups and named according to their pigment composition for over a century of classical systematics. These taxa are in no way equally distributed in the water world. Some phytoplankton are cosmopolitan; others are opportunistic, whereby various pigment groups thrive and proliferate in specific environmental niches. Pigment composition can be detected by absorption and/or fluorescence spectra, or by combining a variety of excitation and emission wavelengths. The FCM method of pigment group clustering has been termed *ataxonomy*, and the resulting knowledge of what groups are present can be used to test various functional-group hypotheses (Fig. 4).

Within the sunlit layer, along with the spectral change/energy of photons, the numbers of photons decrease exponentially from the surface. This results in light-limiting conditions (classically considered the 1% light depth), which can occur anywhere between 1 m in highly turbid waters to well over 100 m in clear waters. This variability is in part due to attenuation

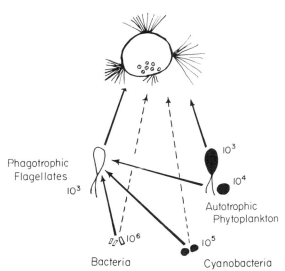

FIG. 3. Conceptual diagram of the microfood web in the marine environment. Autofluorescent cells are shaded. Numbers indicate approximate cell densities per milliliter in coastal waters. From Yentsch and Pomponi (1986).

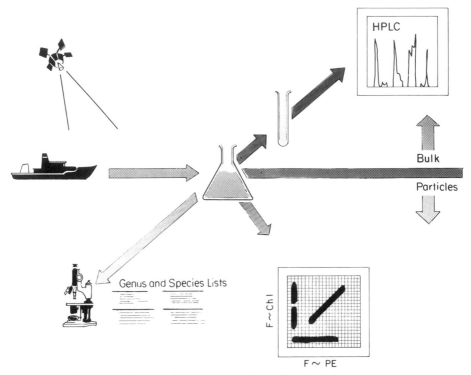

FIG. 4. Schematic diagram of ataxonomic analysis. Bulk analysis (top) results in extracts and sophistication of HPLC (high-performance liquid chromatography) to separate and quantify various pigments. Individual particle analysis (bottom) results in distributions as either genus and species lists or automated characterization into pigment/functional groups. It is clear that one water mass can be identified from another using FCM characterization. F ~ Chl, Chorophyll autofluorescence; F ~ PE, phycoerythrin autofluorescence.

by the water itself, but the major cause is the presence of particles in the water—the living cells, detritus (decaying organic matter), and suspended sediments. The particles in water absorb and scatter sunlight and therefore reduce the penetration of light into the water column. The size and numbers of particles present in water have a profound and predictable effect on the transmission of light throughout a water column (Fig. 5), and in turn, the available light has a profound and predictable effect on the growth and cell division rate of the living particles—in particular the photoautotrophs or phytoplankton.

The measurement of light penetration has been routine for the aquatic sciences during the past century. Common tools used to measure light penetration include depth at which visualization of a standard-sized white disk

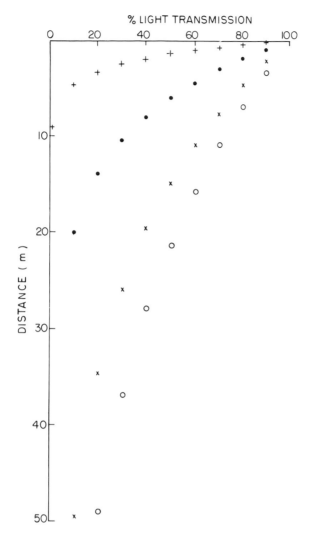

FIG. 5. Percentage of light transmitted at various depths for particles at concentrations of 10^4 per 1 ml. A refractive index of 1.03 relative to seawater with a 5-μm diameter is represented with a (+). A refractive index of 1.01 and 5-μm diameter is represented with a (●). A refractive index of 1.03 and 2-μm diameter is (×) and a refractive index of 1.01, 2 μm diameter is represented with a (○). These values are for scattering contribution only; absorbing particles such as phytoplankton would decrease transmission further. Courtesy of R. Spinrad; from Yentsch and Yentsch (1984).

is obscured (the secchi disk), the submersible UV and visible spectroradio-meter, the submersible transmissometer, and, to measure bioluminescent forms, the bathyphotometer. Unfortunately, none of these instruments has the capacity to provide individual particle measurements on all living particles. Accordingly, when the Coulter counter was first taken to sea in the mid-1970s there was the hope that by measuring all the individual par-ticles in a given volume of water, populations and subpopulations could be distinguished. Unfortunately, the Coulter counter was incapable of discri-minating between living cells (either photoautotrophs or microhetero-trophs, or both), detritis, and/or suspended sediments. Turbid water, containing suspended sediments from river runoff or tidal scouring of the bottom sediments, gave misleading cues in efforts to reconcile phytoplank-ton's contribution to the optical budget and estimate global aquatic pro-ductivity.

III.   The Advent of Fluorescence Techniques

A.   Autofluorescence

Chlorophyll fluorescence, which appears as a brilliant red, was first observed by a Scottish clergyman, Brewster, in 1833 (as reported in Yentsch and Pomponi, 1986). He was observing plant cell material through a microscope using window light, which fortuitously passed through a bottle of copper sulfate, providing blue excitation. It was not until the late 1950s and early 1960s, however, that Chl fluorescence and other types of autofluorescence[1] were exploited as important discriminators for particle suspensions in water. Traditional bulk optical measurements were then routinely normalized to Chl *a* content. As a result, an index of phytoplank-ton biomass (vs total particulates) contributing to the water turbidity was possible, a real breakthrough for the aquatic researchers. To discriminate among various pigment groups, high-performance liquid chromatography (HPLC) (Mantoura and Llewellyn, 1983) or bulk spectral fluorescence signatures were endorsed (Yentsch and Yentsch, 1979).

Yet the dilemma of a thorough optical accountability of light in water remained. This was primarily because light "loss" terms are affected by the size, shape, refractive index, and absorption of the particles. Thus no "bulk" index could adequately describe the packaging. It has been learned the hard way that one must honor the integrity of the cell. Accordingly, the

---

[1] Not only does Chl *a* fluoresce strongly, but so do many of the other accessory pigments such as phycobilin pigments, luciferin, and other flavins.

coupling of all of the techniques presented to date has still failed to yield a complete impression of light and life in water, and there remains little hope that the aquatic light budget can be solved in this way.

The flow cytometer is able to progress past this barrier. While the advent of FCM into oceanography is still in its infancy (the first instrument was introduced into an aquatic science laboratory in 1983), the potential is enormous. To date, all attempts at a photon budget or light closure ("Does the total equal the sum of the parts?") are puzzling. It must be recognized that errors are as likely in the bulk measurements as in the individual cell measurements. Considerable energies must be expended over the forthcoming years to produce robust models that can serve the aquatic scientist as optical budgets. The intercalibration of instruments is a necessity but merely the beginning of the effort.

The promise of FCM goes beyond the light budget and closure. Many aquatic scientists are biologists who are interested in the signal processing within the cell—asking fundamental biological questions using aquatic organisms as model systems. The autofluorescent pigments are not merely dyes but integral components of photosystems, and thus fluorescence is related to the kinetics of photosynthesis and not merely the concentration of the pigments present. This natural wonder can be a complication for the scientists. Yet this characteristic can also be exploited to gain information about what processes are going on within the cell. Autofluorescent pigment methods that help to decipher packaging and processing will be covered in detail.

A few precautions must be discussed. As mentioned, an active photosystem is being processed. It was determined (Rivkin et al., 1986; Cullen, unpublished data) that photosynthetic rates of phytoplankton became significantly lower after FCM analysis with a high-powered laser or mercury–xenon lamp. Exposure to the laser beam during the sorting process caused significant physiological damage. The cellular content of a radiolabel accumulated prior to FCM was not affected by FCM. Although it may not be possible to use FCM to isolate cells for further physiological studies, samples may be incubated with stable or radioactive isotopes and then analyzed by FCM.

To elaborate, as a cell moves through FCM, it undergoes physical shear within the liquid stream, laser exposure, and electric current (EC). To discriminate the physiological effects of each, carbon uptake for *Ditylum brightwelli* and *Alexandrium tamarensis* (formerly *Protogonyaulax tamarensis*) were measured under the following conditions: (1) unsorted, (2) fully sorted (at a 25-mW laser exposure), (3) shear only (no EC or laser), (4) shear plus EC (no laser), (5) shear plus a 25-mW laser exposure (no EC), and (6) shear plus a 500-mW laser exposure (no EC). During these experiments, the charged droplets in the stream were collected in a single collection container rather than being deflected into separate collection tubes. Results from Rivkin *et al.* (1986) are given in Table I.

The influences of shear, EC, and laser irradiation on carbon uptake are significant. Although there were species differences in magnitude, the trends were similar for both the diatom *Ditylum brightwelli* and the dinoflagellate *Protogonyaulax tamarensis*. After all treatments except shear, the carbon uptake by cells was significantly lower than it was for cells that were not subjected to FCM. Shear plus EC reduced carbon uptake by 15–20%. Carbon uptake for both the fully sorted samples and the samples subjected to shear plus a 25-mW laser exposure (no EC) was ~50% of that for unsorted samples, whereas carbon uptake for samples subjected to shear plus a 500-mW laser exposure (no EC) was ~12% of that for unsorted samples. Since, in the absence of laser irradiation, photosynthesis was reduced by a maximum of 20%, it appears that exposure to the laser beam caused significant physiological damage. Increasing the power of the laser beam increased the physiological damage; carbon uptake was five to six times lower after a 500-mW laser exposure than after a 25-mW laser exposure.

TABLE I

RATES OF CARBON UPTAKE BY *ALEXANDRIUM TAMARENSIS* AND *DITYLUM BRIGHTWELLI* PRIOR TO FCM (UNSORTED) AND AFTER VARIOUS TREATMENTS[a,b]

| | Carbon uptake at indicated incubation irradiance (pg of C/cell/hour) | | | |
| | *A. tamarensis* | | *D. brightwelli* | |
Treatment	208	20	208	20
Unsorted	105 ± 7	30 ± 5	210 ± 18	35 ± 3
Fully sorted	58 ± 8	15 ± 2	150 ± 15	25 ± 3
Shear only	98 ± 4	28 ± 4	190 ± 13	32 ± 2
Shear + EC	87 ± 8	24 ± 4	180 ± 12	28 ± 3
Shear + 25-mW laser exposure	55 ± 4	18 ± 3	160 ± 12	22 ± 2
Shear + 500-mW laser exposure	12 ± 4	3 ± 1	29 ± 9	5 ± 3

[a] Samples were incubated at photosynthesis saturating and limiting irradiances (in microeinsteins per square meter per second) for 1–2 hours. Each value represents the mean $\pm$ SD ($N = 4$).

[b] From Rivkin *et al.* (1986).

Despite such damage, however, the recovery of even the most delicate cell occurs within hours, as Haugen *et al.* (1987) discovered using cell growth and cell division as an index of damage due to FCM. Damage done to especially delicate marine phytoplankton cells (*Chroomonas salina*, 3C; *Micromonas* sp., IB4; *Tetraselmis* sp., BE50; *Gyrodinium* sp., 94GYR) by passage through a Coulter EPICS V flow cytometer was assessed. The cells did not distort or lyse after exposure to fluidics or to laser light up to 1000 mW. However, the cells did sustain damage as evidenced by temporary growth rate depressions. As expected, some pigment groups were more susceptible to excitation wavelengths than were others. In general, the more closely "matched" the absorption and excitation wavelengths used, the greater the extent of the damage. Thus, for sorting prior to physiological experiments, we choose low-power and suboptimal excitation wavelengths. The four clones tested eventually resumed control growth rates after growth lags of as long as 48 hours (Haugen *et al.*, 1987).

B.   Fluorescent Probes

One area in which aquatic scientists have lagged behind their biomedical counterparts is in the use of cytochemical stains for FCM. Such stains do permit the detection and sizing of aquatic bacteria and microheterotrophs (Hobbie *et al.*, 1977; Haas, 1982; Button and Robertson, 1989; Sieracki *et al.*, 1989a,b) and are routinely used with these cells. With phytoplankton the issue of stain addition is complicated by the fact that many parts of the spectrum are already "filled" with autofluorescent pigments. Thus appropriate stains are sought to match available spectral voids or "windows." What might be feasible for one pigment group might not be possible for another. In other cases, involving work with preserved versus living cells, methods have been devised to extract or photobleach the autofluorescent pigments (Yentsch *et al.*, 1983a; Olson *et al.*, 1983). Variable cell wall composition and complex membrane or cellular osmoregulation mechanisms cause additional challenges. Nonetheless, cytochemical stains in aquatic sciences have been fruitful to studies of cell cycling, cell metabolic activity, the chemical composition of proteins and lipids, and the scavenging of organics by sediments—to name a few. Cytochemical methods will be summarized in detail.

The advent of immunofluorescent probes is as promising in the aquatic sciences as it is in any other science (Yentsch, *et al.*, 1989b). Designer reagents have been produced to identify specific organisms of particular interest and against specific intracellular compounds including toxins. Immunochemical methods will be summarized in some detail.

		MANUAL light microscope
Conclusion: **30 μm**	30 μm cell diameter	**AUTOMATED** Flow Vision Analyzer Coulter volume
photosynthetic	brilliant red auto- fluorescence from chlorophyll ex488 nm em680 nm	flow cytometer
living	brilliant yellow-green fluorescence with FDA stain ex488 nm em530 nm	flow cytometer
	green auto-fluorescence from luciferin-luciferase ex488 nm em530 nm	flow cytometer
hv	bioluminescence flash stimulated with pH, motion, or light change	photometer
bioluminescent **cell**	fluorescein, rhodamine or phycoerythrin tag to antibody to luciferase; α L-ase	flow cytometer

FIG. 6. Manual light microscope versus automated detection for the microalga *Alexandrium tamarensis*, the New England toxic red tide dinoflagellate (clone GT-429). Comparison of cell discrimination based on parameters of cell diameter, chlorophyll autofluorescence, live/dead metabolic activity with fluorescein diacetate (FDA) stain, bioluminescence potential based on luciferin autofluorescence, bioluminescence flash detection by bathyphotometers and immunochemical reagents, antiluciferase. From Yentsch and Yentsch (1989).

IV. General Methods Using Flow Cytometry

Flow cytometry is used in a variety of ways with sizing, autofluorescence, cytochemical stains, and immunofluorescent probes. An example of many parameters used to study one cell type appears as Fig. 6 for *Alexandrium tamarensis*, clone GT-429, the New England red tide dinoflagellate. The cell is both photosynthetic and bioluminescent. Not all autotrophic cells are so complex. "Primitive" cells such as cyanobacteria are little more than bacteria with photosynthetic capabilities (Fig. 7). A scheme for characterization of aquatic particles using multiple parameters was originally published in 1983, and is being increasingly used (Fig. 8).

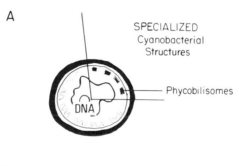

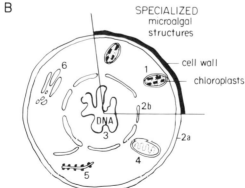

Fig. 7. Basic features of (A) prokaryotic and (B) eukaryotic cells, which can be exploited for fluorescence detection by image analysis and/or FCM. (A) Prokaryotes are characterized by phycobilisomes, autofluorescence, membrane potential, and DNA stains. (B) Eukaryotes are characterized by (1) chloroplasts, (2) cell membranes, (3) DNA, (4) mitochondria, (5) ribosomes on endoplasmic reticulum and (6) Golgi apparatus. Cell walls of a variety of composition are present in plant eukaryotes. From Yentsch and Pomponi (1986).

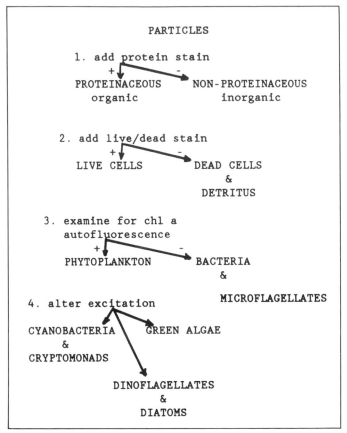

FIG. 8. Scheme for characterizing aquatic particles in which (1) protein stain is added to differentiate between proteinaceous and nonproteinaceous particles; (2) live/dead stain FDA is added to detect percentage living particles; (3) autofluorescence is examined for using excitation at 488 nm to detect autotrophs; and (4) excitation is altered to differentiate various color groups. From Yentsch *et al.* (1983b).

A. Cell Sizing and Light Scatter

The size and shape of cells have always been of intrigue to the aquatic scientist, but it was not until the work of ecologist R. Margalef and students that a true appreciation of life forms of phytoplankton as survival alternatives in an unstable environment occurred. At least a few cell types seem to have evolved for every need. This evolution includes a seemingly infinite variety of both morphological and physiological features. There are features for both buoyancy and sinking. There are life history strategies that depend on mixing of the water column for adequate nutrients and other strategies that exploit stratified stable environments. Some cells thrive in high light, others in low light. Some cells adapt to sporadic nutrient pulses, and others to low continuous recycled nutrient flux. More than two-thirds of the variability in response is attributable to body/cell size alone (Conrad, 1983). Indeed there is a tremendous ecological significance to body/cell size in both terrestrial and aquatic environments. Hypotheses developed and tested in the terrestrial environment are now being tested in the aquatic environment.

As mentioned, due to contaminating signals from detritus and suspended sediments, particle sizing and counting by the Coulter counter is but one parameter in isolation and therefore of limited usefulness. Yet particle sizing, when used in combination with Chl fluorescence as in FCM, is of tremendous value. Cell volume measurement features are present on the pre-IFA and IFA models of FACS analyzers made by Becton Dickinson (BD). Unfortunately, both Coulter and BD do not have direct volume output on their latest models, the EPICS Profile and the FACScan. Partec does have a cell volume option.

Figure 9A is a bifurcation model presented by Legendre and LeFèvre (1989). Figure 9B shows actual FACS analyzer data in support of this model. The left panel of Fig. 9B is from coastal Gulf of Maine waters, where large-sized cells are present superimposed onto the smaller sized cell populations. The right panel is from the open ocean Sargasso Sea, where only the smaller sizes are present. The small-sized cells appear almost as "background" throughout the world's oceans. The cubes in Fig. 10 represent 100 μl of water. (Autofluorescent cells are shaded, while suspended sediments and other nonfluorescent material are not shaded.) The size and numbers of particles are representative of the various situations and are drawn to scale.

With the smaller (<10 μm cell diameter) spherical cells, light scatter can be used to approximate cell size. When this is done with clonal cultures, the volume relationship is accurate. However, most natural populations are complex mixtures of sizes, shapes, and refractive indices. Cell wall composition is also highly variable. These factors complicate the signal and confuse the interpretation.

Forward-angle and 90° light scatter are interesting as parameters in their own right. Again, using multiple parameters, one can segregate the light scatter coming from Chl-containing cells versus detritus and suspended sediment, and thus the information on each subset is useful in attempts to construct the photon budget of the water column. In fact, it is argued that

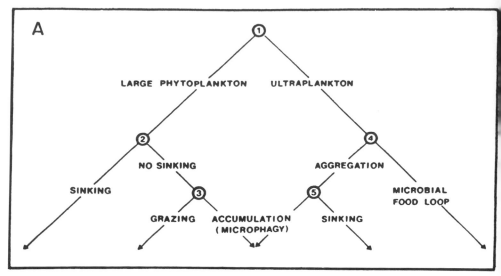

FIG. 9. (A) Schematic of the bifurcation model of the particle size paradigm advanced by Legendre and LeFèvre (1989) and (B) supporting FCM data depicting comparison of size distributions of chlorophyll-containing cells in the 3–53 μm size range. (B) Left panel is from the Gulf of Maine and the right panel is from the Sargasso Sea. Nutrient input to the Gulf of Maine is primarily nitrate, which supports "new" production, while nutrients in the open Sargasso Sea are "recycled," and smaller sized cells are favored. Data from Yentsch and Phinney (1989).

as the number of angles sampled is increased, light scatter measurements will become even more useful. Backscatter measurements most resemble the satellite or remotely sensed signal from phytoplankton and other aquatic particles. Spinrad (1984) and Ackleson and Spinrad (1988; Ackleson et al., 1988, 1989) have pioneered the synthesis of important information from phytoplankton light scatter FCM measurements.

Optical packaging is critical to understanding the individual cell's total light budget as well as the light budget in the water column. Much of the conceptualization is summarized by Bricaud and Morel (1986) and Kirk (1983). While others have used microspectrophotometry (Iturriaga and Siegel, 1988). Perry and Porter (1989) have attempted the determination of the cross-section absorption coefficient of individual phytoplankton cells by analytical FCM.

Sediments play an important part in the transport and scavenging of substances in the aquatic environment. Baier et al. (1989) have initiated the use of FCM in studies of adsorption of yellow substance onto clay particles.

Table II is a summary of ongoing efforts using cell sizing and light scatter parameters.

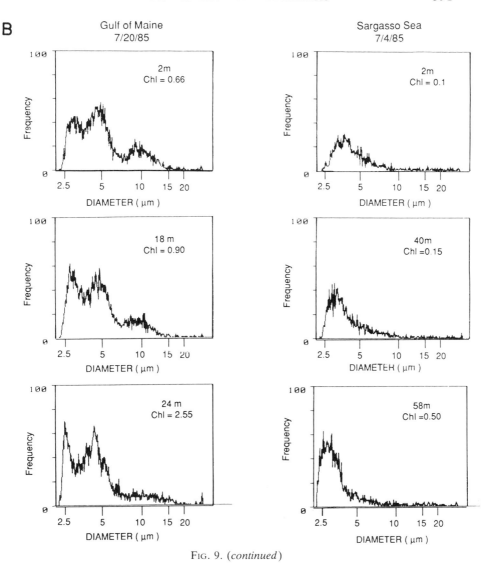

FIG. 9. (*continued*)

B. Autofluorescence

Table III is a summary of autofluorescence studies. Autofluorescence can be regarded as unfortunate or advantageous, depending on your point of view. Biomedical researchers constantly try to override the noise of autofluorescence from NADH, flavins, and other cellular constituents to optimize the signal of interest. On the other hand, aquatic science

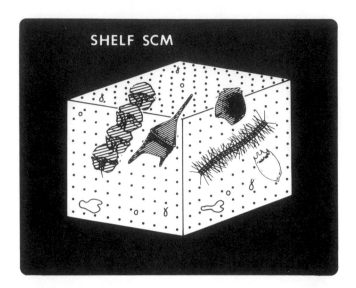

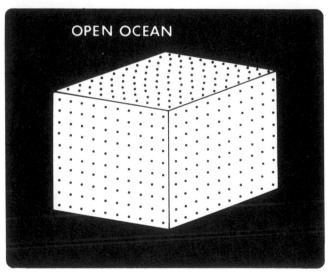

FIG. 10.   Schematic representation of the size and numbers of particles in various water types. Autofluorescent particles are shaded; nonfluorescent particles are not. The size and numbers of particles are drawn to scale, and each cube represents 10 μl. Cubes represent the shelf subsurface chlorophyll maximum (SCM) and the open ocean. From Yentsch and Spinrad (1987).

TABLE II

Individual Particle Analysis: Cell Sizing and Allometry

Parameter	Instrument(s)	Reference researchers
Cell size		
Cell diameter	Microscopy	Blasco et al. (1982); Margalef (1978)
Cell volume	Coulter counter	Sheldon and Parsons (1967); Sheldon et al. (1972)
Cell volume	FACS Analyzer	Yentsch et al. (1986a); Yentsch and Spinrad (1987)
Forward-angle light scatter	EPICS V	Ackleson and Spinrad (1988); Ackleson et al. (1988)
	ORTHO	Button and Robertson (1989)
	EPICS Profile	Azam and C. M. Yentsch (unpublished data)
Optical packaging		Spinrad (1984); Spinrad and Brown (1986); Spinrad and Yentsch (1987); Bricaud and Morel (1986); Ackleson and Spinrad, (1988); Ackleson et al. (1988)
Sediment studies		Baier et al. (1989)

researchers have sought to exploit autofluorescence. The sources are numerous and varied and some poorly understood.

In photosynthetic organisms, the main source of autofluorescence is a single pigment–protein complex, photosystem II (PSII). Only a small number of the protons absorbed by PSII are reemitted as fluorescence (the fluorescence yield). The majority is either used for photosynthesis (oxygen evolution) or converted to heat. The fluorescence yield of PSII *in vivo* is not a constant. Variability in fluorescence yield can create problems if trying to relate fluorescence to concentration. Yet, PSII yield does change in response to, and is an indicator of, the phytoplankter's environment (Neale et al., 1989; Demers et al., 1989).

Autofluorescence in nonphotosynthetic aquatic organisms is poorly understood and has only recently become the subject for investigations. The provocation of excitation photons transforms a rather dull suite of browns, greens, and grays into a "carnival under the sea" (Catala, 1964; Mazel, 1988). A mixture of phytoplankton appears multicolored. The red is Chl fluorescence in phytoplankton; the yellow-green, presumably from flavins, including luciferin, is the substrate for bioluminescence; and the orange is from phycoerythrin (PE) in phytoplankton. Natural population

TABLE III

INDIVIDUAL PARTICLE ANALYSIS: AUTOFLUORESCENCE[a]

Type of study	References[b]
Taxonomy/pigments/epifluorescence microscopy	Waterbury et al. (1979); Johnson and Sieburth (1979, 1982)
Ataxonomy	
Photosynthetic pigments	
Chl a	C. S. Yentsch and Yentsch (1979)
Accessory pigments: phycobilins ⎤	Chisholm et al. (1988a); Trask et al.
Phycoerythrin ⎟	(1982); Frankel et al. (1989)
Phycocyanin ⎟	Wood et al. (1985);
Phycourobilin ⎦	Olson et al. (1985, 1988); Li and
Allophycocyanin ⎤	Wood (1988); Li (1989)
Accessory pigments: carotenoids ⎟	Olson et al. (1989),
Fucoxanthin ⎟	C. S. Yentsch and C. M. Yentsch
Peridinin ⎟	(unpublished data); Cullen et al.
Photoprotective pigments ⎟	(1988); Yentsch and Brown
Diatoxanthin ⎟	(unpublished data); Yentsch
Diadinoxanthin ⎟	(unpublished data)
Ultraviolet screens ⎦	
Flavins	Yentsch et al. (1989b)
Riboflavin, flavin coenzymes	
Luciferin	
Others	
NADH	
Aromatic amino acids	
Physiology	
Chl a fluorescence per unit chlorophyll	Demers et al. (1989)
Photoinhibition of Chl a fluorescence	Neale et al. (1989); Cullen et al. (1988); Sakshaug et al. (1987)
Tracking fate of cells	Stoecker et al. (1986); Cucci
Sinking	et al. (1985, 1989); Shumway et al.
Grazing	(1985); Shumway and Cucci (1987); Gerritsen et al. (1987); Porter (1988); Shumway et al. (1989)

[a] Chl a, Chlorophyll a.
[b] Omission of references in some examples is deliberate.

mixtures usually contain red, orange, and yellow-green. Multiple-wavelength fluorescence detection is essential to separate out the signal from one type of cell versus another in a highly complex heterogeneous mixture. Flow cytometry plus multiple-parametric analysis is proving highly useful for this challenge. But the work has just begun.

A medley of fluorescent colors can also be observed in the macroscopic organisms using UV light. The red fluorescence is believed to be from the

Chl-containing endosymbionts zooxanthellae; the source of the other colors is still unknown. Dissociation of cells is required prior to FCM analysis (Yentsch and Pomponi, 1986). Image analysis is a promising tool for the study of the ecology and physiology of intact macroscopic organisms.

1. TAXONOMY/PIGMENTS/EPIFLUORESCENCE MICROSCOPY/FLOW CYTOMETRY

It was not until the late 1970s, when epifluorescence microscopes were taken to sea that numerous small bacterial-sized autofluorescing cells were noted for the first time (Waterbury *et al.*, 1979; Johnson and Sieburth, 1979). These cells, now known to be primitive prokaryotes, are cyanobacteria (blue-green algae). They fluoresce orange from heavy pigmentation by the phycobilins, primarily PE and phycocyanin (PC). Fortunately, these cells are at nearly ideal concentrations for FCM analysis, some 10^5–10^6 cells per 1 ml. Before long, small eukaryotes were also found to be common (Shapiro and Guillard, 1986). The advent of FCM revealed the discovery of abundant small primitive prochlorophytes (Chisholm *et al.*, 1988b). Olson, a researcher at the Woods Hole Oceanographic Institution, is the pioneer of seagoing FCM, recently advancing the characterization and verification of a numerous primitive prokaryotic life form with a unique pigment complement (Chisholm *et al.*, 1988). These prochlorophyte cells have also been documented by Li and Wood (1988).

The use of flow cytometry/cell sorting was critical in this discovery. First of all, the cells are exceedingly small and have weak fluorescence that fades in seconds, rendering epifluorescence microscopy almost useless. Fortunately, cells are only in the laser light of the flow cytometer for $\sim 10^{-5}$ seconds. Once the signals were observed, cells were sorted for (1) electron microscopy (2) HPLC pigment analysis, and (3) establishing cultures. Culturing success will permit intensive study of these fascinating organisms.

2. ATAXONOMY

The clustering of FCM data based on pigments alone is reasonably crude, yet highly useful. It has been termed "ataxonomy," because it bears little resemblance to classical taxonomy. While all phytoplankton contain Chl *a*, small cells with predominantly PE or PC fluorescence appear in an envelope on an FCM bivariate plot as cyanobacteria. These organisms can be further subdivided, if desired, by changing the excitation wavelength and emission filters (Wood *et al.*, 1985). Cells with strong fluorescence of both PE and Chl are cryptomonads. Cells with only Chl fluorescence are the diatoms, dinoflagellates, coccolithophores, and so on. These cell types can be further subdivided by light scatter methods (Trask *et al.*, 1982; Olson *et al.*, 1989) or dual-wavelength excitation (Olson *et al.*, 1988). Clustering methods for phytoplankton data analysis via a neural network system have been described by Frankel *et al.* (1989).

3.   PHYSIOLOGY

As noted earlier, pigment content is able to change in response to the light environment. Time scales vary from seconds to days (Fig. 11).

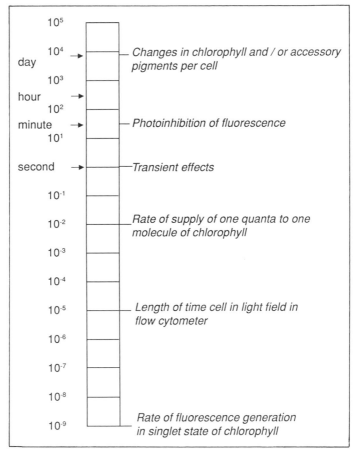

FIG. 11.   Time scale (as fraction and multiples of a second) of fluorescence changes noted for photosynthetic pigment in microalgae. After G. Harris; from Yentsch and Pomponi (1986).

Yentsch *et al.*, (1985) used FCM to indicate the degree of photoadaptation (i.e., increase of Chl concentration per cell over the cell cycle) in light-limited cells. Sakshaug *et al.* (1987), Cullen *et al.* (1988), and Neale *et al.* (1989) have used FCM methods to study photoinhibition of Chl fluorescence (time scale of seconds to minutes) at light levels that approach full sun. Sosik *et al.* (1989) have attempted to characterize the FCM signal obtained from Chl *a* fluorescence.

It is known that the fluorescence yield from PSII does change in response to, and is therefore an indicator of, the phytoplankter's environment. Several studies are under way; (Sakshaug *et al.*, 1987; Cullen *et al.*, 1988; Neale *et al.*, 1989; Demers *et al.*, 1989).

This photoinhibition of fluorescence is coupled with photoprotective pigments, namely the carotenoids diatoxanthin and diadinoxanthin. Lesser (1989) has studied UV protection in autofluorescent endosymbionts present in sea anemones. Data are from isolated cultured cells and cells in the intact association exposed to UV (Fig. 12). Yentsch *et al.* (1989a) have demonstrated that monitoring the 515–530 nm window of autofluorescence on marine particles can be used as an index of luciferin, and therefore bioluminescence capacity. Coleman (1988) discovered a phylogenetic enigma with the autofluorescent flagellum.

4.   TRACKING THE FATE OF CELLS

Autofluorescence is a feature that has been exploited as a tracer of the fate of cells. The fate of phytoplankton in the aquatic environment is of keen interest. The size, number, and percentage of cells that sink to the bottom forming pools of organic carbon in the sediments is tracked, versus that grazed and transferred as energy flow throughout the food web. Grazing experiments have been strongly augmented by the incorporation of FCM techniques (Stoecker *et al.*, 1986; Cucci *et al.*, 1985; Shumway and Cucci, 1987; Shumway *et al.*, 1989; Gerritsen *et al.*, 1987; Porter, 1988). One review (Cucci *et al.*, 1989) indicates that these methods can now be used effectively on natural populations from the field as well as mixtures of clonal cultures (Fig. 13).

C.   INDUCED FLUORESCENCE, CYTOCHEMICAL PROBES

Aquatic scientists have been slow to incorporate fluorescent cytochemical stains into their research. The initial use was merely to make nearly invisible things become visible. Methods are evolving to include image analysis and FCM approaches for quantitation of a variety of cell types,

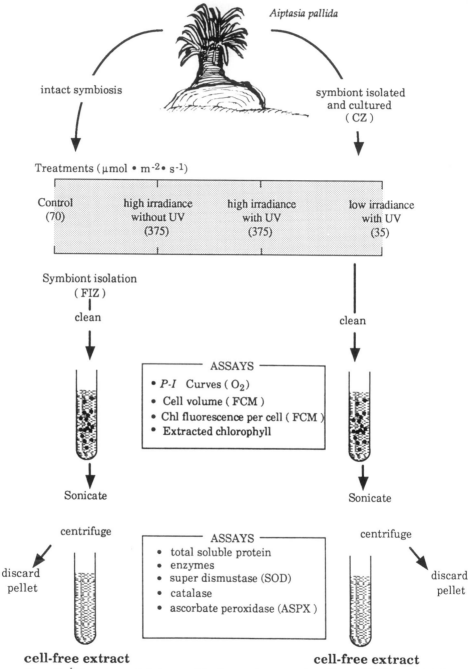

FIG. 12. Experimental design used by Lesser for measuring effects of UV on symbiotic zooxanthellae in sea anemone. CZ, cultured zooxanthellae; FIZ, freshly isolated zooxanthellae; P–I, photosynthesis–irradiance. From Lesser (1989).

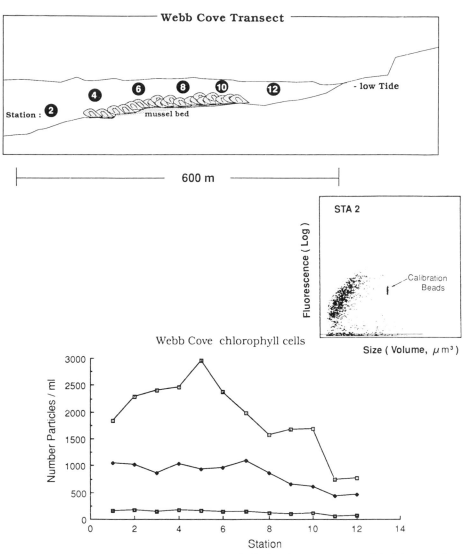

FIG. 13. Selective feeding measured in a field situation over a mussel bed in coastal Maine. Lower figure is a plot of Webb Cove Chl cells of various sizes by station location and number of particles per 1 ml. Sizes: 3–5 μm (▫); 5–8 μm (●); 8–10 μm (■). From Cucci *et al.* (1989).

TABLE IV

INDIVIDUAL PARTICLE ANALYSIS: INDUCED FLUORESCENCE CYTOCHEMICAL PROBES

Type of study	References
Cell types	Caron (1985); Hobbie et al. (1977); Porter and Feig (1980)
Cell molecular constituents	
DNA	Hobbie et al. (1977); Porter and Feig (1980); Chisholm et al. (1986); Yentsch et al. (1983a) Dortch et al. (1985)
Proteins	Dortch et al. (1985)
Lipids	Cooksey et al. (1987)
Toxins	Yentsch (1981)
Cell organelles	
Nucleus	
Chloroplasts	Yentsch et al. (1985)
Mitochondria	
Flagella	Coleman (1988)
Cell processes and physiology	
Photosynthesis	Neale et al. (1989);
Cell cycling	Chisholm et al. (1986b); Vaulot et al. (1986); Vaulot and Chisholm (1987a,b); Armbrust et al. (1989)
Membrane potential	Yentsch and Horan (unpublished data)
Metabolic activity	Dorsey et al. (1989); Yentsch et al. (1989a)

molecular constituents, organelles, and cell processes that had previously eluded the investigators. Table IV is a summary of ongoing studies using cytochemical probes.

1. CELL TYPES

Fluorescent DNA stains have been used to visualize bacteria in preserved samples for many years. Stains have included 4′,6-diamidino-2-phenylindole (DAPI), Hoechst (HO), and acridine orange (AO), and many studies (e.g., Hobbie et al., 1977) have demonstrated that such an approach is highly useful in aquatic microbial ecology. Button and Robertson (1989) have extended the utility of stains to FCM methods. Using the EPICS Profile in a three-color setup (Chl PE, and fluorescein), Yentsch and Azam (unpublished data) have demonstrated the utility of the metabolic indicator, fluorescein diacetate (FDA) for live marine bacteria simultaneously with phytoplankton. Revelante and Gilmartin (1983) and Porter and Feig (1980) have used cytochemical stains, and Sherr *et al.* (1987) have developed double-staining techniques on microheterotrophs.

2. CELL MOLECULAR CONSTITUENTS

Cytochemical probes are most useful when optimized to quantitate specific molecular constituents within cells using FCM. DNA stains have been used by Yentsch *et al.* (1983a) and Chisholm *et al.* (1986). Dortch (unpublished data) worked on methodology of proteins and amino acids; and Cooksey *et al.* (1987) have developed Nile red as a lipid stain useful for phytoplankton. Carbohydrate quantitation has been more elusive to date, in part because a suitable probe is yet to be realized.

Some cells produce "unique" natural products, such as the neurotoxins synthesized by certain species of dinoflagellates. Some of these toxins, while not fluorescent in the native state, become fluorescent upon the addition of hydrogen peroxide; thus their detection by FCM is possible (Yentsch, 1981).

3. CELL ORGANELLES

At times, cytochemical probes are used to visualize various organelles or localize particular constituents. Fluorescent stains have long been employed in this manner for detection of the nucleus and following the chromosomes through mitosis. DNA stains are also useful for localizing and quantitation of DNA in the chloroplasts and mitochondria. Chloroplasts can be studied by autofluorescence alone, or in combination with other stains. There is now a possibility for the study of mitochrondria using rhodamine 123 or membrane potential stains.

4. CELL PROCESSES AND PHYSIOLOGY

The most intriguing aspect of the employment of FCM into aquatic sciences goes beyond quantitation of various cell products. Experiments can be designed to investigate the dynamic processes and physiology of the cell.

As mentioned, photosynthesis and fluorescence from Chl *a* are linked, and Neale and co-workers (1989) are making headway at interpreting complex signals. This method can be used in combination with stains. Cell cycling studies have been highly productive, with investigations by

Chisholm *et al.* (1986), Vaulot *et al.* (1986, 1989), Vaulot and Chisholm (1987a,b), and Armbrust *et al.* (1989).

Getting a handle on cell growth as a process somewhat independent from cell division is an important undertaking. In field measurements, the incorporation of carbon-14 isotope during the process of photosynthesis has been the standard method to assess "population growth rate." Yet one does not gain information on the cell-to-cell variability. Cytochemical stains for pH and membrane potential have been attempted, but to date have yielded inconsistent results. At the time of this writing, the most promising approach appears to be the viability stain FDA used in a quantified manner to estimate metabolic activity (Dorsey *et al.*, 1989). It is useful for phytoplankton, and in general, FDA data correspond reasonably with carbon-14 data (Fig. 14). Data from a field population are given in Fig. 15. Added utility is gained using a three-color setup (Fig. 16) and five-intensity Immunobrite standard beads with intensities expressed as molecular equivalents of fluorescein (Fig. 17). Accordingly, cytotoxicity testing can be measured by esterase activity (FDA cleavage), PSII damage by change in Chl fluorescence; PE fluorescence intensity is yet another index of damage. Damage to each system is on a different time scale.

D.   Immunofluorescent Probes and Immunocytochemistry

The thrust and promise of immunochemical reagents has been awakened in the aquatic science community. The earliest reagents were prepared in the late 1970s by Ward and Perry (1980). Again, the probes have been used to ask questions about what cell types are present, to investigate molecular constituents, and to study cell organelles. In many instances, image analysis is as important as FCM approaches. In combination, they offer even greater benefits. Table V is a summary of ongoing efforts using immunocytochemical probes.

1.   CELL TYPES

The first antibodies produced for marine organisms were to nitrifying bacteria, a group of organisms with a unique function. This specialization makes their localization and pattern of distribution in the world's oceans important. Ward and Perry (1980) and Ward (1984) have pioneered the use of immunocytochemistry by the development of these probes and protocols. Campbell *et al.* (1983) and Campbell (1988) followed the technology by producing several antisera to various clones of cyanobacteria. Her effort was then to study the distribution patterns of the various "serotypes." Antibodies have also been produced to various eukaryotic

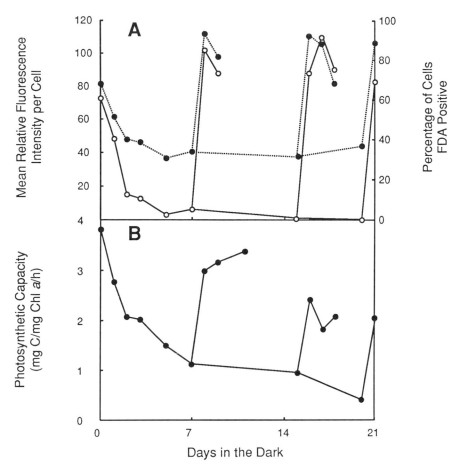

FIG. 14. Comparison of (A) FDA assay results and (B) photosynthesis maximum (P_{max}, maximum ^{14}C uptake) in dark-survival experiment, in collaboration with J. J. Cullen and H. MacIntyre. Data are for clone Omega 48–23. Batch culture grown in F/2 medium on 14 : 10 light/dark cycle, is placed in the dark. Data show similar response of P_{max} and FDA assay [both percentage of cells exhibiting fluorescein fluorescence (○) and mean fluorescence intensity (●)] to light limitations. The two offshootlike curves are responses of cells when placed back into low-light environments after 7, 14, and 21 days in the dark. From Dorsey et al. (1989).

cells based on pigment types or cell wall composition (Lee et al., 1988; Shapiro, 1988; Shapiro et al., 1989).

In some cases, the questions of interest are concerned with one particular species, not to a group or clustering. Shapiro et al. (1990) have had considerable success with genus-, species-, and even clone-specific antibodies.

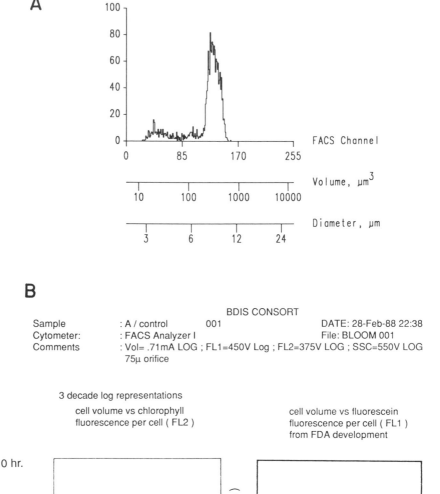

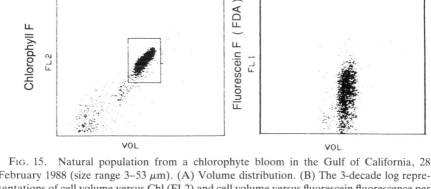

FIG. 15. Natural population from a chlorophyte bloom in the Gulf of California, 28 February 1988 (size range 3–53 μm). (A) Volume distribution. (B) The 3-decade log representations of cell volume versus Chl (FL2) and cell volume versus fluorescein fluorescence per cell (FL1) from FDA assay. Here nearly all of the cells are metabolically active. In most waters studied to date, < 50% of the cells show a positive response in a 15-minute assay, suggesting that many cells are inactive or senescent. Note the considerable variability in metabolic activity, while the Chl fluorescence is a tight cluster. Cell number analyzed was 1978.

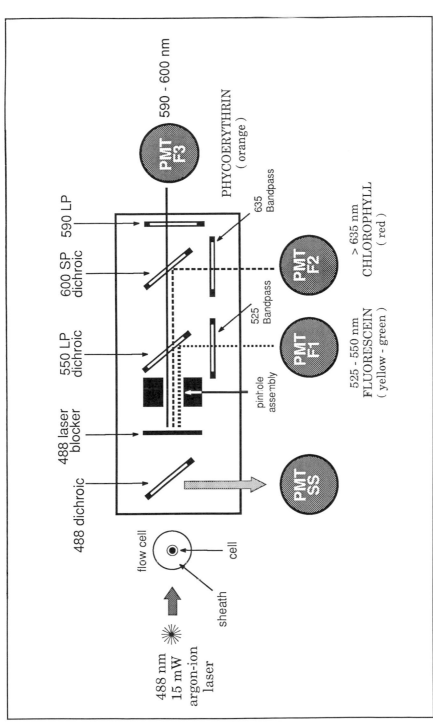

Fig. 16. Optical setup for three-color fluorescence using the EPICS Profile. With this setup, PE Chl, and fluorescein fluorescence can be measured simultaneously. Courtesy of Doug Rick.

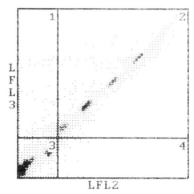

FIG. 17. Immunobrite beads (Coulter) run on the EPICS Profile for the three-color assay. The lowest intensity is a "negative."

TABLE V

INDIVIDUAL PARTICLE ANALYSIS: IMMUNOFLUORESCENCE
IMMUNOCYTOCHEMICAL PROBES

Type of analysis	References
Cell types	
Ataxonomy	
α-Nitrifying bacteria	Ward (1984); Ward and Perry (1980)
α-Cyanobacteria	Campbell et al. (1983); Campbell (1988)
α-Eukaryotes	
Pigment types	Lee et al. (1988)
Cell surface types	Shapiro (1988)
Specific cell types	
genus-, species-,	Shapiro et al. (1989)
clone-specific antibodies	
Cell molecular constituents	
α-RUBISCO	Orellana et al. (1988)
α-Nitrate reductase	Balch et al. (1988)
α-Luciferase	DeMey, (1983)
α-Saxitoxin	Guire et al. (1988)
α-Brevetoxin	Baden et al. (1988)
Cell organelles	
Golgi bodies	
Cytoskeleton	
Scintillons	Nicolas et al. (1988)

2.   CELL MOLECULAR CONSTITUENTS

Certain cellular constituents are of particular interest to aquatic scientists in an attempt to understand balanced and unbalanced growth and to predict noxious bloom situations. Key enzymes were the first targets. Antibodies have been developed to ribulose-1,5-biphosphate carboxylase/oxygenase (RUBISCO) (Orellana *et al.*, 1988) to obtain individual cell measurements related to carbon fixation. Antibodies to nitrate reductase have been developed to assess the capacity of individual cells to reduce external cell nitrate (Balch *et al.*, 1988). There is also an antibody to luciferase, the enzyme responsible for bioluminescence.

Antibodies have been developed in attempts to quantify toxins as well. Successful attempts include antisaxitoxin by Guire *et al.* (1988). antibrevetoxin by Baden *et al.* (1988), and anticiguatoxin by Hokama *et al.* (1988). Specific applications especially designed as rapid assay methods are under development.

3.   CELL ORGANELLES

While cytologists per se have made remarkable progress using immunochemical probes for Golgi bodies and the cytoskeleton, at the time of this writing, these have not been used in microbial ecology studies. Bioluminescence is of considerable interest to aquatic scientists. The one organelle unique to bioluminescent cells, the scintillon, was localized using antiluciferase (Nicolas *et al.*, 1988).

V.   Summary

Flow cytometry/cell sorting in aquatic sciences has been driven in two directions. The frontier directions are on shipboard and shore-based. On the one hand, the rapid analytical technique has been taken on shipboard to provide a real-time assessment of the particles and phytoplankton in water masses. These data also give information on the amount of vertical mixing and advection, and denote fronts between two or more water masses. There is an optical characterization (based on sizes, numbers, and pigment groups) of the individual primary producers, as well as detritus and suspended sediments. An optical-closure question is being addressed: "Does the total optical signal equal the sum of the parts?" Additionally, associations with chemical and physical oceanographic features are readily accomplished. A "census" of thousands of phytoplankton cells is obtained and can be mapped. Scientists are able to identify "who is where?" Such data are critical to understand the optical-feedback loop or the so-called photon-budget-in-the-sea, which in turn controls the rates at which growth processes occur in nature.

On the other hand, an in-depth understanding is sought as to how particle size, shape, refractive index, nutritional status (nutrient and/or light limitation), growth dynamics, and cell cycle combine to control the optics (light scatter and fluorescence at the moment, and ideally absorption as well) or the photon-budget-of-the-cell. For this purpose, a shore-based facility associated with a diverse collection of phytoplankton is ideal. The development at Bigelow Laboratory of the Jane J. MacIsaac Facility is to provide services for the oceanographic community. Association and co-location with the Provasoli–Guillard Center for Culture of Marine Phytoplankton is key. Visitors are trained and given access to state-of-the-art instrumentation. Visiting investigators have available "the tropical, temperate, and polar seas" in concentrated form, as marine phytoplankton isolated worldwide and maintained as living clonal cultures. In this way, frontline cell biology questions can be addressed.

The relentless exploration of standards and controls appropriate for the aquatic community must be continued. An intercalibration effort is a vital step. It is only with the widespread acceptance of particular reference materials and uniform optical filters among research groups utilizing FCM that comparable data sets describing aquatic particle distributions will be possible. For a global science, this strategy is imperative.

ACKNOWLEDGMENTS

I thank the National Science Foundation and the Office of Naval Research for instrumental and research support. Both the NSF and the ONR have been supporters of the application of new technologies to aquatic science problems. Thanks, too, to Terry Cucci, Bridget Holligan, and David Phinney for technical and conceptual development. Janet Campbell, John Cullen, Paul Horan, Katherine Muirhead, and Charles Yentsch are ongoing sources of inspiration. Fran Scannell processed the manuscript and Jim Rollins drafted the illustrations.

REFERENCES

Ackleson, S. G., and Spinrad, R. W. (1988). *Appl. Opt.* **27,** 1270–1277.
Ackleson, S. G., Spinrad, R. W., Yentsch, C. M., Brown, J., and Korjeff-Bellows, W. (1988). *Appl. Opt.* **27,** 1262–1269.
Ackleson, S. G., Balch, W. H., and Holligan, P. M. (1989). *Oceanography* **1,** 18–22.
Armbrust, E. V., Bowen, J. D., Chisholm, S. W., and Olson, R. J. (1989). *Appl. Environ. Microbiol.*, 425–432.

Baden, D. G., Mende, T. J., and Szmant, A. M. (1988). *In* "Immunochemical Approaches to Coastal Estuaries and Oceanographic Questions" pp. 134–144. Springer-Verlag, Berlin.

Baier, R., Cucci, T. L., and Yentsch, C. M. (1989). *Limnol. Oceanogr.* **34,** 947–952.

Balch, W. M., Yentsch, C. M. Reguera, B., and Campbell, W. (1988). *In* "Immunochemical Approaches to Coastal Estuaries and Oceanographic Questions" pp. 263–276. Springer-Verlag, Berlin.

Blasco, D., Garfield, P., and Packard, T. (1982). *J. Phycol.* **18,** 58–63.

Bricaud, A., and Morel, A. (1986). *Appl. Opt.* **25,** 571–580.

Button, D. K., and Robertson, B. R. (1989). *Cytometry* **10**.

Campbell, L. (1988). *In* "Immunochemical Approaches to Coastal Estuaries and Oceanographic Questions" pp. 208–229. Springer-Verlag, Berlin.

Campbell, L., Carpenter, E. J., and Iacono, V. J. (1983). *Appl. Environ. Microbiol.* **46,** 553–559.

Caron, D. A. (1985). *Appl. Environ. Microbiol.* **46,** 491–498.

Catala, R. (1964). "Carnival Under The Sea." Sicard, Paris.

Chisholm, S. W. (1981). *Can. Bull. Fish. Aquat. Sci.* **21,** 150.

Chisholm, S. W., Morel, F. M. M., and Slocum, W. S. (1980). *In* "Primary Productivity in the Sea" pp. 281–299. Plenum, New York.

Chisholm, S. W., Armbrust, E. V., and Olson, R. J. (1986). *Can. Bull. Fish. Aquat. Sci.* **214,** 343–369.

Chisholm, S. W., Olson, R. J., Zettler, E. J., Waterbury, J., Goericke, R., and Welschmeyer, N. (1988). *Nature (London)* **334,** 340–343.

Cho, B. C., and Azam, F. (1988). *Nature (London)* **332,** 441.

Cleveland, J. S., and Perry, M. J. (1987). *Mar. Biol.* **94,** 489–498.

Coleman, A. W. (1988). *J. Phycol.* **24,** 118–120.

Conrad, M. (1983). Adaptability: The Significance of Variability from Molecule to Ecosystem. Plenum, New York.

Cooksey, K. E., Guckert, J. B., Williams, S. A., and Callis, P. R. (1987). *J. Microbiol.* **6,** 333–345.

Cucci, T. L., Shumway, S., Newell, R. D., Selvin, R., Guillard, R. R. L., and Yentsch, C. M. (1985). *Mar. Ecol. Prog. Ser.* **24,** 201–204.

Cucci, T. L., Shumway, S. E., Yentsch, C. M., and Newell, C. (1989). *Cytometry* **10,** 659–669.

Cullen, J. J., Yentsch, C. M., Cucci, T. L., and MacIntyre, H. L. (1988). *Proc. SPIE* **925,** 149–156.

Dorsey, J., Yentsch, C. M., Mayo, S., and McKenna, C. (1989). *Cytometry* **10,** 622–628.

DeMey, J. (1983). *In* "Immunocytochemistry" (J. Polak, and S. Van Norden, eds.), pp. 82–112. Wright-PSG, Bristol.

Dorsey, J., Yentsch, C. M., Mayo, S., and McKenna, C. (1989). *Cytometry* **10**.

Dortch, Q., Clayton, Jr., J. R., Thoresen, S. S., Cleveland, J. S., Bressler, S. L., and Ahmed, S. I. (1985). *J. Mar. Res.* **43,** 437–464.

Frankel, D. S., Olson, R. J., Frankel, S. L., and Chisholm, S. W. (1989). *Cytometry* **10,** 540–550.

Gerritsen, J., Sanders, R. W., Bradley, S. W., and Porter, K. G. (1987). *Limnol. Oceanogr.* **32,** 691–698.

Guire, P. E., Duquette, P. H., Amos, R. A., Behrens, J. C., Josephson, M. W., and Chambers, R. P. (1988). *In* "Immunochemical Approaches to Coastal Estuaries and Oceanographic Questions" pp. 145–154. Springer-Verlag, Berlin.

Haas, L. W. (1982). *Ann. Inst. Oceanogr.* **58,** 940–946.

Haugen, E. M., Cucci, T. L., Yentsch, C. M., and Shapiro, L. P. (1987). *Appl. Environ. Microbiol.* **53,** 2677–2679.

Hobbie, J. E., Daley, R. J., and Jasper, S. (1977). *Appl. Environ. Microbiol.* **33,** 1225–1228.

Hokama, Y., Honda, S. A. A., Kobayashi, M. N., Nakaguwa, L. K., Shirai, L. K., and Miyahara, J. T. (1988). *In* "Immunochemical Approaches to Coastal Estuaries and Oceanographic Questions" pp. 155–166. Springer-Verlag, Berlin.

Iturriaga, R., and Siegel, D. C. (1988). *Ocean Opt. IX, SPIE* **925,** 277–287.

Johnson, P. W., and Sieburth, J. McN. (1979). *Limnol. Oceanogr.* **24,** 928–935.

Johnson, P. W., and Sieburth, J. McN. (1982). *J. Phycol.* **18,** 318–327.

Kirk, J. T. O. (1983). "Light and Photosynthesis in Aquatic Ecosystems." Cambridge Univ. Press, London.

Kornberg, W., ed. (1988). *Natl. Sci. Found.* **19,** 34–113.

Lee, J. J., Chan, Y., and Lagziel, A. (1988). *In* "Immunochemical Approaches to Coastal Estuaries and Oceanographic Questions" pp. 230–241. Springer-Verlag, Berlin.

Legendre, L., and LeFèvre, J. (1989). "Productivity of the Ocean: Present and Past." Wiley, Chichester.

Lesser, M. P. (1989). *Cytometry* **10,** 653–658.

Li, W. K. W. (1989). *Cytometry* **10,** 564–579.

Li, W. K. W., and Wood, M. (1988). *Deep-Sea Res.* **34,** 1615–38.

Mantoura, R. F. C., and Llewellyn, C. A. (1983). *Anal. Chim. Acta* **15,** 297–314.

Margalef, R. (1978). *Oceanol. Acta* **1,** 493–509.

Mazel, C. (1988). pp. 264–279. International Oceanographic Foundation/Univ. of Miami.

Morel, A. (1988). *J. Geophys. Res.* **93,** c9:10, 749–10, 768.

Muirhead, K. A., Schmitt, T. C., and Muirhead, R. A. (1983). *Cytometry* **3,** 251–256.

Neale, P. J., Cullen, J. J., and Yentsch, C. M. (1989). *Limnol. Oceanogr.* submitted.

Nicolas, M.-T., Bassot, J.-M., Johnson, C. H., and Hastings, J. W. (1988). *In* "Immunochemical Approaches to Coastal Estuaries and Oceanographic Questions" pp. 278–282. Springer-Verlag, Berlin.

Olson, R. J., Frankel, S. L., Chisholm, S. W., and Shapiro, H. M. (1983). *J. Exp. Mar. Biol. Ecol.* **68,** 129–144.

Olson, R. J., Vaulot, D., and Chisholm. S. W. (1985). *Deep-Sea Res.* **32,** 1273–1280.

Olson, R. J., Chisholm, S. W., Zettler, E. R., and Armbrust, E. V. (1988). *Deep-Sea Res.* **35,** 425–440.

Olson, R. J., Zettler, E. R., and Anderson, K. O. (1989). *Cytometry* **10,** 636–643.

Orellana, M. V., Perry, M. J., and Watson, B. A. (1988). *In* "Immunochemical Approaches to Coastal Estuaries and Oceanographic Questions" pp. 243–262. Springer-Verlag, Berlin.

Perry, M. J., and Porter, S. M. (1989). *Limnol. Oceanogr.* **34,** 1727–1738.

Phinney, D. A., and Cucci, T. L. (1989). *Cytometry* **10,** 511–521.

Platt, T. (1989). *Cytometry* **10,** 500.

Porter, K. G. (1988). *Hydrobiologia* **159,** 89–98.

Porter, K. G., and Feig, Y. S. (1980). *Limnol. Oceanogr.* **25,** 943–948.

Revelante, N., and Gilmartin, M. (1983). *RAPP REUN. CIESM.* **28,** 121–122.

Rivkin, R. B., Phinney, D. A., and Yentsch, C. M. (1986). *Appl. Environ. Microbiol.* **52,** 935–938.

Robertson, B. R., and Button, D. K. (1989). *Cytometry,* **10,** 558–563.

SAC (1988). *Int. Conf. Soc. Anal. Cytol., 13th, Sept.*

Sakshaug, E., Demers, S., and Yentsch, C. M. (1987). *Mar. Ecol. Prog. Ser.* **41,** 275.

Shapiro, L. (1988). *In* "Immunochemical Approaches to Coastal Estuaries and Oceanographic Questions" p. 242. Springer-Verlag, Berlin.

Shapiro, L. P., and Guillard, R. R. L. (1986). *Can. Bull. Fish. Aquat. Sci.* **214,** 371–389.

Shapiro, L. P., Campbell, L., and Haugen, E. M. (1989). *Mar. Ecol. Prog. Ser.* **57,** 219–224.

Sheldon, R. W., and Parsons, T. R. (1967). *J. Fish. Res. Board Can.* **24,** 909–915.

Sheldon, R. W., Prakash, A., and Sutcliffe, Jr., W. H. (1972). *Limnol. Oceanogr.* **17,** 327–340.

Sherr, B. F., Sherr, E. B., and Fallon, R. D. (1987). *Appl. Environ, Microbiol.* **53,** 958–965.

Shumway, S. E., and Cucci, T. L. (1987). *Aquat. Toxicol.* **10,** 9–27.

Shumway, S. E., Cucci, T. L., Newell, R. C., and Yentsch, C. M. (1985). *J. Exp. Mar. Biol. Ecol.* **91,** 77–92.

Shumway, S. E., Newell, R. C., Crisp, D. J., and Cucci, T. L. (1990). Particle selection in filter-feeding bivalve molluscs: A new technique on an old theme. *In* "The Bivalvia. Proceedings of a Symposium in Memory of Sir Charles Maurice Yonge, Edinburgh, 1986" (B. Morton, ed.). Hong Kong University Press, Hong Kong.

Sieracki, M. E., and Webb, K. L. (1989a). "Protozoa and Their Role in Marine Processes." Springer-Verlag, Berlin, in press.

Sieracki, M. E., Viles, C. L., and Webb, K. L., (1989b). *Cytometry* **10,** 551–557.

Silvert, W., and Platt, T. (1978). *Limnol. Oceanogr.* **23,** 813–816.

Silvert, W., and Platt, T. (1980). *Am. Soc. Limnol. Oceanogr. Spec. Symp.* **3,** 754–763.

Sosik, H. M., Chisholm, S. W., and Olson, R. J. (1989). *Limnol. Oceanogr.* **34,** 1749–1761.

Spinrad, R. W. (1984). *SPIE–Ocean Opt. VII* **489,** 355–342.

Spinrad, R. W., and Brown, J. (1986). *Appl. Opt.* **25,** 1930–1934.

Spinrad, R. S., and Yentsch, C. M. (1987). *Appl. Opt.* **26,** 357–362.

Stoecker, D. K., Cucci, T. L., Hulburt, E. M., and Yentsch, C. M. (1986). *J. Exp. Mar. Biol. Ecol.* **95,** 113–130.

Trask, B. J., van den Engh, G. J., and Elgerhuizen, J. H. B. W. (1982). *Cytometry* **2,** 258–264.

Van de Hulst, H. C. (1957). "Light Scattering by Small Particles." Wiley, New York.

Vaulot, D., and Chisholm, S. W. (1987a). *J. Plankton Res.* **9,** 345.

Vaulot, D., and Chisholm, S. W. (1987b). *J. Phycol.* **23,** 132–137.

Vaulot, D., and Ning, X. (1990). *Contam. Shelf Res.* in press.

Vaulot, D., Olson, R. J., and Chisholm, S. W. (1986). *Exp. Cell Res.* **167,** 38–52.

Vaulot, D., Courties C., and Partensky, F. (1989). *Cytometry* **10,** 629–635.

Ward, B. B. (1984). *Limnol. Oceanogr.* **29,** 402–410.

Ward, B. B., and Perry, M. J. (1980). *Appl. Environ. Microbiol.* **39,** 913–918.

Waterbury, J. B., Watson, S. W., Guillard, R. R. L., and Brand, L. E. (1979). *Nature (London)* **277,** 293–294.

Winfield, D. L. (1987). "Advanced Flow Cytometry. Space Applications and Terrestrial Spinoffs." Report of a workshop held at the Les Alames Natl Laboratory, June 1–3, 1987. NASA JSC Technology Utilization Office. L. B. Johnson Space Center, Houston.

Wood, A. M., Horan, P. K., Muirhead, K., Phinney, D. A., Yentsch, C. M., and Waterbury, J. B. (1985). *Limnol. Oceanogr.* **30,** 1303–1315.

Yentsch, C. M. (1981). *Toxicon* **19,** 611–621.

Yentsch, C. M., and Pomponi, S. A. (1986). *Int. Rev. Cytol.* **105,** 183–243.

Yentsch, C. M., and Spinrad, R. W. (1987). *Mar. Technol. Soc. J.* **21,** 58.

Yentsch, C. M., and Yentsch, C. S. (1984). *Oceanogr. Mar. Biol. Annu. Rev.* **22,** 55–98.

Yentsch, C. M., and Yentsch, C. S. (1989). "Recent Advances in Microbial Ecology," Proc. Fifth Int. Symp. on Microbial Ecology, ISME5, *Business Center of Academic Societies, Japan*, pp. 707–711.

Yentsch, C. M., Mague, F. M., Horan, P. K., and Muirhead, K. (1983a). *J. Exp. Mar. Biol. Ecol.* **67,** 175–183.

Yentsch, C. M., Horan, P. K., Muirhead, K., Dortch, Q., Haugen, E., Legendre, L., Murphy, L. S., Perry, M. J., Phinney, D. A., Pomponi, S. A., Spinrad, R. W., Wood, M., Yentsch, C. S., and Zahuranec, B. J. (1983b). *Limnol. Oceanogr.* **28,** 1275–1280.

Yentsch, C. M., Cucci, T. L., Phinney, D. A., Selvin, R., and Glover, H. E. (1985). *Mar. Biol.* **89,** 9–20.

Yentsch, C. M., Cucci, T. L., and Phinney, D. A. (1986a). *In* "Tidal Mixing and Plankton Dynamics" pp. 414–448. Springer-Verlag, New York.

Yentsch, C. M., Cucci, T. L., Phinney, D. A., and Topinka, J. A. (1986b). *In* "Tidal Mixing and Plankton Dynamics" pp. 414–448. Springer-Verlag, New York.

Yentsch, C. M., Losee, J., Milan-Nunez, E., Cucci, T. L., Alvorez-Borrego, S., Lara-Lara, R., and Lavin, M. (1989a). Submitted.

Yentsch, C. M., Mague, F. C., and Horan, P. K. (1988). *In* "Immunochemical Approaches to Coastal, Estuarine and Oceanographic Questions" Springer-Verlag, Berlin.

Yentsch, C. S., and Phinney, D. A. (1989). *Limnol. Oceanogr.* **34,** 1698–1709.

Yentsch, C. S., and Yentsch, C. M. (1979). *J. Mar. Res.* **37,** 471–483.

Chapter 51

Flow Microsphere Immunoassay for the Quantitative and Simultaneous Detection of Multiple Soluble Analytes

MACK J. FULWYLER AND THOMAS M. McHUGH

Department of Laboratory Medicine
University of California
San Francisco, California 94143

I. Introduction

A quantitative and highly sensitive assay, named flow microsphere immunoassay (FMIA), permitting the simultaneous quantification of several analytes in a sample has been developed.[1] Illustrated in Fig. 1, the method employs several discrete size classes of plastic microspheres of 5, 7, 10, and 15 μm diameter (Fig. 1A), each class coated with a specific capture reagent (CR1–CR4, Fig. 1B). The four size classes of coated microspheres are mixed together and added to the sample under analysis (Fig. 1C). Each microsphere acts as an accumulator that removes from the sample and concentrates specific analyte for later measurement. Following incubation in the sample, the microspheres are washed to remove unbound sample and labeled with a fluorescent reagent that will react only with the microsphere-accumulated analyte, labeling only those microspheres (Fig. 1D). Following washing, the microsphere suspension (Fig. 1E) is analyzed using a flow cytometer that measures microsphere size and fluorescence. As it

[1] Abbreviations used in this chapter: CMV, cytomegalovirus; CV, coefficient of variation; EIA, enzyme immunoassay: FCM, flow cytometric; FITC, fluorescein isothiocyanate; FMIA, flow microsphere immunoassay; HIV, human immunodeficiency virus; HSV, herpes simplex virus; IFA, slide, immunofluorescence assay; LA, latex agglutination; MFC, mean fluorescence channel; PE, phycoerythrin; SD, standard deviation; WB, western blot.

613

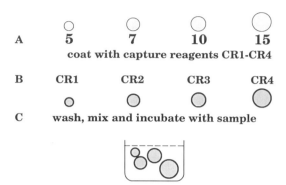

C   wash, mix and incubate with sample

D   wash, incubate with fluorescent reagent

E   wash and analyze in cytometer

FIG. 1.   Flow microsphere immunoassay (FMIA) diagrammed.

passes through the sensor region of a cytometer, the size of each micro-sphere is measured identifying the microsphere class and, thereby, the capture reagent. Simultaneously, fluorescence of the microsphere is mea-sured; the amplitude of the signal indicates the quantity of the analyte in the sample. In our work we denote as positive, microsphere staining that results in the mean fluorescence channel (MFC) of the positive peak being significantly removed from the MFC of the negatively staining micro-spheres. As will be discussed later, the positivity threshold must be set for each assay based on analysis of multiple positive and negative samples.

Invented by Fulwyler and patented in the United Kingdom by Coulter Electronics, Inc. (1976), the method was described by Horan *et al.* (1977). Unappreciated for many years, the method is currently enjoying renewed attention and has been applied to the detection of antibodies to *Candida albicans* (McHugh *et al.*, 1989) and to the detection of human im-munodeficiency virus (HIV) antigen-containing immune complexes and total immune complexes (McHugh *et al.*, 1986, 1988a). It has also been applied to the detection of cytomegalovirus (CMV) and herpes simplex virus (HSV) antibodies (McHugh *et al.*, 1988b). The method has been used for the detection of antibodies to four different HIV antigens and is under development as a routine screening assay for use in blood banks (Scillian *et al.*, 1989).

II. Instrumentation and Materials

A. Flow Cytometry Instrumentation Requirements

The minimum flow cytometer capabilities required to utilize FMIA are the ability to measure microsphere size (diameter, area, or volume) and the ability to measure fluorescence at one color. The capability of distinguishing small differences in size ultimately determines the number of microsphere sizes that can be used and thereby determines the number of analytes that can be simultaneously assayed. We have successfully used the Becton Dickinson FACS Research analyzer, a volume, 90° light scatter, and two-color fluorescence sensing instrument, a Becton Dickinson FACScan analyzer, a laser-based, forward and 90° light scatter, and three-color fluorescence sensing instrument and a Coulter Electronics EPICS-C, a laser-based, forward and 90° light scatter and two-color fluorescence sensing instrument. If the microsphere classes are well separated in size, volume or forward light scatter sensing will be sufficient; however, if the microsphere sizes are similar, the combination of forward and 90° light scatter measurement is helpful. Most of the currently available commercial cytometers have size resolution capability adequate to use four or more size classes.

Typically, fluorescence signals are amplified and displayed logarithmically. The MFC can be used as the unit of measure for data interpretation.

B. Microspheres

In our work, we have used polystyrene microspheres of 4–15 μm in diameter for the noncovalent adsorption of proteins. This noncovalent adsorption of proteins onto polystyrene is based on electrostatic interactions and/or van der Waals forces. These microspheres should be stored at 4°–22°C; storage at lower temperatures may produce irreversible flocculation of the microspheres. For long-term storage, a preservative should be added (i.e., 0.1% sodium azide or thimerosal). The microspheres should be resuspended before use by shaking or by gentle sonication.

For the covalent coupling of ligands to the microsphere, chemically reactive groups can be incorporated into the polystyrene surface. Amino (NH_2)- and carboxylate (COOH)-modified microspheres are most commonly used; however, many other surface groups are available. Storage conditions are described later.

Microspheres can also be obtained with incorporated visible and fluorescent dyes. See Section VI,B (Appendix) for a list of microsphere suppliers.

III.   Microsphere Coating and Assay Protocols

A.   Preparation of Microspheres for Coating

Microspheres are supplied in an aqueous suspension containing a low concentration of surfactant, which may reduce or prevent the adsorption of protein to the microsphere surface (both ionic and nonionic surfactants reduce the adsorption of protein onto polystyrene). This surfactant can be removed by the following procedure.

Washing in Buffer or Water

1. Dilute the microspheres in an equal volume of buffer or distilled water (dH$_2$O) and centrifuge for 5 minutes at 3000 g.
2. Discard the supernatant and resuspend the pellet in buffer or in water and centrifuge for 5 minutes at 8000 g (repeat wash step twice). Note that removal of the surfactant makes the microspheres hydrophobic and may cause them to be captured in the liquid surface film. To prevent the loss of microspheres during this washing step, a small amount of the protein (~ 1 μg/ml) that will be used to coat the microspheres can be added to the buffer or water used for washing.
3. The final wash step should use the buffer to be used in the protein coating of the microspheres.
4. In place of centrifugation:
   a. Filter the microspheres through a small filter (i.e., 0.45-μm syringe filter). To resuspend the microspheres, vigorously flush the filter taking care not to break or tear the filter. Refilter the microspheres three or four times using clean buffer or water for each wash. Ensure that the last wash is in the coating buffer.
   b. Dialyze the microspheres over 24 hours at 4°C using dH$_2$O followed by the coating buffer. To prevent clumping and incomplete washing of the microspheres, a low concentration (~ 1 μg/ml) of the protein to be coated can be added to the microspheres. Be sure to use dialysis tubing of a pore size smaller than the size of the added protein.
   c. Large batches of microspheres can be cleaned by ultrafiltration (Amicon or Millipore).
   d. Microspheres can be cleaned by ion exchange (Bio-Rad).

Determination of Microsphere Concentration

This is accomplished by first counting and then adjusting to 10^6 microspheres per 1 ml of buffer.

B.   Noncovalent Coating of Microspheres

Because of its ease, adsorption of the capture reagent onto the microspheres is often used. Additionally, this method of coating causes little conformational change in the protein—although proteins coated in this manner do tend to lose some function that is, a diminution of the ability of an antibody to bind antigen. The coating process is dependent on the pH and the ionic strength of the buffer and on the molecular weight and isoelectric point of the protein.

1.   Using a large-diameter glass container, to 10^6 microspheres in 1 ml of coating buffer add 0.1 ml of the protein to be coated onto the microspheres. The optimal concentration of protein should be determined in dilution experiments. If poor adsorption of the protein occurs with the standard coating buffer, try coating the microspheres in PBS at pH 7.0 or in Tris buffer at pH 8.4 (see Section VI). To determine the optimal protein concentration, coat microspheres with 0.1, 1, 10, and 25 μg protein per 1 ml.

2.   Mix the microspheres and protein for 3 hours at 37°C, then transfer to 4°C for 18 hours with constant rotation. Note that the microspheres can be adequately coated without continual rotation, although the coefficient of variation (CV) of the resulting fluorescence distribution histogram is lower when continual rotation is used.

3.   Transfer the mixture to a plastic centrifuge tube, add PBS–BSA–Tw and centrifuge for 5 minutes at 500 g.

4.   Remove the superanatant and resuspend the pellet in PBS–BSA–Tw. Incubate for 1 hour at 37°C. This step serves to block extra polystyrene sites with BSA; other blocking reagents, such as casein hydrolysate, gelatin and irrelevant serum, can be used.

5.   Centrifuge the microspheres for 5 minutes at 500 g and resuspend the pellet in PBS–BSA–Tw at a concentration of 10^6 microspheres per 1 ml. Note that some microspheres are lost in the washing steps, so readjust the concentration to assure accuracy. For capture reagents that are difficult to adsorb to the microsphere surface, pretreatment of the microspheres with glutaraldehyde or poly-L-lysine (0.025%–1%) followed by the adsorption technique just described may help.

C.   Covalent Coupling of Protein to the Microspheres

This method of protein attachment provides a stable protein–microsphere complex and allows for orientation of proteins (i.e., antibody can be bound with the active binding sites away from the microsphere). Numerous methods exist for coupling to polystyrene, and the optimal method is determined by the nature of the compound to be attached. For proteins, carbodiimide coupling using microspheres with COOH or NH_2 groups on their surface is best. A general coupling protocol follows.

1. Wash the microspheres as for noncovalent coupling.
2. Incubate the microspheres in 0.1% (w/v) N-ethyl-N'-(3-dimethyl-aminopropyl)-carbodiimide for 4 hours at 22°C.
3. Wash the microspheres twice and add protein in MES buffer (Section VI); incubate for 24 hours at 22°C with constant gentle mixing.
4. Wash the microspheres and resuspend in 0.1 M ethanolamine for 1 hour at 22°C.
5. Wash the microspheres twice and resuspend final pellet in PBS–BSA–Tw; incubate for 1 hour at 37°C.
6. Centrifuge and adjust microsphere concentration to 10^6 per 1 ml.

D.   Storage of Coated Microspheres

Once microspheres are coated they can be stored for long periods before use. Note that maximum storage times for optimal performance of coated microspheres should be determined for each coating reagent.

1. Adjust concentration to 10^7 microspheres per 1 ml in storage buffer with glycerol (see Section VI).
2. Mix well and store at −20°C in 0.5–1.0 ml aliquots.
3. Prior to use in assay, warm the vial of coated microspheres at 30°C for 5–10 minutes.
4. Dilute to 15 ml in cold PBS–BSA–Tw and centrifuge for 5–10 minutes at 500 g.
5. Wash microsphere pellet twice in cold PBS–BSA–Tw.
6. Resuspend final pellet in PBS–BSA–Tw at 10^6 microspheres per 1 ml.

E.   Assay Procedures

Detection of Antibody with Antigen-Coated Microspheres

1. To 100 μl (10^5 microspheres) of the antigen-coated microspheres add 100 μl of antibody sample.
2. Incubate for 0.5–2 hours at 37°C.
3. Wash microspheres twice with PBS–BSA–Tw using centrifugation (or filtration) between washes.
4. Resuspend the microsphere pellet from the final wash in 100 μl of fluorescent anti-antibody (if human serum is used as the primary

antibody to the antigen on the microsphere, use fluorescently labeled anti-human Ig).

5. Incubate as in step 2 and wash; resuspend the final microsphere pellet in sheath fluid for flow cytometric (FCM) analysis.

6. For the simultaneous detection of more than one antibody, coat separate-sized microspheres (use microspheres that differ by $\sim 1 \mu m$ in diameter to ensure ability to discriminate the different populations) in separate tubes, wash and block as described previously; then mix the microspheres in equal proportions. Add the sample to the mixed-microsphere population and incubate; follow procedure as outlined before. Note that in analyzing for more than one antibody the optimal sample dilution needs to be determined to avoid prozone or over dilution effects.

7. Once the reaction is complete the labeled microspheres are stable for at least 8 hours before analysis. Samples are best stored at 4°C in the dark (some analytes may be stable for longer periods).

Detection of Antigen with Antibody-Coated Microspheres

Note that the coating determines the amount of functioning antibody on the microsphere (some buffers may partially inactivate the coating material), and it is prudent to evaluate the function of the coated antibody using different coating buffers.

1. To 100 μl of the antibody-coated microspheres, add the antigen-containing sample.

2. Incubate 0.5–2 hours at 37°C. Note that if the antigen concentration is $< 10^{-12}$ M, longer incubation periods may be needed.

3. Wash the microspheres twice and resuspend the final pellet in a second antibody to the antigen. Ensure that these two antibodies, the one on the microsphere surface and the soluble one, do not block one another; usually the use of a polyclonal antibody on the microsphere and a soluble monoclonal or second polyclonal works well.

4. If the soluble antibody is not fluorescently labeled (or biotinylated), a third antibody (or avidin/streptavidin fluorochrome) will be needed. Resuspend the pellet in 100 μl of a fluorescent antibody directed toward the second antibody (i.e., goat antibody to the antigen on the microsphere + soluble antigen + unlabeled mouse monoclonal to the antigen + fluorescent anti-mouse Ig), and incubate for 0.5–2 hours at 37°C.

5. Wash as before and resuspend the final pellet in sheath fluid for analysis.

IV. Results

A. Flow Microsphere Immunoassay Quantification of Analyte

In contrast to latex agglutination (LA), enzyme immunoassay (EIA), immunofluorescence assay (IFA), western blot (WB), and similar methods that are of qualitative or semi quantitative character, the FMIA method, like RIA, is by its nature quantitative. The quantitation of antibody in a sample is best represented in relative terms rather than in weight/volume. This is due to the observation that variations in antibody affinity can affect the amount of antibody bound to the antigen. However, standard samples should be used in each assay to control for variability and as reference points for unknown samples. For quantification of antigen using antibody-coated microspheres, it is possible, using a standard curve, to report results in weight/volume. This assumes the coating antibody is a standard lot used for each run, thereby eliminating the effect of variation in antibody affinity. Results for antigen detection can be compared between laboratories, although the analytical sensitivity (minimum detectable level) may vary because of assay conditions.

The FMIA has been used to quantitate antigen (immune complexes) in micrograms per 1 ml of serum using a standard curve (McHugh *et al.*, 1986, 1988a). Other applications have concentrated on the detection of antibody and have reported results in fluorescence units (MFC). In these applications, antibody levels determined by FMIA have been compared to LA, IFA, EIA, and WB assays. The FMIA has been shown to be more sensitive and to be a quantitative assay of antibody. Important discoveries were the false negative results of CMV, HSV, and HIV antibodies by LA, IFA, and EIA, respectively, which were detected using FMIA (McHugh, *et al.*, 1988b; Scillian *et al.*, 1989). Additionally, the prognostic value of quantitative antibody levels to HIV p24 using FMIA has been documented (Scillian *et al.*, 1989).

B. Sensitivity of the Method

Positively staining microspheres will have higher MFC than negatively staining microspheres. How large a difference in MFC is required to represent a positive signal must be determined for each assay. For the detection of serum antibody, the MFC obtained with samples known to lack the specific antibody should determine the negative range above which a MFC is considered positive. Similarly, for detection of antigen, samples known to lack the antigen will be needed to determine a positive threshold.

Figure 2 shows the fluorescence histogram seen when analyzing a negative (dashed line) and positive (solid line) sample for analyte using a single-size microsphere population. Clear separation is seen between the positive sample and the negative sample. The ordinate is the relative number of events, and the abscissa is increasing fluorescence. In this case the signal-to-noise ratio is approximately $2000/2 = 1000$.

Figure 3 shows the result when microspheres of four different sizes are coated separately, then mixed and added to a sample. The ordinate represents increasing fluorescence and the abscissa forward light scatter (size). The first and third microsphere populations are below the threshold for

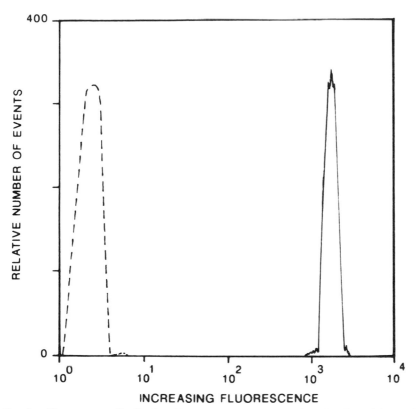

FIG. 2. Fluorescence distribution histogram. Fluorescence (logarithmic scale) on the abscissa and relative number of events (linear scale) on the ordinate. The broken line represents the fluorescence distribution histogram of a single class of antigen-coated microspheres reacted with a sample that does not contain antibody to the antigen, followed by a fluorescent anti-antibody. The solid line is the FMIA fluorescence distribution histogram of microspheres reacted with a serum containing high levels of antibody followed by the fluorescent anti-antibody. The signal/noise ratio is $\sim 2000/2 = 1000$.

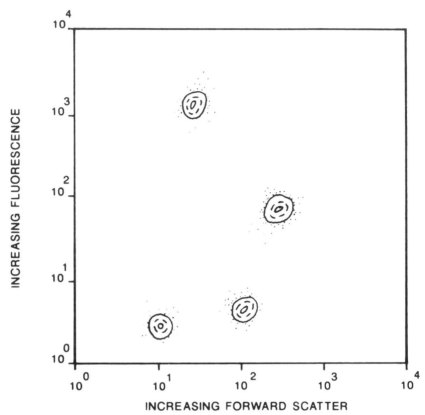

FIG. 3. Forward-angle light scatter and fluorescence distribution histogram. Forward-angle light scatter (size) is shown on the abscissa and fluorescence on the ordinate. The z axis (upward out of the page) represents microsphere frequency producing steep peaks represented by the circular, isocount contour lines. Four microsphere size classes can be seen; the second microsphere class is strongly fluorescent, therefore antibody is present. The fourth and largest microsphere class is also positive but to a lesser degree. The smallest (first) microsphere class and the third microsphere class are both nonreactive (only nonspecific background fluorescence). The slight increase in fluorescence of the third microsphere class over the first is presumably due to increased nonspecific staining.

positivity (MFC = 9). The second microsphere population is strongly fluorescent; the fourth is as well—though less so.

Evaluation of Sensitivity Using Antigen-Coated Microspheres

1. Microspheres (5 μm) were coated with purified human IgG at 10 μg/ml.
2. Serial dilutions of phycoerythrin (PE)–goat, anti-human IgG

(TAGO, Inc., Burlingame, CA) were used to determine relative sensitivity.

3. Shifts in the MFC were compared to the controls (10 μg BSA per 1 ml coated 5 μm microspheres reacted with PE–goat anti-human IgG and 10 μg human IgG per 1 ml coated 5-μm microspheres reacted with preimmune PE–goat IgG). A MFC 3 SD over the control sample MFC was considered significant. Using this positivity threshold we were able to detect a significant signal from a solution of 75 pg of goat anti-human IgG per 1 ml. Note that determination of absolute sensitivity of antibody binding to antigen-coated beads is dependent on the affinity of the antibody and should best be done comparing radiolabeled antibody to the fluorescent antibody both of the same lot. However, this quantitation does not apply to other sources of antibody the affinity of which may be different. Sensitivity measurements of polyclonal antibody-binding assays are best represented in relative terms and not in weight/volume. The sensitivity of the FMIA relative to standard immunoassays has been described in previous studies. In all of the studies to date, the FMIA has been shown to be more sensitive than LA, EIA, WB, and IFA.

Figure 4 shows data comparing the sensitivity of the FMIA to EIA for detecting antibody in a sample. Purified antigen was coated onto the 5-μm microspheres and onto the wells of a 96-well plate. Using PE–goat anti-human IgG for the FMIA and peroxidase–goat anti-human IgG for the EIA and optimal assay conditions, serum samples known to contain the specific antibody were serially diluted from 10^{-2} through 10^{-8}. The results were plotted as FMIA or EIA signal versus sample dilution. The data in Fig. 4 demonstrate that the FMIA was able to detect specific antibody at perhaps 14-fold greater dilution than was EIA able to detect antibody.

Evaluation of FMIA Sensitivity Using Antibody-Coated Microspheres

1. Microspheres (5 μm) were coated with 5 μg/ml of goat anti-human IgG.
2. Serial dilutions of purified human IgG were added, followed by 5 μg PE–goat anti-human IgG per 1 ml.
3. The resulting shift in the peak value of MFC was compared to the MFC value of a control consisting of microspheres coated with preimmune goat antibody with all the subsequent steps and goat antibody to human IgG-coated microspheres without the human IgG.
4. Experiments show that 15 pg/ml of human IgG could be detected using this system; at this concentration a fluorescence signal > 3 SD over the negative control was seen.

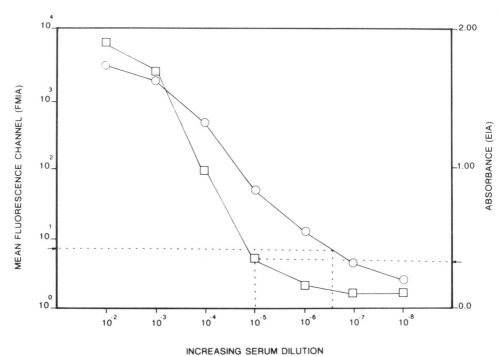

FIG. 4. Sensitivity comparison of FMIA versus EIA. A comparison of the sensitivities of FMIA and EIA for the detection of antibody using microspheres and 96-well plates coated with antigen. (○) Values seen with the FMIA; (□) values obtained with the EIA. The abscissa is serum dilution (log scale) and the ordinate is absorbance (linear scale) for the EIA, and MFC (log scale) for the FMIA. The broken lines begin at the threshold for positivity (marked by arrows) for FMIA or EIA and indicate serum dilution above which the assay was positive. The detection threshold for EIA is a dilution of $\sim 10^{-5}$ and for FMIA $\sim 3 \times 10^{-6}$

Note that sensitivity determinations are dependent on reagent quality and assay parameters. The sensitivity experiments of polyclonal antibody determination used a 2-hour incubation with the PE-conjugated antibody; most assays will not use a fluoresceinated first antibody (i.e., antibody in test solution) but will use a second fluoresceinated antibody that increases the sensitivity. The assay for soluble human IgG used a 2-hour incubation; extension to an overnight incubation could increase the ability to detect lower levels of analyte.

Ultimately, the sensitivity of analyte detection depends on the ability to distinguish a positively staining microsphere from a negatively staining microsphere. Within a given sample, there will be only a limited amount of analyte available for detection, and that available analyte will be distributed uniformly over the coated microspheres added to the sample. Obviously, with a given number of microspheres, the larger the microspheres, the greater the surface area per microsphere available for collecting analyte. We have some evidence that the fluorescence intensity of positively staining microspheres is closely related to microsphere surface area.

Likewise, the smaller the number of microspheres over which the available analyte is distributed, the greater the amount of analyte available per microsphere and the larger the fluorescence signal obtained. We recommend that ≥ 5000 microspheres of each class be analyzed in order to obtain a reliable microsphere population distribution histogram. And we suggest that the capture reagents for the less abundant analytes be placed on the largest microspheres used.

The highest sensitivity obtained to date made use of the mercury arc light source of the Becton Dickinson Research Analyzer, and the highly fluorescent label, PE, in a biotin–streptavidin amplification procedure (McHugh et al., 1988a,b). The mercury arc light was filtered to provide illumination at 546 nm, which maximally excites PE while producing little nonspecific autofluorescence giving an increased fluorescence signal-to-noise ratio. Argon ion laser excitation using the 488- or 514-nm line is less effective in exciting PE but 488-nm excitation of PE is still superior to fluorescein isothiocyanate (FITC).

It should also be noted that the sensitivity of the method may be enhanced by the use of the "fluorescent small microsphere" amplification method of Saunders et al. (1985). Using fluorescent 0.1-μm microspheres, this group has shown an amplification of ~ 10-fold over that obtained with FITC labeling.

C. Dynamic Range

The dynamic range of the test represents the range of analyte concentration that can be assayed using a single dilution of sample. Enzyme immunoassay typically has a 40-fold dynamic range assuming a background or negative sample providing an absorbance value of 0.05 and a maximum positive sample producing an absorbance of 2.0. The dynamic range of FMIA is between 200 and 2000, calculated using a negative-sample MFC of 5 and a positive-sample MFC of 1000 or 10,000. The increase in dynamic range over that of EIA is at least 5-fold and can be as high as 50-fold. The dynamic range possible in RIA can be similar to FMIA, since a greater amount of surface area can be used (coated paper disks) to increase the resulting signal; however, it is unusual for the dynamic range in RIA to be > 100, assuming a background or negative sample with 100 cpm and a maximum positive at 10,000 cpm. Other positive/negative (binary) assays, such as slide IFA and LA, require serial dilution of the sample to obtain a semiquantitative result; thus, dynamic range is not a relevant parameter. With the FMIA one can quantitate antibody or antigen concentration from a single sample dilution. Additionally, the signal in the FMIA is discrete, is time-independent, and is a more accurate measure of analyte concentration than is the continuous, time-dependent enzyme–substrate reaction used in EIA.

D. Potential Applications

The FMIA has been used as an immunoassay but is not restricted to antibody–antigen reactions. In principle, any soluble analyte for which a capture reagent exists can be assayed with the FMIA method. As used here, the term "soluble analyte" refers to antibodies, to antigens, to molecular aggregates such as immune complexes, or to small particles such as viruses or bacteria. The only constraints on the assay are the ability to bind a specific capture reagent to the microsphere and the availability of a fluorescently labeled analyte indicator.

E. Extension of the Method to 10 or More Analytes

The FMIA method has been applied to the simultaneous detection of five analytes. The number of analytes that can be assayed increases to perhaps seven or eight with the judicious selection of microsphere diameter.

The number of microsphere classes usable could be doubled by adding a fluorescent dye to some of the microsphere classes. If we have 5 size classes of uncolored microspheres, into another group of the same 5 size classes a fluorescent dye, which does not interfere with measurement of analyte, can be incorporated, yielding 10 classes of microspheres distinguishable by measurement of size and presence or absence of microsphere-fluorescent dye. Note that the use of a second spectrally separated fluorophore as suggested here requires that the cytometer measure a second fluorescence color.

We have observed that the attachment of capture reagent(s) to the surface of dyed microspheres is the same as for uncolored microspheres.

F.   Limitations

As with any immunoassay, the results are dependent on the quality of the materials used. The principal limitations of the FMIA method are (1) the need for an expensive flow cytometer and (2) the time involved in analyzing these microsphere assays. With automated devices introducing the samples, these microsphere assays could be much more quickly analyzed and would require much less operator time. Additionally, large-scale testing can be time-consuming because of the multiple centrifugation washing steps. The use of filtration for washing the microspheres would be more efficient in time and in thoroughness of the wash. However, to date filtration washing steps when using undiluted serum have been problematic because of plugging of the membranes with serum proteins. Filtration has worked using serum diluted 1 : 40 or using sources of analytes that contain low amounts of protein. Progress will certainly be made in this area and will reduce the amount of time needed to perform these assays. The final limitation remains the calibration and standardization of flow cytometers, which makes it difficult to compare results between laboratories and difficult to define a standard unit for reporting antibody levels (also a problem with other antibody assays). However, using standard curves for antigen testing, results can be reported in weight/unit volume and, if the same reagents are used in different laboratories, results can be compared. Clearly intraassay and interassay (within run and between runs) variability needs to be determined with any new assay. This determination is best accomplished using frozen aliquots of samples that contain either high or low levels of the analyte in question. Analysis of these aliquots over time has indicated the reproducibility of the FMIA.

V.   Conclusion

Flow microsphere immunoassay (FMIA) represents a significant development in immunoassays. It is now possible to measure simultaneously multiple analytes with sensitivity superior to that of current methods. The ability to measure multiple analytes simultaneously is especially advantageous where sample volume is limiting.

Simultaneous measurement also reduces the number of separate assays that need to be performed, reducing reagent expense and the time needed to carry out the assay. The use of FCM to detect the signal is of benefit to those laboratories where FCM is currently used because the application of

FMIA broadens the utility of this instrument. However, because flow cytometers are expensive and complex, laboratories without access to a flow cytometer may find FMIA impractical. Additionally, the assay result is objective as there is no operator interpretation as for slide-based IFA or WB.

Increased sensitivity over current immunoassays has been demonstrated for FMIA. This increase in sensitivity coupled to the ability to measure multiple analytes will propel the use of FMIA into numerous new areas.

VI.   Appendix: Solutions and Suppliers

A.   Solutions

1. PBS–BSA–Tw: for blocking and washing microspheres; phosphate-buffered saline + 0.5% bovine serum albumin + 0.1% Tween 20.
2. Na_2CO_3 buffer: for noncovalent protein attachment; 1.06 g sodium carbonate (Na_2CO_3); add dH_2O to 1 liter (0.01 M); adjust pH to 9.5.
3. Tris buffer: for noncovalent protein attachment; 7.88 g Tris HCl ($C_4H_{12}ClNO_3$), add dH_2O to 1 liter (0.05 M); adjust pH to 8.4.
4. MES buffer: for covalent protein attachment; 4.88 g 2-(N-morpholino)-ethanesulfonic acid ($C_6H_{13}NO_4S$), add dH_2O to 1 liter (0.025 M); adjust pH to 6.0.
5. Storage buffer (adapted from Wilson and Wotherspoon): 50 mM HEPES, 0.01% sodium azide, 5% BSA, 2.5 mM EDTA, 2 mM phenylsulfonyl fluoride, 2 mM iodoacetic acid, and 5 units/ml aprotinin in PBS. Make appropriate volume for use and store at 4°C for up to 1 month; just before use add glycerol (40%, v/v) to the volume required. Do not store with glycerol added.

B.   Microsphere Suppliers

1. Bangs Laboratories
   6 Sue Springs Ct.
   Carmel, Indiana 46032
   (317) 844-7176
2. Duke Scientific
   P.O. Box 50005
   Palo Alto, California 94303
   (800)334-3883
   (415)962-1100

3. Flow Cytometry Standards Corporation
P.O. Box 12621
Research Triangle Park, North Carolina 27709
(919) 967-9345
4. Polysciences
400 Valley Road
Warrington, Pennsylvania 18976-2590
(215)343-6484
5. Seradyn
P.O. Box 1210
1200 Madison Avenue
Indianapolis, Indiana 46206
(800)428-4007
(317)266-2915

REFERENCES

Fulwyler, M. J. (1976). UK patent #1561042.
Horan, P. K., Wheeless, L. L. (1977). *Science* **198,** 149–157.
McHugh, T. M., Stites, D. P., Casavant, C. H., and Fulwyler, M. J. (1986). *J. Immunol. Methods* **95,** 57–61.
McHugh, T. M., Stites, D. P., Busch, M. P., Krowka, J. F., Stricker, R. B., and Hollander, H. (1988a). *J. Inf. Dis.* **158,** 1088–1091.
McHugh, T. M., Miner, R. C., Logan, L. H., and Stites, D. P. (1988b). *J. Clin. Microbiol.* **26,** 1957–1961.
McHugh, T. M., Wang, Y. J., Chong, H. O., Blackwood, L. L., and Stites, D. P. (1989). *J. Immunol. Methods* **116,** 213–219.
Saunders, G. C., Jett, J. H., and Martin, J. C. (1985). *Clin. Chem.* **31,** 2020–2023.
Scillian, J. J., McHugh, T. M., Busch, M. P., Tam, M., Fulwyler, M. J., Chien, D. Y., and Vyas, G. N. (1989). *Blood* **73,** 2041–2048.
Wilson, M. R., and Wotherspoon, J. S. (1988). J. *Immunol. Methods* **107,** 225–30.

Chapter 52

On-Line Flow Cytometry: A Versatile Method for Kinetic Measurements

KEES NOOTER,* HANS HERWEIJER,*† RICHARD JONKER,* AND GER VAN DEN ENGH††

*TNO Institute of Applied Radiobiology and Immunology
2280 HV Rijswijk
The Netherlands

†Daniel den Hoed Cancer Center
Rotterdam
The Netherlands

††Biomedical Sciences Division
Lawrence Livermore National Laboratory
University of California
Livermore, California 94550

I. Introduction

Since the early 1970s, flow cytometry (FCM) has become a highly developed cell analysis technique. Its use is widespread among a great variety of disciplines, and new areas of application are being exploited continuously. The FCM technique is ideally suited for the quantitation of the uptake of virtually any fluorescent substance by cells or cell organelles. With classical FCM, quantitative analysis of the uptake of fluorescent dyes by cells has to be done by examining cell samples at intervals. This method becomes very difficult, if not impossible, to perform when the speed of the kinetic phenomenon of interest is very fast (seconds instead of minutes). Then the most appropriate approach is to analyze a single sample continuously. We here describe an FCM technique that enables the study of real-time, kinetic processes in biological materials in time periods of seconds up to a few hours.

631

The RELACS-III (three-laser Rijswijk experimental light-activated cell sorter) was modified to allow for on-line measurements. The term *on-line* is used here to describe conditions where the cells from an individual sample are passing the excitation light beam continuously for a predefined time, and that all necessary additions of reagents are performed during this uninterrupted measurement. To perform on-line measurements, the flow cytometer must be equipped with an adequate sampling system and a dedicated combination of data acquisition electronics and software. The method of on-line FCM offers especially the great advantage to visualize the very early and rapid phases of kinetic processes. Furthermore, compared to classical kinetic measurements, no intersample variations or irregular delays between time of addition of reagents and time of measurement are of influence.

In this article a technical description will be given of the flow cytometer used to perform real-time kinetic studies, including the software necessary to acquire and analyze data. This is followed by a description of two applications of on-line FCM. First, the real-time analysis of chromatin structure changes by enzymatic digestion of the DNA is shown. Changes in the fluorescence intensity of bound DNA stains occurred within minutes. These experiments therefore demonstrate the applicability of on-line FCM to study fast kinetic changes. Second, an assay is described to measure uptake kinetics of anthracycline drugs. Uptake profiles and changes therein after addition of membrane transport modulators can show the presence of typical multidrug-resistant cells. This assay is much more accurate and sensitive then classical FCM.

II. Instrumentation

A. On-Line Flow Cytometer

Figure 1 is a flow diagram of an on-line flow cytometer. The cells (or particles) of interest are in a reaction tube, which is surrounded by a water jacket that can be maintained at a constant temperature (usually 37°C). The cells are kept in suspension and the added reagents are mixed by a small magnetic stirrer. The stopping of the reaction tube has two inlets and one outlet. Adjustable air pressure on one of the inlets forces the cells to pass the tubing of the outlet toward the flowthrough cuvet of the flow cytometer. At any desired time reagents can be added to the cell suspension by the second inlet, using a syringe. From the very moment of addition, effects of the reagents on the measured parameter(s) can be

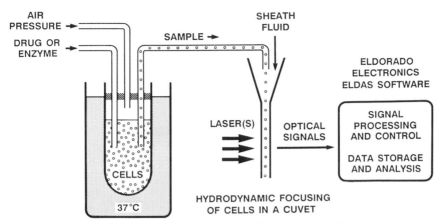

FIG. 1. A schematic representation of an on-line flow cytometer. Cells (or particles) are suspended in a thermostated tube and forced to the measuring point of the flow cytometer by means of air pressure. Additives can be mixed into the suspension at any time via an extra inlet in the reaction tube. Cells are kept in suspension and reagents are mixed by a magnetic stirrer.

studied. The time for cells to travel from the sample tube to the measuring point of the flow cytometer is on the order of 5—20 seconds, depending on the flow rate of the sample and the length of the sample-line tubing.

The RELACS-III flow cytometer is equipped with three lasers: two argon-ion lasers (Coherent Innova 90-5, Palo Alto, CA, and Spectra Physics 2020/5, Mountain View, CA) and one helium–neon laser (Shanghai Institute of Laser Technology, Shanghai PRC), thus enabling all standard excitation wavelengths to be used in single- double-, or triple-wavelength excitation experiments. The flowthrough cuvet used (Hellma, Müllheim, FRG) has a diameter of 250 μm. A maximum of eight parameters per cell can be measured.

B. Electronics

The electronics of the RELACS flow cytometers have two unique features. Neither is an absolute requirement for kinetic studies, but both do increase accuracy and ease of operation. The first of these features is a complete parallel processing system (van den Engh and Stokdijk, 1989). In a multiparameter flow cytometer, the incoming signals are handled most efficiently when each detector is equipped with its own electronics for pulse processing and analog-to-digital conversion (ADC). The input channels can then operate in parallel, and all signals can be processed simultaneously. The several pulses measured from one particle may reach the detectors at different times (multilaser measurements). The RELACS electronics allow for parallel processing of these in time-separated signals by storing the data temporarily in first-in/first-out (FIFO) buffers. After a particle has been seen by all detectors, the stored values are combined and transferred from the FIFO units over a bus to the acquisition computer. The data are transferred via a 16-bit parallel interface, and stored under direct-memory access (DMA) in the computer memory. This allows for an effective transfer of 1 Mbyte of data per second, corresponding to a list-mode file of 128,000 four-parameter events. As a result of the scheme of parallel-pulse processing, the dead time of the system is independent of the number of parameters measured, or the number and time separation of the excitation light beams. The instrument has a cycle time of 5 μsec, which corresponds to a nominal throughput rate of 200,000 events per second.

The second feature of the RELACS electronics is the measurement of general experimental conditions. During data sampling, eight DC values are read periodically (10 times per second) and placed in between the list-mode data. These eight DC channels can contain parameters like experiment time, sample temperature or pH, laser output power, and background fluorescence. The channels can also function as counters, for instance to determine the total number of processed events. For the determination of the kinetics of fluorescence changes, the time marker is used. The time stamps within each block of data at the several time points of a measurement allow for the construction of uptake or release curves of the parameter of interest versus the actual measurement time. Experimental errors, like a temporary clogging, can be filtered out (see also Watson, 1987).

C. Software

The ELDAS (eight-parameter list-mode data analysis system) software package is composed and implemented to allow for acquisition and analysis of data from the RELACS flow cytometers. The ELDAS package consists of several modules capable of handling data-acquisition (ACQUIS); analysis of list-mode data (ANNA); analysis and graphic representation of single-parameter histograms (HISTO), bivariate histograms (BIVAR), and kinetic data (KINET). The program further serves modules for standard operations as data storage and data management. The package is implemented on Hewlett-Packard (HP) 9000 microcomputers (series 200 and 300). Data acquisition is done with HP 9000-220 computers equipped with 15–20 Mbyte hard disks for immediate data storage. Analysis of data is done with a HP 9000-330 computer, equipped with a hard disk, a color monitor, a printer, and a plotter for graphic representation of data. At present, three acquisition and three analysis computers are connected in a network system with hard disk and tape back-up storage. The software is written in HP BASIC, with all calculation-intensive subroutines in HP PASCAL.

1. DATA ACQUISITION

The data acquisition module (ACQUIS) is used to sample data from the RELACS electronic bus and write the data to disk in a list-mode file. For kinetic measurements the program asks in advance for the total duration of the experiment, the number of sample points (blocks) in this time (usually 35), the number of cells to be measured at each sample point (usually 2000), and the distribution of the sample points over the time (linear or logarithmic). Also, some experimental details like the number of parameters, a parameter identification, a description of the sample and the additives, and a description of the experimental parameters used (like time and sample temperature) are added as a text file to the measurement data.

After alignment of the flow cytometer, and adjustment of the amplification for all parameters, a kinetic measurement starts by sampling the first block of (generally) 2000 cells. This data set is used in the analysis to set windows, and is a reference because no changes have occurred yet. Thereafter, a reagent is injected through the inlet. This can be the fluorescent dye of interest, or some substrate that influences an already present fluorochrome. Upon addition, the kinetic acquisition program is activated. The program then immediately starts sampling data blocks according to the predefined specifications. It is also possible to set up sampling profiles for multiple addition of subtrates, each time with its own distribution of the sample points over the measurement time period. The RELACS electronics put the eight experimental (DC) parameters on the bus 10 times per second. These are acquired by the computer like normal parameters, and included in the list-mode file. All parameters carry an identification code in their 16-bit data: 4-bit identification + 12-bit parameter value = 16-bit data. With these identification codes, DC parameters can be recognized apart from "normal" parameters.

Besides these basic acquisition features, the program also continuously displays the measured data. This allows for immediate interpretation of the experiment, and even more importantly, this serves as a quality control monitor. The computer displays three dot plots and two histograms (Fig. 2). The parameters of the dot plots and the histogram can be set or changed at any time (except during storage of data, which has always the highest priority). Usually, histogram 1 is used to display the parameter of

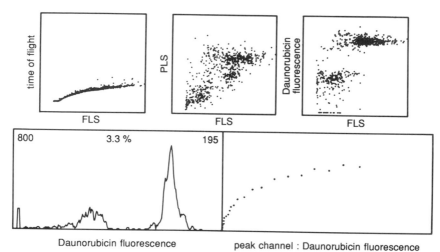

Notebook: HH 881208   Page: 3   Kinetic experiment of 35 blocks

Waiting for sampling time nr. 28   Time before sampling: 41 seconds

Sampling ready in: 1180 seconds

FIG. 2.   Data display of the acquisition program while running a kinetic experiment. The displayed parameters of the three dot plots (top) and two histograms (bottom) can be freely chosen. Usually, histogram 1 displays the data of the kinetic parameter. The peak position of this histogram is plotted in histogram 2 at each time point, thus creating a provisional kinetic curve. The total number of cells in the histogram (800), the CV (3.3%), and the peak channel of the histogram (195) are plotted at the top of the histogram 1 panel.

interest; the peak channel of this histogram at each sample point can be plotted in histogram 2 (see Fig. 2). This already shows the kinetic plot during the measurement, although it is rather inaccurate (no windowing or averaging of data). The dot plots are refreshed after a preset number of events, the histograms after each sample point.

2. DATA ANALYSIS

The analysis of the stored list-mode data is performed with the program module ANNA. Like all list-mode data analysis software, windows can be set on all parameters (even on the kinetic parameter if wanted). To obtain kinetic curves, the data are reduced to a set of values versus the time (sample points). That is, for each data block the cells in window are selected and the values of the kinetic parameter are reduced to one value. The modal, mean, and median fluorescence can be computed together with the coefficient of variation (CV). During these calculations, logarithmically amplified parameter values can be recalculated to linear values or vice versa, based on the known amplifier characteristics. The data set of one value per time point thus obtained is stored in a new file, which can be used in the KINET module. It is also possible to store a complete histogram of the events in window per time point. The first time marker in a data block is taken as the measurement time of the time point. Excessively long sampling times for a particular point (as will be the result of a clogging of the sample line) can be noticed, since the duration of sampling (= sample rate) is calculated using all time markers in a data block. These time points may be excluded from further analysis (Watson, 1987). The other stored experimental parameters can also be used to control the quality of the measurements.

Construction of kinetic curves is performed with the KINET program module. This allows for plotting (overlay) of data in curves of the fluorescence (or some other kinetic parameter) versus the measurement time (see Fig. 3 for an example). The program allows for scaling of both axes, plotting of several curves in one graph (see Fig. 5), and plotting of histograms of selected time points (see Fig. 6). Fitting of kinetic curves has not been implemented yet. The data structure, however, is well suited for these calculations.

D. Standardization

We calibrate our flow cytometer with fluorecent microspheres (Fluoresbrite, Polysciences, Inc. Warrington, PA). Fixed laser powers are used in all experiments (set in light control modes). Within one type of experiment the same photomultipliers (PMT), filters, and (pre)amplifiers are used for each parameter each time. Using fixed amplifier settings, the PMT voltages are adjusted to measure the mean peak of the particle signals in the same channel number. This way, measurements performed on different days can be compared. The CV of the fluorescence histogram is calculated (see Fig. 2, bottom left, histogram 1), and is used as a quality control for the alignment of the flow cytometer.

III. Applications

A. Real-Time Analysis of Chromatin Structure Changes

DNA *in situ* is highly organized with nuclear proteins bound to it. The local structure of this chromatin is a function of gene activity, and differentiation state of the cell. Chromatin structure can be investigated by deoxyribonuclease I (DNase I) digestion of the DNA. The sensitivity of DNA for DNase I appears to depend on the chromatin structure, and is increased by damage caused by ionizing radiation. To probe for DNase I sensitivity, the DNA can be first stained to equilibrium with DNA-specific fluorochromes. Changes in DNA stainability are then measured during the enzymatic digestion (Roti Roti *et al.*, 1985).

We have attempted to find a relationship between the different binding modes of several DNA-binding dyes and changes in the fluorescence of these dyes during digestion of the DNA with DNase I. Darzynkiewicz (1979) showed that digestion of DNA can be determined by monitoring the fluorescence of the DNA-binding dye acridine orange (AO). The presence of AO had no effect on the digestion of the DNA by DNase I. Therefore, the measured differences in fluorescence were mainly attributed to DNase I activity and not to unwinding of the two DNA strands by the fluorochrome. Roti Roti *et al.* (1985) showed that at the time of DNase I addition to the cells there already was an increase in fluorescence. We attributed this to a delay between sampling and measuring of the nuclei. Therefore, we have chosen to perform the study of chromatin structural changes with on-line FCM, allowing for real-time measurements.

Fluorochromes with different binding modes to the DNA have been tested. Acridine orange (AO) has two binding modes. It emits green fluorescent light upon intercalation in the double helix. This requires the stacking of several dye molecules together (Kapuscinski *et al.*, 1983). Acridine orange emits red light upon external binding to single- and double-stranded nuclei acids. Propidium iodide (PI) and ethidium bromide (EB) intercalate in the DNA. Hoechst 33258 (HO258) and Hoechst 33342 (HO342) bind externally, inside the minor groove where they generally are hydrogen-bonded to four adjacent AT base pairs (Teng *et al.*, 1988).

Mouse thymocytes were used in this study because these represent a population with uniform DNA content and chromatin structure. Thymocytes were fixed by use of formaldehyde and ethanol. RNA was removed with RNase A and cells were stained with one of the fluorochromes. Six million cells in 3 ml DNase I buffer were measured with the RELACS-III flow cytometer. The sample rate was 500 cells per second and the temperature of the reaction mix was kept at 37°C. Besides dye fluorescence(s), forward light scattering and time-of-flight (TOF; scatter pulsewidth) were also measured. The TOF parameter was used in the data analysis to remove doublets of thymocytes. DNase I (10 μg/ml = 720 units/ml final concentration) was added to the reaction mix at time zero.

In Fig. 3 the kinetics of changes in AO fluorescences are shown. At each time point the mean red and green fluorescence was computed and plotted versus the time after addition of DNase I. Within 2 minutes, the red

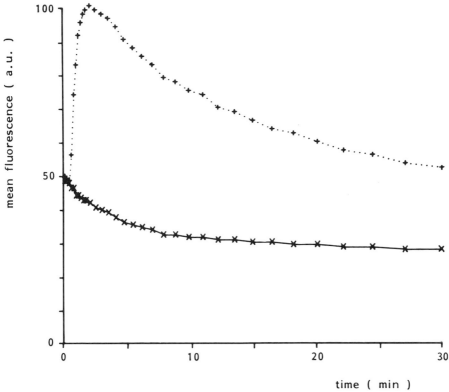

FIG. 3. DNase I digestion of mouse thymocytes: mean fluorescence of AO-stained (7.5 μg/ml) thymocytes as a function of time after addition of 10 μg/ml DNase I. The laser was set to emit 488-nm light (500 mW). Red fluorescence ($+ \ldots +$) was filtered through two RG610 long-pass filters (Schott). Green fluorescence ($\times$——$\times$) was filtered through a 530 ± 10 nm bandpass filter (Melles Griot).

fluorescence (external binding) of the thymocytes increased >2-fold, whereas the green fluorescence (intercalative binding) decreased at the same time. After 2 minutes the red fluorescence also decreased in time. These steadily decreasing fluorescence signals are caused by the breakdown of the DNA by prolonged DNase I activity.

Fluorescence from PI and EB showed the same sudden increase as the red AO fluorescence, although the effect is somewhat smaller (20% increase). In Fig. 4, PI fluorescence histograms of DNase I-treated thymocytes are plotted versus the measuring time. At the maximum mean fluorescence intensity, discrete subpopulations of cells with original and with increased fluorescence were measured, indicating that the process of

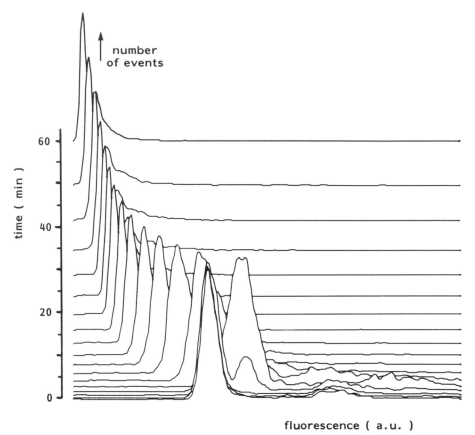

FIG. 4. DNase I digestion of mouse thymocytes: PI fluorescence histograms as function of incubation time after addition of 25 μg/ml DNase I. Thymocytes were stained with 10 μg/ml PI. The laser was set to emit 488-nm light (500 mW). PI fluorescence was filtered through a KV550 long-pass filter (Schott). The heights of the peaks indicate the number of events.

DNase I digestion is fast and complete within one nucleus. When HO258 or HO342 were used as the DNA stain, DNase I digestions showed a directly decreasing fluorescence intensity (data not shown).

The fluorescence maximum proved to be independent of the DNase I concentration. The time necessary to reach the maximum fluorescence level was longer with lower DNase I concentrations. Concentrations of as low as 0.1 μg/ml DNase I gave a 20% increase in PI fluorescence after 15 minutes. After washing away the histones with 0.1 N HCl, the initial PI fluorescence was 40% higher. When these pretreated cells were digested with DNase I, there was still an additional increase (10%), indicating that HCl only washes away proteins, leaving most of the DNA intact.

DNase I produces single-strand breaks at potentially 10-intervals (Sollner-Webb et al., 1978). These so-called nicks allow for unwinding of the DNA, thus enabling more fluorochromes to intercalate into the DNA. We attribute the increase in fluorescence of the intercalating dyes (PI, EB) to this effect. The denaturation of DNA decreases the number of external binding sites for the Hoechst dyes HO342 and HO258. These dyes require an intact minor groove for binding. The unwinding of the DNA also decreases the possibility of the stacking of AO molecules in the DNA helix, resulting in a decreased green fluorescence. The increase in red AO fluorescence, caused by external binding to single- and double-stranded nucleic acids, must be attributed to the unwinding of the DNA helix and the increased single-strandedness, resulting in more accessible external binding sites for AO.

Roti Roti et al. (1985) showed that the increase in EB fluorescence can also be obtained after large amounts of irradiation. Since chromatin structure and accessibility of DNA is very important in the distribution of radiation damage (Chiu et al., 1982), this measuring system can give further insight into the relation between chromatin structure and radiation sensitivity.

B. Detection of Multidrug-Resistant Cells

Mammalian cell lines which have been selected *in vitro* for resistance to cytotoxic drugs like the *Vinca* alkaloids or the anthracycline antibiotics display the so-called multidrug-resistant (MDR) phenotype (for a review on MDR, see Bradley et al., 1988). The functional change in MDR cells is a defect in drug accumulation due to activity of an energy-dependent rapid drug efflux pump. There is accumulating evidence that MDR can also occur in human cancer (Fojo et al., 1987; Goldstein et al., 1989; Nooter et al., 1990; Herweijer et al., in press). This is potentially of therapeutic importance, since a variety of substances have been identified that can overcome MDR in *in vitro* model systems (Tsuruo et al., 1982; Nooter et al., 1989). The current hypothesis about reversing MDR for some of these agents (e.g. verapamil) is that they have an affinity for the intracellular drug efflux pump, and thus can compete for outward transport and thereby restore the cellular cytotoxic drug accumulation. For that reason the MDR-reversing agents are here referred to as membrane transport-modulating agents (MTMA).

Since in model systems MTMA have also been able to overcome MDR *in vivo*, agents are available now with which clinical trials can be started in typical MDR patients. Therefore, it is of great importance to determine the occurrence of MDR in human cancer. Thus, accurate, sensitive, and rapid assays for the detection of MDR cells are highly desirable. We here describe an FCM assay for the detection of MDR cells in heterogeneous cell populations that is based on the ability of MDR cells to respond to MTMA (Herweijer *et al.*, 1989).

The human ovarian carcinoma A2780 anthracycline-sensitive (2780/S) and the anthracycline-resistant 2780AD (2780/R) cells were used as a model system for MDR. Daunorubicin uptake kinetics have been determined for both types of cells. The daunorubicin content of individual cells can be measured by FCM. The method makes use of the fluorescent properties of the anthracycline drugs (Nooter *et al.*, 1983; see Krishan, Chapter 43; this volume). Upon excitation with 488-nm laser light (close to the absorption maximum), anthracyclines fluoresce, with an emission maximum around 600 nm.

For the kinetic measurements, the cells were suspended in HEPES-buffered Hanks' balanced salt solution (HHBSS) at a final concentration of 2×10^5 cells/ml. Uptake kinetics of daunorubicin (2 μM final concentration, added in 2–8 ml of cell suspension) were determined over a time period of 90 minutes. Besides anthracycline fluorescence, forward and perpendicular light scattering were also measured. Daunorubicin fluorescence was filtered through a KV550 long-pass filter (Schott, Mainz, FRG) and measured with a S20-type PMT (Thorn EMI, Fairfield, NJ). The signal was logarithmically amplified (4 decades full scale). Scattering light was filtered through 488-nm bandpass filters (Melles Griot, Irvine, CA) and linearly amplified.

Daunorubicin net-uptake curves of 2780/S and 2780/R cells are plotted in Fig. 5. These curves were computed from the list-mode data by recalculating the logarithmically measured value to the corresponding linear value (which is possible because the amplification characteristics of the amplifiers are known) and, subsequently, averaging all values at each time point. It is clear that steady-state kinetics are reached after ~60–90 minutes for the 2780/S cells and 30 minutes for the 2780/R cells. The sensitive cells take up 5-fold more daunorubicin compared to the resistant cells. The typical MDR phenotype of the 2780/R cells was demonstrated by on-line addition of a MTMA to these cells after 30 minutes of daunorubicin accumulation. Figure 5 shows the effects of the addition of different (final) concentrations of cyclosporin A to the cells. Almost immediately after addition an increase in daunorubicin accumulation could be measured. The effect of cyclosporin A is concentration-dependent. Full restoration of daunorubicin uptake could not be obtained, even after addition of 10 μM cyclosporin

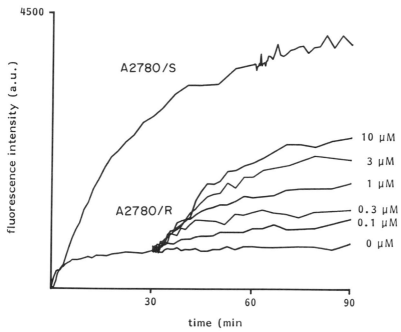

FIG. 5.   Daunorubicin fluorescence intensity (in arbitrary units, a.u.) of 2780/S and 2780/R cells versus the drug exposure time. At time zero, daunorubicin (2 μM final concentration) was added to the cells, which were incubated at 37°C. After 30 minutes, cyclosporin A was added to the 2780/R cells. Daunorubicin net uptake was measured for another 60 minutes thereafter. Several samples were measured, to which different concentrations of cyclosporin A were added. After 60 minutes, HHBSS was added as control to the 2780/S cells.

A. Higher concentrations of cyclosporin A led to acute cell death. Addition of this MTMA to the sensitive 2780/S cells did not result in changes in the daunorubicin accumulation.

To obtain an indication of the sensitivity of the on-line FCM assay for the detection of MDR cells, mixing experiments were performed. Sensitive A2780/S and resistant A2780/R cells were mixed in different ratios, and subsequently the daunorubicin uptake kinetics and modulation by cyclosporin A were measured. Plotting of the fluorescence histograms measured at several time points after addition of daunorubicin allowed for the detection of small subpopulations of MDR cells. As few as 2.5% MDR cells could readily be detected in the mixture of 2780/S and 2780/R cells (Fig. 6).

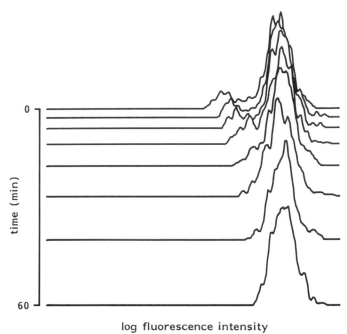

log fluorescence intensity

FIG. 6. Histograms of daunorubicin fluorescence intensity (logarithmic scale) of a mixture of 2780/S and 2780/R cells (24 : 1) measured at several drug exposure times, plotted in front of each other. Cells were incubated with daunorubicin (2 μM) at 37°C for 60 minutes, prior to addition of 10 μM cyclosporin A at time zero. Daunorubicin accumulation was measured for another 60 minutes.

ACKNOWLEDGMENTS

Funds for the development of the RELACS electronics and ELDAS software were provided by the Dutch research organizations TNO and ZWO. The multidrug-resistance research program is supported by the Dutch Cancer Society, grant number RRTI 88-8.

REFERENCES

Bradley, G., Juranka, P. F., and Ling, V. (1988). *Biochim. Biophys. Acta* **948,** 87–128.
Chiu, S. M., Oleinick, N. L., Friedman, L. R., and Stambrook, P. J. (1982). *Biochim. Biophys. Acta* **699,** 15–21.
Darzynkiewicz, Z. (1979). *In* Flow Cytometry and Sorting'" (M. R. Melamed, P. F. Mullaney, and M. L. Mendelsohn, eds.), pp. 285–316. Wiley, New York.
Fojo, A. T., Ueda, K., Slamon, D. J., Poplack, D. G., Gottesman, M .M., and Pastan, I. (1987). *Proc. Natl. Acad. Sci. U.S.A.* **84,** 265–269.
Goldstein, L. J., Galski, H., Fojo, A. T., Willingham, M., Lai, S. L., Gazdar, A., Pirker, R., Gottesman, M. M., Pastan I. (1989). *J. Natl. Cancer Inst.* **81,** 116–124.

Herweijer, H., van den Engh, G. J., and Nooter, K. (1989). *Cytometry* **10**, 463–468.
Herweijer, H., Sonneveld, P., Baas, F., Nooter, K. (1990). *J. Natl. Cancer Inst.*, in press.
Kapuscinski, J., Darzynkiewicz, Z., and Melamed, M. R. (1983). *Biochem. Pharmacol.* **32**, 3679–3694.
Nooter, K., van den Engh, G. J., and Sonneveld, P. (1983). *Cancer Res.* **43**, 5126–5130.
Nooter, K., Oostrum, R., Jonker, R. R., van Dekken, H., Stokdijk, W., and van den Engh, G. J. (1989). *Cancer Chemother. Pharmacol.* **23**, 296–300.
Nooter, K., Sonneveld, P., Oostrum, R., Herweijer, H., Hagenbeek, A., and Valerio, D. (1990). *Int. J. Cancer* **45**, 263–268.
Roti Roti, J. L., Wright, W. D., Higashikubo, R., and Dethlefsen, L. A. (1985). *Cytometry* **6**, 101–108.
Sollner-Webb, B., Melchior, W., and Felsenfeld, G. (1978). *Cell* **14**, 611–627.
Teng, M., Usman, N., Frederick, C. A., and Wang, A. H. J. (1988). *Nucleic Acids Res.* **16**, 2671–2690.
Tsuruo, T., Iida, H., Tsukagushi, S., and Sakurai, Y. (1982). *Cancer Res.* **42**, 4730–4733.
van den Engh, G. J., and Stokdijk, W. (1989). *Cytometry* **10**, 282–293.
Watson, J. V. (1987). *Cytometry* **8**, 646–649.

Chapter 53

Calibration of Flow Cytometer Detector Systems

RALPH E. DURAND

Medical Biophysics Unit
B. C. Cancer Research Centre
Vancouver, British Columbia V5Z 1L3
Canada

I. Introduction

Flow cytometers can precisely measure fluorescence intensity, but no standard method for quantitative expression of results has yet been adopted. This is largely due to the dependence of these data on instrument alignment, laser power, optical filters, photomultiplier tube (PMT) sensitivity, and display mode (linear or logarithmic) and resolution. With so many factors potentially contributing to the values assigned to any sample intensity (even on a single instrument), it is often most convenient to intercompare experimental samples simply by choosing appropriate machine settings, and collecting a frequency histogram of fluorescence intensity for each sample without changing the measurement conditions. This approach unquestionably produces data with the greatest possible resolution when small differences in sample intensity are expected. However, it suffers from the limitation that only a restricted range of fluorescence intensities (dynamic range) can be observed. In the case of linear amplification, the resolution and dynamic range are dependent on the number of channels selected for analog-to-digital signal conversion (ADC; 2^7 to 2^{10} resolution, corresponding to 128–1024 channels); for logarithmic amplification, the dynamic range is limited by that of the amplifier itself, and overall resolution is decreased.

METHODS IN CELL BIOLOGY, VOL. 33

Since the output signal from all PMTs is a highly predictable and reproducible function of tube voltage over the usual operating range, we have
found it convenient to "calibrate" PMT response as a function of voltage
and wavelength, and to use this information as the basis for a normalization procedure that permits calculation of relative fluorescence intensity
over a wide range of PMT voltages and amplifier gains (Durand, 1981).
Accurate determination of relative fluorescence is then possible over a
wider dynamic range than that for any log amplifier currently available;
however, in view of the inherent convenience of acquiring data with
logarithmic amplification, we have also briefly described a method for
conversion of logarithmic data to linear format. Using this procedure, data
acquired in either the log or linear format can be reduced to the *same*
values for relative fluorescence intensity.

II.   Materials

The PMT calibration procedure subsequently described is not specific to
any particular instrument type, nor are any specialized materials necessarily required. A highly useful accessory, however, is a set of neutral density
(ND) filters (e.g., 0.3, 0.6, 1.0, 2.0, and 3.0) to decrease signal intensity by
known increments. Oriel Corporation (Stamford, CT) sells 1-in.-diameter
filters and provides wavelength-specific calibration curves for each filter.

A.   Linear Amplifier Calibration

To check the absolute (as opposed to nominal, that is, that listed by the
manufacturer) gain of the amplifiers of Coulter or Becton Dickinson cytometers, availability of a precision signal generator is desirable. Use of the
"built-in" signal generator on either instrument is convenient; alternately,
one can also use appropriate fluorescent particles (microspheres, or cells
with high homogeneity).

B.   Photomultiplier Tube Calibration

All PMTs have a response that is highly dependent on the wavelength of
the detected light. As a consequence, calibration for a particular wavelength can be accomplished with "test" signals (e.g., a light-emitting diode,
LED). Since LED and fluorochrome emissions seldom overlap precisely,
we have found it to be preferable to use appropriately stained cells or

fluorescent microspheres, with the same optical filters that would be used under typical working conditions. A separate "calibration" is necessary for each fluorochrome–filter combination.

III. Procedure

Two separate steps are required to complete the basic calibration procedure for Coulter and Becton-Dickinson instruments, and a third step provides calibration of their log amplifiers. Although it is not essential that this suggested sequence be followed, it is suited to laboratories that do not have sophisticated electronic test instruments (oscilloscopes and signal generators) available.

A. Linear Amplifier Calibration

Signal amplification is handled somewhat differently in Ortho, Coulter, and Becton Dickinson instruments. Ortho instruments rely only on PMT voltage; in the EPICS and FACS series, amplifiers nominally adjustable over at least a 32-fold range are provided. However, individual amplifiers (even those for the different channels of a single instrument) may provide actual gains that differ from the nominal gain by as much as 20% (Durand, 1981). Depending on the ultimate accuracy desired in determining the relative fluorescence of experimental samples, the individual investigator must determine whether precise calibration of each amplifier is of benefit.

As indicated in Section II, availability of an electronic signal generator results in maximal precision, but is not essential. Regardless of the signal source, the recommended procedure is as follows:

1. Choose a signal that, when the instrument is operated at maximum gain, produces a mean (mean $= \Sigma n_i x_i / \Sigma n_i$, where n_i is the number of cells in channel x_i) or modal (peak) signal near the maximum channel number of the frequency histogram. Since all events recorded contribute to determination of the mean, good reproducibility can be achieved with less data accumulated when using means rather than modes.

2. For the same signal, record the mean or mode signal position for each reproducible amplifier setting (for amplifiers with continuously adjustable gains, the minimum and maximum of the range are likely to be the only easily reproduced settings).

3. The actual amplification at each setting in steps 1 and 2 can be calculated based on the ratio of distribution means or modes (e.g., means of 161, 81.4, 40.6, 21.2, and 10.0 for amplifier gains of 16, 8, 4, 2, and 1, translate to actual gains (G) of 161/10 or 16.1, 8.14, 4.06, 2.12, and 1). Repetitive measurements with several (slightly) different initial signal levels are recommended; the reproducibility of each calculated gain should be within 1%.

B. Photomultiplier Calibration

Since each PMT is critically sensitive to the wavelength of the light detected, it is recommended that this calibration be undertaken with each fluorochrome and filter combination that will ultimately be employed. Also, bandpass filters are recommended to increase further the accuracy of this procedure. The use of (wavelength-calibrated) ND filters greatly simplifies the following:

1. Select a brightly stained sample, and determine its mean or modal fluorescence (as previously described), using the minimum PMT voltage that produces an easily analyzed frequency distribution with a near-minimum intensity. If a complete set of ND filters is available, any convenient amplification setting can be used; in the absence of ND filters, start with the maximum amplifier gain available.

2. Increase the PMT voltage by a convenient increment (e.g., 50 V), and determine the new mean or mode. This will be the second point on a (log–log) plot of signal mean (or mode) versus PMT voltage.

3. Increase the PMT voltage by another increment, and again determine and plot the mean. In the event that a voltage increment produces an overflow condition on the histogram (intensities greater than full scale), three options are available:

    a. Insert an ND filter in front of the PMT. The calculated mean is then the product of the observed mean and the attenuation factor of the ND filter at the wavelength being used.

    b. Decrease the amplifier gain. The calculated mean is then the product of the observed mean and the ratio of the initial/present amplification (both are accurately known from Section III,A).

    c. In the event that further reductions of amplifier gain are impossible, or that additional ND filters are unavailable, it is possible to substitute a less intensely stained sample at this point. Choose a sample that will give reasonable histograms at two or more amplifier/voltage settings used with the initial sample; this allows normalization of new measurements to those previously taken, and thus expands the calibration range.

4. The appropriate combinations of step 3 and option a, b, or c should be repeated until data have been acquired over the useable voltage range of the PMT (generally 300–1500 V). The log–log plot of signal intensity (calculated means or modes) versus PMT voltage should be linear over most of the range of voltages chosen, indicating that PMT output is proportional to voltage raised to the power indicated by the slope (k) of the line.

Typical data for one of the PMTs of a FACS 440 are shown in Fig. 1. Note that the power function is maintained over at least 6 decades (i.e., over the range indicated by the filled circles).

C.   Derivation of "Calibration Factors"

The procedure described in Section III,A is a method of determining the actual (linear) amplifier gain (G), and in Section III,B, the PMT gain coefficient (k) was determined. It then follows that the relative intensity RI of signals accumulated in channel I is given by:

$$RI = (AI/G)(V_r/V)^k \qquad (1)$$

where V_r is a PMT "reference" voltage, and A is a normalization constant. We have found it convenient to choose a PMT reference voltage near that

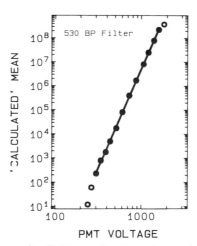

FIG. 1.   Typical response of a PMT as a function of operating voltage. Results were obtained using fluorescent microspheres and ND filters as described in the text, with the calculated mean being the product of observed mean channel number and attenuation factor for the ND combinations used at each experimental point. The response is adequately described by a power function over the range indicated by the closed circles.

normally used for the PMT (e.g., 600 V); this is, however, a completely arbitrary choice, as is the value assigned to A. It is often convenient to choose a value of A that results in unstained cells having a relative intensity of 1.0. Alternatively, if direct intercomparison of results from several instruments is desired, a separate set of calibration constants can be derived for each instrument, so that the fluorescence intensity of a (commercially available) fluorescent microsphere preparation has a defined value.

It should also be noted that this expression for RI can be applied channel by channel as written in the equation, or can be used to normalize previously determined mean or modal fluorescences by simply substituting those values for I in the Eq. (1).

D. Conversion of Logarithmic Data

The advantages of visualizing data having a large dynamic range by using logarithmic amplification have been reviewed by Muirhead *et al.* (1983) and are evident elsewhere in this volume. Mathematical description of such data is, however, accomplished most easily by performing a logarithmic-to-linear transform, and this can be done in a manner consistent with the normalization conditions derived here. Our recommended procedure for this conversion follows:

1. Select a PMT voltage and log amplifier gain that provide a near-maximum mean or modal signal. Record the (log) channel number of the distribution mean (or mode), then determine the RI of the sample using suitable voltage and linear amplifier settings (and the constants previously determined).
2. Repeat step 1 (with log data acquired at the same PMT voltage and amplifier gain as for the initial determination) for several ND filter combinations that reduce the signal intensity (or, alternatively, with cells or microspheres of different intensities).

Data collected in this manner can be plotted directly as in Fig. 2, or if linear conversion without normalization is adequate, the (log) distribution mean or mode can be plotted as a function of relative signal intensity (calculated according to signal attenuation by ND filters). Data obtained using a logarithmic amplifier of a FACS 440 show a dynamic range as large as 5 decades (Fig. 2). Since RI and channel number (I) are obviously related for any curve of Fig. 2 by the expression $RI = a\, e^{bI}$, it follows that the logarithmic data can be converted channel by channel to relative (linear) intensities, and the (weighted) mean fluorescence or other statistical parameters can then be easily calculated.

Data acquired at different PMT voltages produce parallel curves for any

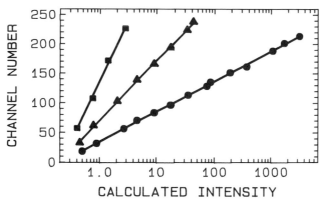

FIG. 2. Typical response of a FACS 440 log amplifier, as a function of the relative fluorescence intensity calculated with linear amplification and the procedures described in the text. Data were accumulated using fluorescent microspheres, with signal intensity varied by interposing appropriate combinations of ND filters. The instrument "multiplier gain" was held at 0.5, with "base gains" as shown, and PMT voltage at 600 V. Note that all curves extrapolate to a common intercept, and that the relative slopes of the curves were essentially equal to the nominal gains shown. Gain 1 (●); gain 2 (▲); gain 4 (■).

amplifier gain, and are displaced by the same factor $(V_r/V)^k$ previously determined for PMT dependence on operating voltage (at a particular wavelength). Similarly, the slopes of the curves in Fig. 2 differ by exactly the same amplifier gain factor (G) previously determined for linear amplification. Thus, it follows that the general expression:

$$RI = a(V_r/V)^k e^{bI/g}$$

will convert log data acquired in channel I at any amplifier (relative) gain (g) and PMT voltage V (note that g is the gain relative to that used to define the "standard" curve, with slope b, at voltage V_r). After performing this channel-by-channel conversion, the mean of the linearized distribution can easily be calculated.

IV. Comments

The PMT calibration procedure requires ≤ 30 minutes for each filter–fluorochrome combination; we have found the procedures to be sufficiently useful that we now express all our quantitative measurements only in terms of relative intensities. To facilitate these calculations further, it is useful to monitor PMT voltages and amplifier settings electronically in a manner that permits automated recording and use of these values by software packages.

It should be obvious that the procedures just described allow standardization of fluorescence measurements, but do not substitute for the usual techniques of instrument alignment, or for the necessity of internal controls in high-precision studies. These procedures do, however, permit quantitation of fluorescence intensity (relative to an arbitrary standard) over the entire useable range of the detectors, without ADC or amplifier limitations.

ACKNOWLEDGMENTS

This work was supported by grant CA37775 from the U.S. Public Health Service.

REFERENCES

Durand, R. E. (1981). *Cytometry* **2**, 192–193.
Muirhead, K. A., Schmitt, T. C., and Muirhead A. R. (1983). *Cytometry* **3**, 251–256.

Chapter 54

Spectral Properties of Fluorochromes Used in Flow Cytometry

JAN KAPUSCINSKI AND ZBIGNIEW DARZYNKIEWICZ

Cancer Research Institute of the
New York Medical College at Valhalla
Valhalla, New York 10595

I. Introduction

Proper selection of the excitation wavelength as well as the choice of emission filters and dichroic mirrors of flow cytometers require the knowledge of spectral properties of the fluorochrome to be used. This information is generally provided by the authors describing individual methods or probes. Thus, when needed, the reader can search for these data by going through the original publications. Nearly all methods described in this volume contain information regarding excitation and emission properties of the dyes described in these chapters.

Often, however, as in designing new staining reactions or selecting combinations of multiple dyes for different cell constituents in a single staining mixture, there is a need for rapid review of spectral properties of numerous probes. Several such reviews have been published (Latt, 1979; Steinkamp, 1984; Crissman and Steinkamp, 1987; Shapiro, 1988; Waggoner et al., 1989; Tsien, 1989; Arndt-Jovin and Jovin, 1989; Haugland, 1989). Among them, the most comprehensive is the catalog of fluorochromes compiled by Haugland (Haugland, 1989). The latter source, describing chemical properties, affinities, and spectral characteristics of numerous dyes as well as providing extensive literature, is a specially valuable contribution for the researchers searching for new dyes for flow cytometry (FCM).

The scope of this chapter does not allow us to present the spectral properties of what is presently an excessively large number of fluorochromes.

655

Fortunately, however, most of the reagents that may have application in FCM are derivatives of a smaller group of dyes such as fluorescein, rhodamine, coumarin, acridine, xanthene, or thiazine. Many groups of reagents thus share similar spectral properties with these dyes.

Chemical modifications of these few, common fluorochromes to obtain a multitude of useful probes serve three main objectives. The first is to alter spectral characteristics of the dye so that the probe can better match the instrumental or analytical requirements (e.g., optimal excitation wavelength, efficient fluorescence detection, detection of the resonance energy transfer). This is achieved by modification of the fluorophore via addition of groups that induce hipsochromic or bathochromic shifts.

The second objective is to introduce a reactive chemical group capable of forming a permanent (covalent) bond with the target molecule(s). This modification is most successful in designing probes toward proteins. By addition of a haloacetyl- or maleimide group, the fluorochrome gains the ability to bind sulfhydryl groups in most proteins. The amino groups are targeted by introduction of an isothiocyanate substituent into the probe.

The third objective of the modification is to alter the permeability of the probe. Generally, charged probes poorly cross the cell membrane. Neutralization of the probe's charge by its reduction or by addition of a protective group (e.g., acetylation or acetoxymethylation) enhances its ability to penetrate the membrane. Once inside the cell, the protective group is removed by esterases or is oxidized, which results in charge restoration and entrapment of the probe.

Chemical modifications can change spectral characteristics of the probe not only by shifting the excitation and/or emission band in either direction but also by altering the luminescence quantum yield. Caution therefore should be exercised in projecting spectral properties of the derivatives based on spectral data of the parent dye.

This chapter presents, in shorthand notation for easy review and practical application, the excitation (absorption) and emission spectra of 67 fluorochromes with affinities to different cell constituents. The selected probes are either already well established in the field of FCM or have a potential for wider use. The data are presented in the form that may be more relevant for applications in FCM. Namely, because significant differences exist between spectral properties of the free dye compared to those of the dye bound to the substrate, when possible, the spectra of the dyes complexed with the respective cellular substrates are presented, rather than those of pure dyes (often in nonaqueous solutions) as is commonly practiced. The fluorochromes were subdivided into three groups: probes of nucleic acids, of proteins, and of other cell constituents. The latter group includes probes of membrane potential, polarity, pH, ion concentration, and viability (Figs. 1–3).

The information provided in this chapter complements other chapters of this volume. It may be useful in selecting new probes or probe combinations, in designing new applications, as well as in adapting the methods to different instruments that require alterations of optical filters or excitation wavelengths.

II. Fluorescent Probes of Nucleic Acids

For DNA and RNA molecules there are few chemical groups suitable for covalent labeling. Therefore, most fluorescent probes of nucleic acids (e.g., all dyes listed in Fig. 1) form nonbinding complexes with these polymers. This fact bears important implications for the use of these probes as a quantitative analytical tool. Typically, such interactions are reversible and the amount of bound dye depends not only on concentration of the free dye but also on the concentration of electrolytes and on temperature. These properties should be considered in designing optimal staining conditions

---

FIG. 1.   (*Overleaf.*) Complexes of probes with nucleic acids. Approximate positions of excitation (shaded bars) and emission (solid bars) spectra. The positions of the maxima are marked by narrow bars; the long bars represent the wavelength range (in nanometers) in which the fluorescence intensity values are not lower than 50% of the main peak amplitude. Abbreviations used are as follows. Probes: ss, single-stranded; ds, double-stranded; NA, nucleic acids; DAPI, 4′, 6-diamidino-2-phenylindole · 2HC; FMA, fluorescein mercuric acetate; LD700, rhodamine 700; L585, 6-benzothiazolyl-3-ethyl-2-(4-dimethylamino-β-styryl)-benzothiazolinium p-toluenesulfonate; VL772, 6-(dimethylamino)-2-[2,4-dimethyl-(1-phenyl-1H-pyrrol-3-yl)ethenyl]-1-methyl quinolinium methosulfate. Specificity: A-T and G-C indicate the base specificity of the ligand in term of fluorescence properties of the its complex with NA rather than on favorable thermodynamics of the interaction as described in the text.

References (far-righthand colum): 1. Gill *et al.* (1975); 2. Kapuscinski and Darzynkiewicz (1987a); 3. Kapuscinski *et al.* (1982); 4. Kapuscinski *et al.* (1983); 5 Darzynkiewicz *et al.* (1975); 6. Pachmann and Rigler (1972); 7. Crissman *et al.* (1978); 8. Stohr *et al.* (1978); 9. Chien *et al.* (1977); 10. Crissman *et al.* (1979); 11. Kapuscinski and Skoczylas (1977); 12. Kapuscinski and Szer (1979); 13. Darzynkiewicz *et al.* (1984); 14. Kennedy *et al.* (1987); 15. Bell (1988); 16. Gigli *et al.* (1988); 17. LePecq and Paoletti (1967); 18. Latt and Stetten (1976); 19. Weisblum (1974); 20. Shapiro (1981); 21. Takeuchi and Maeda (1976); 22. Takeuchi and Maeda (1979); 23. Shapiro and Stephens (1986); 24. Latt *et al.* (1984); 25. Kapuscinski and Darzynkiewicz (1987b); 26. Darzynkiewicz *et al.* (1987); 27. Lee *et al.* (1986); 28. Terstappen and Loken (1988); 29. Shapiro (1988); 30. Sage *et al.* (1983). K. Kapuscinski-unpublished data.

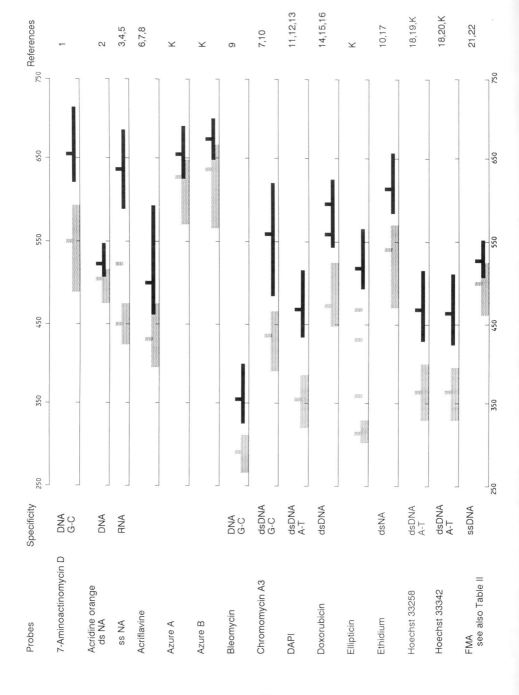

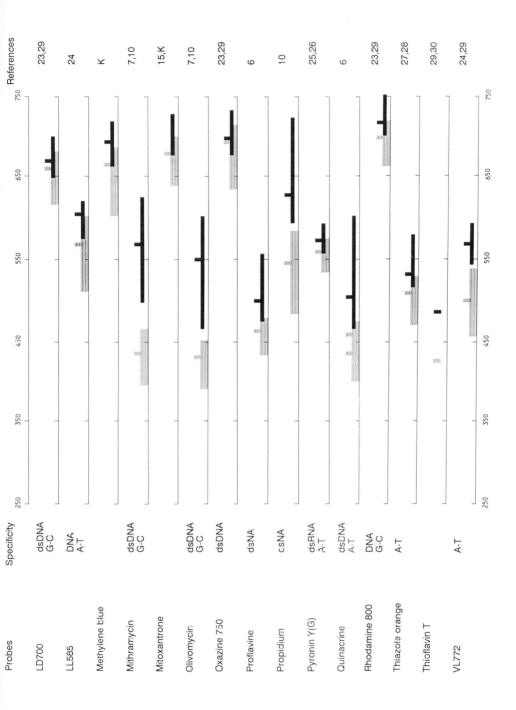

659

The majority of dyes listed in Fig. 1 are intercalators. To this group belong classical stains such as acridine orange, acriflavine, azure dyes, ethidium, proflavine, propidium, pyronin, and quinacrine. However, in addition to their well-known intercalative mode of binding to ds nucleic acids (Waring, 1981; Wilson and Jones, 1982), most of these compounds can also interact with ss nucleic acids; the products of such interactions undergo condensation and precipitation (Kapuscinski et al., 1982; Kapuscinski and Darzynkiewicz, 1984a). Furthermore, at higher concentrations most intercalators denature ds nucleic acids, subsequently forming condensed complexes with the resulting ss sections (Kapuscinski and Darzynkiewicz, 1984a,b, 1990; Darzynkiewicz et al., 1976, 1985). The stoichiometry and spectral properties of the condensed complexes differ from the intercalative products (Kapuscinski and Darzynkiewicz, 1984b, 1990), and this phenomenon explains the metachromatic properties of acridine orange and the specificity of pyronin Y(G) (Kapuscinski and Darzynkiewicz, 1984a, 1987a,b; Darzynkiewicz et al., 1987).

The antibiotics mithramycin, chromomycin, and olivomycin bind to the exterior of the DNA double helix. However, 7-aminoactinomycin D, bleomycin, doxorubicin, and the synthetic antitumor agent mitoxantrone have aromatic systems that can intercalate into the double helix and also have substituents to the aromatic systems that can bind into grooves of the double helix.

Hoechst 33258, Hoechst 33342, and DAPI have been considered classical narrow-groove binders. However, strong evidence has been presented that DAPI, and perhaps other dyes of this category, can also intercalate (Wilson et al., 1989). In addition, these three dyes, at higher concentrations, form condensed complexes with nucleic acids and other polyanions. Both excitation and emission of the dye in such complexes is shifted to the red as compared to the spectrum of these same dyes in the unbound form or in complexes with DNA at low binding density. The fluorescent efficiency of the dyes in the condensed complexes is generally higher compared to the efficiency of free dyes but is still lower than in complexes with DNA at low binding density (unpublished data from this laboratory).

It should be emphasized that the specificity of the probes as marked in Fig. 1 reflects the *preferential fluorescent properties* of its complex with particular nucleic acids or base sequences rather than its affinity. Thus, for instance, pyronin Y(G) and quinacrine bind more strongly to GC-rich than to AT-rich sequences of RNA or DNA, respectively. Guanine, however, quenches the fluorescence of these dyes. In effect, only AT-rich sequences of nucleic acids form fluorescent complexes and, therefore, these two probes have presently been classified as AT-specific.

From the above short description one can see the complexity of the mechanisms of interactions of noncovalent probes with nucleic acids. New methods and strategies have been developed in molecular biology based on the use of covalent ligands (photochemically activated or amino-active probes), biotinylated labels, enzymatic reactions, immunoprobes and cross-linking reagents (Matthews and Kricka, 1988). These probes are not yet fully exploited but may have potential value in FCM.

III. Complexes of Fluorescent Probes with Proteins

Figure 2 presents the spectral properties of several fluorescent probes used for quantitative assays of proteins. The major step in development of these probes was chemical derivation of fluorochromes via introduction of substituents that selectively react with proteins, predominantly with their amino or sulfhydryl groups. The fluorochromes, depending on the type of group(s) introduced, often have altered spectral characteristics compared to the parent dye. This is evident, for instance, from the comparison of the spectra of four derivatives of coumarin listed in Fig. 2, namely BDCM, coumarin-ITC, coumarin-MI, and CPI.

The introduction of protein-reactive fluorochromes opened new strategies for the synthesis of an unlimited number of very specific fluorescent probes. Thus, for example, a high-affinity probe toward actin was

---

Fig. 2. (*Overleaf.*) Complexes of probes with proteins. Approximate positions of excitation (shaded bars) and emission (solid bars) spectra. The positions of the maxima are marked by narrow bars; the long bars represent the wavelength range in which the fluorescence intensity values are not lower than 50% of the main peak amplitude. Abbreviations used are as follows. Probes: AMCA-NHS, 7-amino-4-methylcoumarin-3-acetate; BDCM, (4-brom.methyl)-6,7-dimethoxycoumarin; ITC, isothiocyanate; MI, maleimide; CPI, 7-diethylamino-3-(4'-isothiocyanatophenyl)-4-methylcoumarin; DANS, dansyl chloride; FMA, fluorescein mercuric acetate; LRSC, lissamine rhodamine-B sulfonyl chloride; RITC, rhodamine B isothiocyanate; SITS, 4-acetamido-4'-isothiocyanatostilbene-2,2'-disulfonic acid; SR 101, sulforhodamine 101; TRITC, tetramethylrhodamine isothiocyanate; XRITC, rhodamine X isothiocyanate. Specificity: SH, sulfhydryl-reactive probes; GLT, glutathione; NH, amino-reactive probes; var., selectivity varied depending on the active group attached to the fluorochrome; iodoacetylated and maleimide derivatives are SH-reactive probes; ITC derivatives are amino-selective probes; vital, vital stain.

References (far-righthand column): 1. Haugland (1989); 2. Oi *et al.* (1982); 3. Robinson, (1988); 4. Hiratsuka (1987); 5. Small *et al.* (1988); 6. Stohr *et al.* (1978); 7. Chen and Scott (1985); 8. Cherry *et al.* (1976); 9. Titus *et al.* (1982); 10. Waggoner *et al.* (1989); 11. Hakanson *et al.* (1974); 12. Karush *et al.* (1964); 13. Puri *et al.* (1985); 14. Principe *et al.* (1989); 15. Cohn and Lyle (1960); 16. Shapiro (1988); 17. Hoffman (1988); 18. Crissman and Steinkamp (1982).

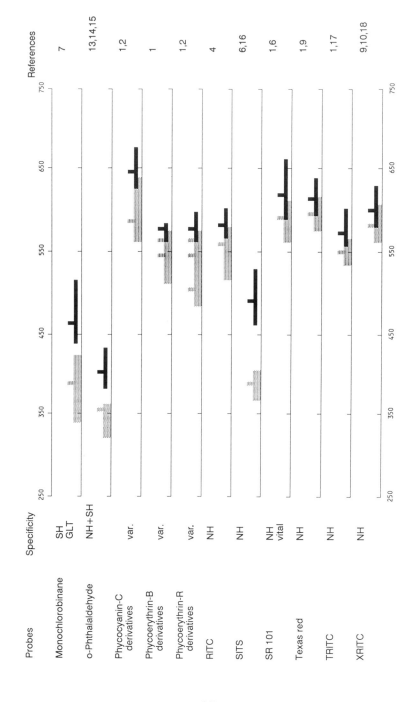

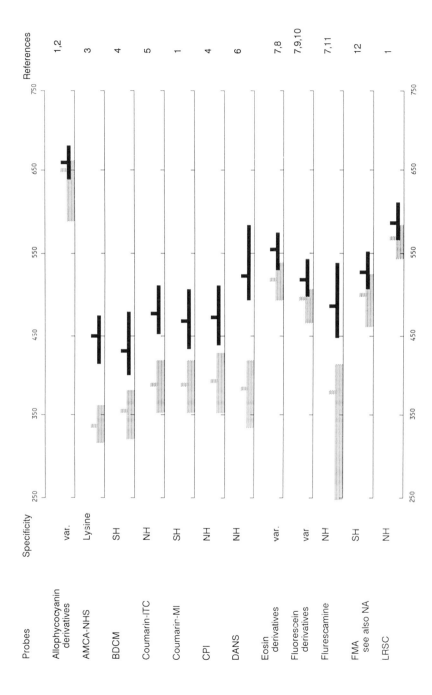

Probes	Specificity						References
Allophycocyanin derivatives	var.						1,2
AMCA-NHS	Lysine						3
BDCM	SH						4
Coumarin-ITC	NH						5
Coumarin-MI	SH						1
CPI	NH						4
DANS	NH						6
Eosin derivatives	var.						7,8
Fluorescein derivatives	var						7,9,10
Flurescamine	NH						7,11
FMA see also NA	SH						12
LRSC	NH						1

663

developed by coupling CPI with the polypeptide toxin phalloidin (Small *et al.*, 1988, and references cited therein). Furthermore, conjugation of the reactive fluorescent labels with monoclonal antibodies (mAb) provided a means for generating highly specific fluorescent probes for individual proteins. A large variety of these labels, with widely different spectral properties, now exists and is used in multiparameter (multicolor) assays to detect several cellular antigens simultaneously. Their usefulness in FCM is discussed in several chapters of this volume (Dolbeare *et al.*, Chapter 21; Baisch and Gerdes, Chapter 22; Bauer, Chapter 24; Stewart, Chapters 39 and 40).

IV.   Probes of Membrane Potential, Polarity, pH, Ion Concentration, and Cell Viability

Fluorochromes of this category are typically nonbinding probes; their mode of interaction with the respective cellular targets involves predominantly electrostatic forces. These probes are often applied in the form of neutral derivatives (e.g., FDA, BCECF) which can penetrate the cell membrane. Then, intracellular enzymes transform them to the ionic form (see Fig. 3).

FIG. 3.   Probes of membrane potentials, polarity, ion concentration, and viability. Approximate positions of excitation (shaded bars) and emission (solid bars) spectra. The positions of the maxima are marked by narrow bars; the long bars represent the wavelength range in which the fluorescence intensity values are not lower than 50% of the main peak amplitude. Abbreviations used are as follows: Probes: BCECF, 2',7'-bis-(2-carboethoxyethyl)-5-(and -6-)carboxyfluorescein; CTT, chlorotetracycline; DiO-C_n-(3), oxacarbocyanine; DiI-C_n-(3), inocarbocyanine; DiI-C_n-(5), inocarbocyanine; FDA, fluorescein diacetate; Fura-2 and Indo-1, selective calcium chelators (pentacarboxylic acids); Mag-indo-1, a selective magnesium chelator (tetracarboxylic acid); PBFI, a selective potassium chelator (crown ether); SBFI, a selective sodium chelator (crown ether); SNARF-1, seminaphthorhodafluor; SPQ, 6-metoxy-N-(3-sulfopropyl)quinolinium. Specificity: m.p., membrane potential; m.i., membrane integrity; vital, vital stain, viability test; polarity, protein polarity (hydrophilicity); mit., mitochondria stain. Note 1: in viable cells, in permeable cells this dye stains NA (see Fig. 1).

References (far-righthand column): 1. Graber *et al.* (1986); 2. Haugland (1989); 3. Caswell (1972); 4. Tsien (1989); 5. Sims *et al.* (1974); 6. Shapiro (1988); 7. Waggoner *et al.* (1989); 8. Grynkiewicz and Tsien, (1985); 9. Greenspan *et al.* (1985); 10. Sackett and Wolff (1987); 11. Greenspan and Fowler (1985); 12. Ratinaud *et al.* (1988); 13. Kapuscinski and Darzynkiewicz (1987b); 14. Darzynkiewicz *et al.* (1986); 15. Darzynkiewicz *et al.* (1987); 16. Steinkamp *et al.* (1979); 17. Darzynkiewicz *et al.* (1982); 18. Illsley and Verkman (1987); 19. Lee *et al.* (1989).

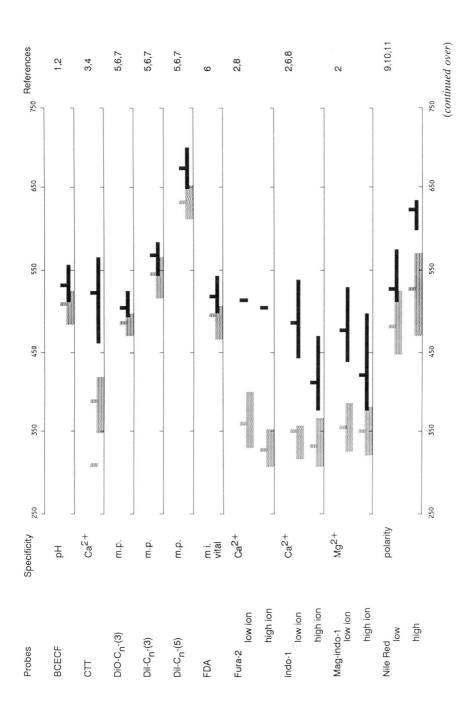

(continued over)

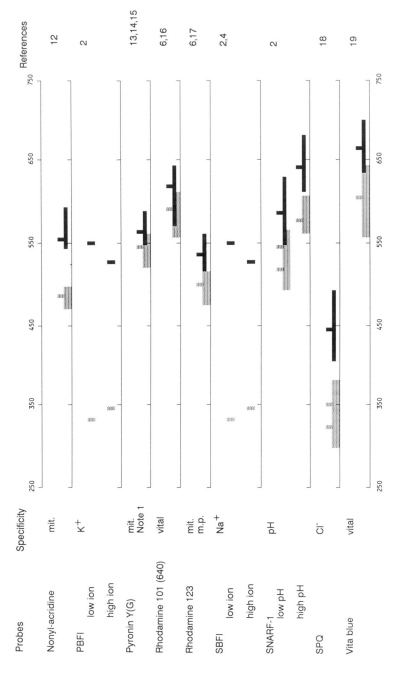

Fig. 3 (continued)

Several positively charged fluorescent probes (carbocyanine dyes, rhodamine 123, pyronin Y) accumulate in viable cells as a result of differences in the potential across cellular and/or mitochondrial membranes. Still others (e.g., nonyl-acridine), have additional binding or structural components that enhance their specificity toward particular targets. New and potentially valuable reagents are the probes of Na^+ and K^+ SBFI and PBFI; their selectivity is based on the size of the chelator (crown ether). At the time of this writing, only the maxima of their excitation and emission were available. Nile red is an interesting fluorochrome that is sensitive to polarity (dielectric constant) of its environments and can be used as an intracellular lipid probe.

Acknowledgments

We would like to thank Dr. Frank Traganos for critical reading of the manuscript. This work was supported in part by NIH grants CA28704 and R37 CA23296.

References

Arndt-Jovin, D. J., and Jovin, T. M. (1989). *Methods Cell Biol.* **30,** 417–448.

Bell, D. H. (1988). *Biochim. Biophys Acta* **949,** 132–137.

Caswell, A. H. (1972). *J. Membr. Biol.* **7,** 345–364.

Chen, R. F., and Scott, C. H. (1985). *Anal. Lett.* **18,** 393–421.

Cherry, R. J., Cogoli, A., Oppliger, M., Schneider, G., and Semenza, G. (1976). *Biochemistry* **15,** 3653–3656.

Chien, M., Grollman, A. P., and Horowitz, S. B. (1977). *Biochemistry* **16,** 3641–3647.

Cohn, V. H., and Lyle, J. (1960). *Anal. Biochem.* **14,** 434–440.

Crissman, H. A., and Steinkamp, J. A. (1982). *Cytometry* **3,** 84–90.

Crissman, H. A., and Steinkamp, J. A. (1987). *In* "Techniques in Cell Cycle Analysis" (J. W. Gray and Z., Darzynkiewicz, eds.), pp 163–206, Humana Press, Clifton, New Jersey.

Crissman, H. A., Stevenson, A. P., Orlicky, D. J., and Kissane, R. J. (1978). *Stain Technol.* **53,** 321–330.

Crissman, H. A., Stevenson, A. P., Kissane, R. J., and Tobey, R. A. (1979). *In* "Flow Cytometry and Sorting" (M. R., Melamed, P. F. Mullaney, and M. L., Mendelsohn, eds.), pp 243–261. Wiley, New York.

Darzynkiewicz, Z., Traganos, F., Sharpless, T., and Melamed, M. R. (1975). *Exp. Cell Res.* **90,** 411–428.

Darzynkiewicz, Z., Traganos, F., Sharpless, T., and Melamed, M. R. (1976). *J. Cell Biol.* **68,** 1–10.

Darzynkiewicz, Z., Traganos, F., Staiano Coico, L., Kapuscinski, J., and Melamed, M. R. (1982). *Cancer Res.* **42,** 799–806.

Darzynkiewicz, Z., Williamson, B., Carswell, E. A., and Old, L. J. (1984). *Cancer Res.* **44,** 83–90.

Darzynkiewicz, Z., Traganos, F., Kapuscinski, J., and Melamed, M. R. (1985). *Cytometry* **6,** 195–207.

Darzynkiewicz, Z., Kapuscinski, J., Carter, S. P., Schmid, F. A., and Melamed, M. R. (1986). *Cancer Res.* **46**, 5760–5766.

Darzynkiewicz, Z., Kapuscinski, J., Traganos, F., and Crissman, H. A. (1987). *Cytometry* **8**, 138–145.

Gigli, M., Doglia, S. M., Millot, J. M., Valentini, L., and Manfait, M. (1988). *Biochim. Biophys. Acta* **950**, 13–20.

Gill, J. E., Jotz, M. M., Young, S. G., Modest, E. J., and Sengupta, S. K. (1975). *J. Histochem. Cytochem.* **23**, 793–799.

Graber, M. L., DiLillo, D. C., Friedman, B. L., and Pastoriza-Munoz, E. (1986). *Anal. Biochem.* **156**, 202–212.

Greenspan, P., and Fowler, S. D. (1985). *J. Lipid Res.* **26**, 781–788.

Greenspan, P., Mayer, E. P., and Fowler, S. D. (1985). *J. Cell Biol.* **100**, 965–973.

Grynkiewicz, G., and Tsien, R. Y. (1985). *J. Biol. Chem.* **260**, 3440–3450.

Hakanson, R., Larsson, L.–I., and Sundler, F. (1974). *Histochemistry* **39**, 15–23.

Haugland, R. P. (1989). "Handbook of Fluorescent Probes and Research Chemicals. Molecular Probes, Eugene, Oregon.

Hiratsuka, T. (1987). *J. Biochem.* **101**, 1457–1462.

Hoffman, R. A. (1988). *Cytometry Suppl.* **3**, 18–22.

Illsley, N. P., and Verkman, A. S. (1987). *Biochemistry* **26**, 1215–1219.

Kapuscinski, J., and Darzynkiewicz, Z. (1984a). *J. Biomol. Struct. Dyn.* **1**, 1485–1499.

Kapuscinski, J., and Darzynkiewicz, Z. (1984b). *Proc. Natl. Acad. Sci. U.S.A.* **81**, 7368–7372.

Kapuscinski, J., and Darzynkiewicz, Z. (1987a). *J. Biomol. Struct. Dyn.* **5**, 127–143.

Kapuscinski, J., and Darzynkiewicz, Z. (1987b). *Cytometry* **8**, 129–137.

Kapuscinski, J., and Darzynkiewicz, Z. (1990). *In* "Structure & Methods" vol. 3 (R. Sarma and M. H. Sarma, eds.), pp. 267–281. Adenine Press, Schenectady, New York.

Kapuscinski, J., and Skoczylas, B. (1977). *Anal. Biochem.* **83**, 252–257.

Kapuscinski, J., and Szer, W. (1979). *Nuclei Acids Res.* **6**, 3519–3534.

Kapuscinski, J., Darzynkiewicz, Z., and Melamed, M. R. (1982). *Cytometry* **2**, 201–211.

Kapuscinski, J., Darzynkiewicz, Z., and Melamed, M. R. (1983). *Biochem. Pharmacol.* **32**, 3679–3694.

Karush, F., Klinman, N. R., and Marks, R. (1964). *Anal. Biochem.* **9**, 100–114.

Kennedy, D. G., Nelson, J., van den Berg, H. W., and Murphy, R. F., (1987). *Anal. Biochem.* **167**, 124–127.

Latt, S. A. (1979). *In* "Flow Cytometry and Sorting" (M. R. Melamed, P. F. Mullaney, and M. L. Mendelsohn, eds.), pp. 263–284. Wiley, New York.

Latt, S. A., and Stetten, G. (1976). *J. Histochem. Cytochem.* **24**, 24–33.

Latt, S. A., Marino, M., and Lalande, M. (1984). *Cytometry* **5**, 339–347.

Lee, L. G., Chen, C. -H., and Chiu, L. A. (1986). *Cytometry,* **7**, 508–517.

Lee, L. G., Berry, G. M., and Chen, C. -H., (1989). *Cytometry* **10**, 151–164.

LePecq, J. -B., and Paoletti, C. (1967). *J. Mol. Biol.* **27**, 87–106.

Matthews, J. A., and Kricka, L. J. (1988). *Anal. Biochem.* **169**, 1–25.

Oi, V. T., Glazer, A. N., and Stryer, L. (1982). *J. Cell Biol,* **93**, 981–986.

Pachmann, U., and Rigler, R. (1972). *Exp. Cell Res.* **72**, 602–608.

Principe, P., Wilson, G. D., Riley, P. A., and Slater, T. F. (1989). *Cytometry* **10**, 750–761.

Puri, R. N., Bhatnagar, D., and Roskoski, R., Jr. (1985). *Biochemistry* **24**, 6499–6508.

Ratinaud, M. H., Leprat, P., and Julien, R. (1988). *Cytometry* **9**, 206–212.

Robinson, D. (1988). *Biochem. Soc. Trans.* **16**, 11–16.

Sackett, D. L., and Wolff, J. (1987). *Anal. Biochem.* **167**, 228–234.

Sage, B. H., O'Connell, J. P., Jr, and Mercolino, T. J. (1983). *Cytometry* **4**, 222–227.

Shapiro, H. M. (1981). *Cytometry* **2**, 143–159.

Shapiro, H. M. (1988). *In* "Practical Flow Cytometry", pp. 115–198. Liss, New York.

Shapiro, H. M., and Stephens, S. (1986). *Cytometry* **7**, 107–110.

Sims, J., Waggoner, A. S., Wang, C. -H., and Hoffman, F. (1974). *Biochemistry,* **13**, 3315–3330.

Small, J. V., Zobeley, S., Rinnerthaler, G., and Faulstich, H. (1988). *J. Cell Sci.* **89**, 21–24.

Steinkamp, J. A. (1984). *Rev. Sci. Instrum.* **55**, 1375–1400.

Steinkamp, J. A., Orlicky, D. A., and Crissman, H. A. (1979). *J. Histochem. Cytochem.* **27**, 273–276.

Stohr, M., Vogt-Schaden, M., Knobloch, M., and Vogel, R. (1978). *Stain Technol.* **53**, 205–215.

Takeuchi, S., and Maeda, A. (1976). *Biochim. Biophys. Acta* **454**, 309–318.

Takeuchi, S., and Maeda, A. (1979). *Biochim. Biophys. Acta* **563**, 365–374.

Terstappen, L. W. M. M., and Loken, M. R. (1988). *Cytometry* **9**, 548–556.

Titus, J. A., Haugland, R. P., Sharrow, S. O., and Segal, D. M. (1982). *J. Immunol. Methods* **50**, 193–204.

Tsien, R. Y. (1989). *Methods Cell Biol.* **30**, 127–156.

Waggoner, A., DeBiasio, R., Conrad, P., Bright, G. R., Ernst, L., Ryan, K., Nederlof, M., and Taylor, D. (1989). *Methods Cell Biol.* **30**, 449–478.

Waring, M. J. (1981). *Annu. Rev. Biochem.* **50**, 159–192.

Weisblum, B. (1974). *Cold Spring Harbor Symp. Quant. Biol.* **38**, 441–449.

Wilson, W. D., and Jones, R. L. (1982). *In* "Intercalation Chemistry" (M. S. Whittingham and A. J., Jacobson, eds.), pp 445–501. Academic Press, New York.

Wilson, W. D., Tanious, F. A., Barton, H. J., Strekowski, L., and Boykin, D. W. (1989). *J. Am. Chem. Soc.* **111**, 5008–5010.

INDEX

CONTENTS OF
RECENT VOLUMES

Volume 29

Fluorescence Microscopy of Living Cells in Culture

Part A. *Fluorescent Analogs, Labeling Cells, and Basic Microscopy*

Volume 30

Fluorescence Microscopy of Living Cells in Culture

Part B. *Quantitative Fluorescence Microscopy — Imaging and Spectroscopy*

Volume 31

Vesicular Transport

Part A.

Part I. *Gaining Access to the Cytoplasm*

Volume 32

Vescular Transport

Part B.